Fisheries Techniques

Fisheries Techniques

Edited by
Larry A. Nielsen

Department of Fisheries and Wildlife Sciences
Virginia Polytechnic Institute and State University
and
David L. Johnson

Division of Wildlife and Fisheries Management
The Ohio State University

Illustrated by
Susan S. Lampton

Columbus, Ohio

The American Fisheries Society wishes to thank
Conoco, Inc.
for its financial support of this publication.

American Fisheries Society
Bethesda, Maryland

Library of Congress Catalog Card Number: 83-71866

ISBN: 0-913235-00-8

Printed in the United States of America
by Southern Printing Company, Inc., Blacksburg, Virginia

Second printing, 1985

Fisheries Techniques

A special project of the
Fishery Educators Section
American Fisheries Society

Steering Committee

Clarence Carlson, chairman
Bradford E. Brown
Ned E. Fogle
Melvin T. Huish
James K. Mayhew
Robert L. Kendall
Carl R. Sullivan

Editorial Assistant

Sharon K. Nielsen

Dedication

This book has been created for students. The authors' willingness to participate in this project comes largely from their desire to make fishery education as meaningful as possible. It seems appropriate, therefore, to dedicate this book to those educators who inspired the authors when they were students. Each author was asked to name the single educator who most affected his development as a fishery scientist. To these educators we dedicate this book:

Brian F. Allanson
Carl Bond
Bradford E. Brown
Robert S. Campbell
Kenneth D. Carlander
Daniel M. Cohen
Edwin L. Cooper
Kenneth B. Cumming
William D. Davies
B. R. Eddleman
D. W. Hayne
Robert T. Lackey
Ole Mathisen
O. Eugene Maughan
John R. Olive
Harland I. Padfield
George Post
John E. Randall
Carl B. Schreck
R. A. Schrimper
Donald B. Siniff
H. S. Swingle
Howard A. Tanner
J. V. K. Wagar
Dale C. Wallace

Contents

Preface

The objective of this book is to introduce fisheries students to the ways in which fisheries data are collected. We have sought to produce a book that covers most of the techniques that a practicing fishery manager is likely to encounter during his or her career. Almost all of the techniques could be topics of individual books; the reality of space and time has required us to limit the coverage of each to a few pages. It was our intention to describe for each technique the basis for its use, the major options for equipment types and procedures, the advantages and disadvantages, and some practical hints for efficient use. The coverage for each technique is not identical because the nature of each technique and our understanding of each vary. We have taken each topic as far as allowed by the literature, by the willingness of the author, and by our assessment of student familiarity with the technique.

It is just as important to point out what the book was not intended to do, especially because a volume entitled "Fisheries Techniques" can mean many different things. The book does not pretend to be the final authority on the topics covered; the cited references should be consulted for in-depth treatment of each topic. The book should not be used as the sole basis or justification for designing a major data gathering effort; again, the cited references must be consulted. The book does not describe or recommend a "standard" way of collecting data, but rather describes the criteria for selecting a specific method for data collecting. The book does not describe how to manage fisheries; it describes how to collect data, given that such data are needed.

The objective of the book is straight-forward, but turning the objective into a reality has been a complicated process. The book, which undoubtedly has existed as an idea among fisheries professionals as long as there have been such people, began to take form in the heads of the editors in 1979. The idea was pursued through the Fishery Educators Section of the American Fisheries Society, and the book as it now exists represents in large part the collective efforts of the Fishery Educators Section. Through a series of steps, including an educators' workshop, steering committee meetings, input by state and provincial agencies, and manuscript preparation and review, at least 200 people have been involved. Perhaps the greatest impetus driving the book to completion was the grant from Conoco Oil which underwrote the production costs. Carl Sullivan, Executive Director of the American Fisheries Society, acquired the grant, and he should receive much of the credit for turning our objective into printed pages.

Several specific items regarding the content of the book require clarification. We have used the masculine pronoun throughout the book for the sake of readability. Specific products are mentioned by name throughout the book, but the use of such names does not constitute endorsement of any product by the American Fisheries

Society or by the employers of the authors. Many techniques used in fisheries, including those described in this book, are inherently dangerous. We have highlighted safety throughout the book, but the absence of specific warnings for any piece of equipment or procedure does not mean that accidents and injuries will never occur. A fishery professional must always be certain that the specific situation in which he or she is working is free of hazardous conditions.

Each chapter in the book was reviewed in outline and draft form by at least two technical reviewers. We chose reviewers as often as possible from state or federal management agencies to assure that the techniques included in the book were both meaningful and practical. The names of reviewers are as follows:

Chapter 1: Bradford E. Brown and Bruce Shupp
Chapter 2: Johanna M. Reinhart, Robert L. Kendall, and Mercer H. Patriarche
Chapter 3: Lt. Michael Chaplan, Robert Hartmann, and Eric Olson
Chapter 4: Donald A. Duff, Robert L. Hunt, and Martin Marcinko
Chapter 5: Donald F. Amend, Charles E. Hicks, and David W. McDaniel
Chapter 6: W. C. Latta and Larry Mitzner
Chapter 7: Alfred Houser and Robert W. Wiley
Chapter 8: Gary D. Novinger, Larry Olmsted, and Gordon R. Priegel
Chapter 9: C. F. Bryan, Fred W. Vasey, and Ross Rasmussen
Chapter 10: Larry Aggus, Daniel R. Holder, and Leif Marking
Chapter 11: Vaughn Anthony, Don Gabelhouse, Jr. and R. Weldon Larimore
Chapter 12: Clarence Clay and Bert Bowler
Chapter 13: Ronald L. Crunkilton, Patrick H. Davies, and James Mayhew
Chapter 14: Frank Cross and E. J. Crossman
Chapter 15: Robert M. Jenkins and Ronald Schaefer
Chapter 16: Kenneth Carlander, Larry Claggett, and Kim Graham
Chapter 17: James B. Mense and Richard E. Strauss
Chapter 18: George Klontz, Fred P. Meyer, and Robert Murchelano
Chapter 19: Richard A. Ryder, Barry W. Smith, and Roy A. Stein
Chapter 20: C. C. Coutant, Thomas R. Russell, and A. B. Stasko
Chapter 21: James Ryckman, David G. Deuel, and M. D. Norman
Chapter 22: Bradford E. Brown, Gene R. Huntsman, and Ronald Rybicki
Chapter 23: Tommy Brown, Robert T. Lackey, Michael Orbach, and John Charbonneau
Chapter 24: James W. Ayers and Floyd A. Hennagir

This book represents the collected efforts of many individuals, only a few of whose names appear on the title pages. The authors of individual chapters have asked that the following people be recognized for their contribution to this book.

Clarence A. Carlson and Curtis L. Gifford (chapter 2) wish to acknowledge Barbara Aldrich and Marianne Wagers; Donald J. Orth (chapter 4) acknowledges the help of Donald A. Duff, Peter A. Bisson, William S. Platts, and Cole S. Shirvell; Robert R. Stickney (chapter 5) expresses his appreciation for the assistance of Richard L. Noble; and Wayne A. Hubert wishes to thank Kenneth Carlander, John Nickum, and James Mayhew for their help with chapter 6. Murray L. Hayes (chapter 7) acknowledges Jerry E. Jurkovich, Fred Wathne, B. F. Jones, and Marion Hansen; James B. Reynolds (chapter 8) recognizes Terra Shideler; and Darrel E. Snyder expresses his appreciation to Clarence Carlson, Maryann Snyder, the secretarial staff of the Department of Fishery and Wildlife Biology at Colorado State University, the Colorado State University Agricultural Experiment Station, and various members of the Early Life History Section of the American Fisheries Society for their assistance with and support of chapter 9.

William D. Davies and William L. Shelton (chapter 10) acknowledge Larry Aggus; Richard E. Thorne wishes to recognize and thank Clarence Clay for his special help with chapter 12. Donley M. Hill (chapter 13) expresses his appreciation to John Shipp; and Richard O. Anderson and Stephen J. Gutreuter wish to thank Terry R. Finger, W. E. Ricker, Terence P. Boyle, and Daniel W. Coble for their help with chapter 15. Ambrose Jerald, Jr. (chapter 16) expresses his appreciation to Melinda F. Davis, Louise Dery, Sherry Sass, John Ropes, and Gary Shepherd; and Gene S. Helfman (chapter 19) is grateful for the help of E. L. Bozeman, J. B. Heiser, W. N. McFarland, and E. T. Schultz and the partial funding of his effort by NOAA, Office of Sea Grant, Department of Commerce, under Grant NA80AA-D-0091 to the University of Georgia. Jimmy D. Winter (chapter 20) wishes to thank V. B. (Larry) Kuechle and M. Jon Ross of the Cedar Creek Bioelectronics Laboratory, University of Minnesota, for their assistance; Stephen P. Malvestuto (chapter 21) recognizes the help of Paul Eschmeyer; and Fred J. Prochaska and James C. Cato gratefully acknowledge the help of various members of Florida's fishing industry with whom they have worked in an educational and research role.

In addition, the editors wish to acknowledge our department heads, Gerald Cross and Charles Cole, who supported our efforts in many ways. Mary Frye, of the AFS Central Office, handled the financial dealings for the book. Cathy Barker typed the final manuscript. Southern Printing, of Blacksburg, Virginia, composed and printed the book.

L. A. N.

D. L. J.

Chapter 1
Sampling Considerations

DAVID L. JOHNSON AND LARRY A. NIELSEN

1.1 INTRODUCTION

This book is about collecting data. The book is necessary because fisheries professionals spend a major part of their time collecting data with a large variety of techniques for a large variety of reasons. Efficient and effective collection of data can mean the difference between a successful management or research effort and one that ends with inconclusive or useless information.

Collecting the data, however, is only part of the whole process that managers and researchers must perform. Before data are collected, the reasons for the activity and the plan to carry it out must be explicitly defined. After the data are collected, the results must be analyzed, synthesized, and presented to the appropriate audiences. The scope of these "before" and "after" processes is evident in the Lake Barkley study described in Box 1.1. This chapter describes some of the considerations that determine the success or failure of a data collection activity.

Box 1.1. The Barkley Lake Rotenone Study

Cove rotenone sampling is a useful method for gathering information about fish populations in southern reservoirs (*see* chapter 10). Perhaps the most ambitious volunteer sampling study and certainly the largest cove rotenone study in fisheries was conducted on Barkley Lake, Kentucky. The organization of that study illustrates the need for careful planning.

The Reservoir Committee of the Southern Division of the American Fisheries Society decided in June, 1976, to sponsor a large-scale evaluation of the cove rotenone method. Based on their experience with a smaller project ten years earlier, they set the sampling date for September, 1978, leaving 26 months for design and planning.

During the next 12 months, the committee defined the objectives of the project and chose the 85-ha Crooked Creek Bay of Barkley Lake as the sampling location. The project ultimately included ten distinct objectives, addressing not only cove rotenone sampling but also fish kill counting guidelines, mark-recapture techniques, and effectiveness of fish attractors. The incorporation of additional objectives allowed for maximum use of the data with a minimum of extra effort.

After selecting the Barkley Lake site, the Committee sought legal authority and public support for the project. A public meeting, held in September, 1977, at Barkley Lake State Park, attracted 100 local citizens. Description of the sampling and its scientific value by officials of the Kentucky Department of Fish and Wildlife

Continued

Resources converted the anticipated resistance to unanimous public support for the project. The Nashville Office of the Army Corps of Engineers, a participant in the project, prepared the necessary Environmental Impact Assessment, which was subsequently reviewed by appropriate groups and approved by the Environmental Protection Agency.

Field work began at the site in March, 1978, when 15 coves and sites for four fish attractors were chosen. Once the specific sites were identified, decisions could be made regarding the manpower and equipment needed for sampling. Commitments for personnel and equipment were secured from 14 state, 3 federal, and 2 private agencies and from 15 universities. A total of 400 people, 5800 meters of blocknet, 2500 liters of rotenone, 120 boats, 56 sorting tables, 79 scales, 46 rotenone pumps, and 350 kg of potassium permanganate were eventually used.

Sampling activities began on September 18, when two main block nets were sewn into 700-m and 300-m lengths to enclose the entire Crooked Creek Bay. During the following two days, the nets were set, and sites for 13 other block nets, electrofishing zones, and fish processing stations were marked. For three nights, fish were captured by electrofishing, tagged, and returned to the bay. A largemouth bass fishing tournament was held in the bay on September 23-24 as the recapture sample for the mark-recapture abundance estimate.

Most of the volunteers arrived on Sunday, September 24; equipment was assigned and moved to pre-marked staging areas. On Monday, block nets were set and crews received final instructions. Electrofishing crews captured, tagged, and released fish in each blocked cove on Monday night. Rotenone was applied to all areas on Tuesday morning and fish were collected throughout the next two days. In all, over 3,000,000 fish, weighing more than 90 tons, were collected. After fish were processed, they were taken to pre-arranged waste disposal sites or were plowed into the ground to fertilize wildlife food plots in the area.

Data recorders, assigned to each sampling area, received blank data forms each morning and returned completed forms to project leaders each evening. Forms were proofread, checked for legibility, and coded for computer analysis. The project produced a total of 3178 data sheets. By midday Thursday, September 28, block nets had been removed and field crews had headed for home.

Within one year, data were analyzed, and completed reports of various aspects of the project were published in the *Proceedings of the 33rd Annual Conference of the Southeastern Association of Fish and Wildlife Agencies*. From first commitment to final report, the project took 39 months; *data collection took two days.*

1.1.1 Sampling Justification

As professionals in public agencies, we must always be ready to justify what we do. National news of unprofessional use of research funds and useless data sets make us uncomfortably aware of our vulnerability. It is vital, therefore, that we ensure continued public support of our activities by following some simple guidelines.

Field research must address an identified problem that cannot be solved using existing information. If a particular question is considered important, the first effort should be directed at historical records and scientific literature. Only if the answers are unavailable and the question is of high priority do we undertake a field investigation.

The clientele we serve must recognize the problem along with us and understand our approach. Regular communication of problem statements and research objectives to client groups will help weld that partnership. This is as true for a state agency and a bass club as it is for a graduate student and his advisory committee.

1.1.2   Defining the Problem

Research activities do not arise from a vacuum, but rather from a need for knowledge. Depending on the situation, the need may be for basic research, for long-term applied research, or for rapid assessment of an immediate problem. In a practical profession like fisheries, most needs are for knowledge to solve management problems. The process of defining the problem proceeds through a series of steps (Box 1.2). The steps vary with the situation, but each situation will likely have the basic steps described below.

Box 1.2.   The Problem at Lotsa Luck Lake

Step 1:
Problem Suggested
by Clients

A letter sent to the State Fisheries and Wildlife Department complained about poor fishing in a 50-hectare lake named Lotsa Luck. The major complaint was that there were too many small bluegill. This letter was answered by a biologist who stated that a survey of the lake five years earlier had revealed a good bluegill population. Later, the biologist received two more letters complaining about the fishing. One of the letters was from Senator Fish who lives on the lake. The Lotsa Luck Lake Association also complained about poor fishing.

Step 2:
Preliminary
Investigation

Upon inquiring, the biologist found that reliable anglers, the county game protector, and a marina operator all agree that poor fishing is a problem.

Step 3:
Historical Data
Searched

The biologist examined historical data, finding that the last survey was five years old and not very useful.

Step 4:
Causes Postulated

The biologist recognized several possible causes. New housing developments, all with septic tanks which could be adding excess nutrients to the system, had been built since the last survey. Abundant aquatic plants could be causing the problem of a stunted bluegill population. Over-harvest

Continued

---

	of largemouth bass, poor bass recruitment, or the presence of a new, more preferred prey species for the bass could have the same result.
Step 5: Deciding to solve the problem	Senator Fish heads the State Advisory Committee for the Fisheries and Wildlife Department, the lake represents an important economic component of the county, there are few other fishing opportunities nearby, and the predator-prey relationship in the lake appears to be out of balance. Also, managing the fishery of the lake is part of the mandate of the agency. For these reasons, the biologist rated the problem as a high priority. The Department decided that the District had sufficient funds and personnel to investigate the problem, and, further, that a field investigation was required because no recent information was available.

Identifying the problem. The clients for a fisheries manager include all people in his jurisdiction and subsets of special interest groups; the clients for a researcher are other researchers and fisheries managers. Specific problems are suggested by clients through letters, personal contact, or popular articles, and scientific journals. We must be alert constantly for the concerns expressed by our clients, whenever and however they are expressed.

Verifying the problem. There are several ways to follow up on a problem identified by a client. A trip to the site will allow first-hand examination of the situation and opportunity to interview other people in the vicinity. Contact with persons directly involved in the problem area, such as area managers or county game protectors, can help place the problem in its broadest context.

Data may be already available from previous work on the problem. A preliminary search for information in local data files and in published literature will give perspective to the problem. Such a search can help determine if historical or published information are a sufficient basis for solving the problem. This decision must be made with great care because depending on past information, which may be outdated or of poor quality, can lead to wrong conclusions.

Postulated causes. Frequently, the problems identified by clients are really symptoms of underlying problems. It is necessary, therefore, to list some potential causes of the identified problem. If the hypothesized causes are outside the investigator's area of knowledge, it may be necessary to include others with the appropriate skills and knowledge. This is also the point at which the investigator can begin to evaluate his ability to remove some of the suspected causes of the problem.

Deciding to solve the problem. Each identified problem is surrounded by various political, social, and biological issues. Each problem must be examined in light of overall agency or institutional responsibilities and goals. If the problem is not within previously established mandates of the agency, then the mandate must be redefined or the problem must be referred to the appropriate unit. If the problem falls within the agency mandate, a decision must be made about the importance of solving the problem. Ecological, social, and political factors must be examined closely to evaluate how critical the problem is. For example, the reported illegal snagging of coho salmon in a stream may be of little or no importance biologically because the fish will die after the spawning anyway, but the social pressures by landowners and anglers may dictate further investigation of the problem. If a field sampling activity is approved, the research sampling plan is begun.

1.2 PLANNING THE INVESTIGATION

The planning process has many steps; the exact process followed depends on the situation and the personal preferences of the planners. The basis of all planning approaches, however, must be to define explicitly the purpose of collecting data. A general format for the planning process steps would be to (1) define the goal, (2) choose objectives that lead to the goal, (3) devise strategies to meet the objectives, (4) identify activities to implement the strategies, (5) lay out a timetable, (6) request resources, and (7) calculate costs.

1.2.1 Goals

A goal is a general statement of purpose that is usually long-term in nature, not necessarily very precise, and may not even be measurable. Using the problem identified in Box 1.2, a goal might be "to determine why fishing is poor for largemouth bass and bluegill in Lotsa Luck Lake." Goals give a general shape to the subsequent planning and create a framework for applying specific techniques. Defining goals provides a good starting point for interaction with clients. It is unwise for the manager or researcher to define goals without client interaction. It is equally dangerous, however, to allow clients to choose goals independently. Fisheries professionals must advise clients about the feasibility and projected consequences of alternative goals.

1.2.2 Objectives

Although all steps of a planning process are important, the most critical step is creating objectives. Objectives are explicit statements of exactly what an activity will accomplish and when the accomplishments will occur. It is impossible to over-emphasize the importance of high quality objectives. Objectives must be precise, measurable, compatible with goals, and realistic in relation to the time, funds, and manpower available. An objective for the problem in Box 1.2 might be "to determine in the first year of the study the annual growth rate of bluegill and largemouth bass in Lotsa Luck Lake." Because objectives define where we are going, good objectives greatly enhance the probability of project success. If adequate progress is not being made past stated milestones, then we can quickly identify the reasons and revise the project accordingly. In this sense, project planning is an ongoing process culminating in the final product of our field effort.

1.2.3 Strategies and Tactics

The implementation of objectives requires the choice of strategies and tactics (sometimes called sub-objectives) for doing all the appropriate activities and no inappropriate ones. The top-down planning process (Phenicie and Lyons 1973) allows for the logical consideration of tasks to be accomplished (questions answered) in an explicit manner.

In the top-down process, planners answer the question, "What must we find out in order to answer that question or reach that objective?" Another way of asking that question is, "We can attain our objective *if and only if* we know the answers to these questions." The process continues through succeedingly lower levels until the answers to the questions are obviously attainable with current techniques. If one constructs a top-down plan properly, using the "if and only if" logic, then sequentially answering the questions from the bottom up will result in resolution of the central objective. The planning occurs from the top down, but activities proceed from the bottom up. Top-down planning for the Lotsa Luck Lake problem is shown in Figure 1.1.

The top-down plan shows that the objectives for the first year of this study should be to determine the growth, recruitment, and mortality of bluegill (from questions *1, 2,* and *3* in Fig. 1.1) and largemouth bass (questions *4, 5,* and *6*), the fish community composition (question *7*), the dissolved oxygen concentration and temperature profiles through one season (questions *8* and *9*), the pH, alkalinity, and vegetation abundance through one season (questions *10, 11,* and *12*), and to produce a contour map of the lake (question *13*).

1.3 SAMPLING

The purpose of sampling is to make judgments about the whole by looking at only a portion of the whole. Sampling is necessary for most fisheries data because it is either impractical or impossible to observe any aspect of a fishery completely. We collect a sample of fish, take a scale sample, or sample the habitat. How these samples are collected determines how well the sampling data help achieve the objectives of the activity.

If field data are to be useful, they must be acquired according to established statistical procedures. These procedures involve experimental design, sampling design, and data analysis. A detailed discussion of these topics is beyond the scope of this chapter; our main intent is to introduce the topic of sampling design. The entire discipline of statistics is relevant, however, and Table 1.1 lists commonly cited statistics texts that should be used as references on the topic.

Two characteristics of sampling data—accuracy and precision—describe the value of the data. Accuracy refers to how well the sample represents the whole. If we wish to know the average size of rainbow trout in a lake, and we collect fish with a gill net that captures large fish more effectively than small fish, the sample will over-estimate the average size. Collecting accurate, or unbiased, data is extremely difficult in fisheries. Most of this book describes the biases of data collection techniques and ways to avoid them.

Precision of sampling data refers to repeatability. Statistics like standard deviations and quartile ranges are measures of precision. In general, the narrower the "confidence interval" around an estimate made from a sample, the more likely you are to get a similar result if you take another sample in the same way. Collecting highly precise data usually requires extensive sampling effort as well as a careful design.

1.3.1 Sampling Location

Where to collect data is a decision that affects the quality of the data and the ease with which they can be collected. If data are being collected to describe conditions at a given water body, for example, a stream that will be impounded by a reservoir, then

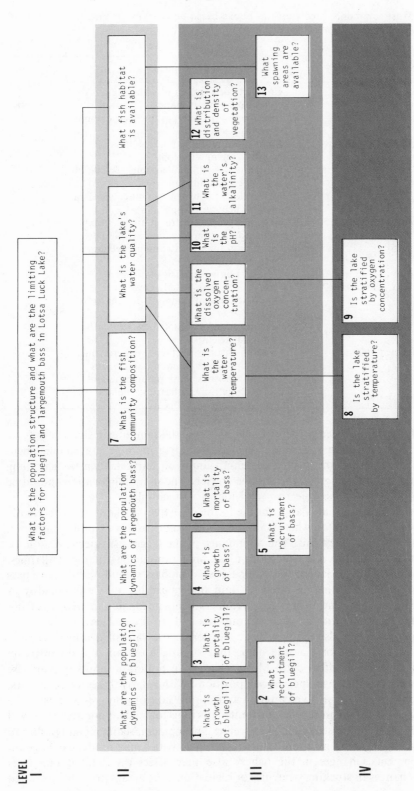

Figure 1.1. An example of top-down planning for Lotsa Luck Lake. The central question (the uppermost box) can be answered "if and only if" the questions in the second row are answered. The questions in the second row can be answered "if and only if" the questions in the third row are answered, and so on. The numbered questions refer to questions that are answerable with established methods.

Table 1.1. Most commonly cited statistics books based on citations in Volume 109 and 110, *Transactions of the American Fisheries Society.*

Authors (Year)	Principal Contents
Cochran (1977)	Complete description of sampling designs, emphasizing mathematics
Conover (1971)	Introduction to non-parametrics, emphasizing application and assumptions
Draper and Smith (1966)	Simple and multiple linear regression, emphasizing application
Elliot (1977)	Sampling design, emphasizing application to biological field data
Helwig and Council (1979)	Instructions for use of computerized data analysis programming (SAS)
Hollander and Wolfe (1973)	General non-parametric statistics, emphasizing application to experimental data
Nie et al. (1975)	Instruction for use of computerized data analysis program (SPSS)
Siegel (1956)	General non-parametric statistics
Snedecor and Cochran (1967)	General statistics, briefly covering experimental and sampling designs, non-parametrics
Sokal and Rohlf (1969)	General parametric statistics, written for biological scientists
Steel and Torrie (1960)	General statistics, including experimental and sampling design, non-parametrics
Zar (1974)	General statistics, both parametric and non-parametric, emphasizing biology

the study location has been predetermined and only the sampling design needs to be chosen. If, however, data are being collected to study a more general management situation, for example, the impact of length limits on harvest, then the sampling location and the sampling design must be chosen. The sampling location should be chosen on the basis of the objectives and logistics of sampling.

Choose a location that is appropriate to the objectives of the study. Frequently convenience alone dictates where sampling will occur, and a sampling program sometimes fails because of this. The abundance of the target fish species should be appropriate for the study, and conditions for sampling should be suitable at all times. For example, a study of length limits for largemouth bass should occur where bass densities are high, the fishery is intensive, and anglers can be interviewed readily. A study to determine the relation between stream gradient and abundance of brook trout must include locations with a range of gradients and trout densities.

Meeting objectives also requires that conditions in the watershed or in other parts of the water body do not bias the data. In a stream, upstream and downstream conditions always will affect data. Major changes in pollution inputs or land use during a study will change species composition. Within large lakes or reservoirs, similar problems can occur. Even after an appropriate lake or reservoir is chosen, the large surface area may require selecting limited portions for study. Shoreline areas will vary greatly in characteristics such as shoreline development, sediment and pollution inputs, and angling concentration. Other research activities may produce changes in the environment. Changes in the fishery also may affect the data; if managers anticipate changes in stocking strategies or regulations, the area is probably unsuited for long-term sampling.

Logistic considerations are of critical importance in selecting a sampling location. If several suitable locations are available, choose locations that optimize sampling effectiveness. Choose a site close to the work force, so that the time and cost of transportation are reduced. A nearby site also makes sampling more flexible, which is desirable because weather and equipment problems inevitably disrupt schedules. The work site should be easily and safely accessible for all sampling times; muddy roads, impassable streams, and high waves can cause missed sampling dates. Also consider working conditions because uncomfortable workers collect poor data and make bad decisions in the field. Consider conditions for taking the samples, processing the catch (for example, picking fish from gill nets or sorting seine catches), and recording data.

Although choosing the best sampling location is often a subjective decision, use as much objective information as possible. Begin by listing all possible locations that match the objectives, and omit those that cannot be used because of legal or political constraints. The remaining locations can then be compared explicitly to choose the best site. Examine publications and reports, and talk to everyone with experience at these locations. Visit the locations and, if possible, fly over them (chartering a plane costs very little in relation to the information gained). Choose preferred locations and alternates, in case the preferred locations become unavailable. Most importantly, record the entire process so that the choice can be explained fully if the need arises.

No location is perfect, either in terms of logistics or sampling objectives. With proper planning, however, the objectives and sampling design often can be modified to eliminate problems. The entire planning process, including choice of objectives, locations, and sampling design, must be dynamic so that the eventual data collection yields the highest return for the time and money expended.

1.3.2 Sampling Design

Once a location is chosen, the actual sites of data collection must be selected. Because the entire surface area of a waterbody or every fish cannot be examined, a subset (or sample) must be selected according to a sampling design. A great variety of sampling designs exist, all of which have advantages and disadvantages. We describe basic designs that are appropriate in most situations.

Random sampling. Random sampling is the most basic design for collecting data. Every possible sample has an equal chance of being collected. For example, if sampling plans call for seining bluegills in six 10-m sections of shore along a 300-m shoreline, each possible set of six sections would be equally likely to be chosen. In practice, the shoreline would be divided with markers into 30 sites and the sites numbered consecutively. Then, using a random method, such as choosing numbers from a random number table or picking numbered papers from a hat, six sampling sites would be chosen. In the example in Figure 1.2, sections *2, 3, 7, 15, 19,* and *28* were chosen from a random number table.

Random sampling prevents the investigator from using his judgment to pick sampling sites. Therefore, no biases, intentional or unintentional, affect the choice of sampling units. More complicated sampling designs are described below, but random sampling is used at some stage in each design to guarantee that the accuracy of the data has not been affected by the preferences of the sampler.

Stratified sampling. In order to increase the precision of random sampling, the possible sampling units (statisticians call this the "population") can be divided into groups (statististicians call them "strata") that tend to be homogeneous. Data are

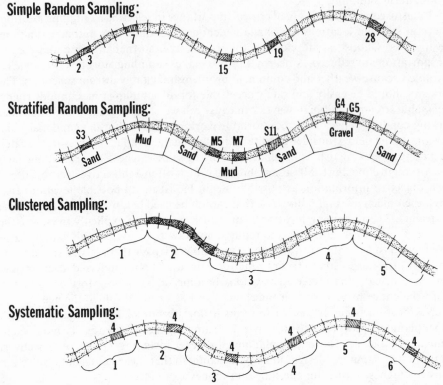

Figure 1.2. Example of various sampling designs for the selection of six seining locations along a continuous shoreline. All numbers were chosen from a random number table.

collected from chosen sampling units within each stratum, then data are analyzed by stratum, and the results are combined to yield a final answer. Using the previous example, the shoreline might be divided into three strata according to the composition of the bottom. Figure 1.2 shows that the mud stratum has 8 possible sites, sand has 15, and gravel has 7. Two seine sampling sites might be collected from each stratum, again chosen randomly.

The advantage of stratified sampling is that differences among sampling locations or times are recognized and accounted for in the sampling procedure. The idea is to divide a heterogeneous sampling environment into several subsets, each of which is relatively homogeneous. The variance of sampling data within each stratum will be low, and, when combined for all strata, the total variance will be lower than for simple random sampling. Because fisheries sampling occurs in an uncontrollable environment, stratification is almost always desirable.

The three most common types of stratification coincide with major dimensions of the aquatic ecosystem. First, differences in physical habitat within a water body will affect data and, therefore, should be stratified. Examples of habitat strata include shallow and deep waters, hypolimnion and epilimnion, exposed and sheltered shorelines, substrate types, and pools, riffles, and runs (*see* chapter 4). Second, the annual climatic cycle imposes a seasonal stratification on data. Therefore, sampling might be stratified according to periods of growth, spawning, or migration for data about fish or according to periodicity in water flow rates or thermal stratification for habitat data. Third, diurnal variations in environment require attention to time of day. For example, water quality, fish species composition, and stomach contents change throughout the day. Sampling is most often stratified into light and dark periods.

Stratification involves a series of choices concerning how many and what kinds of strata to create, and how many units (sites) to sample in each stratum. Such decisions will vary according to the objectives of sampling, including whether variance or cost of the data is to be minimized. Formal procedures for selecting strata are found in most statistical design textbooks; Cochran (1962) is the most frequently cited.

Clustered sampling. In some situations, both random and stratified random sampling are impractical or impossible. For example, in the case of choosing seining locations along a shoreline, it may be too expensive to identify every possible 10-m site in order to choose six sites. Instead, the area may be divided into groups of six sites each, which can be identified more economically. By this process, the shoreline in Figure 1.2 would be divided into five clusters. The cluster to be sampled would be chosen randomly, the six sites within the cluster would be staked off, and each site would be seined to collect the data.

Cluster sampling is appropriate if having adjacent sites will not bias the data. This is likely to occur where the habitat is either very uniform or very heterogeneous. Coral reefs and small streams are habitats in which adjacent sites are likely to be dissimilar and where cluster sampling could be used. Sampling 25 fish from a large catch to estimate average length is another likely situation for cluster sampling. Because the lengths of successive fish removed from the catch are not likely to be related, it probably is unnecessary to randomly sample the catch. Instead, a random number can be chosen to start the sample (e.g., the 18th fish) and the next 25 fish comprise the sample.

Systematic sampling. Systematic sampling, also known as uniform sampling or list sampling, is a type of cluster sampling. It consists of dividing the sampling population into sections, numbering the sampling units in each section consecutively, and sampling the same numbered unit in each section. Using the seining example, the shoreline would be divided into six sections (one for each site to be seined) of five sites each, numbered 1 to 5. A random number between 1 and 5 would be chosen (for example, *4* in Figure 1.2), and the site with the number in each section would be sampled. The seining locations are located uniformly along the shore, 50 meters apart. This is a form of cluster sampling in which the clusters are all sites with the same number rather than adjacent sites.

Systematic sampling provides an easier way to select sampling units than random sampling. Where the numbering of the sampling units is unrelated to the desired data (that is, the units are numbered randomly), this process is equivalent to random sampling. For example, choosing fishermen to interview from an alphabetical list or choosing fish to weigh from a conveyor belt probably will sample the population randomly with systematic sampling. Systematic sampling also may increase precision when the population is not randomly arranged. If adjacent samples tend to be like each other (for example, if adjacent seining sites tend to have similar fish abundance), then systematic sampling, which spreads sites out evenly, will eliminate "clumps" of sampling sites. Accuracy may decrease with systematic sampling, however, if the sampling environment changes in a repeated pattern that coincides with the distance between sampling units. For example, if the shoreline tends to have gravel substrate every 50 meters, the six seining sites could all occur in similar habitats and the data would not represent the entire shoreline. While unlikely for the small distances in this example, a regular pattern in habitat change may exist over longer distance in reservoirs, where coves occur periodically, and in streams, where gradient may alternate between steep and shallow sections.

Permanent sampling stations. Fisheries data are often collected to monitor a situation over time rather than to compute numbers that absolutely describe the situation at one instant. Such data are called indices, and the relative change in the index is the information of interest. To collect such data, permanent sampling sites are selected and sampling is conducted in the same way and at the same time each year. For the seining example, we would establish six permanent sites along the shoreline, perhaps randomly but probably in a subjective way. We would then calculate an index of "all bass caught at the six locations on June 1." Comparison of the index for successive years would indicate if abundance of bass was increasing, decreasing, fluctuating, or stable.

Randomness is abandoned in this case, and the data are not expected to represent accurately the average situation that would be revealed with random sampling. The use of the data is to monitor change, and the only assumption is that changes seen at permanent sampling stations reflect overall changes. The data must be interpreted cautiously, however, because the trend at the permanent stations may change relative to the rest of the area. A sampling design which includes mostly permanent sites plus a few randomly chosen sites allows the validity of the index to be checked constantly.

The advantage of permanent sampling sites is that sampling can be arranged to optimize logistic concerns. For example, sites can be established close to access areas, where sampling conditions are optimal, or at times when sampling is convenient or economical. It is imperative that the complete details of sampling (including site, time, equipment, and manner of sampling) and environmental conditions are all carefully described so that sampling can be conducted identically in all years or so that extreme changes in the data can be interpreted.

Altering the sampling design. It is inevitable that a pre-determined schedule for field sampling will need to be altered at times. Trees fall into water, barges sink along the shore, trucks break down, and storms occur. While careful consideration of sampling location can anticipate and avoid most of these problems, virtually no sampling schedule will ever be completed without some changes. There are no theories that describe what to do when an assigned place, time, or organism is impossible to sample. The field worker's best judgment is necessary in all instances.

Three general options are available if a sample cannot be collected. First, the specific sampling unit may be omitted if there are a large number of sampling points in total. It may be desirable to include more sampling units than needed in the original design so that the precision of the data is not reduced. Second, it may be possible to add another randomly selected site or time by the same process as the original samples were chosen. Probabilities of selecting from the remaining units will be higher than in the original process. Third, the adjacent site or time may be selected to replace the missing sampling unit. This is most appropriate if the date of sampling needs to be changed, especially if the problem is related to an equipment failure. This is less appropriate if the sampling site needs to be changed because differences between the chosen and adjacent site may bias the data.

Whether these or other options are used for adjusting a sampling design, two rules must be followed. First, an explicit (that is, written) plan for dealing with sampling problems must be created before sampling begins. This maintains the objectivity of sampling, assures consistent changes, and avoids poor decisions in the field. Second, every departure from the sampling design must be thoroughly described in a field notebook or on the data sheets; why the change was made, how it was made, and differences in conditions between the original and alternate sampling unit all need to be recorded.

Size and number of sampling units. The example used in the previous descriptions assumed that sampling would include six sites, each 10 m long. For an actual sampling program, however, the number and size of sampling units also must be chosen. As with the selections of strata, the definition of sampling units is a complex task beyond the scope of this chapter. Formulas for calculating the optimum size and number of sampling units can be found in statistical design texts.

The number and size of sampling units, either the physical dimensions or the number of organisms to observe, will be a compromise between the precision and cost of the data. Generally, more sampling units mean lower variance in the data and, therefore, higher precision. Because more small units can be sampled than large units, sampling many small units generally improves the data. Cost of sampling increases with smaller units, however, because traveling between sites, setting up equipment, and other logistic tasks require more time. With a fixed amount of time or money available for sampling, the total sample size (that is, number of sampling units times size) will decline as the size of individual sampling units declines.

Because the number of samples collected largely determines the value of the data, fishery agencies frequently develop standardized sampling schemes for specific situations (Table 1.2). Such designs consolidate the experience of the agency, the general uses of the data, and the constraints on the manager's time. Fixed designs like these are compromises between assuring that sampling will be accurate for all situations and optimizing sampling in each water body.

Table 1.2. Examples of standardized sampling schedules set up by Texas Parks and Wildlife Department for long-term sampling of fish communities in reservoirs.

Sampling technique	Size of reservoir (acres)	Number of sampling units	Where to collect	When to sample
Cove Rotenone	< 10,000	3 coves. > 1 acre each	randomly selected	between June and Sept., all within 30 days
	≥ 10,000	3 coves, > 5 acres total		
Gill Netting	<100	≤ 5 netnights	permanently selected in upper, middle, and lower section	May
	101-5000	5 netnights		
	5001-10,000	10 netnights		
	>10,000	15 netnights		
Seining	< 5,000	5 100-ft stations	permanently selected, various habitats	May, July, and Sept. during daylight
	5,001-10,000	10 100-ft stations		
	>10,000	15 100-ft stations		
Trap Netting	<100	5 netnights	permanently selected, various habitats	once between April and June, once between September and November
	101-5,000	5 netnights		
	5001-10,000	10 netnights		
	>10,000	15 netnights		

1.3.3 Information Gathered

What information to gather in the field should be a logical outgrowth of the objectives set for the study. Every piece of information has an associated price tag, and the field crew must know what information to gather before arriving on the scene. Collecting too little information in inappropriate units endangers the entire project while collecting excess or exceedingly precise data because they "might be useful" reduces the crew's capability to collect pertinent information.

Cost-benefit in information gains. Generally, low cost activities produce low benefit information. The investigator must decide what quality of information he needs. This information level will largely determine the cost of the project. For each subdiscipline within fisheries we could produce a hierarchical cost-benefit list of investigative activities. This cost-benefit hierarchy is illustrated for fish community and population studies in Box 1.3. Similar continua exist for habitat and user studies. Many of these cost-benefit decisions are discussed in appropriate chapters of this book. We wish to emphasize that costs and benefits are related and that the quality of information needed relative to the available resources must be chosen before launching a field study.

Measurement. Units of measurement also are a function of study objectives. Less precise measurements can be taken more quickly and efficiently than more precise ones and, thus, may save considerable time in the field. For example, Proportional Stock Densities (Anderson 1976) can be calculated for a fish species with only three length categories, and measurements do not need to be extremely precise. If body length-scale length regressions are necessary for establishing a back-calculation formula, however, each length measurement must be quite precise.

Many state and federal biologists and most university researchers now take all measurements in metric units. Metric units have the advantage of easy conversion to larger or smaller units because the entire system is based on units of 10. Our recommendation is that all field data be taken in metric units. Reports of a technical nature should retain the metric units, while popular publications in the United States should include both metric units and the English equivalents.

Two researcher-imposed errors in measurement affect the reporting of data. The first is the tendency to exceed least significant digits when reporting means. For example, a mean length of 384.14 mm for a group of fish cannot result from measurements of the individual fish to the nearest 10 millimeters. The second error is digit bias which occurs when a worker unintentionally favors certain numbers when making measurements. For example, some field team members will report many more even measurements than odd, exhibiting a preference for even numbers.

1.4 DATA MANAGEMENT

1.4.1 Field Recording Sheets

Accurate and rapid recording of data and subsequent protection of the records is essential for successful field work. Use waterproof paper for field sheets, and record data in pencil to ensure they are readable after getting wet. Using waterproof binders also helps prevent damage to the data sheets. There are no all-purpose field data sheets because each study is different; however, for each particular study, pre-printed forms with locations for all necessary information will help prevent oversights during field

Box 1.3. A Cost-benefit Hierarchy for Information About Fish Populations and Communities

The lowest level of information available about fish communities is how many species are present and susceptible to your sampling gear. That information is relatively easy to get but may be of limited value. At the other end of the scale are radio tracking and food habit information; both activities are very expensive, but the resultant information may be extensive. Between these extremes are a series of activities that will provide more information at increasing cost. A list of these activities and an estimate of their relative cost follows: each activity is compared to the first activity. Each activity will be affected by sampling bias, a topic discussed in this book for each gear type.

Activity	Information	Relative Cost	Comments
Species enumeration	Number of species present	1	Useful in sampling
Numbers of fish caught of each species	Relative abundance of the species present	x2	Usually the minimal level of information needed
Length of fish	Length distributions can give relative year-class strength, growth, and mortality, especially for young fish; Proportional Stock Densities (see chapter 16)	x4	A great deal of helpful information added, particularly for fast growing, temperate latitude
Weight of fish	Length-weight curves, condition factors, relative weight (see chapter 16)	x12	Most field scales are not accurate enough. Must construct special shelter or move indoors. Condition factors need both accurate lengths and weights
Age determina-ation - hard part analysis	More accurate than length distributions for calculating year-class strength, age distribution, growth history, and mortality (see chapter 17)	x120	Must have accurate length measurements, requires extra handling of fish (see chapter 6) and laboratory time for analysis
Radio or sonar tagging	Exact information about fish location and mea-surements. May be com-bined with information about water depth, temperature (see chapter 21)	x1200	Much information gained about movements of relatively few fish. Equipment cost and maintenance can be quite high

Continued

If an investigator draws up a list of this sort, the options become much more clear. Optimization is the key, i.e., a balance between the cost of an activity and the new information provided. The information gained must meet the study objectives and costs must stay within the budget. If these two criteria are not met, then either the objectives or the budget must be modified.

sampling. Design the data sheets to follow the order in which data are gathered and to eliminate the need for using several different sheets at the same time. Design field sheets so that other workers can use the data with a minimum of rewriting because errors are common when transcribing data onto summary sheets or keypunch forms. Furthermore, design the data sheets to be compatible with the data storage and retrieval system being used.

1.4.2   Data Storage and Retrieval

If data are not available when needed for decision making, then the gathering of those data was futile. Limited amounts of data can be stored in paper files, if the files are carefully organized for easy searching (e.g., by lake, stream, watershed) and maintained in good order. This storage method, however, is usually inefficient. Paper files require much space, and the entire system must be searched manually each time a portion of the data is needed. Usually data stored this way are so hard to find and put together that integrative studies using historical data are seldom attempted.

Data storage and retrieval using computer files is an efficient alternative to paper files. Computer files eliminate the problem of storage space, can be searched easily, and can be edited quickly, accurately, and economically. The major drawback of computer files is the lack of confidence many people have when faced with a computer terminal. This problem is amplified when computer programmers write storage, retrieval, and analysis packages without the consultation of the ultimate users (e.g., managers). Whenever new systems are developed, fisheries and computer professionals must cooperate so that the problems and anxieties of both groups are eliminated.

Data storage can be accomplished using two basic formats (Fig. 1.3). The first, and currently most common, is the rectangular data set. This system requires that all data pertaining to an observation be placed on one card. A great many entries can be placed on a single card, and cards can be placed in any order within the data set. The second method, the hierarchical format, requires more cards, and the cards must be maintained in correct sequence. The disadvantages of the hierarchical format are becoming less important as many systems switch to direct data input from keypunch to tape without cards. The advantages of the hierarchical format include elimination of data crowding and compatibility with alpha numeric input. The input is very similar to field data sheets so that field personnel can enter data into the computer with no coding. The input is much easier to check for errors, and reductions in processing time and core storage space by the elimination of data duplication cuts computer costs. A Catch Analysis Program using this system has been developed by Hansley (1978).

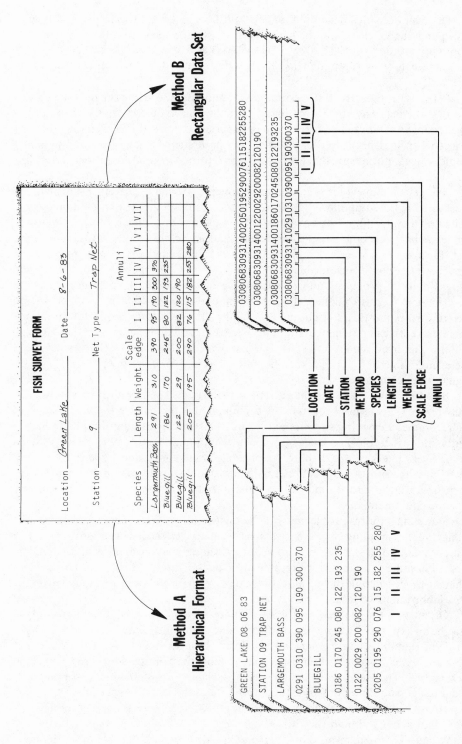

Figure 1.3. Comparisons of rectangular and hierarchical formats for storing data from a typical field data sheet.

1.5   LOGISTICS OF SAMPLING

Once all decisions have been made about how and what to sample, and how to record and analyze the data, the next step is to execute the data collection activities. Proper preparation for the daily field activities is as necessary as the project planning that came before. A sampling trip that is well-organized and well-equipped will be successful.

Perhaps the most frustrating situation in sampling is to arrive at a field site only to find that the nets have holes, the battery for the boat motor is dead, or the electro-shocker does not work. The first axiom of fisheries could read, "Everything is broken or lost when it is needed." Because fisheries equipment usually is shared by many people, you cannot be sure that everything was returned in good condition after its last use.

The first step in avoiding equipment problems is to make a careful list of all needs by visualizing the sampling activities as you read through the written sampling plans. Write down everything! For any item that can fail or break, like backpack shockers, dip nets, chemicals, and sample containers, add a backup to the list. Include spare parts for items that wear out in the field; for example, carry extra spark plugs, shear pins, batteries for electronic devices, and fuses. Assign specific people with the responsibility of procuring and maintaining each piece of equipment needed.

The next step is to collect the equipment and make sure each item works. Test batteries, start motors, stretch out nets, and calibrate meters. Look for situations where preventive maintenance will be needed soon, and do it before sampling. Repair all items that need repair, and start well in advance of the sampling date; what looks like a small hole in a net may take hours to re-weave, and getting parts for a small engine may require weeks.

Be prepared to make emergency repairs in the field. The second axiom of fisheries could read, "If it was not broken in the beginning, it will break while you are using it." We recommend carrying supplies and tools in a surplus ammunition box, which is nearly indestructable and waterproof. As a minimum, carry a slot and Phillips screwdriver, pliers, a spark plug wrench, electrical tape, duct tape, baling wire, electrical wire, and first aid supplies; add additional supplies for special equipment. Carry operating and repair manuals for specialized equipment even if you are familiar with the equipment.

Sampling requires people, and much of fisheries sampling requires many people. Managing the sampling crew is critical to efficient and effective collection of data. The first consideration is having enough people. Although having extra people should be avoided in some situations (for example, if a boat is over-crowded), the usual problem is having too few people during parts of the task. Plan for a crew large enough to do the most labor-intensive task efficiently.

Organizing the workers at the site is necessary for successful sampling. There must be an explicit hierarchy of authority and responsibility among the field workers. A rule-of-thumb is that one person can effectively supervise five others; therefore, it will be necessary to assign supervisory roles to more than one person when a large field crew is used. The leader must assign specific tasks to each individual and preferably keep each person doing the same tasks throughout the sampling period. In this way, the activity will be done consistently, and the responsible person can answer questions about that activity should they arise later.

Whenever possible, only one person should record all data. If more than one person is needed, each should be told exactly how to record, including instructions about units of measurement and number of decimal places. Recording data is the only link between the field and the final results, so an experienced and capable worker should do this job.

The one person who functions as leader of the field activity should not have a specific job assignment. The leader assures that sampling occurs smoothly and according to design. The leader should oversee the set-up of the field activity, assign and explain tasks, check to see that workers are following the design and instructions, and assist where extra help is needed.

The last step before sampling is informing relevant people about sampling plans. The appropriate law enforcement offices should always be notified so that they can answer calls from casual observers. State fishery agencies and nearby universities should be notified so that other sampling or projects in the area will not be affected. Fishing groups and outdoor writers should be notified so that they can observe (maybe even help) and inform their members or readers about the work. Good relations with the public are as important for fishery management as good data.

1.6   COMMUNICATING THE RESULTS OF SAMPLING

Responsibilities of fisheries professionals do not end with data collection and analysis. A project is not complete until the information is communicated to the appropriate audience. The communication process takes several forms depending on the clients and the purpose of the investigation. It is important to realize that no single report or visual aid will serve all audiences. In fact, each set of clients normally will require a separate preparation carefully tailored to their needs and interests. No matter what the purpose of the written or oral presentation, peer review of the presentation is critically important.

The most common outline for a report or presentation is (1) abstract/executive summary, (2) problem statement, (3) introduction or background, (4) objectives, (5) methods and site description, (6) results, (7) discussion, and (8) summary and recommendations.

1.6.1   Written Communications

Written communications take a variety of forms depending on the audience. A common, but often frustrating, assignment for fisheries researchers and managers is the production of popular articles for newspapers or magazines. These outlets generally require rather short articles that must be written in simple language with adequate background about why the investigation was done and what the results mean to the readers.

Technical publications are generally written for colleagues within an agency. They have limited distribution and contain little interpretation of data; instead they stress communication of results or new techniques. Many of these reports become part of the "fugitive" literature because they are almost impossible to find except by personal contact. Some agencies and institutions, however, produce series of technical publications that reach a wider audience and are indexed by abstracting services.

The greatest exposure for research or management activities and results comes from publishing in a journal or symposium. Each journal has its own format and style which must be followed precisely. The submitted paper is usually reviewed by a panel of peers selected by the journal editor. Rejection rates are high for the major journals, and a period of 6-12 months between submission and publication is common.

A general characteristic of journal articles is concise, efficient writing. Titles must be carefully worded to include the important terms in a paper yet remain short. The increasing use of computerized literature searches based on titles makes title construction critically important. Abstracts should emphasize results, not methodology, because abstracts are often available through computer searches and because readers of journals often read abstracts to choose papers for further study.

1.6.2 Figures and Visual Aids

The figures used in a communication, whether written or oral, provide the focus for the audience. Ideally, figures for written publications should be prepared by professional artists. Clarity and economy of design are critical. Most published graphics will be reduced in size by 1/3 to 1/2 and, therefore, should be drawn with very large letters and symbols. Design tables to present information without repetition and arrange the entries in a vertical format to reduce space required.

Most verbal presentations should include supporting visual aids. Projection of 5 x 5-cm slides is the most effective method of graphic presentation for large audiences. If the audience is small, paper charts or overhead projections of large transparencies also may be useful. Use only a single medium for visual aids whenever possible. If more than one medium is needed, switch among types carefully to avoid confusion and audience discomfort associated with abruptly changing levels of room light.

A table of data or a figure prepared for print is seldom appropriate for illustrating an oral presentation. Such figures are generally too complex, and the lettering is too small. Each visual aid for an oral presentation should concentrate on a single topic and be easily interpreted. Each visual aid should have a purpose; never use slides for fillers. Plan ahead so that the necessary graphics are ready in time for practicing. As with printed graphics, seek help from professional photographers.

1.7 REFERENCES

Anderson, R. O. 1976. Management of small warm water impoundments. *Fisheries* 1(6):5-7, 26-28.

Cochran, W. G. 1977. *Sampling techniques,* third edition. John Wiley and Sons, New York, New York, USA.

Conover, W. J. 1971. *Practical nonparametric statistics.* John Wiley and Sons, New York, New York, USA.

Draper, N. R., and H. Smith. 1966. *Applied regression analysis.* John Wiley and Sons, New York, New York, USA.

Elliot, J. M. 1977. *Some methods for the statistical analyses of benthic invertebrates.* Freshwater Biological Association Scientific Publication 25, second edition. The Ferry House, Ambleside, England.

Hansley, J. E. 1978. *The catch analysis program (CAP): An interactive computer program for the management and analysis of fisheries data.* Master's thesis. The Ohio State University, Columbus, Ohio, USA.

Helwig, J. T., and K. A. Council. 1979. *SAS User's Guide.* SAS Institute, Cary, North Carolina, USA.

Hollander, M., and D. A. Wolfe. 1973. *Nonparametric statistical methods.* John Wiley and Sons, New York, New York, USA.

Nie, N .H., C. M. Hull, J. G. Jenkins, K. Steinbrenner, and D. H. Bent. 1975. *SPSS: Statistical package for the social sciences,* second edition. McGraw-Hill, New York, New York, USA.

Phenicie, C. K., and J. R. Lyons. 1973. *Tactical planning in fish and wildlife management and research.* Bureau of Sport Fisheries and Wildlife Resource Publication 123.

Siegel, S. 1956. *Nonparametric statistics for the behavioral sciences.* McGraw-Hill, New York, New York, USA.

Snedecor, G. W., and W. G. Cochran. 1967. *Statistical methods,* sixth edition. Iowa State University Press, Ames, Iowa, USA.

Sokal, R. R., and F. J. Rohlf. 1969. *Biometry.* W. H. Freeman, San Francisco, California, USA.

Steel, R. G. D., and J. H. Torrie. 1960. *Principles and procedures of statistics.* McGraw-Hill, New York, New York, USA.

Zar, J. H. 1974. *Biostatistical analysis.* Prentice-Hall, Englewood Cliffs, New Jersey, USA.

Chapter 2
Finding Literature and Reports

CLARENCE A. CARLSON AND CURTIS L. GIFFORD

2.1 INTRODUCTION

Everyone in the fisheries profession eventually needs information available only in scientific literature. Students must learn to search this literature to prepare term papers and research proposals. Fisheries managers and administrators can improve their planning or sampling by considering research results in scientific books and journals. Researchers should be aware that science proceeds only by addition of new information to knowledge established by earlier workers and that much duplication of effort can be avoided by familiarity with the literature. Wilson (1952), in his classic book on scientific research, referred to two important and feasible goals of a literature review: to find out if the information sought is already available and to acquire a broad general background in the given field.

This chapter introduces the literature of fish and fisheries and the techniques for locating it. This chapter appears before the chapters on sampling because a literature search is needed prior to data collection in fisheries research or management programs.

2.2 MODERN LITERATURE SEARCHING

Literature searches should progress logically according to established procedures. They are usually conducted by examining printed indexes or by using computers to access databases (continuously updated information files) created from these indexes.

2.2.1 Beginning a Search

Although the beginning point of a literature search depends on the knowledge and experience of the researcher, all should start with an outline of objectives and boundaries which may be modified as the search progresses. If the subject is unfamiliar, it is helpful to read an authorative summary in a textbook, encyclopedia, or handbook. A good general source is the *McGraw-Hill Encyclopedia of Science and Technology*.

Another approach, perhaps for more knowledgeable researchers, is to review a library's card catalog and the *Monthly Catalog of United States Government Publications*. (These publications usually are not cataloged with other holdings.) *Library of Congress Subject Headings* can be consulted for the exact words or phrases to use in searching the subject part of the card catalog. Because books rarely provide the most up-to-date information, they should usually be supplemented by articles found in serials or journals. Review articles in these journals can be extremely valuable in providing orientation and key references.

To locate journal articles, use published or computerized indexes and abstracts discussed later in this chapter. These often lead to reports of original research, but the user needs to know enough about a subject to reduce it to a set of descriptive terms (keywords). Finding and reading recent articles and following their references to earlier works is the penultimate step in a literature review.

Finally, it is important to locate new papers by known specialists on specific topics and to learn of research in progress. Current-awareness publications, the Smithsonian Science Information Exchange (both discussed later in this chapter), and personal contacts are most useful for achieving these goals. Biologists have long been dissatisfied with their ability to keep up to date; the best products of modern information retrieval technology have not yet satisfied the conscientious authority.

Veteran researchers recognize, and novices are advised, that the quality of information discovered in a comprehensive search will vary considerably. Information from such sources as the popular press, progress reports, books, and peer-reviewed journal articles is not all equally reliable. The latter obviously deserve most credence, but they also vary in quality because of differences in standards of referees and editors.

2.2.2 Manual Searches

Moore (1980) discussed developing a file of index cards as a literature search progresses. Most experts advise placing bibliographic citations on the cards according to the style prescribed by a major journal in one's field. We, however, recommend recording verbatim information, at least for articles from less-familiar journals and government documents. Major North American fisheries journals have different requirements regarding title abbreviations. Moreover, certain United States Forest Service publications require the full spelling of even the first names of authors. An Interlibrary Loan always requires the full title, as well as verification of its source in a standard indexing or abstracting tool. For this reason we recommend writing the title, year, volume, page, and/or abstract number of each source on index cards. File completed cards alphabetically by author's surname under keyword categories. Complete citations require a little extra time and patience but can minimize repeated visits to the library to look up bibliographic details.

Because most libraries have copying facilities, many researchers choose to copy relevant pages from books, journals, or microfiche as they are found or obtained on loan. Copy only in compliance with provisions of the Copyright Revision Act of 1976 (CRA, Public Law 94-554) and international copyright laws. Fair use provisions of CRA establish rights of persons other than copyright owners; these are vague, but there is general agreement that single copies of book chapters or journal articles can be made and kept in a personal file for use in research, study, or lesson planning. Teachers should consult Anonymous (1977), Johnston (1978), or Miller (1981) for limitations on copying by students or for classroom use. Securing reprints of journal articles (and other material), organizing personal collections, and retrieving documents were discussed by Jahoda (1970) and Mosby and Burns (1980). Many journals, abstracts, and indexes include addresses of authors to facilitate reprint acquisition. The user must weigh costs in time and dollars in deciding whether to write for a reprint or make a copy.

2.2.3 Computer-assisted Searches

Early reports on computer-assisted bibliographic retrieval systems for biologists considered timeliness and comprehensiveness of databases, overlap between databases, and costs of their use. Although concerns about these and other limitations persist (Doszkocs et al. 1980), we are convinced that computer-assisted literature searches will become increasingly important to fisheries biologists. Mayer and Waters (1981) urged fisheries scientists to ". . . throw away your 3 X 5 cards and send your bookcases back to the library to make room in your office for a computer terminal . . ." to get ready for the information flood. The following introduction to computerized searches is based on their report, that of Cuadra et al. (1981), and personal experience.

Computerized bibliographic search facilities vary from specialized in-house systems to large national or international systems composed of subsystems managed by vendors; their stored information ranges from titles to complete documents. The development of online systems (by which database information is transmitted directly to a user) has allowed several commercial services to provide computer-assisted searches at nominal cost (Fig. 2.1). In the past decade several specialized databases have been consolidated under master search systems. DIALOG Information Services, Inc., associated with Lockheed Corporation since 1972, is one of the largest of these. It contains over 250 databases and over 55 million records; the user selects the most productive databases for his search. Many university libraries and research agencies are affiliated with these systems and have equipment necessary to communicate with them.

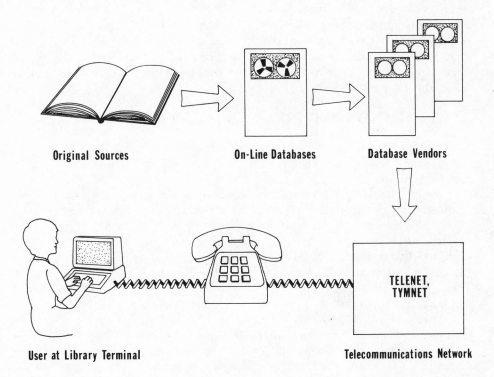

Original Sources On-Line Databases Database Vendors

TELENET, TYMNET

User at Library Terminal Telecommunications Network

Figure 2.1. Access to bibliographic information by computer.

To use computer databases in a literature search, the user must have the level of familiarity with a subject that is necessary for using abstracting or indexing journals. Consult a librarian or bibliographer familiar with the techniques of computer-assisted literature searching after carefully compiling a keyword list. Select the most appropriate database for the search on the basis of such factors as cost, availability, and system capabilities. The next steps are to log on to the computer, call up the database file, and request information in interactive steps proceeding from general to more specific keyword descriptors. After each request, the computer displays the number of records in the database relating to the designated subject. After narrowing the search sufficiently, i.e., reducing the number of citations to a reasonable number, obtain a display of them on the terminal. The references usually contain complete bibliographic descriptions and may include an abstract or keywords. After reviewing these records and modifying the strategy as necessary, issue a final command to display and print the results of the search at the terminal or at the location of the database. Some vendors will automatically update searches on specific subjects on a continuing basis. Copies of relevant papers can be ordered through some systems. If the full text relating to a citation is not available through the database vendor, obtain it from a library.

Most commercial and governmental computerized information retrieval systems charge a service fee based on an hourly connect-time rate, the number of citations or pages printed, and (if applicable) telecommunication costs. Connect-time rates range from about $25 to $300 per hour and average about $65 per hour. Users can minimize costs through careful refining of keywords before logging on to the system. The average charge for offline printing is about $0.15 per citation. A search of a single bibliographic database generally takes less than 15 minutes and costs about $30. Although database use has rarely been subjected to cost-benefit analysis, user reports leave no doubt that this is a valuable resource. In selecting methodologies, the researcher must weigh the cost of a computer-assisted search against the extra time required for a manual search.

2.3   FISH AND FISHERIES-RELATED LITERATURE

Literature on fish and fisheries includes separate and collected works, fisheries statistics, and bibliographic guides. The latter are used for navigation in a sea of published information.

Several guides to the literature of biology and zoology are available, including the relatively recent guide of Kirk (1978). Typical studies of literature of field-oriented biological disciplines are those of Gorham (1968) and Hein (1967). Burns' (1971) chapter "Using the Literature on Wildlife Management" provided a comprehensive guide which was updated by Moore (1980). Although more limited in scope, our chapter is recommended as a complementary treatment of a related subject. Throughout the rest of this chapter we have created tables to summarize information and have used text sparingly to clarify or supplement the tables.

Artedi's (1738) *Bibliotheca ichthyologica. . .*, the first bibliography of fishes, was expanded and corrected by Walbaum (1788-93). Dean's (1916-1923) *A Bibliography of Fishes* is widely recognized as a major contribution to the development and history of ichthyology. Scattergood (1953) reviewed bibliographic sources for fisheries researchers, and Lagler (1956) provided a guide to literature of fish and fisheries. The annotated bibliography by Kelts and Bressler (1971) is a noteworthy listing that particularly warrants consideration. Brigham (1974) and Nielsen and Summers (1979) studied sources of literature cited in fisheries journals.

2.3.1 Monographic Literature

Monographs are learned treatises on circumscribed areas of learning. A library's card catalog, as an index to the books in its collection, may not lead you to new publications in a field of specialization. One way to accomplish this is to peruse serials for lists of new books. Many leading journals in aquatic sciences contain reviews and listings of these books.

Table 2.1 summarizes sources which list availability of books, including scientific and technical monographs. Subject guides to *Book in Print* and *Forthcoming Books,* the most frequently used sources, are published annually as separate companion volumes. *American Book Publishing Record* is supplemented by cumulatives for the periods 1876-1949 and 1950-1977. *Libros en Venta,* a Spanish-language list of books in print, and *Les Livres Disponibles: French Books in Print* are representative of foreign language guides. (A complete list of titles available in French is published annually by Edi-Quebec.) We suggest consulting a reference librarian for information on current books outside North America.

Recently it has also become possible to identify a book's availability through online computer databases such as BOOKS IN PRINT (BBIP), ISI/ISTP & B (Index to Scientific & Technical Proceedings & Books), and BOOKSINFO (Books Information). Abstracts and indexes, discussed later in this chapter, often cover new books as well as journal articles. Guides to the availability of out-of-print books or reprint editions include *A B Bookmans Weekly* (published by Moses Pitt, Clifton, New Jersey), *American Book Trade Directory* (Bowker), *Books in Series in the United States* (Bowker), and *Books on Demand* (reprint guide published by University Microfilms, Ann Arbor, Michigan).

Locations of many important books can be determined by consulting the *National Union Catalog,* which lists the works cataloged by the Library of Congress and by other North American libraries contributing to its cataloging programs.

2.3.2 Serial Literature

Serials are defined as publications issued in successive parts bearing numerical or chronological designations and intended to be continued indefinitely. Articles in serials are the primary sources of information in fisheries science. Lists of serials or serials bibliographies which contain fisheries periodicals include *Ulrich's International Periodicals Directory, Irregular Serials and Annuals: An International Directory, World List of Periodicals for Aquatic Sciences and Fisheries, Serial Sources for the BIOSIS Data Base,* and *World List of Scientific Periodicals, 1900-1960.* The online computer database ULRICH'S INTERNATIONAL PERIODICALS DIRECTORY corresponds to the printed *Ulrich's International Periodicals Directory* and other Ulrich's publications.

Serials used and cited by fisheries professionals are far too numerous to present here. Table 2.2, a list of some commonly-cited serials, is based on a recent study of citations in fisheries journals and on a compilation of serials surveyed by the most comprehensive abstracts for fisheries publications. *Serial Sources for the BIOSIS Data Base* is recommended as a source of abbreviations of serial titles. Important new

Table 2.1. Guides to new books and books in print

Title	Publisher	Scope	Indexes	Frequency of issue
American Book Publishing Record	R. R. Bowker Co., New York	Monthly cumulation of *Weekly Record* listings arranged by subject according to Dewey Decimal classification	Author and title	Monthly, with annual cumulations
Book in Print and Books in Print Supplement	Bowker	Books on all subjects if available for sale in U.S. *Supplement* lists changes and books missed	Author, title, and information on publishers (and subject in *Supplement*)	Annual
Books in Series in the United States	Bowker	Titles in popular, scholarly, and professional series	Author, series, subject, and title	Annual
Canadiana	National Library of Canada, Ottawa	Publications of Canadian origin or interest arranged by Dewey Decimal classification	Author, series, and subject	11/year; July and August are combined
Canadian Books in Print	University of Toronto Press	Books published in Canada, all subjects	Author, title, and publisher	Annual
Cumulative Book Index	H. W. Wilson Co., New York	International bibliography of books published in the English language	Author, subject, and title	Monthly except August with annual cumulations
Forthcoming Books in Print	Bowker	Cumulative list of forthcoming titles	Author, title, and publisher information	Annual
Library of Congress: Books: Subjects	Library of Congress	Continuing and cumulative subject bibliography of recent works received by Library of Congress and other U.S. libraries participating in their cataloging	None	Quarterly with annual and quinquennial cumulations

Table 2.1. (Continued)

Title	Publisher	Scope	Indexes	Frequency of issue
National Union Catalog	Library of Congress	Complements *Library of Congress Books: Subjects*; worldwide compilation of books received, arranged by author	None	9/year; monthly except March, June and September. Variable cumulations
Scientific and Technical Books and Serials in Print	Bowker	Subject listing of books on science and technology	Author, title, and publisher information	Annual
Weekly Record	Bowker	Current American books and foreign books distributed in U.S. listed by author (or title for multi-authored books). Cumulated in *American Book Publishing Record*	None	Weekly

Table 2.2.   Journals commonly cited in fisheries literature (according to Nielsen and Summers 1979) and where they are indexed. Journals are listed in order of citation frequency, based on lists of serials reviewed (Anonymous 1976, 1980, 1981; Schoumacher 1980).

Journal	Indexed in			
	Aquatic Sciences & Fisheries Abstracts	Biological Abstracts	Sport Fishery Abstracts	Zoological Record
Canadian Journal of Fisheries and Aquatic Sciences (continues Journal of the Fisheries Research Board of Canada)	X	X	X	X
Transactions of the American Fisheries Society	X	X	X	X
Copeia	X	X	X	X
U.S. National Marine Fisheries Service Fishery Bulletin (continues U.S. Fish and Wildlife Service Fishery Bulletin)	X	X	X	X
Progressive Fish-Culturist	X	X	X	X
Comparative Biochemistry and Physiology (continued in parts as Comparative Biochemistry and Physiology, [A] Comparative Physiology, [B] Comparative Biochemistry, and [C] Comparative Pharmacology)	X	X	X	X
Journal of Fish Biology	X	X	X	X
Ecology	X	X	X	X
Limnology and Oceanography	X	X	X	X
Science (Washington, D.C.)	X	X	X	
Canadian Bulletin of Fisheries and Aquatic Sciences (continues Bulletin of the Fisheries Research Board of Canada)	X	X	X	X
Canadian Journal of Zoology (continues Canadian Journal of Research, Section D, Zoological Sciences)	X	X	X	X
Proceedings of the Annual Conference Southeastern Association of Fish and Wildlife Agencies (continues Proceedings Annual Conference Southeastern Association of Game and Fish Commissioners)	X	X	X	X
American Midland Naturalist	X	X	X	X
Biological Bulletin (Woods Hole)	X	X	X	X
Journal of Animal Ecology	X	X	X	X
California Fish and Game	X	X	X	X
Journal of Experimental Zoology	X	X	X	X
Estuaries (supersedes Chesapeake Science)	X	X	X	X
NOAA (National Oceanic and Atmospheric Administration) Technical Report NMFS (National Marine Fisheries Service) SSRF (Special Scientific Report Fisheries) (continues U.S. Department of Commerce National Marine Fisheries Service Special Scientific Report Fisheries)		X	X	X
California Department of Fish and Game Fish Bulletin	X	X	X	X

Table 2.2. (Continued)

	Indexed in			
Journal	Aquatic Sciences & Fisheries Abstracts	Biological Abstracts	Sport Fishery Abstracts	Zoological Record
Journal of Wildlife Management	X	X	X	X
New York Fish and Game Journal	X		X	X
U.S. Fish and Wildlife Service Research Report (continued as U.S. Bureau of Sport Fisheries and Wildlife Research Report. Reverted to U.S. Fish and Wildlife Service Research Report)	X	X	X	

fisheries journals include *Fisheries Research* and *North American Journal of Fisheries Management.*

The *Union List of Serials in Libraries of the United States and Canada* and its supplements (*New Serial Titles*), published by the Library of Congress, are useful in determining which libraries hold a particular journal. The on line database UNION corresponds to the *Union List of Scientific Serials in Canadian Libraries.*

2.3.3 Indexes and Abstracts

Indexing and abstracting tools providing information on a particular subject are basic to any literature search. Abstracts provide brief summaries of publications in addition to bibliographic descriptions contained in indexes. Published versions of indexes and abstracts are usually supplemented by computer databases, and many of the most useful bibliographic services for fisheries professionals exist in both formats.

Published versions. Aquatic Sciences and Fisheries Abstracts (ASFA) is likely to be of most value to fisheries professionals. It is published by Cambridge Scientific Abstracts in cooperation with the Food and Agriculture Organization of the United Nations (FAO), the Intergovernmental Oceanographic Commission (IOC), and the United Nations Ocean Economics and Technology Office (UNOETO). Since 1978, *ASFA* has been published in two parts. *ASFA 1: Biological Sciences & Living Resources* covers biology and exploitation of aquatic organisms and related legal, policy, and socio-economic considerations. It includes a fisheries section which covers fishing methods, statistics, aquaculture, food technology, and marketing. *ASFA 2: Ocean Technology, Policy & Non-living Resources* covers pollution, oceanography, and limnology. Over 5,500 journals, books, and reports are monitored in compiling each monthly issue; nearly 30,000 abstracts are published annually.

Moore (1980) described several other published abstracting and indexing services that might also be useful to fisheries professionals, including *Bibliography of Agriculture, Biological Abstracts, Biological and Agricultural Index, Chemical Abstracts, Dissertation Abstracts International, Forestry Abstracts, Key-Word Index of Wildlife Research, Wildlife Review,* and *Zoological Record.* We have summarized data on published indexes and abstracts that are most likely to be valuable to fisheries biologists in Table 2.3. The best way to gain familiarity with these is to read their guides and to practice using them.

Table 2.3. Important published bibliographic indexes and abstracts for fisheries biology (* = Described by Moore 1980)

Title	Publisher	Scope	Dates Covered	Types of Indexes	Abstracts
Aquatic Sciences & Fisheries Abstracts	Cambridge Scientific Abstracts	Science, technology, and management of freshwater and marine systems	1971-Present	Author, subject, geographic (and taxonomic in ASFA 1)	Yes
Biological Abstracts*	BIOSIS	Published serial and monographic biological research literature	1927-Present	Author, biosystematic (phylum to family), generic (genus and/or species), concept (broad biological), and subject (key words)	Yes
Biological Abstracts/RRM	BIOSIS	Biological reports, reviews, and meetings	1965-Present	Same as BA	No
CIS/Index	Congressional Information Services (CIS)	Publications of the U.S. Congress (except *Congressional Record*)	1970-Present	Bill number, document number, report number, subjects and names, and titles	Yes
Current References in Fish Research	Victor Cvancara, University of Wisconsin - EC, Eau Claire, WI	Annual title listing of published fisheries research papers	1976-Present	Author, keyword, and scientific name	No
Dissertation Abstracts International*	University Microfilms International	Doctoral dissertations from cooperating institutions, U.S. and Canada	1938-Present	Author, keyword, title, and classified	Yes
EIS: Digest of Environmental Impact Statements	Information Resources Press, Arlington, VA	Environmental Impact Statements	1977-Present	Agency/organization, EIS number, geographical, legal, and subject	Yes
Fishery Publication Index	U.S. Fish and Wildlife Service (Circulars 36 and 296)	Publications of U.S. Bureau of Fisheries and U.S. Fish and Wildlife Service	1920-54 and 1955-64	Author, publication series, and subject	No

Table 2.3. (Continued)

Title	Publisher	Scope	Dates Covered	Types of Indexes	Abstracts
Index and List of Titles, Fisheries Research Board of Canada and associated publications	Fisheries Research Board of Canada	Publications of the Fisheries Research Board and its predecessors and associated publications	1900-1973	Author and subject (and list of titles)	No
Index to Scientific and Technical Proceedings	Institute for Scientific Information (ISI) Philadelphia, PA	Published proceedings in scientific and technical disciplines, worldwide	1978-Present	Author/editor, corporate, meeting location, sponsor and subject	No
Index-Transactions of the American Fisheries Society	American Fisheries Society	Transactions of AFS	1872-1976	Author, subject, and species	No
Monthly Catalog of United States Government Publications*	U.S. Government Printing Office	Most U.S. Government publications	1895-Present	Author, title, subject, series/report, stock no., classification no., and title keyword	No
Science Citation Index*	Institute for Scientific Information	Worldwide scientific literature, with emphasis on journal articles	1961-Present	Author (cited names and citing names) and subject	No
Sport Fishery Abstracts	U.S. Fish and Wildlife Service	Current literature in sport fishery research and management	1955-Present	Author, geographical, systematic, and subject	Yes
Zoological Record*	BIOSIS and The Zoological Society of London	Zoological literature by year (most recent, volume 115, on literature of 1978)	1864-Present	Author, geographical, paleontological, subject, and systematic	Yes

Indexes to *Biological Abstracts* (BA) were changed in 1979 and are not as described by Moore (1980). *Biological Abstracts/RRM* (Reports, Reviews, Meetings), a supplement to *Biological Abstracts,* replaced *BioResearch Index* and should be searched separately. Both *BA* and *Biological Abstracts/RRM* are published by BioSciences Information Services (BIOSIS) of Philadelphia.

EIS: Digests of Environmental Impact Statements is supplemented by an annual *EIS Cumulative* for years after 1976 and by *EIS Retrospective 1970-1976,* which includes indexes and abstracts to all final environmental impact statements issued between 1969 (the date of enactment of the National Environmental Policy Act) and 1977.

The two volumes of the *Fishery Publication Index* by Aldous (1955) and George Washington University (1969) are supplemented by McDonald's (1921) and Aller's (1958) indexes to publications of the United States Bureau of Fisheries. Carter's (1968, 1973) indexes and lists of titles may be updated by consulting indexes in individual volumes of the *Journal of the Fisheries Research Board of Canada* or its successor, the *Canadian Journal of Fisheries and Aquatic Sciences.* The same means may be used to complement the indexes to the *Transactions of the American Fisheries Society* by Chapin (1929) and Beckman (1955, 1980). Several indexes to other journals which contain articles on fish and fisheries are available; 20-year indexes for *Ecology* and for *Limnology and Oceanography* were issued recently. Many journals have annual indexes.

Cumulative indexes of the *Monthly Catalog* are now published annually and semiannually. The *Cumulative Subject Index to the Monthly Catalog of United States Government Publications (1900-1971)* should be used for starting subject searches prior to 1971. *Microlog Index* is a comprehensive guide to federal, provincial, and local Canadian government publications. The *Catalogue: Government of Canada Publications* provides spotty coverage of fisheries literature. *SciTech Publications,* published by the Scientific Information and Publications Branch of the Department of Fisheries and Oceans, Canada, may be consulted for publications of that department.

Garfield (1979) comprehensively discussed theory and application of citation indexing. He provided a rationale for and described *Science Citation Index* in 1964 (Garfield 1964). It now consists of three related portions: the *Source Index,* the *Citation Index* and the *"Permuterm" Subject Index. Sport Fishery Abstracts,* available for a modest subscription fee, is the most easily accessible abstracting journal for managers with limited opportunity to visit libraries. The printouts of computerized bibliographic analysis of ichthyological literature prepared by Atz (1971,1973) developed into the "Pisces" section of *Zoological Record* after the 1970 edition.

Commercial Fisheries Abstracts (1948-1973) was succeeded by *Marine Fisheries Abstracts* (1973-1974); these provided information on progress in fishery research and technology. *World Fisheries Abstracts* was published by the Food and Agriculture Organization of the United Nations from 1950 to 1972. *Oceanic Abstracts,* published by Cambridge Scientific Abstracts since 1964, covers worldwide technical literature on the oceans; it is overlapped considerably by *ASFA.* Other indexes and abstracts that may be of interest to fisheries biologists are *Animal Behavior Abstracts, Ecological Abstracts, Ecology Abstracts, Energy (Information) Abstracts, Entomology Abstracts, Environmental Abstracts, Environmental Health and Pollution Control* (abstracts), *Environmental Periodicals Bibliography, Pesticides Abstracts, Pollution Abstracts, Review of Applied Entomology, Selected Water Resources Abstracts,* and *Toxicology Abstracts.*

Monthly Checklist of State Publications, a record of state documents received by the Library of Congress, is the primary listing of state publications.

Computer databases. Databases are likely to remain in a state of flux for sometime as the trend toward online literature searching continues, and we recommend consulting librarians, bibliographers, or listings of databases for new developments. Table 2.4 was constructed primarily from information presented by Cuadra et al. (1981, 1982). *ASFA,* the primary database for fisheries biologists, was described by Adams and Walter (1981).

Online bibliographic databases not shown in Table 2.4 which might be of interest to fisheries professionals include AGRICOLA (corresponds to *Bibliography of Agriculture),* AGRIS (*Agricultural Research Information System of the United Nations),* CNEXO (literature of oceanography), ENERGYLINE (*Energy Information Abstracts),* ENVIROLINE (*Environmental Abstracts),* ENVIRONMENTAL BIBLIOGRAPHY (*Environmental Periodicals Bibliography),* LIFE SCIENCES COLLECTION (*Ecology Abstracts, Entomology Abstracts, Toxicology Abstracts* and others), OCEANIC ABSTRACTS, POLLUTION ABSTRACTS, SELECTED WATER RESOURCES ABSTRACTS, and STATE PUBLICATIONS (contains citations to a broad range of publications of state agencies and universities).

2.3.4 Technical Reports

Technical reports describe results of research usually supported by the federal government; they are issued in tremendous numbers in a variety of formats. Although they may contain significant information, they are often limited in distribution and, compared to journal articles, are less likely to be discovered in a literature search. Some abstracts and indexes make special efforts to include technical reports on fish or fisheries; *Aquatic Sciences and Fisheries Abstracts* and *Sport Fishery Abstracts* are notable examples. The biweekly *Government Reports Announcements & Index* (GRA&I), produced by NTIS (National Technical Information Service), is an excellent source of information on United States government technical reports. Its bibliographic citations are arranged by subject, and it contains indexes of authors, keywords, and contract, grant, or report numbers. The online computer database NTIS, which corresponds to the printed version of GRA&I, can also be used to access government technical reports.

The Fish and Wildlife Reference Service operates under a contract between the Denver Public Library and the United States Fish and Wildlife Service to provide access to selected technical reports generated by the Anadromous Sport Fish Conservation Program, the Federal Aid in Fish and Wildlife Restoration Program, the Endangered Species Grants Program, and the Cooperative Fish and Wildlife Research Units. The Reference Center's computerized information retrieval system is particularly useful in accessing Dingell-Johnson (DJ) reports. The database STATE PUBLICATIONS can also be used to locate state technical reports.

2.3.5 Fisheries Statistics

Fisheries statistics constitute a rather unique form of literature in this field, and they range from data on fish hatchery production to tons of food fish landed in various ports around the world. Publications dealing exclusively with fisheries statistics include the *FAO Yearbook of Fishery Statistics, U.S. National Marine Fisheries*

Table 2.4. Online computer databases for retrieval of bibliographic information for fisheries biologists

Title	Producer	Online Service	Scope	Coverage
AQUACULTURE	National Oceanic and Atmospheric Administration Environmental Science Information Center	DIALOG	Citations and some abstracts on marine, brackish, and freshwater aquaculture, international	1970 to present
AQUALINE	Water Research Center	DIALOG	Citations and abstracts on water, wastewater, and the aquatic environment, international	1974 to present
ASFA (Aquatic Sciences and Fisheries Abstracts)	Cambridge Scientific Abstracts	DIALOG & QL Systems, Ltd.	Citations and abstracts; corresponds to printed *ASFA*	1978 to present
BIOSIS PREVIEWS	BIOSIS	Bibliographic Retrieval Services (BRS), Canada Institute for Scientific and Technical Information (CISTI), DIALOG, and System Development Corporation (SDC)	Citations (and abstracts since 1976 on some services); corresponds to printed *BA* and *BA/RRM*	1970 to present (except on CISTI)
CANADIAN ENVIRONMENT (CENV)	Environment Canada, Inland Waters Directorate (WATDOC)	QL Systems, Ltd.	Citations and abstracts of water-related and other environmental literature pertaining to Canada	1970 to present
COMPREHENSIVE DISSERTATION INDEX (CDI)	University Microfilms International, Inc.	BRS & DIALOG	Citations; corresponds to *Dissertation Abstracts International* and other printed indexes	1861 to present (1962 to present for master's theses)

Table 2.4. (Continued)

Title	Producer	Online Service	Scope	Coverage
CIS/INDEX	Congressional Information Service (CIS)	DIALOG & SDC	Citations and abstracts; corresponds to printed *CIS/Index*	1970 to present
EIS: Digest of Environmental Impact Statements	Information Resources Press	BRS	Abstracts of environmental impact statements	1970 to present
ISI/ISTP & B (Index to Scientific & Technical Proceedings & Books)	Institute for Scientific Information	ISI Search Network	Citations; corresponds to *Index to Scientific and Technical Proceedings*	1978 to present
GPO MONTHLY CATALOG	U.S. Government Printing Office	BRS, DIALOG & SDC	Citations; corresponds to printed *Monthly Catalog*	1976 to present
SCISEARCH	Institute for Scientific Information	BRS & DIALOG	Citations; corresponds to printed *Science Citation Index*	1974 to present on DIALOG; 1978 to present on BRS

Service Current Fisheries Statistics (including annual preliminary reports entitled *Fisheries of the United States)*, and *U.S. National Marine Fisheries Service Statistical Digest* (including annual summaries of final data entitled *Fishery Statistics of the United States)*. Kelts and Bressler (1971) described several statistical sources. Statistics Canada, an agency of the Canadian government, is responsible for Canadian fisheries statistics.

Other publications contain fishery statistics as well as those on a great variety of other subjects. *Statistics Source* is a guide to statistical information sources from the United States and foreign countries. *American Statistics Index,* published by the Congressional Information Service (CIS), includes citations and abstracts of documents containing statistical data collected and analyzed by the United States government since 1972. The online database ASI corresponds to the printed index. *Statistical Abstract of the United States* (published annually since 1878) and *Historical Statistics of the United States, Colonial Times to 1970* (published by the United States Bureau of the Census) are available for access to earlier summaries of data from governmental and private sources. CIS began publication of the *Statistical Reference Index* in 1980 to abstract and index sources of statistical information other than the United States government (including state government agencies, universities, and independent research centers). The *Statistical Yearbook* of the United Nations, produced by the Statistical Office, Department of International Economic and Social Affairs, also contains summaries of fisheries statistics.

2.3.6 Current-Awareness Tools

Fisheries biologists may keep up with what is currently being published in their field by subscribing to a number of journals or by perusing current journals in a library. A convenient and inexpensive alternative involves use of published aids designed to provide rapid dissemination of titles of journal articles or books (Table 2.5). These usually consist of sets of reproduced tables of contents issued at regular intervals. *Current Contents: Agriculture, Biology & Environmental Sciences,* for example, presents titles of papers in 1,025 journals. It features such categories as fishery science and technology, marine biology, ecology, zoology, environmental studies, oceanology, conservation, wildlife management, and water research and engineering. *Current Contents: Life Sciences* covers journals of cytology, endocrinology, genetics, microbiology, pathology, and physiology. A cumulative index to journal issues is published in both of these aids three times a year, and a *Quarterly Index* to each is available. Garfield (1982) discussed ways to use *Current Contents* as a personal reference tool.

Index to Scientific and Technical Proceedings, which is published monthly, and its online equivalent ISI/ISTP&B can be used as current-awareness aids. The Smithsonian Science Information Exchange is another means of learning about research in progress. In a 1976 survey of recipients of *Sport Fishery Abstracts,* 59 percent of respondents reported that they used this journal primarily to scan current literature.

2.3.7 Other Bibliographic Aids

Finally, we wish to direct readers to miscellaneous science reference sources. Kelts and Bressler's (1971) annotated bibliography remains the best compilation of such references for fisheries biologists, and we can only begin to update their listings in

Table 2.5. Important current-awareness publications for fisheries biologists

Title	Publisher	Scope	Format	Indexes	Frequency of issue
Current Advances in Ecological Sciences	Pergamon Press	Journals and books in ecology	Lists of titles, authors (with addresses), and bibliographic details by subject or habitat	Author	Monthly
Current Contents: Agriculture, Biology & Environmental Sciences [a]	Institute for Scientific Information	Journals, worldwide	Tables of contents reproduced	Author (with addresses), journal, and subject	Weekly
Current Contents: Life Sciences [a]	Institute for Scientific Information	Journals, worldwide	Tables of contents reproduced	Author (with addresses), journal, and subject	Weekly
FAO Documentation - Current Bibliography	FAO of the U.N. Documentation Center	Documents published by or available from FAO	List of accession number with bibliographic details and synopses or abstracts	Author, geographic, project, and subject	Bimonthly
Freshwater and Aquaculture Contents	ASFIS	Core journals	Tables of contents reproduced; schedule of future meetings	None	Monthly
Marine Science Contents Tables	ASFIS	Core journals	Tables of contents reproduced; schedule of future meetings	None	Monthly
Translated Tables of Contents of Current Foreign Fisheries, Oceanographic, and Atmospheric Publications [b]	U.S. Dept. of Commerce, NOAA, National Marine Fisheries Service	Foreign journals	Tables of contents translated	None	Irregular

[a] Described by Moore (1980)
[b] Translation time and irregular issue dates result in less "currency" than others

the space available here. Several general guides to scientific reference works (including bibliographies, dictionaries, encyclopedias, yearbooks, handbooks, directories, catalogs, biographies, histories, and translations) are available, including the guide to reference sources by Malinkowsky and Richardson (1980).

The best general guide to bibliographies is *Bibliographic Index, a Cumulative Bibliography of Bibliographies* (published three times a year by H. W. Wilson Co., New York). Andriot's (1980) *Guide to U.S. Government Publications* provides an annotated description of series and periodicals published by United States government agencies, including reference publications within series. In addition to *Index to Scientific and Technical Proceedings*, guides to conference proceedings include *Bibliographic Guide to Conference Publications* (published by G. K. Hall & Co., Boston) and *InterDok, Directory of Published Proceedings*, Series SEMT (published by InterDok Corp., Harrison, New York).

Significant general guides to translations are *Index Translationum* (published annually by the United Nations Educational, Scientific and Cultural Organization, Paris, France) and *Translations Register-Index* (published 12 times yearly by the National Translations Center, Chicago). The *Fisheries Research Board of Canada Translations Series* ceased publication in 1974, was continued as the *Fisheries and Marine Service Translation Series,* and since 1980 has been the *Canadian Translations of Fisheries and Aquatic Sciences.* The United States National Oceanic and Atmospheric Administration publishes *Received or Planned Current Foreign Fisheries, Oceanographic, and Atmospheric Translations. Journal of Ichthyology* and *Hydrobiological Journal* are translated from Russian and published by Scripta Technica.

Especially useful fisheries handbooks include Bagenal (1978), Carlander (1969 and 1977), *FAO Manuals in Fisheries Sciences,* and Ricker (1975). The *FAO Fisheries Documents Catalog* is available from UNIPUB, New York. The most up-to-date directory of fisheries biologists in North America is included in the *American Fisheries Society Membership Directory and Handbook.* FAO has published international directories of fishery limnologists and marine scientists. The *Directory of Marine Scientists in Canada—1981* (updated annually) lists names, addresses, specialties, and activities of marine scientists in public and private sectors. The *Conservation Directory* (published by the National Wildlife Federation) is an invaluable source of addresses, names, and telephone numbers.

2.4 REFERENCES

Adams, G. H., and R. R. Walter. 1981. Aquatic Sciences and Fisheries Abstracts (ASFA). *Database,* March 1981:37-53.

Aldous, I.D. 1955. *Fishery publication index, 1920-54.* United States Fish and Wildlife Service Circular 36, Washington, District of Columbia, USA.

Aller, B. B. 1958. *Publications of the United States Bureau of Fisheries 1871-1940.* United States Fish and Wildlife Service Special Scientific Report - Fisheries 284, Washington, District of Columbia, USA.

Andriot, J. L., editor. 1980. *Guide to United States Government publications. Documents Index,* volumes I and II, McLean, Virginia, USA.

Anonymous. 1976. *Aquatic Sciences and Fisheries Abstracts - List of periodicals monitored.* Food and Agriculture Organization of the United Nations ASFIS-ML-1 (rev 0), Rome, Italy.

Anonymous. 1977. *The new copyright law; questions teachers and librarians ask.* National Education Association, Washington, District of Columbia, USA.

Anonymous. 1980. List of serials indexed for this volume. *The Zoological Record 112 (1975 literature).* The Zoological Society of London, Wetherby, West Yorkshire, England.

Anonymous. 1981. *Serial sources for the BIOSIS data base,* volume 1981. BioSciences Information Service, Philadelphia, Pennsylvania, USA.

Artedi, P. 1738. *Bibliotheca ichthyologica seu historica litteraria ichthyologiae* (Edited by Carolus Linnaeus). Lugduni Batavorum, Leyden, Holland.

Atz, J. W. 1971. *Dean bibliography of fishes.* 1968. The American Museum of Natural History, New York, New York, USA.

Atz, J. W. 1973. *Dean bibliography of fishes.* 1969. The American Museum of Natural History, New York, New York, USA.

Bagenal, T., editor. 1978. *Methods for assessment of fish production in fresh waters,* 3rd edition. Blackwell Scientific Publications, Oxford, England.

Beckman, W. C. 1955. *Index-Transactions of the American Fisheries Society 1929-1952, volumes 59-82.* American Fisheries Society, Bethesda, Maryland, USA.

Beckman, W. C. 1980. *Index-Transactions of the American Fisheries Society 1953-1976, volumes 83-105.* American Fisheries Society, Bethesda, Maryland, USA.

Brigham, W. U. 1974. Journal coverage by North American fishery biologists. *Transactions of the American Fisheries Society* 103:387-389.

Burns, R. W. 1971. Using the literature on wildlife management. Pages 13-45 *in* R. H. Giles, editor. *Wildlife management techniques,* 3rd edition, revised. The Wildlife Society, Washington, District of Columbia, USA.

Carlander, K. D. 1969. *Handbook of freshwater fishery biology,* volume 1, 3rd edition. Iowa State University Press, Ames, Iowa, USA.

Carlander, K. D. 1977. *Handbook of freshwater fishery biology,* volume 2. Iowa State University Press, Ames, Iowa, USA.

Carter, N. M. 1973. *Index and list of titles, Fisheries Research Board of Canada and associated publications, 1900-1964.* Fisheries Research Board of Canada Bulletin 164.

Carter, N. M. 1973. *Index and list of titles, Fisheries Research Board of Canada and associated publications, 1965-1972.* Fisheries Research Board of Canada Miscellaneous Special Publication 18.

Chapin, M. K. 1929. *Index-Transactions of the American Fisheries Society 1872-1929, volumes I-LVIII.* American Fisheries Society, Bethesda, Maryland, USA.

Cuadra, R. N., D. M. Abels, and J. Wanger. 1981. *Directory of online databases,* volume 3, number 1. Cuadra Associates, Inc., Santa Monica, California, USA.

Cuadra, R. N., D. M. Abels, and J. Wanger. 1982. *Directory of online databases,* volume 3, number 2. Cuadra Associates, Inc., Santa Monica, California, USA.

Dean, B. 1916-23. *A bibliography of fishes....* Enlarged and edited by Charles Rochester Eastman. American Museum of Natural History, New York, New York, USA.

Doszkocs, T. E., B. A. Rapp, and H. M. Schoolman. 1980. Automated information retrieval in science and technology. *Science* 208:25-30.

Garfield, E. 1964. "Science Citation Index"-a new dimension in indexing. *Science* 144:649-654.

Garfield, E. 1979. *Citation indexing - its theory and application in science, technology, and humanities.* John Wiley and Sons, New York, New York, USA.

Garfield, E. 1982. On the 25th anniversary of *Current Contents/Life Sciences* we look forward to the electronic online microcomputing era. *Current Contents/Agriculture, Biology and Environmental Sciences* 13:5-11.

George Washington University. 1969. *Fishery publication index, 1955-64.* United States Fish and Wildlife Service Circular 296, Washington, District of Columbia, USA.

Gorham, E. 1968. Journal coverage in the field of limnology. *Limnology and Oceanography* 13:366-369.

Hein, D. 1967. Sources of literature cited in wildlife research papers. *Journal of Wildlife Management* 31:598-599.

Johoda, G. 1970. *Information storage and retrieval systems for individual researchers.* Wiley-Interscience, New York, New York, USA.

Johnston, D. F. 1978. *Copyright handbook.* R. R. Bowker Company, New York, New York, USA.

Kelts, L. I., and J. I. Bressler. 1971. Fish and fisheries literature resources: an annotated bibliography. *Transactions of the American Fisheries Society* 100:403-422.

Kirk, T. G. 1978. *Library research guide to biology: illustrated search strategy and sources.* Pierian Press, Ann Arbor, Michigan, USA.

Lagler, K. F. 1956. *Freshwater fishery biology*. W. C. Brown Company, Dubuque, Iowa, USA.

Malinkowsky, H. R., and J. M. Richardson. 1980. *Science and engineering literature: a guide to reference sources,* 3rd edition. Libraries Unlimited, Inc. Littleton, Colorado, USA.

Mayer, D. L., and B. F. Waters. 1981. *Computerized information management in the fisheries sciences.* Paper presented at Symposium on Acquisition and Utilization of Aquatic Habitat Inventory Information, Portland, Oregon, 28-30 October 1981. (Available from B. F. Waters, 3400 Crow Canyon Road, San Ramon, California 94583, USA.)

McDonald, R. M. E. 1921. *An analytical subject bibliography of the publications of the Bureau of Fisheries.* United States Bureau of Fisheries Document 899, Washington, District of Columbia, USA.

Miller, J. K. 1981. *U.S. copyright documents: an annotated collection for use by educators and librarians.* Libraries Unlimited, Inc., Littleton, Colorado, USA.

Moore, J. L. 1980. Wildlife management literature. Pages 7-38 *in* S. D. Schemnitz, editor. *Wildlife management techniques manual,* 4th edition, revised. The Wildlife Society, Washington, District of Columbia, USA.

Mosby, H. S., and R. W. Burns. 1980. Developing and maintaining a small personal reprint library. Pages 39-43 *in* S. D. Schemnitz, editor. *Wildlife management techniques manual,* 4th edition, revised. The Wildlife Society, Washington, District of Columbia, USA.

Nielsen, L. A., and P. B. Summers. 1979. Journal use in fisheries as determined by citation analysis. *Fisheries* 4:10-11, 21.

Ricker, W. E. 1975. *Computation and interpretation of biological statistics of fish populations.* Fisheries Research Board of Canada Bulletin 191.

Scattergood, L. W. 1953. Bibliographic sources for fishery students and biologists. *Transactions of the American Fisheries Society* 83:20-37.

Schoumacher, R., editor. 1980. List of publications screened. *Sport Fishery Abstracts* 25:333-339.

Walbaum, J. J. 1788-93. *Peitri Artedi Renovati pars I. et II. (-V.) i.e. bibliotheca et philosophia ichthyologica.* Cura Iohannis Iulii Walbaumii. Grypeswaldiae, Germany.

Wilson, E. B. 1952. *An introduction to scientific research.* McGraw-Hill Book Company, Inc., New York, New York, USA.

Chapter 3
Safety in Fishery Fieldwork

CHARLES R. BERRY, JR., WILLIAM T. HELM,

AND

JOHN M. NEUHOLD

3.1 INTRODUCTION

The fishery biologist conducting a field study may be exposed to hazardous working conditions. Working on water or ice and manipulating specialized equipment and boats, sometimes in inclement weather, may lead to situations that increase the likelihood of an accident involving personnel or equipment. However, the risks can be reduced by following certain safety procedures for fishery fieldwork. The equipment list for any fishery field trip might include items such as boats, motors, gasoline, nets, sample jars, and data sheets. At the top of the list should be the word SAFE, reminding the biologist that he is being negligent unless he has the *Skills, Attitudes, Facts,* and *Equipment* needed to travel and work safely.

Skills are obtained by enrolling in boating, first aid, and other safety courses. Driving, towing a trailer, boating, and other skills can be learned by practice. With observers present and safety precautions taken, trailer maneuvering skills can be practiced in an empty parking lot, boating skills on rough water can be practiced near shore on a windy day, and skills in avoiding injury in falls while one is wading can be gained in a swimming pool. Experience and skills gained in the practice sessions will remove some of the fears and uncertainties that hamper performance in an emergency situation. Attitudes about working safely, however, are the most important part of a safety program. Accidents do not just happen—they are caused by deficiencies in one's state of mind (carelessness, fatigue, laziness, disobedience, forgetfulness, or poor judgment) and by faulty conditions (poor lighting, improper equipment, poor maintenance, or inclement weather). Facts are the third necessity when assembling the SAFE equipment. Facts are gathered from courses of instruction, films, and written material. A wide variety of training aids are readily available from a number of sources and are usually free for the asking (Box 3.1). The last item on the SAFE list is Equipment. First aid kits, fire extinguishers, other standard boat, auto, and field equipment, and a knowledge of how to use them must be available to help prevent an accident or to mitigate the effects of an accident that does occur.

The information in this chapter briefly covers safety equipment and skills specifically needed by the fishery biologist; however, the information presented here is not adequate training material. It is presented only as an impetus for the untrained person to seek training and as a refresher for trained individuals.

Box 3.1   Major Sources For Safety Information, Pamphlets, and Films

Source	Address
American Red Cross[a]	National Headquarters, Washington, D.C. 20006
Consumer Information Center[b]	Pueblo, CO 81009
Council on Family Health	633 Third Avenue, New York, NY 10017
Health and Safety Education Division	Metropolitan Life Insurance Company, One Madison Avenue, New York, NY 10010
National Fire Protection Association	470 Atlantic Avenue, Boston, MA 02210
National Safety Council	Membership Service Bureau, 444 North Michigan Avenue, Chicago, IL 60611
United States Coast Guard District[c]	Commander 1st Coast Guard District 150 Causeway Street, Boston, MA 02114
	Commander 2nd Coast Guard District 1430 Olive Street, St. Louis, MO 63103
	Commander 3rd Coast Guard District Governors Island, New York, NY 10004
	Commander 5th Coast Guard District Federal Building, 431 Crawford Street, Portsmouth, VA 23705
	Commander 7th Coast Guard District Federal Building, 51 S.W. 1st Avenue, Miami, FL 33130
	Commander 8th Coast Guard District, Hale Boggs Federal Building 500 Camp Street, New Orleans, LA 70130
	Commander 9th Coast Guard District 1240 East 9th Street, Cleveland, OH 44199
	Commander 11th Coast Guard District, Union Bank Building, 400 Oceangate Blvd., Long Beach, CA 90822
	Commander 12th Coast Guard District 630 Sansome Street, San Francisco, CA 94126
	Commander 13th Coast Guard District, 915 2nd Avenue, Seattle, WA 98104
	Commander 14th Coast Guard District, Prince Kalanianaole Federal Building, 300 Ala Moana Boulevard, 9th Floor, Honolulu, HI 96850
	Commander 17th Coast Guard District, PO Box 3-500, Juneau, AK 99802
United States Department of Agriculture[d]	Agriculture, Engineering, Science and Education Administration—Extension Washington, DC 20250

Continued

[a]An American Red Cross Chapter is located in most cities.
[b]Ask for the guide to Federal Information Centers which will direct you to safety information on many topics.
[c]Coast Guard Auxiliary units are located in many other cities not included in this list.
[d]A safety extension office is located in the land grant university of each state.

3.2 FIRST AID

First aid skills are necessities for fishery biologists because their work is frequently done far from sources of medical care. Many of us have been led to believe that saving lives is a complex task that should be left to the medical profession. This is not always true. In some emergencies time counts more than medical expertise. If a life is in your hands, you must act. First aid is not complicated; the biggest obstacles are ignorance, fear, and nervousness. When time is crucial and you are working in the field far from help, your simple action may save a life. It is foolish and negligent to be unprepared for an emergency. First aid skills can be obtained by enrolling in standard first aid courses conducted by the American Red Cross. Information on courses is available from any local chapter.

3.2.1 Drowning and Electrical Shock

Drowning or near-drowning and electrical shock are accidents to which fishery biologists are sometimes exposed and which can cause the victim to stop breathing (Eastman 1974). Breathing must be restarted in 4-6 minutes. However, recent medical breakthroughs in treating drowning victims in cold water (less than 21 C) have amended this general rule. Now rescuers are advised to aggressively resuscitate anyone who has been submerged for *up to an hour*. First aid for all drowning victims means a good understanding of cardiopulmonary resuscitation (CPR) procedures. Formal training in CPR techniques is available from the Red Cross, American Heart Association, universities, local hospitals, and others. Every field crew should include a person trained in CPR, and every crewman should be familiar with the technique. Restarting breathing is the first step in the CPR procedure. First turn the victim face down. Allow several seconds for water to drain from the lungs. If the victim coughs and sputters and starts breathing without assistance, he will get rid of the remaining water. If the victim is not coughing, position the victim on his back and check for signs of breathing. Begin mouth-to-mouth breathing if necessary (Fig. 3.1). Continue it until the victim can breathe unassisted; then seek professional medical help at once to prevent the onset of pneumonia. Once CPR is begun, the rescuer is legally obligated to continue it until help arrives.

3.2.2 Temperature Hazards

Working outdoors exposes the biologist to emergencies caused by excessive heat or cold. During physical exertion in hot weather, much of the circulation is directed to blood vessels in the skin to enable heat to be radiated from the body surface. When temperature is excessively high or when exposure to high temperature is continued for excessively long periods, the circulatory and nervous systems may be unable to control body temperature and, thereby, cause hyperthermia. A mild disturbance is termed

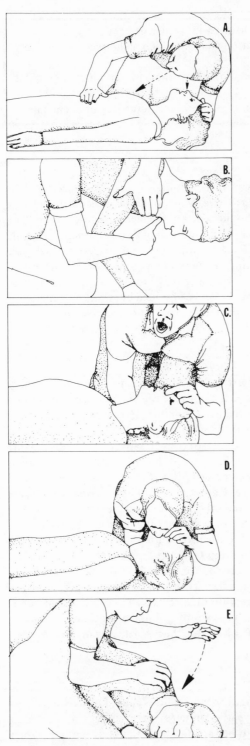

A. Awaken the victim by shaking and shouting. If there is no response, position the victim on his back, gently lift the neck with one hand to open the air passage. Listen and feel for breath while watching for movement in the victim's chest and stomach.

B. If there is a question about whether the victim is breathing, begin rescue breathing. Clear the mouth of foreign objects by making a hook with your index finger and sweeping the throat.

C. Place your hand on the victim's forehead and pinch the nose shut with your fingers, while holding the forehead back and lifting on the neck with the other hand.

D. Take a deep breath and place your mouth over the mouth of the victim (if you are hesitant, mouth-to-mouth breathing can be done adequately through a handkerchief). Blow air into the victim until you see the chest rise. Remove your mouth from the victim's. Turn your head to the side and watch the chest fall while you listen for air escaping from the victim's mouth. Repeat every 5 seconds (12 breaths per minute). Often only a few artificial breaths are needed to restart breathing in the victim.

E. Foreign bodies stuck in the airway may prevent air from entering the victim's lungs. Pull the victim on his side and apply four blows to the back with the hand, in rapid succession. Sweep the mouth with the index finger. Reposition the body, tilt the head to open the airway, and again attempt to ventilate the victim's lungs. Repeat the back blows if necessary.

Figure 3.1. Steps in determining whether a person subject to near-drowning or electric shock is breathing, and steps in rescue breathing techniques.

heat exhaustion; the extreme condition is heatstroke (Table 3.1). Hypothermia describes the physical and mental collapse that occurs when the body is chilled to the core (Hirsh 1975). Wetness, wind, and exhaustion aggravate the chilling, which occurs most often when the air temperature is below 10 C. Cold kills in two distinct steps: (1) exposure and exhaustion, and (2) hypothermia. The victim may fail to realize what is happening. The potential victim feels fine while moving, but exposure and exhaustion are debilitating. When the victim stops moving, body heat production instantly drops, violent shivering may begin, and the onset of hypothermia occurs in a matter of minutes (Table 3.1).

3.3 BOATING SAFETY

Fishery biologists work on a wide variety of waters while carrying out tasks ranging from electrofishing and trawling to running rivers and conducting creel censuses. The variation in waters and the variation in tasks requires a variety of boats and rafts, each outfitted for a particular task. Small, open boats are certainly the most frequently used and are often heavily laden with equipment and passengers.

3.3.1 Power and Capacity Ratings

A capacity label showing the maximum load in equipment and passengers and the maximum horsepower suitable is displayed on most boats (Fig. 3.2). The load capacity and number of passengers must be calculated if the capacity plate is missing or if heavy equipment is permanently attached to the boat (USCG 1980b). Five pieces of data are needed: (1) maximum length, (2) maximum width, (3) transom height, (4) steering style, i.e., remote or direct, and (5) bottom shape (V-shaped or flat). The recommended motor size can then be calculated as shown in Box 3.2. The recommended passenger load and weight capacity can be calculated as shown in Box 3.3.

Do not overload the boat. The load information found on the capacity label or from calculations should be considered as having been recommended for calm water conditions only. In rough water, the safe carrying capacity is decreased. When the boat is loaded, the load should be distributed evenly and kept low in the boat.

3.3.2 Safe Boating Procedures

Proper equipment and skills are needed to operate a boat safely. Fishery biologists should take one of the courses in boating skills such as those offered by the United States Coast Guard Auxiliary and United States Power Squadron. The courses cover seamanship, rules of the road, aids in navigation, piloting, safe motor boat operation, and boating laws. A self-instruction course is also available (USCG 1972). Safe boating literature is available from the Coast Guard and from most state governments (Fleming et al. 1975). Safe boating is based on good sense, courtesy, and respect for property and life. Operating a boat under control at reasonable speeds, letting the other skipper know your intentions, and staying away from swimmers and dangerous areas all make good sense and require the exercise of courtesy and respect. Passing other boats safely is the skill most often required when under way (Fig. 3.3). Although the maneuver is simple, it does require communication between the two boats.

A boat should be outfitted with the safety gear appropriate for its size and use.

Table 3.1. Symptoms and treatment for hyperthermia and hypothermia, from USCG (1980a)

Condition and symptoms or signs	Treatment
Heat Exhaustion 1. Faintness (with a sense of pounding of heart) 2. Nausea, vomiting 3. Headache 4. Restlessness 5. Unconsciousness (the victim who has collapsed and is perspiring surely has heat exhaustion—perspiration rules out the diagnosis of heat stroke).	1. Move victim to a cool place 2. Keep victim lying down; treat for shock 3. If the victim is conscious, water to which 1/2 teaspoon of salt per glass has been added, or stimulants such as iced coffee or tea may be given freely.
Heat stroke 1. Headache, dizziness 2. Extreme elevation of body temperature (105-109°F) 3. Frequent desire to urinate 4. Irritability 5. Disturbed vision or unconsciousness 6. Skin hot and dry 7. Pupils constricted 8. Pulse full, strong, pounding	1. Place victim in shade 2. Reduce body temperature as rapidly as possible by: (1) pouring cold water over the body, (2) rubbing the body with ice, or (3) covering the body with sheets soaked in ice water 3. Remove clothing 4. Lay victim with head and shoulders slightly elevated 5. Give cool drinks after consciousness returns 6. Do not give stimulants
Hypothermia 1. Shivering 2. Reduction in or lack of judgment and reasoning power 3. Vague, slow, slurred speech 4. Memory lapse 5. Incoherence 6. Fumbling hands 7. Drowsiness 8. Stumbling, lurching gait 9. Exhaustion (inability to get up after resting)	1. Believe the signs; a victim often denies the problem 2. Shield the victim from wind and rain 3. Remove wet clothing (replace them with dry clothes, a blanket, or sleeping bag) 4. Provide warm drinks 5. Keep the semiconscious victim awake 6. Place semiconscious victim between two warm heat donors

State regulations differ (OBCA 1970a, 1970b, 1974). Take precautions to be certain your boat is properly equipped and meets legal safety regulations. A courtesy motor boat examination by a qualified Coast Guard Auxiliary member incurs no obligation. The examiner advises on correcting deficiencies; if none are needed the boat is given a "Seal of Safety."

Fire control is most important. Gasoline-powered inboard motor boats of all sizes should be outfitted with flame arresters and ventilation systems (USCG 1980b). Outboard motor boats do not require flame arresters, but their enclosed fuel compartments must be properly ventilated. Fire extinguishers classifed B-I or B-II (suitable for gasoline, oil, or grease fire) are a necessity.

Personal flotation devices (PFDs) should be on board, and many agencies require that they be worn while the boat is afloat. The PFDs (or life jackets) must be Coast Guard approved and are classified according to performance (Table 3.2). Whistles or

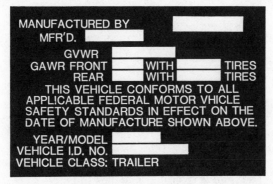

Figure 3.2. Representative capacity labels found on boats (top) and boat trailers (bottom). Specific information will be found stamped in the blank areas of the metal labels. GVWR = gross vehicle weight rating; GAWR = gross axle weight rating.

horns are suggested for all craft, but they are optional for boats shorter than 18 feet. Signaling devices, such as red hand-held flares, are easily obtained and are approved for either night or day use (USCG 1981). Lights are required at night. Motor boats less than 40 feet long and using inland waters must display a green port light and a red starboard light that are visible for one nautical mile. A white light at the stern must be visible for two nautical miles.

3.3.3 Storms on Water

Coast Guard stations, yacht clubs, and many public harbors display flags describing weather conditions. Fishery biologists should know how to read the flags and also how to read the clouds (NOAA 1980). Locally severe storms may not be predicted. Watch the sky for cumulonimbus clouds (thunderheads)—typically, towering clouds with an anvil-shaped leading edge at the top and a churning, rolling bottom edge. Frequently the dark area between the cloud and the earth is where rain, hail, or lightning occurs. Spend some time learning or reviewing the cloud types and sky characteristics that may be used as indicators of potential weather problems (NJMP 1979).

Lightning is a universally feared feature of thunderstorms. The best course of

Box 3.2. Power Factors and Horsepower Ratings

Power factors for boats of various sizes and corresponding horsepower ratings for outboard motors that can be safely used on each boat. The power factor is determined by multiplying the length (in feet) by the transom width (in feet). Horsepower listed is that safe for use with V-bottom boats. For flat-bottom boats, the next lower horsepower is suggested.

Power factor (F)	Outboard motor size (horsepower)
0-35	3
36-39	5
40-42	7.5
43-45	10
46-52	15

Example 1: A boat 13.25 feet long with a transom width of 3.75 feet has an F value of approximately 50; 13.25 x 3.75 = 49.7. The above table shows that maximum permissible horsepower is 15.

For power factors greater than 52 and if the transom height is 20 inches or greater and steering is remote, horsepower (HP) is calculated as follows:

$$HP = 2F - 90 \text{ (rounded to the nearest multiple of 5)};$$

if the transom height is less than 20 inches or the steering is direct,

$$HP = 0.8F - 25 \text{ (rounded to the nearest multiple of 5)}.$$

Example 2: A boat 17.50 feet in length, a transom width of 5.09 feet, a transom height of 20 inches and remote steering yields an F of 17.50 x 5.09 = 89 for which the maximum recommended horsepower is (2 x 89) - 90 = 90.

action to minimize lightning danger is to stay off or get off the water during thunderstorms. Large boats that have flag masts, radio antennas, or other vertical projections should be outfitted with grounding systems (straight, high capacity electrical conductors from the highest point on the boat to a submerged ground plate). Engines and fuel tanks must also be grounded. All parts of the boat contained within an imaginary cone formed by a 60° angle from the vertical on all sides of the highest grounded point are protected against lightning (USCG 1979a). This highest grounded point can be regulated to include all extremities of the boat in this zone of protection.

Although lightning is probably most feared, strong winds and rough seas are actually more dangerous features of thunderstorms. Reactions to storm emergencies depend on the size of the boat, water, and storm (Henderson 1972). Start for the harbor if in doubt about weather conditions. If reaching the dock is not possible, seek a sheltered location. Do not leave an uncomfortable but safe location to reach the dock until the storm has passed. Making headway in heavy seas requires experience and skill. When heading into waves, proceed slowly at a 45° angle to the direction of the waves. Vary the angle of the attack into the waves to attain the best combination of progress and smooth ride. Running in the same direction as large waves requires precise timing to hold the boat in the trough ahead of the following wave. Slight changes in lateral direction, rather than throttle adjustment, will keep the boat ahead of the following wave and provide a smooth, safe ride with the waves. Never force the

---

Box 3.3   Passenger Load and Weight Capacity

Formulae for determining passenger load (A) and weight capacity (B) of a boat when capacity information is unavailable. An example for a 19-ft boat is included.

A.   Passenger Load (*PL*): (overall length) x (maximum width)/15 = number of passengers
         Note: length and weight are in tenths of a foot
B.   Weight capacity (*WC*): *WC* = *PL* x 141 + 32
         Note: *WC* is in pounds
C.   An Example: a boat 19 feet long and 7 feet wide will have a passenger load of:

$$PL = \frac{19 \times 7}{15} = 9,$$

and the weight capacity will be
         WC = 9 x 141 + 32 = 1,301 pounds.

---

boat up the back of a wave and over the crest; never allow the following wave to catch up and wash over the transom.

A last resort in heavy seas is to use a sea anchor—a conical canvas sheet that is deployed at the end of a rope attached to the bow of the boat (Fig. 3.4). Rope length may vary from about 50 to 200 feet for boats about 16 to 30 feet long. Rope length should provide a small angle between the rope and the plane of the water surface (Shewman 1980, 1981). Never attach the sea anchor to the stern. A sea anchor holds the bow of the boat into the waves while the boat drags the anchor slowly through the water. Motion is less violent than when the boat is under way, the potential for hull damage is reduced, and there is less danger of capsizing. It is advisable to devise a means of attaching the sea anchor line to the boat quickly and safely in the event of a sudden storm. Practice deploying the anchor under moderate wave conditions to determine whether your methods will work under severe conditions.

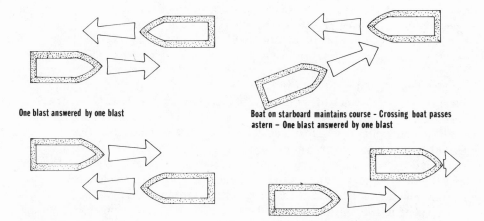

One blast answered by one blast

Boat on starboard maintains course - Crossing boat passes astern – One blast answered by one blast

Two blasts answered by two blasts

Pass on starboard – One blast answered by one blast

Figure 3.3 Rules of the road when two boats are passing, showing positions and describing communications recommended.

Table 3.2. Personal flotation devices (PFDs or life jackets)

Type and size	Buoyancy (pounds)	Positioning
I		Positions an unconscious person from
Adult	22	a face down to an upright or slightly
Child	11	backward position
II		Turns the wearer to a vertical or
Adult	15.5	slightly backward position in water
Child	11	
III		Wearers can place themselves in a
Adult	15.5	vertical or backward position in the
Child	11	water
IV	—	Designed to be thrown to a person in the water and grasped and held by the user
V	—	Approved for restricted use only

When the boat is swamped, it will probably remain floating and provide some security to the victims. Always stay with the boat because it can be more readily located than a lone swimmer. Staying with the boat also enables the victims to conserve body heat by climbing partially onto it to help reduce the amount of body heat lost in cold water. Water conducts heat away from the body 25 times faster than air (USCG 1980c). Swimming increases the circulation rate and pumps warm water away from the spaces between your body and layers of clothing, and, thus, further increases the rate of body heat loss. Remain as still as possible in cold water.

The laws of the sea require that a boater come to the aid of another in distress. If an accident has happened or distress signal fired, take immediate and positive action. Several rescue skills should be noted. When going to the aid of people in the water, approach *from* downwind to prevent wind and wave forces from pushing your boat into the capsized boat or over people in the water. Disengage your propeller while retrieving people from the water. In strong winds your boat will drift rapidly, requiring repositioning after each person is retrieved. Do not overload the rescue boat.

Occasionally, it is necessary to tow another boat. In calm waters an alongside tow works well (Chambers 1981). When one boat is tightly secured beside another (with bumpers in place), the two boats respond as a unit. In choppy waters, however, the disabled boat should be towed behind the tow boat. A towing bridle is the key to the towing operation. Attach the bridle to the towing craft as far forward as possible and run the bridle through or around stern fittings. Attach the towline to the bridle with a metal ring or similar device that will allow the towline to center itself during the tow. The other end of the towline should be attached to the bow eye of the disabled boat. The rougher the water, the longer the towline should be. The reactions of the towed boat can be uncertain, and passengers should wear life jackets while being towed. Towing should be conducted at a speed which will not strain the towrope or bridle and which will permit complete control. If the towed boat begins to swing from side to side, reduce speed and move weight in the towed boat to the stern. When approaching the dock, a crew member in the towing boat should step onto the dock with the towrope in hand and pull the towed boat to the dock.

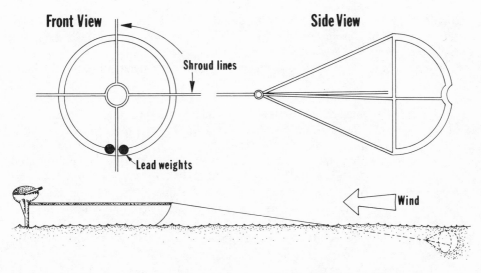

Figure 3.4 Schematic diagram of an open sea anchor and position in the water after proper deployment. Note the small angle between anchor rope and water surface.

3.4   SAFETY ON THE ROAD

Safety requires the adoption of an attitude toward operating a motor vehicle that is known as "defensive driving." The defensive driver is not timid or overcautious, but is determined to take every reasonable precaution to prevent traffic mishaps. The standard accident prevention formula on which the principles of defensive driving are based involves three interrelated steps: (1) *see the hazard* by thinking about what might happen as far ahead as possible; (2) *understand the defense* for specific hazards and apply the correct defensive maneuver; and (3) *act in time* once the hazard is recognized and the defense determined. All federal and many state employees are required to take defensive driving courses; however, everyone should be familiar with the technique. Information on defensive driving courses is available from the National Safety Council and often from municipal or county law enforcement agencies.

3.4.1   Towing a Trailer

The fishery biologist frequently has a boat trailer and boat, equipment trailer, or camp trailer in tow. With the trailer attached, the likelihood of being involved in a single-car accident increases by a factor of four, especially during long-distance travel (Dark 1971). Safe trailer towing has four requirements: (1) the proper trailer and load, (2) the right hitching equipment, (3) a towing vehicle with adequate weight and power, and (4) a driver familiar with trailer towing techniques (Bottomly 1974).

Surveys have shown that 10% of the trailers on the highway are overloaded, 30% of the towing vehicles are overloaded, and 35% of the trailers have excessive weight on the hitch and tongue (Dark 1971). All trailers have capacity information—the gross trailer weight rating and the gross axle weight rating—displayed on a capacity label (Fig. 3.2). Two weights must be known—the gross trailer weight and the tongue weight. The tongue weight should be 10-15% of the gross trailer weight. These weights can be determined by weighing the trailer at a public scale when the trailer is loaded for a normal trip. The gross trailer weight is the weight of the loaded trailer by itself, unhitched and supported on a jack. To determine the gross axle weight, move the tongue and jack off the scale and reweigh. Determine tongue weight by subtracting gross axle weight from gross trailer weight.

Wheel and tire condition and size also determine the capacity of a trailer. Under-inflated tires are the most common trailer towing problem. They become hot, sway more easily, wear out more quickly, and carry less load safely. The size of the tires needed to carry the load for which the trailer is rated is also listed on the capacity plate. Trailers subject to immersion should have bearing protectors (spring loaded caps that pressurize the bearing grease to protect the wheel bearings).

Trailer brakes, brake lights, and turn signals extend the braking and signaling ability of the towing vehicle to the trailer. The connection and operation of lights should be checked often during a trip. The useful life of lights is increased if they are detachable and are removed before the boat is launched. State laws differ in requirements for trailer brakes. Most states require brakes on trailers above a certain gross weight and some specify the types that are acceptable. Check to ensure that the trailer is in compliance with the laws in your state.

The hitching equipment consists of the coupler, safety chains, hitch, and ball. The coupler is the mechanism that attaches the trailer to the ball. The size of the coupler and the size of the ball are determined by gross trailer weight (Table 3.3). Safety chains should be attached to the towing vehicle at points separate from the bracket that holds the ball. The hitch size and weight depend on the gross trailer weight and tongue weight.

Safe trailer towing requires the driver to have certain knowledge and skills. The driver should be certain that the trip begins with safe equipment by completing a multi-point check list (Box 3.4). Lack of towing experience is the most important factor causing the high accident rate of vehicles towing trailers. The two most important skills required for the driver are backing up and correcting fishtailing. When backing, place one hand on the bottom of the steering wheel and move the wheel in the direction the trailer must go. Learn how to avoid and correct fishtailing—a warning signal and the first symptom of impending disaster. When a trailer begins to fishtail, it begins to wag the towing vehicle. When the swaying goes beyond a certain point, both usually flip over (Anonymous 1977). Gusty winds, passing trucks, high speeds (especially downhill), and insufficient tongue weight all may cause fishtailing. Avoid fishtailing when going downhill by slowing down at the top of the hill and shifting to a lower gear. If the trailer starts to fishtail, let up on the gas, but do not touch the brake. Keep the wheel straight and hold the wheel tightly. Compensatory steering with a trailer in tow often exaggerates the swaying.

When under way, never forget that a trailer is following. Start slowly, proceed more slowly than usual, and slow down for bumpy areas. Maintain a greater following distance than normal, and think twice about passing. Pull over and check every hour for high temperatures in the wheel bearings, loose tie-downs, defective lights, incorrect tire pressure, and high vehicle engine temperature.

Launching a boat is a critical part of trailer towing (USCG 1979b). "Ramp courtesy" requires preparation of the boat for launching away from the ramp. Check with knowledgeable persons about unusual hazards at the ramp. Raise the engine to avoid scraping, install the drain plug, release the tie-downs, and disconnect or remove the lights. Allow the trailer hubs to cool before entering the water. Sudden cooling may crack or chip the bearings or suck them full of water. On the ramp, never turn the

Table 3.3. Matching trailer hitching equipment to trailer load[a]

Trailer class	Gross trailer wt (lbs)	Hitch class	Hitch type[b]	Ball diameter (inches)	Ball shank diameter (inch)
I	2,000	I	WC or WD	1-7/8	3/4
II	3,500	II	WC or WD	1-7/8	3/4
III	5,000	III	WD	2	1
IV	> 5,000	IV	WD (with anti-sway)	2-5/6	1[c]

[a]From USCG (1979b).
[b]WC=weight carrying (holds all trailer weight), WD=weight distribution (redistributes the trailer weight to all six wheels of the towing vehicle and trailer).
[c]Shank diameter should increase to 1-1/4 inches if gross trailer weight is about 10,000 lbs, and to 1-3/8 inches if gross trailer weight is about 13,000 lbs.

engine off. Set the parking brake and use tire stops. Only the driver should be in the towing vehicle; others should direct and assist the launch by having tire stops ready and holding a line to the boat.

3.4.2 Trucks and Campers

Field fishery work often requires the use of trucks, vans, and campers with which the worker may be unfamiliar. Unfamiliarity with equipment is a chief cause of accidents. The biologist should be aware of the differences between driving a large vehicle and the family car and should be familiar with the safe use of the life support systems on trailers and campers (USDOT 1977). Safe driving begins with having the right equipment. Make sure the total weight of the camper and contents does not exceed the gross vehicle weight and tire rating. If in doubt, weigh the vehicle on a

---

Box 3.4.   Safe Trailer Towing
   Pre-departure check list for safe trailer towing.

TOWING VEHICLE
   Radiator coolant
   Transmission fluid
   Engine oil
   Shocks/springs
   Vision/auxiliary mirrors
   Emergency equipment
   Tire pressure
TRAILER
   Tire wear
   Wheels, lugs, and bearing protectors
   Lights
   Load balance
   Tongue weight
   Tiedowns

HITCHING
   Safety chains
   Locked coupler
   Ball-coupler match
DRIVER
   Backing skills
   Ramp courtesy
   Defensive driving

public scale. About 20% of the recreational vehicles on the highway have loads in excess of tire capacity (Dreiske 1971). Overloaded vehicles are unstable, have steering and braking difficulties, and are subject to overheated tires that may catch fire or blow out. Mirrors are important auxiliary equipment. Both standard and convex mirrors provide the best vision down the sides of the vehicle. Finally, check the tie-down mechanisms on the camper for tightness and make sure standard safety equipment is loaded.

Behind the wheel, driving habits should be changed by making a deliberate and conscious effort to "think big" and "think slow." The vehicle's size, weight, and center of gravity cause it to be more unstable than a passenger car. Handling, acceleration, turning radius, and stopping are affected (NSC 1978). The surface area of the vehicle exposed to the wind is substantial, compared with its weight. During windy conditions the side of the camper acts like a sail. The driver normally compensates, but trouble comes from gusts (natural or from other vehicles) or when the wind is cut off (by a passing truck or when one enters a tunnel). Anticipate these conditions, hold the wheel firmly, and move slowly. Drivers frequently overcompensate and veer out of control. Crowding the center lane is a dangerous habit of beginners. Watch the shoulder and keep to the right. Backing up may be a problem when only mirrors are used. Practice backing, turning, and other driving skills in an empty parking lot.

Several points should be remembered when stopping for fuel. Be alert for overhanging structures that may be lower than the height of the vehicle. When refueling, extinguish pilot lights in a camper. Fill dual tanks and use fuel from each for a short time to avoid overflow. When stopped, check the engine fluid levels, tire pressure, tie-downs, and load.

At the camp, the life support system of a trailer or camper (electricity, gas) must be used safely—accidents are deadly (NCS 1978). Read the owner's manual thoroughly. The liquid petroleum gas system is the most hazardous aboard the vehicle because of the possibility of fire, explosion, or asphyxiation caused by leaking gas (DMV 1979). Equipment must be serviced regularly by trained personnel. Know where the gas shut-off valves are located. The gases (propane, butane) have chemical additives that produce a strong odor. If the odor of gas is detected, immediately turn off valves and air the vehicle. Check for leaks by observing bubbles from joints coated with soapy water. Tighten loose fittings, but if there is any doubt about whether repairs have been effective, have the system serviced. When the unit is infrequently used, spiders build webs in appliances—so vacuum clean all appliances well. All liquid gas appliances must use outside air for combustion and release exhaust gases outside the vehicle. Never block outside vents of gas furnaces and refrigerators. Never use open flames (oven, stove, catalytic heater) for supplementary heat—even with a vent or window open—while you sleep. Open flames in a vehicle or even a tent soon deplete the oxygen and fill the space with carbon monoxide. Sleeping persons have become victims of carbon monoxide poisoning even with roof and side vents of a camper open. Initial symptoms are headache and sleepiness. Once victims fall asleep, they die if the gas is not ventilated. One expert (Hirsh 1977) pointed out that propane and butane heaters have caused deaths and suggested a warm sleeping bag instead of relying on heating devices while sleeping.

3.5 SAFE WADING

Most freshwater streams are paved with slippery rocks and have uneven bottoms and water currents that make wading hazardous. A bad fall on a slick, rock-studded shoreline may result in injury, and loss of footing in a swift stream may threaten life. As with many other measures, practice can help mitigate damages in case of an accident. It can be simple, informative, and enjoyable to practice surviving a fall while wearing waders and dummy equipment. Safe wading requires certain equipment, skills, and knowledge. Waders should fit the foot and allow leg mobility for stepping on or off rocks and climbing banks. A belt around the middle of the waders helps maintain the fit and will prevent water from pouring into the waders should a fall occur. Waders with standard soles are appropriate for slowly flowing streams with mud, sand, or pebble substrates. Felt soles or other types of non-slip soles are almost a necessity on slippery rocks. Lug soles can be readily converted by bonding unbacked outdoor carpeting to the soles with contact cement. Sandals with spikes or aluminum bars may be required for very slippery moss-covered rocks or slippery soapstone, shale, or slate. A wading staff (a "third foot") is a valuable aid in crossing fast water. It should be held on the upstream side and planted firmly before the feet are moved. Flotation devices are frequently overlooked if a boat is not in use; however, a life jacket should be on the equipment list if the wading of large streams is part of the job. If you are carrying equipment or an electro-shocker on your back, the shoulder and waist straps should be equipped with quick release buckles, similar to those on safety belts in automobiles or airplanes.

The secret of safe wading is to always think "one foot at a time." First, enter a stream at a point of relatively slow-moving water. Plant the feet firmly on the bottom. Place your weight on the downstream foot and slide the upstream foot forward carefully. Find a spot for the front foot, transfer weight to it and slide the rear (downstream) foot to it. Anchor the downstream foot and again seek a secure position for the upstream foot. While keeping weight on the ball of the foot, take slow, short steps to maintain balance. In crossing a fast-water stream, it is best to wade on a slight downstream angle. Avoid the main current by seeking slack water behind rocks. Keep sideways to the current to minimize the force of water. Learn to "read" water to identify obstacles, drop-offs, and other hazards in the water before you reach them. When turning around, always turn into the current.

Most falls are caused by carelessness, e.g., moving too fast, taking long steps, wading directly downstream, or jumping off the bank. If you fall in water of moderate depth and current velocity, first get into a head-upstream, face-down position to take advantage of the buoyancy of the upper body. Water pressure will assist when you attempt to rise to a standing position.

Wearing waders while carrying heavy equipment poses special problems if the fall is into a deep hole while wading or if the fall is from a boat (Spurr 1980). Waders full of water do not drag the victim to the bottom. Boots and waders weigh less in water than out and the water that fills the boots adds no extra weight. If the fall is feet first, most air is forced out of the waders but some is trapped in clothing. The trapped air plus that in the lungs (take a deep breath prior to immersion) will provide sufficient flotation to return to the surface. Raising the arms, cupping the hands, and pulling the arms downward will provide enough upward force to return to the surface. One disadvantage of waders is the fact that they streamline the legs and make kicking for propulsion useless. Do not waste energy kicking. Use the side, breast, or back strokes to swim to nearby safety. If you fall headfirst, the air trapped in the boots and clothing will cause you to float. After rolling over on your back, bend the knees slightly so sufficient air can be retained in the legs to provide ideal flotation. By sculling with the hands, breathing is easy, and progress can be made toward safety.

Chest waders can be quickly converted to an emergency life preserver if safety is not nearby. The waders must be removed and filled with air by one of two methods. Remove the waders and bring them to the surface feet first so that most of the water drains out. Stream the waders behind you while treading water, and with both hands hold the top above the surface. Swing the waders rapidly overhead with the top held open and push them down into the water. You will probably submerge briefly while doing this, but usually more than enough air is then trapped in the legs and feet of the waders to keep your head out of the water (Spurr 1980). Another method of adding air to the waders is to hold the top open just below the surface and splash down into the waders with one hand. Air carried into the water with the splash will become trapped in the waders and they will begin to bulge. Shift the air to the legs by pulling the wader down into the water. After filling waders with air by either method, hold the top closed underwater, and slide between the legs—now emergency water wings.

3.6 SAFETY ON ICE

Ice on lakes and rivers can be classified according to formation and safety (Table 3.4). It is safe to work extensively only on new ice that is clear—the strongest type of ice. The popping and cracking sounds heard when walking on new ice more than three inches thick actually indicate that the ice is good and solid. As a rule of thumb, ice three inches thick will support one man on foot, ice six inches thick will support a group moving single file, ice eight inches thick will support an automobile, and ice twelve inches thick will support a heavy truck. However, it is impossible to make generalizations. While three inches of ice on a farm pond poses little danger, the same thickness on a moving stream or a lake with springs, stumps, plant life, and water currents may be dangerous. River ice is especially treacherous because water currents may erode the undersurface, creating dangerously thin ice in proximity to thick, safe ice (Ashton 1979). On both rivers and lakes, there may be trouble spots even on new ice. Ice near shore may be unsafe because the proximity to land hastens melting. Winds may force ice onto the bank, leaving cracks and broken ice near shore, or it may blow

Table 3.4. Types of ice, and appearance, structure, and relative strength of each type

Ice type	Appearance	Structure	Relative strength
Lake			
New	Shiny, clear	None apparent, clear cracking sounds indicate good ice	Highest strength depending on thickness— 6 inches will support a group of persons in single file
Slush or snow	Opaque-white	Grainy, may have entrained air	About one-half as strong new ice
Floe	Rough, edges protruding at angles	Broken pieces of new or slush ice frozen together	Unsafe
Rotting	Dark, dull	Vertical crystals, soft	Unsafe
River	Variable in appearance	Variable in structure	About 15% as strong as new lake ice; unsafe because water currents beneath cause great variation in thickness

ice away from the shoreline. Pressure ridges in the ice sheet may have strips of open water between the upper and lower ice sheets. Springs in a lake may create thin and potentially dangerous ice. Each trouble spot is made more dangerous by snow-cover.

The first and most important ice-safety tool is an ice spud or needle bar. An ice spud is a metal staff about five feet long with a chisel welded onto the end. On a needle bar, the chisel is sharpened to a point. These tools can be used to test ice thickness by forcefully striking ice ahead of you as you walk. When in doubt, the spud can be used to dig a test hole. The tool can also be used to keep from falling through a hole should an accident occur. When first testing ice thickness, an observer should be standing by with a pole and at least 50 feet of rope.

If someone falls through the ice, the victim and the observers must follow specific rescue techniques. The victim should place his arms on the ice and roll the upper body onto the ice by pulling and kicking. Biologists who frequently work on ice carry spikes to gain a handhold when crawling from a hole. Once on the ice, the victim should roll away from the hole. The rescuer should lie down to distribute weight when approaching the hole. A rope, belt (or several fastened together), branch, or other life line can then be passed to the victim. The rescuer serves as an anchor on one end while the victim pulls himself out of the hole. Both rescuer and victim should roll away from the hole to avoid breaking through again. In most cases, the way you walked in is the best way out.

3.7 REFERENCES

Anonymous. 1977. Put safety behind you. *Family Safety Magazine* 36(2):28-30.

Ashton, G.D. 1979. River Ice. *American Scientist* 67:38-45.

Bottomly, T. 1974. *The complete book of boat trailering.* Stock No. 1887-7. Association Press, New York, New York, U.S.A.

Chambers, V. 1981. Towing—doing it properly and safely. *Pennsylvania Angler* 50(7):16-17

Dark, H.E. 1971. Know before you tow. *Family Safety Magazine* 30(2):24-26.

DMV (Division of Motor Vehicles). 1979. *RV Safety.* Commonwealth of Virginia, Division of Motor Vehicles, Form DL17, Richmond, Virginia, USA.

Dreiske, P. 1971. Have home, will travel. *Family Safety Magazine* 30(3):24-26.

Eastman, P.F. 1974. *Advanced first aid afloat.* Cornell Maritime Press, Cambridge, Maryland, USA.

Fleming, J.P., B. Harris, M. Avery, and M. Roche. 1974. *National directory of boating safety materials.* National Safety Council, Chicago, Illinois, USA.

Henderson, R. 1972. *Sea Sense.* Stock No. 1860-5. Association Press, New York, New York, USA.

Hirsh, T. 1975. The chill that kills. *Family Safety Magazine* 34(1):5-6.

Hirsh, T. 1977. The last campout. *Family Safety Magazine* 36(1):8-9.

NJMP (New Jersey Marine Police). 1979. *Squall line.* New Jersey Marine Police, Trenton, New Jersey, USA.

NOAA (National Oceanic and Atmospheric Administration). 1980. *Marine weather services.* National Ocean Survey, NOAA/PA 70029, Riverdale, Maryland, USA.

NSC (National Safety Council). 1978. *RV camping guide.* National Safety Council, Stock No. 08302, Chicago, Illinois, USA.

OBCA (Outdoor Boating Club of America). 1970a. *Handbook of boating laws—Northeastern states.* Outdoor Boating Club of America, Chicago, Illinois, USA.

OBCA (Outdoor Boating Club of America). 1970b. *Handbook of boating laws—Southern states.* Outdoor Boating Club of America, Chicago, Illinois, USA.

OBCA (Outdoor Boating Club of America). 1974. *Handbook of boating laws—Northcentral states.* Outdoor Boating Club of America, Chicago, Illinois, USA.

Shewman, D.C. 1980. *Comparison of Mill-Spec Sea Anchor to Shewman, Inc. sea anchor of similar size.* Shewman, Inc., Dunedin, Florida, USA.

Shewman, D.C. 1981. *Sea anchors, drogues, and salvage bells.* Shewman Inc., Dunedin, Florida, USA.

Spurr, E. 1980. Worthy notes for wise waders. *Ducks Unlimited Magazine* 40(4):43-45.

USCG (United States Coast Guard). 1972. *The skipper's course; a self-introduction program.* United States Government Bookstore, Pueblo, Colorado, USA.

USCG (United States Coast Guard). 1979a. *Lightning: cone of protection.* United States Government Printing Office, 631-875/11, Washington, District of Columbia, USA.

USCG (United States Coast Guard). 1979b. *Trailer-boating: A primer.* United States Government Printing Office, 1979-623-210, Washington, District of Columbia, USA.

USCG (United States Coast Guard). 1980a. *First aid for boatsmen.* United States Coast Guard Auxiliary, CG-525 (Old Aux. 206), Boston, Massachusetts, USA.

USCG (United States Coast Guard). 1980b. *Federal requirements for recreational boats.* United States Coast Guard, Washington, District of Columbia, USA.

USCG (United States Coast Guard). 1980c. *A pocket guide to cold water survival.* United States Government Printing Office, COMDTINST 3131.5 (Old CG 473), Washington, District of Columbia, USA.

USCG (United States Coast Guard). 1981. *Visual distress signals for recreational boaters.* United States Coast Guard, G-BEL-4, Washington, District of Columbia, USA.

USDOT (United States Department of Transportation). 1977. *Travel and camper trailer safety.* United States Department of Transportation, National Highway Safety Administration, Washington, District of Columbia, USA.

Chapter 4
Aquatic Habitat Measurements

DONALD J. ORTH

4.1 INTRODUCTION

Aquatic habitat measurements play an integral role in fisheries management. Their use is critical in predicting such things as the impacts of habitat alterations, potential fish production and probable limiting factors, and the success of a species introduction. Habitat measurements also make it possible to classify aquatic habitats into similar groups so that research and management results may be generalized.

The number and types of species and their relative abundance is largely determined by the habitat. In stream systems, for example, there is a pattern of longitudinal zonation in which the headwaters are usually inhabited by only a few species, but more species are found and average size of fish increases downstream (Thompson and Hunt 1930; Huet 1959; Harrel et al. 1967; Sheldon 1968). This phenomenon is associated with the greater diversity of habitat (Gorman and Karr 1978) in downstream reaches and more variable physical-chemical conditions in the intermittent headwaters (Matthews and Styron 1981). Measurements of habitat-related factors have been used to develop predictive equations to estimate standing stocks in aquatic habitats (Jenkins 1976; Aggus and Lewis 1977; Binns and Eiserman 1979). In many situations an environmental assessment must be made to determine the impact of alteration of the habitat as a result of impoundment, channelization, or other proposed land use. This requires habitat measurement procedures to estimate present habitat conditions and to predict future conditions (Stalnaker and Arnette 1976; Stalnaker 1979; United States Fish and Wildlife Service 1980a, 1980b, 1981; United States Department of Agriculture Forest Service 1982; Bovee 1982).

Habitat measurements can be used in conjunction with knowledge of a species' habitat preferences to determine if habitat parameters are limiting and which type of habitat improvement may be beneficial. Hunt (1971, 1979) has described many habitat improvement techniques used in Wisconsin trout streams. Alterations in reservoir operation and design also have been used to improve fish habitat (Jenkins 1970; Groen and Schroeder 1978; Benson 1973, 1980). Habitat measurements are also used to classify fishery habitats (Platts 1980). Habitat classification systems are being refined to integrate them with landform and vegetative type classification systems (Olsen et al. 1981; Lotspeich and Platts 1982).

Fisheries scientists recognize aquatic habitat as including all nonliving aspects of the aquatic ecosystem and often measure parameters over a wider geographic scale than simply a species' distribution. Therefore, a distinction needs to be made between microhabitat and macrohabitat. Microhabitat is the localized area in which a species occurs. Microhabitat measurements, therefore, depend on where the species lives. For example, if a fish lives within the top 2 meters of a 15 meter-deep reservoir, its microhabitat is the area less than 2 meters deep. Habitat measurements taken are usually referred to as macrohabitat measurements. For example, the entire reservoir is the macrohabitat of the fish species described above. The decision to make measurements of microhabitat or macrohabitat depends on the objectives of the habitat survey.

The purpose of this chapter is fourfold. First, the sources of information and techniques used to measure watershed hydrology and geomorphology are discussed (section 4.2.1). Second, procedures for field mapping of aquatic habitats are presented (section 4.2.2). Third, the habitat parameters that are usually measured for fisheries management are described. Fourth, techniques, equipment, and biases associated with these habitat measurements are presented. This information is organized by river and stream habitats (section 4.2.3) and lake and reservoir habitats (section 4.2.4). Measurements of fish habitat in marine environments are not specifically described.

4.2 MEASUREMENT OF HABITAT COMPONENTS

4.2.1 Topographic Map Interpretation

Much information about the watershed can be obtained from a good topographic map. In addition, important features and sampling locations can be located for future reference using geodetic coordinates (latitude and longitude), the Universal Transverse Mercator (UTM) system coordinates, or, in the United States, the Public Lands System. Most highway and United States Geological Survey topographic maps contain lines of latitude and longitude along the map borders. The UTM coordinate system locates coordinates based on the distance (in meters) from the equator and from an arbitrary reference meridian.

The Public Lands System, established in 1784, divided the United States into townships six miles square, using parallel and meridian lines (Fig. 4.1). Each township is identified based on its distance north or south from the parallel base line and east or west from the principal meridian. For example, township 3 south, range 2 west (T3S, R2W) is the third one south of the parallel base line and the second one west of the principal meridian (Fig. 4.1). Each township is then divided into 36 sections, each 1 square mile. These are numbered from right to left in the first row, then from left to right in the second row, and so forth. Finally, each section is divided into four corners: NW, NE, SW, and SE, each of which can be further subdivided into corners.

Topographic maps contain landmarks and locations of aquatic habitats and provide contour lines of equal elevations. Surface areas of lakes (existing or proposed) and their watersheds can be easily determined from topographic maps. The boundaries of a watershed can be drawn by looking for ridges that divide adjacent watersheds. The outline for new reservoirs can be determined based on the surface elevation of the dam's spillway. A new contour line drawn at the spillway level would represent the outline of the proposed reservoir.

Acreage (dot) grids and planimeters are most frequently used to estimate surface areas. An acreage grid consists of a transparent grid overlay with one dot per grid cell. The dot grid is placed over the area to be measured, and the number of dots included in the area is counted. Every other dot that falls on the boundary line is counted. Number of dots can then be converted to an area. For many fishery applications, the acreage grid may be accurate enough, although the polar planimeter is preferred for applications that require greater accuracy. A polar planimeter (Fig. 4.2) is an instrument that is used to trace the outline of an area and convert the measurement to

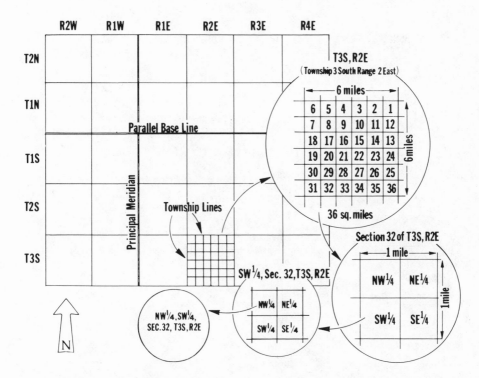

Figure 4.1 United States Public Land System

an area. The use of the planimeter is explained in Box 4.1. The surface area estimates obtained represent the area projected on a flat surface; true surface area, if needed, would have to be obtained by using a correction factor related to the amount of slope.

The size of streams can be determined from topographic maps. Width may be measured on larger rivers using small-scale topographic maps. Stream length can be determined with a map measurer or cartometer (Fig. 4.2). To measure distance, set the tracer wheel at zero and trace the stream. The true distance can then be determined by using the scale of the map.

Stream gradient (vertical drop per horizontal distance) can be determined from topographic maps between two points where the contour lines cross the stream. Average stream gradient between these points is the contour interval divided by the stream length between these points. Gradient may be expressed in m/km, ft/mi, or percentages.

Stream sinuosity is the ratio of stream length to watershed length. Watershed length is measured along the valley axis. The stream sinuosity index reflects the amount of meandering and, hence, the diversity of habitat types (Zimmer and Bachmann 1978).

Stream order is a useful classification based on branching of streams (Strahler 1957). On a topographic map showing all intermittent and permanent streams, the smallest unbranched tributaries are designated order 1 (Fig. 4.3). Where two first order streams join, a second order stream segment is formed; where two second order stream segments join, a third order stream segment is formed; and so forth (Fig. 4.3). The usefulness of the stream order classification depends on the assumption that order

Map Measurer   **Polar Planimeter**

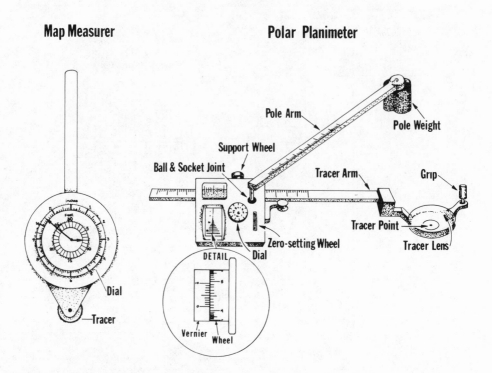

Figure 4.2 Polar planimeter and map measurer

number is directly proportional to stream size, discharge, and watershed area. These relationships generally hold within a drainage (Harrel et al. 1967; Platts 1979), although one should not expect streams of a given order in different drainages to have similar physical features. Within similar drainage basins, stream order can be used to estimate approximate stream size and fish population numbers and diversity (Platts 1979; Barila et al. 1981).

The major question concerning stream order is what constitutes a first-order stream. United States Geological Survey topographic maps distinguish between intermittent and perennial (permanent) streams, but the smallest tributary stream drawn on these maps depends on the scale of the map. For consistency, most workers follow Leopold et al. (1964) and define a first-order channel as the smallest unbranched tributary that appears on a 7½ minute quadrangle map (1:24,000 scale). The definition is often modified to include perennial streams (i.e., those with sufficient flow to develop biota).

Three other geomorphic parameters that are useful for characterizing watersheds are the drainage density, constant of channel maintenance, and the relief ratio. Drainage density is the sum of all stream channel lengths divided by the watershed area. The constant of channel maintenance is the amount of surface area (square meters) required to maintain 1 meter of stream channel. This constant can be estimated from a log-log plot of basin area versus cumulative stream length (Strahler 1957). The relief ratio is defined as the ratio between (1) the difference in elevations of the river mouth and the watershed summit and (2) the watershed length, measured along the longest dimension of the drainage area. The relief ratio indicates the overall

Box 4.1   Use of the Planimeter

The planimeter is used to measure the surface area of an irregularly shaped area. This instrument consists of a measuring wheel that rotates, a tracing arm and point, and a pole arm that rotates around a fixed point or pole weight (Figure 4.2). The planimeter units are read from divisions on the measuring wheel and a dial that counts the number of revolutions of the measuring wheel. Conversion of the planimeter units to units of area depends on the length of the tracer arm. If the tracer arm length is fixed, the values read from the dial and measuring wheel convert directly to units of area, and the conversion factor is supplied by the manufacturer. If the tracer arm length is adjustable, a calibration device is supplied by the manufacturer to convert to units of area. Alternately, you can draw a square of known area and carefully trace this area three times. The average of the three planimeter readings represents the number of planimeter units that equal this known area.

To determine the surface area from a map, follow these procedures:

1. Spread the map on a smooth, hard surface and fasten it to the surface with tape.
2. Determine if the area to be measured is too large to be traced without moving the fixed point. If so, divide the area into sections and measure each section separately.
3. Place the pole weight outside the area to be measured, and make certain that all parts of the map can be reached with the tracer point. Also make certain that the angle between the tracer arm and pole arm stays between 15 and 165 degrees.
4. Mark a starting point with pencil and zero the measuring wheel.
5. Beginning at the starting point, move the tracer point accurately along the outline of the area to be measured, proceeding in a clockwise direction and returning to the starting point.
6. Record the number of planimeter units. The dial shows the number of revolutions of the measuring wheel, and the measuring wheel is divided into 100th intervals. Make a second tracing and average the two readings.
7. Convert the planimeter units to units of area using the conversion factor as described above.

The planimeter shown in Figure 4.2 has a mechanical readout system. Digital readout and computing planimeters are available to automatically compute areas in any units or for any scale map. The principal advantage of the mechanical readout is its low cost. For frequent users, however, the digital and computing planimeters are worth the extra cost.

slope of the watershed and may be correlated with sediment losses. *See* Leopold et al. (1964) for more detailed accounts of river and watershed geomorphology.

4.2.2   Mapping Techniques

Lakes and ponds can be mapped easily if aerial photographs are available. The scale of the map can be determined by measuring the length of a building or other object in the picture. If aerial photographs are not available, however, field surveys are

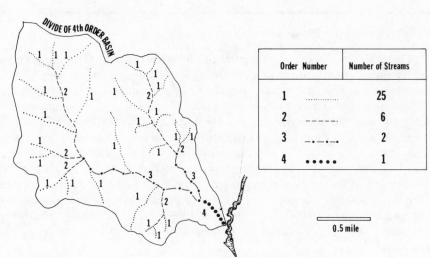

Figure 4.3 Stream order designation according to Strahler (1957)

usually necessary for developing shoreline maps. Welch (1948) provides detailed descriptions of traverse, stadia, two transit, plane-table, transverse measurement, triangle, and rectangle surveys for mapping lakes, and deflection-angle traverse and plane-table traverse surveys for streams. In this chapter the plane table-alidade, traverse, and deflection-angle traverse methods are described.

The plane table-alidade (or base line) method is most appropriate for smaller lakes and regular shorelines. It is described in greater detail by Welch (1948), Lind (1979), and Wetzel and Likens (1979). Minimum equipment required are a plane table and tripod, alidade, measuring tape, stakes, range (or sighting) poles, and paper. Drive numbered stakes along the water's edge at major changes in shoreline configuration (Fig. 4.4a). Establish a base line along a relatively straight section of shoreline so that each stake can be sighted from two locations to triangulate its location. Drive stakes at each end of the base line. Set up the plane table directly over the end of the base line (point A), and use a plumb bob to center the table over the stake. Attach paper to the table and level the table. Draw a base line on the map by sighting with the alidade to point B. Measure and record on the map the distance between A and B to determine the scale of the map. One person then takes the range pole and holds it vertically at each stake while another person sights with the alidade and draws in the appropriate lines of sight (Fig. 4.4 a). After all lines of sight are drawn, center the plane table over point B and align it in the proper direction by sighting with the alidade back to point A. Draw in lines of sight from point B to all stakes. The intersections of lines of sight represent the shoreline locations for the numbered stakes. Draw in the shoreline outline, and make measurements of irregular features.

The traverse method is more suitable for larger bodies of water than the plane table-alidade method. This method (Fig. 4.4 b) requires an engineer's transit and a person knowledgeable in its operation. Starting at point 1, make a series of distance measurements and angle measurements using the transit and a measuring tape. A stadia rod can be substituted to measure distances if the transit is equipped with stadia hairs. Use the field measurements (angles and distances) to plot the shoreline outline in the laboratory.

The plane table-alidade and traverse methods can be used to map medium to large-size stream segments; however, for smaller streams neither method may be

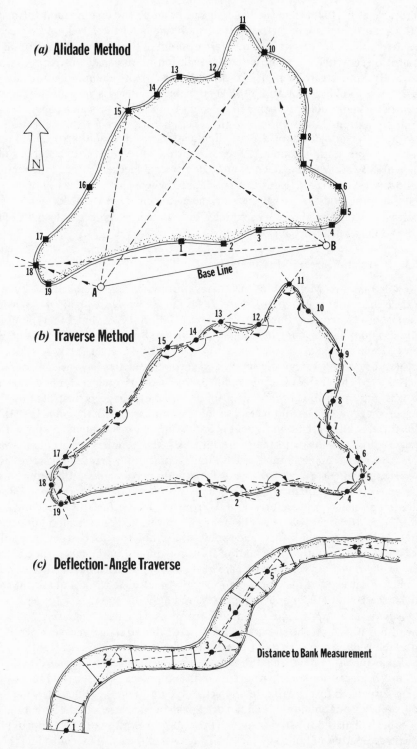

Figure 4.4 Mapping methods: *(a)* plane table - alidade (or base line method, *(b)* traverse method, and *(c)* deflection-angle traverse method.

suitable. Welch (1948) described the defection-angle traverse method for mapping small streams (Fig. 4.4 c). A transit with stadia hairs, a stadia rod, and a measuring tape are required. Drive a stake at the starting point of the survey (station 1), and set up a transit directly over it. Choose the next station (2) as far downstream as possible while still permitting a direct line of sight within the stream channel, between stations 1 and 2. Drive a stake at station 2 also. At appropriate intervals, make measurements of the distance right and left of the line of sight to the bank or water's edge. Move the transit to station 2, and choose station 3 as before. Record the deflection angle between the prolongation of the first line of sight (joining stations 1 and 2) and the second (joining stations 2 and 3). Measure distances between with a stadia rod. The principal advantage of the deflection-angle traverse method is that long sections of small, canopy-covered streams can be mapped where the use of aerial photographs would be impractical. Also, all measurements are made in the field, but the map is actually drawn in the laboratory so there is no problem, as there is in the field, of trying to fit the map onto the paper.

4.2.3   River and Stream Measurements

Water quality. Methods for measuring the physical and chemical quality of water are described fully in *Standard Methods for the Examination of Water, Sewage and Wastewater* (American Public Health Association 1976) and other limnological methods and pollution control manuals (Gehm and Bregman 1976; Lind 1979; United States Environmental Protection Agency 1979; Wetzel and Likens 1979). Furthermore, many water analysis kits and automatic meters have been manufactured (*see* Box 4.2). The degree of accuracy and precision of these meters is not standardized, and it is important to calibrate them following manufacturers' instructions. Measurements with kits and meters should be compared with the *Standard Methods* to insure that they are accurate and precise enough for the intended application. In general, Boyd (1980) found that water analysis kits were less precise than *Standard Methods* and were not suitable when high accuracy and precision were necessary.

High gradient rivers are typically well-mixed, and stratification or local differences in water quality are not usually problems unless there are point-source pollution discharges. Therefore, samples from a few locations will be accurate for the whole river. In larger, lower gradient rivers, however, there are often differences in water quality between riffles and pools and from top to bottom in pools and backwaters. Also, diurnal fluctuation in water quality parameters, such as carbon dioxide, dissolved oxygen, and pH, are likely in sluggish stream segments. For a more complete discussion of the interpretation of physical and chemical quality of running waters, see Hynes (1970) and Reid and Wood (1976).

Depth. The appropriate techniques for measuring depth in rivers and streams depend on the size of the stream. In small wadeable streams, measure depth with a calibrated wading rod with a flat base. Although these may be easily fabricated, they are also available commercially. In larger, unwadeable streams, measure depth from a boat or a bridge, using some type of sounding equipment—either a rigid rod, sounding weight suspended from a cable, or an electronic depth sounder (Buchanan and Somers 1969). When using sounding cables in deep, fast-flowing streams, the sounding weight and cable will drift downstream, and depth will be overestimated. An approximation to the true depth is obtained by measuring length of the cable extended and the angle between the vertical and the sounding cable. The true depth would be the length of the cable times the cosine of the angle. Electronic depth sounders, which are suitable for large stream surveys, have precision of 15-30 cm (Bovee and Milhous 1978).

Box 4.2. Sources for Equipment and Information

Source	*Items Available*
Ben Meadows Company 3589 Broad St. Atlanta, Georgia 30366	Acreage grids, compasses, current meters, map measurers, planimeters, surveying equipment, tape measures, etc.
Department of Mines and Technical Surveys Map Distribution Office Ottawa, Ontario, Canada	Topographic maps of Canada
Forestry Suppliers, Inc. 205 W. Rankin St. Jackson, Mississippi 39204	Similar to Ben Meadows Company
Hach Chemical Company Box 907 Ames, Iowa 50010	Water quality testing kits and reagents
Hydrolab Corporation P.O. Box 9406 Austin, Texas 78766	Water quality meters and digital recording thermographs
Inter Ocean Systems, Inc. 3540 Aero Ct. San Diego, California 92123	Oceanographic and water quality monitoring systems, current meters, tidal and wave gages
Leeds & Northrup Dickerson Road - MD 105 North Wales, Pennsylvania 19454	Dissolved oxygen continuous monitors and portable meters
Ryan Instruments, Inc. P.O. Box 599 Kirkland, Washington 98033	Continuous recording thermographs
Scientific Instruments, Inc. 518 West Cherry St. Milwaukee, Wisconsin 53212	Current meters
Teledyne Gurley 514 Fulton St., P.O. Box 88 Troy, New York 12181	Current meters

United States Environmental
  Protection Agency
Washington, District of
  Columbia
	STORET - EPA water quality data base

Continued

United States Geological	Topographic maps, state and
Survey Branch of Distribution	county base maps, land use
east of Mississippi River:	and cover maps
1200 S. Eads St.	
Arlington, Virginia 22202	
west of Mississippi River:	
Box 25286, Federal Center	
Denver, Colorado 80225	

United States Geological Survey Branch of Distribution east of Mississippi River: 1200 S. Eads St. Arlington, Virginia 22202 west of Mississippi River: Box 25286, Federal Center Denver, Colorado 80225 — Topographic maps, state and county base maps, land use and cover maps

United States Geological Survey 437 National Center Reston, Virginia 22042 — Water Data Storage and Retrieval System, WATSTORE, contains data collected at stream gaging stations, lakes and reservoirs, water quality, temperature, and sediment stations.

United States Geological Survey Water Resources Division located in most state capitals — Water resources data - publications on water quantity and quality data by state.

Wildlife Supply Company 301 Cass St. Saginaw, Michigan 48602 — Aquatic sampling and water quality testing equipment

Yellow Springs Instrument Co. Box 279 Yellow Springs, Ohio 45387 — Dissolved oxygen meters

Velocity. The simplest way to measure current velocity is to time an object as it floats a known distance downstream. However, there are several problems with this approach. Only the surface velocity, which is the fastest in the water column (Fig. 4.5), is measured. It is difficult to control where velocity is measured with this method. The object may not move in a straight line and, therefore, the distance will be difficult to measure. This method can be used for crude estimates where no other equipment is available. Use an object that floats submerged at the water surface and not above it— an orange works well. Select a straight section of stream with little turbulence. Repeat the measurements of distance and time at least twice, and average the velocities. Convert the surface velocity to a mean water column velocity by multiplying by 0.8 (rough bottom) or 0.9 (smooth bottom).

The Price and the pygmy meters are the most commonly employed meters for measuring river and stream velocity. The Price current meter (Fig. 4.6) was designed by the United States Geological Survey and is used by them to measure discharge at gaging stations. The Price meter consists of a bucket wheel mounted on a vertical axis which revolves when suspended in flowing water. Earphones are used to count the number of revolutions per second, which are converted to current velocity using the

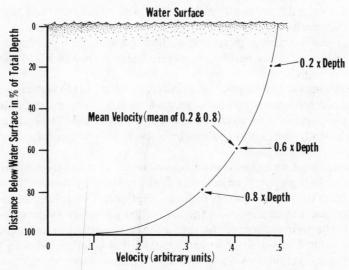

Figure 4.5 Vertical distribution of current velocity in an open channel (adapted from Buchanan and Somers 1969)

calibration charts supplied by the manufacturer. The Price meter is suited to velocities of 0.03-3.38 m/s. A smaller model, the pygmy, is used in small streams and is more accurate at low velocities (0.015-0.914 m/s).

The greatest advantage of the mechanical current meter is that it allows the measurement of velocity at any point in the water column. The mean velocity in the vertical water column can be calculated as the average of measurements taken at many points in the vertical. Typically, however, the mean velocity is approximated by taking velocity at one or two points in the vertical. The 0.6-depth method involves making one measurement at 0.6 of the depth below the surface (Fig. 4.5). This corresponds to the mean velocity in the vertical. The 0.6-depth method should only be used where depths are less than 0.76 m (Buchanan and Somers 1969); at deeper sites use the two-point method. Measure at 0.2 and 0.8 of the depth below the surface, and average these

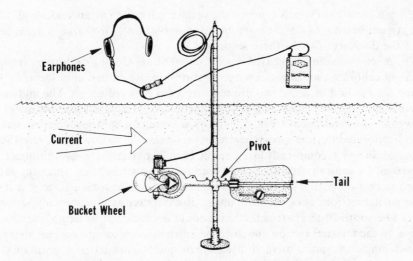

Figure 4.6 Price type current meter shown with wading rod.

two readings to estimate mean velocity (Fig. 4.5). Buchanan and Somers (1969) describe other methods for estimating mean velocity in the vertical.

When using the current meter and wading rod, stand in a position that does not affect the velocity of water passing the meter. The best approach is to hold the meter upstream and to your left.

The methods used to estimate mean current velocity in the vertical water column were developed for estimating discharge, and mean velocity may not always be the most relevant parameter to describe the microhabitat of fish. Depending on the fish species under study and its microhabitat preferences, measurements at other locations may be more meaningful.

Cross-sectional area. The cross-sectional area *(A)* of a stream is the area of the vertical cross section, perpendicular to flow, containing water. The area *(A)* is computed from the width *(w)* times the mean depth *(d)*. Two other parameters useful in characterizing the cross section are the wetted perimeter *(P)* and the hydraulic radius *(R)*. The wetted perimeter is the distance, measured perpendicular to flow, along the bottom and sides of a stream channel that is in contact with water. For rectangular cross sections this is approximately equal to the width plus twice the mean depth. The hydraulic radius is the ratio of cross-sectional area *(A)* and wetted perimeter *(P)*.

Discharge. Discharge can be measured directly using the United States Geological Survey midsection method or the Robins-Crawford (1954) method, or indirectly using Manning's equation or weirs. Before undertaking the task of measuring discharge, check to see if a United States Geological Survey gaging station exists near areas of interest. If so, there is already a daily record of discharge available. If rapid crude estimates are needed, the Robins-Crawford (1954) method is appropriate. Find a cross section in a relatively straight stream segment in which the threads of velocity are parallel to each other and the stream is stable and free of obstructions that would create turbulence. Divide the cross section into three sections. Within each section, measure the width *(w)*, mean depth *(d)*, and the time *(t)* it takes a float to travel a given distance *(l)*. Discharge *(Q)* in each section is estimated from

$$Q = \frac{w \times d \times a \times l}{t} \tag{4.1}$$

where *a* is a coefficient which converts the surface velocity to mean velocity (0.8 for a rough stream bottom to 0.9 for a smooth stream bottom). Discharge is equal to the sum of the discharge for the three sections.

The midsection method is used by the United States Geological Survey at gaging stations to calibrate the stage-discharge relationship. Buchanan and Somers (1969) describe the method in detail and present forms for data collection. The midsection method is similar to the Robins-Crawford method but requires a current meter and more velocity measurements (Fig. 4.7). After selecting an appropriate cross section, stretch a calibrated tag line across the stream perpendicular to flow. The United States Geological Survey recommends using about 25 to 30 partial sections, although fewer may be used for smooth streambeds with even velocity distributions. Space the partial sections so that no partial section has more than 10% of the total discharge in it. Do not use partial sections of equal width unless discharge is distributed evenly across the stream. The width of the partial sections should decrease as the depth and velocity increase. In each partial section, measure the mean velocity using a current meter and the 0.6-depth or two-point method. Measure distances from an initial point on shore *(b)* for each observation point. Discharge *(q)* in each partial section *(i)* is obtained from

$$q_i = v_i \times d_i \times \frac{(b_{i+1} - b_{i-1})}{2} \qquad (4.2)$$

where v=mean velocity and d=depth for each segment i. Total discharge (Q) in the stream at the cross section is

$$Q = \sum_{i=1}^{n} q_i \qquad (4.3)$$

where n is the number of partial sections.

Discharge can also be measured on smaller channels using weirs. A weir is a small dam placed across the stream, which causes the water to pass through a particular opening or notch. The notch may be V-shaped or trapezoidal (broad-crested). The discharge is proportional to the head (vertical distance from the bottom of notch to the water surface), measured upstream from the weir. The relationship between discharge (Q) and the measured head depends on the shape of the notch. Most hydraulics textbooks give appropriate equations for the common notch types.

Manning's equation can be used to estimate discharge and mean velocity in a stream channel indirectly. Discharge equals velocity, estimated by Manning's equation, times cross-sectional area, estimated by techniques described above. *See* Chow (1959) or United States Geological Survey (1976) for guidelines in using this method.

Cover/Shelter. Measurement of cover in streams is extremely important because many stream fishes prefer to remain in cover. What constitutes cover is sometimes ambiguous because cover types are often species-specific (Wesche 1980). Therefore, a species-specific approach which considers the cover type preferences may be required to evaluate the quality of cover in streams. Wesche (1980) and Bovee (1982) describe

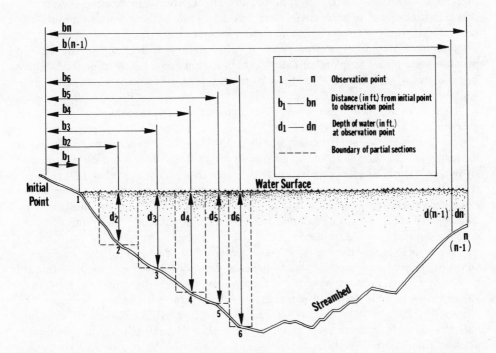

Figure 4.7 United States Geological Survey midsection method for estimating discharge in a stream (adapted from Buchanan and Somers 1969).

methods to evaluate cover quality in stream reaches.

Usually, only the types and quantity of cover are measured. Cover types are based on the object providing cover: boulder, log, root wad, brushpile, vegetation, undercut bank, and overhanging vegetation. Measuring the quantity (surface area coverage) of cover in a stream reach is a problem similar to estimating percentages of various types of vegetation or substrate classes in a stream reach. Therefore, the comments made in this section apply to both substrate and vegetation surveys.

When measurement of the surface area composed of different cover classes is desired, four general methods can be employed. The easiest is a visual estimate of percentage coverage, usually made from one vantage point on the bank. These estimates are usually unreliable because observers are not consistent.

Wright et al. (1981) describe three methods that are more objective: the detailed mapping method, the points sampling method, and the rectangles method. In the detailed mapping method, a grid of small dimension is established over the stream using painted bricks spaced at the corners of the grid or stakes driven at appropriate intervals along the bank and across the stream, with line strung to mark the grid. An observer then maps the type of cover (or substrate or vegetation) in each small grid. The amount of detail in the map is determined by the size of the grid. From these data a detailed map of the spatial distribution of cover types may be constructed. The detailed method is time consuming and often is more detailed then necessary. The points sampling method is quicker than the detailed method. It involves setting up a grid system of larger size and recording the type of cover, vegetation, or substrate at each point in the grid. In practice, the observer determines the dominant type at the point and must use a standard size area at each point to determine dominance or presence/absence of cover. The points sampling method can be used to estimate percentage coverage of each cover type, but it can not be used to make a map. The rectangles method involves setting up rectangles using a grid system and recording the dominant cover type, vegetation type, or substrate in each rectangle. This method provides a simplified map and an estimate of percentage coverage by type. These three methods work best in rivers less than 20 meters wide, where safe wading and good visibility are possible.

Instream vegetation surveys. The distribution and abundance of aquatic macrophytes are important from a water management (flooding) standpoint as well as from an ecological or fisheries standpoint. Maps for general distribution of macrophytes can be made by aerial photography in shallow water (Edwards and Brown 1960) or by the three mapping methods described by Wright et al. (1981) and in the preceding section on cover. The three mapping methods are well suited where estimates of the percentage coverage by different macrophytes are desired. For estimates of biomass (g/m^2) of aquatic vegetation, quadrat grid sampling or water column samplers should be used.

Substrate composition. Substrate composition determines the quality of spawning habitat and cover for many fishes and also influences benthic invertebrate production. In streams the variable of interest may be the percentage composition of various substrate classes in an entire area of the stream bottom. In other cases the variable of interest may be the vertical and horizontal distribution of substrate particles. Therefore, there are a range of methods available. The detailed mapping, points sampling, and rectangles methods described in the section on cover are appropriate for estimating the percentage of cover of the dominant substrate types over the stream bottom. However, stream substrates consist of a variety of particle sizes, and survival of some life stages (e.g., salmon embryos) depends on the relative density of these particle sizes. To measure this, the points sampling method is most appropriate.

There are several methods that may be used at each point in the points sampling method. These include core sampling, mechanical sieving, photographic analysis, and visual techniques.

In order to perform core sampling, insert a stove pipe or large diameter tube (10-30 cm) into the streambed to a standard depth (10-30 cm) and extract the substrate by hand or scoop (Shirazi et al. 1979). Determine the substrate that remains suspended by taking a subsample of the water column inside the core after it has been thoroughly mixed. This manual sampling is suitable when benthic invertebrates are also being sampled and when equipment is limited. However, the disadvantages of manual sampling include the difficulty of inserting the stove pipe in coarse gravel and the difficulty of sampling deep water (Shirazi et al. 1979). An alternative is the frozen core method (Everest et al. 1980) in which a hollow tube is inserted into the substrate, liquid carbon dioxide is poured down the tube to freeze a core of sediment, and the frozen core is then removed. The main advantages of the frozen core method are more complete collection of fine sediments, uniformity, ability to sample deeper water, and ability to analyze vertical distribution of sediment. The added cost and experience required and the difficulty of sampling large size gravel are the main disadvantages (Shirazi et al. 1979).

The percentage composition of the core by size classes is analyzed by sieving (Cummins 1962; Shirazi et al. 1979). The use of standard sieves distributes the particles based on the Wentworth classification (Table 4.1). Based on the sieving, the cumulative weight can be plotted against particle size. Also the geometric mean particle diameter and the percentage of the sediments less than a given size can be determined. Geometric mean particle size (d_g) is calculated as

$$d_g = d_1^{w_1} \times d_2^{w_2} \times d_3^{w_3} \times \ldots d_n^{w_n} \qquad (4.4)$$

where d = midpoint diameter of particles retained by a given sieve, and w = decimal fraction by weight of particles retained by a given sieve. A short cut formula for

Table 4.1 Modified Wentworth classification for substrate particle sizes (Cummins 1962)

Classification	Particle size range (mm)
Boulder	> 256
Cobble (Rubble)	64 — 256
Pebble	32 — 64
	16 — 32
Gravel	8 — 16
	4 — 8
	2 — 4
Very coarse sand	1 — 2
Coarse sand	0.5 — 1
Medium sand	0.25 — 0.5
Fine sand	0.125 — 0.25
Very fine sand	0.0625 — 0.125
Silt	0.0039 — 0.0625
Clay	< 0.0039

geometric mean particle size is

$$d_g = \sqrt{d_{16}\, d_{84}} \qquad\qquad (4.5)$$

where d_{16} is the particle size diameter at the 16th weight percentile, and d_{84} is the diameter at the 84th weight percentile. Platts et al. (1979) recommend the use of the geometric mean particle diameter as a standard measure for substrate characterization. Percent fines is usually calculated as the percentage of sediment by weight that is less than a given diameter, usually from 0.83 to 6.3 mm (Platts et al. 1979). Two other substrate parameters that are useful for evaluating salmonid spawning gravel are the sorting coefficient, a measure of the distribution of particle sizes, and the fredle index, a measure of the quality of gravel for salmonid reproduction (Lotspeich and Everest 1981).

Photographic analysis of substrate composition allows interpretation of only the surface composition in clear water and is not suited for all applications. It does, however, allow rapid evaluation in the field with limited laboratory work. Photographic prints can be analyzed manually (Cummins 1964) or with specialized equipment to count and size particles (Ritter and Helley 1969).

Visual methods are useful where rapid evaluation of the percentage composition of various substrate classes over an entire stream reach is desired. The points sampling method is most appropriate, using either of two basic methods. The simplest method is to measure size of the diameter of the dominant substrate type in a standard-sized area and classify it according to the Wentworth system.

The second method is to compute the substrate score, which is the summation of four numbers, three concerning the size of the substrate particles, and the fourth relating to a measure of embeddedness. At each point visually sampled, measure and assign values as shown in Table 4.2: (1) the size of the predominant substrate; (2) the size of the second most dominant substrate; (3) the size of the material that surrounds these two dominant types; and (4) the extent to which the predominant substrate is covered by finer sediments. The sum of the four numbers is the substrate score. In

Table 4.2 Values for particle sizes and embeddedness used in calculation of substrate scores (Crouse et al. 1981)

Characteristic	Value
Particle type or size	
Organic cover (over 50% of bottom surface)	1
< 1 - 2mm	2
2 - 5 mm	3
5 - 25 mm	4
25 - 50 mm	5
50 - 100 mm	6
100 - 250 mm	7
>250 mm	8
Embeddedness	
Completely embedded	1
3/4 embedded	2
1/2 embedded	3
1/4 embedded	4
Unembedded	5

preliminary observations, the substrate score was positively correlated with the geometric mean particle size, the standard but time-consuming measure (Crouse et al. 1981).

Solar radiation and stream shading. The intensity of solar radiation is measured with a radiometer. Radiometers are called pyrheliometers for measuring direct (parallel beam) solar radiation, pyranometers for measuring direct and diffuse solar radiation, and pyrradiometers for measuring total radiation (solar and reflected). Instruments available are described in United States Geological Survey (1980). The common units used for measurement are calories, kilojoules, ergs, watts (all expressed on a per area-time basis), foot-candles, and lux. Conversion formulas are given in Wetzel (1975) and Wetzel and Likens (1979), along with underwater measurement methods. Government meteorological stations report data collected with pyrheliometers, but where shading may alter the solar radiation reaching the stream, solar radiation should be measured on site.

Because stream shading and surrounding topography affect the amount of solar radiation reaching the stream during a day, measurements of stream shading and the arc of direct sunlight are often made. The canopy shading (percentage shading) is estimated using a densiometer, a convex mirror divided into grids (Murphy et al. 1981). If the boundaries (east and west) that define the arc of direct sunlight can be accurately determined, the angle of this arc can be easily measured with a clinometer (Platts 1982).

Local stream gradient. The stream gradient in a localized area can be measured with a level mounted on a tripod and a leveling rod. The difference in elevation between two sites divided by the horizontal distance between them is the gradient. The location of the upstream and downstream sites is important because gradient changes abruptly rather than smoothly. Stream gradient may also be approximated using a hand level or clinometer.

Habitat types. A wide variety of habitats occur in small to medium-size streams (Box 4.3), but pools and riffles are the types described in most surveys. In streams the pool-riffle ratio or percent pools and riffles is typically estimated visually. These are subjective measurements with poor repeatability because riffles, pools, and other habitat types are not separated by clearly defined boundaries (Platts 1982).

To more objectively differentiate between pools and riffles, Shirvell (1982) created the shape index, calculated as

$$\text{shape index} = (w/d)^{(d/d\,max)} \tag{4.6}$$

where w = width, d = mean depth, and $d\,max$ = maximum depth along a cross section. Pools had shape indices less than 9 and riffles had indices greater than 9.

Streambank measurement. Condition, slope, and vegetative cover of stream banks is important for determining the erosion of sediments during high water and, therefore, the turbidity and sedimentation in streams. Slope of the stream bank is measured accurately using a level mounted on a tripod and a leveling rod. The placement of transects can bias the results, so choose several transects randomly, and measure each. Other riparian measurements, such as percentage streamside cover, stream bank stability, and animal grazing damage, are usually made subjectively and, therefore, are often unreliable. Some rating systems for these measurements are given in the *Aquatic Habitat Surveys Handbook* (United States Forest Service 1982).

4.2.4 Lake and Reservoir Measurements

Water quality. References on methods for measuring water quality are cited in

---

Box 4.3   Terminology of Stream Habitat Types

Terminology and descriptions used for common habitat types in small to medium-size streams (Bisson et al. 1982).

1. Riffles—shallow, turbulent stream segments with higher gradients than pools or glides.
   a. Low gradient riffles—shallow stream reaches with moderate current velocity, some surface turbulence, and gradient less than 4%.
   b. Rapids—higher gradient (greater than 4%) reaches with faster current velocity, coarser substrate, and more surface turbulence.
   c. Cascades—stepped rapids with very small pools behind boulders and small waterfalls.
2. Glides—moderately shallow stream channels with laminar flow, lacking pronounced turbulence.
3. Pools—deeper habitats with slower current velocities.
   a. Secondary channel pools—those that remain along the flood plain when high waters recede.
   b. Backwater pools—slack areas along channel margins, caused by eddies behind obstructions.
   c. Trench pools—slots in stable channel bottom, e.g., bedrock.
   d. Plunge pools—where main flow passes over a complete channel obstruction and drops vertically to scour the streambed.
   e. Lateral scour pools—where flow is deflected by a partial channel obstruction, usually at bends in channel.
   f. Dammed pools—deep water impounded by a channel blockage.

---

section 4.2.3. In lakes and reservoirs vertical stratification is likely, and, therefore, measurements are typically made at intervals from top to bottom. Refer to Wetzel (1975) for stratification patterns to expect in lakes at various times of the year. Water sampling devices designed to take samples at discrete depths are needed, such as the Kemmerer, van Dorn, and Nansen samplers (Lind 1979). Meters are also available to measure some water quality parameters, although, as discussed in section 4.2.3, attempts should be made to determine their reliability before adopting them for standard use. Horizontal differences in water quality are also likely in lakes and reservoirs so several sampling sites are usually selected.

Trophic state. Many attempts have been made to classify lakes by their trophic state (oligotrophic, mesotrophic, eutrophic). The most frequently used index of trophic state for fisheries has been the morphoedaphic index *(MEI)*, calculated as

$$MEI = \sqrt{(TDS)/d} \qquad (4.7)$$

where *TDS* is total dissolved solids and d is mean depth. The *MEI* was originally developed to estimate fish production, but it is also related to trophic state (Ryder et al. 1974; Jenkins 1982). Other trophic state indices in use are reviewed in Maloney (1979).

Solar radiation and temperature. In lakes and reservoirs solar radiation changes with depth of water. Therefore, solar radiation must be measured with a submarine photometer (Lind 1979; Wetzel and Likens 1979). Because of the cost involved, these measurements are not made for most fisheries applications. Instead, because the amount of solar radiation penetrating the water depends on transparency, water transparency is commonly measured with a simple device called a Secchi disk. The disk consists of a circular plate (20 cm diameter), painted black in two opposite quarters and white in the other two. To determine Secchi disk transparency, lower the disk into the water in the shade until it disappears and record the depth. Lower the disk a few feet and then slowly raise it until it reappears, and record this depth. The average of the two readings is the Secchi disk transparency. Make measurements at midday. Secchi disk measurements have been used in empirical estimation of the euphotic depth, the depth at which 99% of the surface radiation has been dissipated (French et al. 1982).

Temperature, which is indirectly related to solar radiation, is the most common habitat measurement. For measuring temperature from top to bottom, electric thermistor thermometers are most appropriate. Other approaches using mercury thermometers, reversing thermometers, or bathythermographs are described in Wetzel and Likens (1979).

Depth. As discussed in section 4.2.3, depth may be measured with sounding cables and weights or electronic echosounders. A continuous record of depth along transects can be obtained using a depth sounder moved across the lake at a constant speed. Because most depth sounders are not accurate in shallow water, sounding cables and weights may be more appropriate. If sounding cables and weights are used on lakes too large to stretch a tag line across, locate soundings along a transect with an alidade. Mark the ends of the transect with colorful flags so that the boat crew can stay on the transect. Use the alidade at each point to locate the point on a map. The frequency of depth soundings and number of transects needed depends on how rapidly depth is changing. In areas where depth changes markedly, more transects and more measurements per transect are needed. Plot the depth data on an outline map so that depth contour lines can be drawn. This map is called a bathymetric or hydrographic map.

Contour maps involve interpolation of the spatial data to connect lines of equal depth (or other habitat parameters). For depth, the shoreline can be used as a guide in drawing the shape of the contour lines between points where data were collected. Contour maps require much data and time to develop and are not always necessary for fisheries applications. Where required, contour mapping can be accomplished more efficiently by digitizing the map data, i.e., expressing the spatial distribution as an array of numbers that a computer can interpret. Several contour mapping programs (e.g., SYMAP, SURFACE II) are available at most university computer centers.

Morphometric measurements. Several morphometric measurements are commonly used to characterize lakes and reservoirs. Volume measurement is important in determining the amount of chemical needed to reach a certain concentration. In relatively uniform basins, depth can be measured at several randomly selected points to estimate mean depth. Volume is approximated by multiplying the mean depth times the surface area (both measured in the same units).

More precise measurements of volume require a good contour map. First, determine the surface area at each contour. Then estimate the volume of water in the stratum between two contour lines *(Vs)* as

$$V_s = \frac{h}{3} (A_1 + A_2 + \sqrt{A_1 A_2}$$

(4.8)

where h = height of the stratum, A_1 = the area of the upper surface of the stratum, and A_2 = area of the lower surface. The total volume is the sum of the volumes for each stratum. Mean depth is estimated simply by dividing the volume by the surface area. Be careful when using old maps for morphometric measurements because erosion, water levels, and sedimentation can alter depth.

The shoreline length is needed to calculate the shoreline development index, a measure of the shape of a body of water and an indicator of the potential amount of littoral zone development. Shoreline length is easily measured from an existing map using a cartometer (Figure 4.2). The shoreline development index *(SDI)* is the ratio of the shoreline length *(L)* to the circumference of a circle of the same area as the lake, calculated as

$$SDI = \frac{L}{2\sqrt{A\pi}} \tag{4.9}$$

A circular lake will have an *SDI* of 1, and the greater the irregularity of the shoreline, the greater the *SDI*.

The mean slope of the lake or reservoir bottom is often of interest since lakes with slight slope tend to be more productive than deep lakes with steep slopes. The mean slope between two contours *(S)* is calculated from

$$S = \frac{L_i + L_{i+1}}{2} \times \frac{h}{A} \tag{4.10}$$

where L_i = length of the contour lines, h = height between contours, and A = area between two contours.

Currents. Current velocity is an important parameter in main stream impoundments. Current velocity and direction can be measured with mechanical current meters available from limnological and oceanographic supply houses. These can be rigged with weights, cables, and buoys to obtain readings at desired depths. Some meters are available with self-contained strip charts or digital recorders for continuous unattended recording.

Flushing rate. The flushing rate in a lake or reservoir is equal to the proportion of reservoir volume discharged per unit of time and is calculated as discharge divided by volume. The flushing rate can be computed on an annual, seasonal, or daily basis. The storage ratio is inversely related to the flushing rate. The storage ratio is the ratio of the average annual volume of the lake to the annual discharge volume, both measured in the same units.

Waves. Wave heights may be important habitat measurements in lakes or reservoirs because of their relation to shoreline erosion. Wave height is measured as the vertical distance from crest of the wave to the lowest point of the water surface. If wave heights are small, they can be measured with a staff gage installed in the area. For continuous measurements of waves under a wider range of climatic conditions, pressure-sensitive wave gages can be permanently installed.

Water levels. Average annual vertical fluctuation in water level is a variable commonly used to characterize the degree of water level fluctuation in a reservoir. Most reservoirs have continuous water level recorders, and data can be obtained from reservoir operators. In lakes where there are no existing recorders, continuous recording gages or staff gages that are manually recorded must be installed.

Cover. Cover can be evaluated in shallow and clear lakes and reservoirs with methods similar to those used in rivers and streams (section 4.2.3). In deeper lakes and reservoirs, cover can be detected using echosounders (chapter 12) to map the location and extent of cover, although the type of cover is not always detectable. Transect sampling is usually most appropriate when using echosounders.

Substrate. In lakes and reservoirs substrate types can be classified using the modified Wentworth scale (Table 4.1), and spatial distribution can be determined using a points sampling method. To obtain substrate samples some type of dredge or core sampler is utilized. The Ekman grab sampler is best suited for sampling soft and finely divided sediments, while the Petersen or Ponar grab samplers are better for sampling larger particle sizes (Lind 1979; Wetzel and Likens 1979). Substrate particle sizes are usually determined in conjunction with benthic invertebrate samples.

Vegetation. In shallow waters aerial photography can be used to record the distribution of aquatic macrophytes (Edwards and Brown 1960). If the vegetation is not too dense, it can be identified and mapped quickly and easily using shoreline surveys from a boat. For densely vegetated areas, the points sampling approach with water column vegetation samplers is more efficient.

4.3   REFERENCES

Aggus, L. R., and S. A. Lewis. 1977. Environmental conditions and standing crop of fishes in predator-stocking-evaluation reservoirs. *Proceedings of the Annual Conference of the Southeastern Association of Fish and Wildlife Agencies* 30:131-140.

American Public Health Association. 1976. *Standard methods for the examination of water, sewage, and wastewater,* 14th edition. American Public Health Association, New York, New York, USA.

Barila, T. V., R. D. Williams, and J. R. Stauffer, Jr. 1981. The influence of stream order and selected stream bed parameters on fish diversity in Raystown Branch, Susquehanna River Drainage, Pennsylvania. *Journal of Applied Ecology* 18:125-131.

Benson, N. G. 1973. Evaluating the effects of discharge rates, water levels, and peaking on fish populations in Missouri River main stem impoundments. *Geophysical Monographs Series* 17:683-689.

Benson, N. G. 1980. *Effects of post-impoundment shore modifications on fish populations in Missouri River reservoirs.* United States Fish and Wildlife Service Research Report 80, Washington, District of Columbia, USA.

Binns, N. A., and F. M. Eiserman. 1979. Quantification of fluvial trout habitat in Wyoming. *Transactions of the American Fisheries Society* 108:215-228.

Bisson, P. A., J. L. Nielsen, R. A. Palmason, and L. E. Grove. 1982. A system of naming habitat types in small streams, with examples of habitat utilization by salmonids during low streamflow. Pages 62-73 *in* N. B. Armantrout, editor. *Symposium on Acquisition and Utilization of Aquatic Habitat Inventory Information,* Western Division, American Fisheries Society, Portland, Oregon, USA.

Bovee, K. D. 1982. *A guide to stream habitat analysis using the Instream Flow Incremental Methodology.* United States Fish and Wildlife Service Instream Flow Information Paper Number 12. FWS/OBS-82/26, Fort Collins, Colorado, USA.

Bovee, K. D., and R. T. Milhous. 1978. *Hydraulic simulation in instream flow studies: theory and techniques.* United States Fish and Wildlife Service Instream Flow Information Paper Number 5, FWS/OBS-78/33, Fort Collins, Colorado, USA.

Boyd, C. E. 1980. Reliability of water analysis kits. *Transactions of the American Fisheries Society* 109:239-243.

Buchanan, T. J., and W. P. Somers. 1969. Discharge measurements at gaging stations. United States Geological Survey, *Techniques of Water-Resources Investigations,* Book 3, Washington, District of Columbia, USA.

Chow, V. T. 1959. *Open channel hydraulics.* McGraw-Hill Book Company, New York, New York, USA.

Crouse, M. R., C. A. Callahan, K. W. Malueg, and S. E. Dominguez. 1981. Effects of fine sediments on growth of juvenile coho salmon in laboratory streams. *Transactions of the American Fisheries Society* 110:281-286.

Cummins, K. W. 1962. An evaluation of some techniques for the collection and analysis of benthic samples with special emphasis on lotic waters. *American Midland Naturalist*

67:477-504.

Cummins, K. W. 1964. Factors limiting the microdistribution of larvae of the caddisflies *Pycnopsyche lepida* (Hagen) and *Pycnopsyche guttifer* (Walker) in a Michigan stream (Trichoptera: Limnephilidae). *Ecological Monographs* 34:271-295.

Edwards, R. W., and M. W. Brown. 1960. An aerial photographic method for studying the distribution of aquatic macrophytes in shallow waters. *Journal of Ecology* 48:161-164.

Everest, F. H., C. E. McLemore, and J. F. Ward. 1980. *An improved tri-tube cryogenic gravel sampler.* United States Department of Agriculture, Forest Service, Pacific Northwest Forest and Range Experiment Station, Research Note PNW-350, Portland, Oregon, USA.

French, R. H., J. J. Cooper, and S. Vigg. 1982. Secchi disc relationships. *Water Resources Bulletin* 18:121-123.

Gehm, H. W., and J. I. Bregman, editors. 1976. *Handbook of water resources and pollution control.* Van Nostrand Reinhold Company, New York, New York, USA.

Gorman, O. T., and J. R. Karr. 1978. Habitat structure and stream fish communities. *Ecology* 59:507-515.

Groen, C. L., and T. A. Schroeder. 1978. Effects of water level management on walleye and other coolwater fishes in Kansas reservoirs. *American Fisheries Society Special Publication* 11:278-283.

Harrel, R. C., B. J. Davis, and T. C. Dorris. 1967. Stream order and species diversity of fishes in an intermittent Oklahoma stream. *American Midland Naturalist* 78:428-436.

Huet, M. 1959. Profiles and biology of western European streams as related to fish management. *Transactions of the American Fisheries Society* 88:155-163.

Hunt, R. L. 1971. *Responses of a brook trout population to habitat development in Lawrence Creek, Wisconsin.* Wisconsin Department of Natural Resources Technical Bulletin Number 48, Madison, Wisconsin, USA.

Hunt, R. L. 1979. *Removal of woody streambank vegetation to improve trout habitat.* Wisconsin Department of Natural Resources Technical Bulletin Number 115, Madison, Wisconsin, USA.

Hynes, H. B. N. 1970. *The ecology of running waters.* University of Toronto Press, Toronto, Ontario, Canada.

Jenkins, R. M. 1970. The influence of engineering design and operation and other environmental factors on reservoir fishery resources. *Water Resources Bulletin* 6:110-119.

Jenkins, R. M. 1976. Prediction of fish production in Oklahoma reservoirs on the basis of environmental variables. *Annals of the Oklahoma Academy of Science* 5:11-20.

Jenkins, R. M. 1982. The morphoedaphic index and reservoir fish production. *Transactions of the American Fisheries Society* 111:113-140.

Leopold, L., M. G. Wolman, and J. P. Miller. 1964. *Fluvial processes in geomorphology.* Freeman Company, San Francisco, California, USA.

Lind, O. T. 1979. *Handbook of common methods in limnology.* 2nd edition. C. V. Mosby Company, St. Louis, Missouri, USA.

Lotspeich, F. B., and F. H. Everest. 1981. *A new method for reporting and interpreting textural composition of spawning gravel.* United States Department of Agriculture, Forest Service, Pacific Northwest Forest and Range Experiment Station, Research Note PNW-369, Corvallis, Oregon, USA.

Lotspeich, F. B., and W. S. Platts. 1982. An integrated land-aquatic classification system. *North American Journal of Fisheries Management* 2:138-149.

Maloney, T. E. editor. 1979. *Lake and reservoir classification systems.* United States Environmental Protection Agency, Office of Research and Development, EPA-600/3-79-074, Corvallis, Oregon, USA.

Matthew, W. J., and J. T. Styron, Jr. 1981. Tolerance of headwater vs. mainstream fishes for abrupt physicochemical changes. *American Midland Naturalist* 105:149-158.

Murphy, M. L., C. P. Hawkins, and N. H. Anderson. 1981. Effects of canopy modification and accumulated sediment on stream communities. *Transactions of the American Fisheries Society* 110:469-478.

Olsen, J. E., E. A. Whippo, and G. C. Horak. 1981. *Reach file phase II report: a standardized method for classifying status and type of fisheries.* United States Fish and Wildlife Service, FWS/OBS-81/31, Washington, District of Columbia, USA.

Platts, W. S. 1979. Relationships among stream order, fish populations, and aquatic geomorphology in an Idaho river drainage. *Fisheries* 4 (2):5-9.

Platts, W. S. 1980. A plea for fishery habitat classification. *Fisheries* 5 (1):2-6.

Platts, W. S. 1982. Stream inventory garbage in—reliable analysis out: only in fairy tales. Pages 75-84 *in* N. B. Armantrout, editor. *Symposium on Acquistion and Utilization of Aquatic Habitat Inventory Information*, Western Division, American Fisheries Society, Portland, Oregon, USA.

Platts, W.S., M. A. Shirazi, and D. H. Lewis. 1979. *Sediment particle sizes used by salmon for spawning with methods for evaluation.* United States Environmental Protection Agency, EPA-600/3-79-043, Corvallis, Oregon, USA.

Reid, G. K., and R. D. Wood. 1976. *Ecology of inland waters and estuaries*, 2nd edition. Van Nostrand Company, New York, New York, USA.

Ritter, J. R., and E. J. Helley. 1969. Optical method for determining particle size of coarse sediment. United States Geological Survey, *Techniques of Water-Resources Investigation*, Book 5, Washington, District of Columbia, USA.

Robins, C. R., and R. W. Crawford. 1954. A short accurate method for estimating the volume of stream flow. *Journal of Wildlife Management* 18:366-369.

Ryder, R. A., S. R. Kerr, K. H. Loftus, and H. A. Regier. 1974. The morphoedaphic index, a fish yield estimator—a review and evaluation. *Journal of the Fisheries Research Board of Canada* 31:663-688.

Sheldon, A. L. 1968. Species diversity and longitudinal succession in stream fishes. *Ecology* 49:193-198.

Shirazi, M.A., D. H. Lewis, and W. K. Seim. 1979. *Monitoring spawning gravel in managed forested watersheds, a proposed procedure.* United States Environmental Protection Agency, EPA-600/3-79-014, Corvallis, Oregon, USA.

Shirvell, C. S. 1982. Objective differentiation of pool and riffle habitat based on transect data. Unpublished Manuscript. (Available from Department of Fisheries and Oceans, Freshwater and Anadromous Division, P.O. Box 550, Halifax, Nova Scotia, Canada, B3J 2S7).

Stalnaker, C. B. 1979. The use of habitat structure preferenda for establishing flow regimes necessary for maintenance of fish habitat. Pages 321-337 *in* J. V. Ward and J. A. Stanford, editors. *The ecology of regulated streams.* Plenum Press, New York, New York, USA.

Stalnaker, C. B., and J. L. Arnette, editors. 1976. *Methodologies for the determination of stream resource flow requirements: an assessment.* United States Fish and Wildlife Service, Washington, District of Columbia, USA.

Strahler, A. N. 1957. Quantitative analysis of watershed geomorphology. *Transactions American Geophysical Union* 38:913-920.

Thompson, D. H., and F. D. Hunt. 1930. The fishes of Champaign County: a study of the distribution and abundance of fishes in small streams. *Illinois Natural History Survey Bulletin* 19 (1):1-101.

United States Environmental Protection Agency. 1979. *Methods for chemical analysis of water and wastes.* Office of Research and Development, EPA-600/4-79-020, Washington, District of Columbia, USA.

United States Fish and Wildlife Service. 1980a. *Habitat as a basis for environmental assessment.* Division of Ecological Services, ESM 101, Washington, District of Columbia, USA.

United States Fish and Wildlife Service. 1980b. *Habitat evaluation procedures (HEP).* Division of Ecological Services, ESM 102, Washington, District of Columbia, USA.

United States Fish and Wildlife Service. 1981. *Standards for the development of habitat suitability index models.* Division of Ecological Services, ESM 103, Washington, District of Columbia, USA.

United States Forest Service. 1982. *Aquatic habitat survey handbook (FSH 2609.23) for the General Aquatic Wildlife System (G.A.W.S.).* United States Department of Agriculture, Forest Service, Region 4, Ogden, Utah, USA.

United States Geological Survey. 1967. *Roughness characteristics of natural channels.* United States Geological Survey, Water Supply Paper 1849, Arlington, Virginia, USA.

United States Geological Survey. 1980. Radiation. Office of Water Data Coordination, United States Geological Survey, *National handbook of recommended methods for water-data acquisition,* Chapter 10.H, Reston, Virginia, USA.

Welch, P. S. 1948. *Limnological methods.* McGraw-Hill Book Company, New York, New York, USA.

Wesche, T. A. 1980. *The WRRI trout cover rating method. Development and application.* Water Resources Research Institute, Water Resources Series Number 78, University of Wyoming, Laramie, Wyoming, USA.

Wetzel, R. G. 1975. *Limnology.* W. B. Saunders Company, Philadelphia, Pennsylvania, USA.

Wetzel, R. G., and G. E. Likens. 1979. *Limnological analyses.* W. B. Saunders Company, Philadelphia, Pennsylvania, USA.

Wright, J. F., P. D. Hiley, S. F. Ham, and A. D. Berrie. 1981. Comparison of three mapping procedures developed for river macrophytes. *Freshwater Biology* 11:369-380.

Zimmer, D. W., and R. W. Bachman. 1978. Channelization and invertebrate drift in some Iowa streams. *Water Resources Bulletin* 14:868-883.

Chapter 5
Care and Handling of Live Fish

ROBERT R. STICKNEY

5.1 INTRODUCTION

Many activities undertaken by fisheries workers require the handling and transporting of live fish. Data gathering and tissue collection are often the most stressful form of handling over short periods of time. Individual fish require handling when they are measured or weighed; when scales, blood, or biopsy samples are taken; when they are tagged or marked; when tags are recovered or marks read; and when they are examined for diseases (Fig. 5.1).

Live hauling can be accomplished without severe stress if the proper facilities and conditions exist. Fish may be moved in buckets or tubs from one place to another on a hatchery or in the field, hauled distances of a few kilometers to several hundred kilometers in live hauling tanks on trucks, or shipped around the world in plastic bags into which pure oxygen has been injected.

Holding before, during, and after handling is also required in many instances.

Figure 5.1. Careful handling is necessary to avoid unneeded stress to fish. Examine fish at time of handling to check for diseases and general condition.

Holding facilities may feature flow-through water systems (raceways or cages), static water systems (containers of standing water which may receive aeration), or closed recirculating systems, in which water flowing through the holding tank is treated and reused. The type of system used depends on the species of fish being held, the density of fish per unit volume of holding water, and the amount of holding time that is required. In both holding and hauling it is necessary to maintain the best water quality possible until the fish are removed from the high stress environment.

This chapter presents information related to the care and handling of live fish under a variety of circumstances. The material presented is of a summary nature; specifics may vary considerably from one fish species to another and from region to region depending upon climate and ambient water quality. The most important concept associated with handling and hauling fish is to avoid, as much as possible, conditions which will cause stress. This goal is pervasive and must be kept in mind at all times. If stressful conditions cannot be avoided, fish health will be adversely affected. Mortality may occur very rapidly, or may be delayed for several days (up to two weeks) while disease organisms become abundant enough to cause epizootics. Many of the techniques discussed in this chapter are based upon common sense and the experience of workers who have handled fish for decades. Others have been developed in response to specific needs by engineers and those not routinely involved with fisheries.

5.2 MORTALITY RESULTING FROM FISH HANDLING

5.2.1 Physical Injury

Direct mortality may occur when fish are handled roughly or kept out of the water for extended periods. There is no hard and fast rule about how long a fish can be kept from the water before irreversible damage occurs. Some fish can take considerable abuse without apparent long-term effects, while the mere removal of some species from the water leads to almost instant mortality, for example, the bay anchovy, *Anchoa mitchilli*. As a general rule, fish should never be kept from the water any longer than is necessary to accomplish the task for which they were captured.

The manner in which fish are captured has an impact on how well the animals adapt to holding and handling. All fishing methods cause stress and may lead to immediate or delayed mortality. Rough fishing gears, such as trawls and gill nets, may lead to high mortality and considerable injury to fish that are recovered alive. Injury can be reduced, although not eliminated, by using short duration trawl hauls (5 to 10 minutes maximum) and short gill net sets (check the net hourly). Trot lines should be checked about every hour, and fish should be removed from seines as quickly as possible after the haul is completed.

Delayed mortality is often associated with a disease epizootic which occurs from 24 hours to 14 days after handling. If a fish is injured during handling, disease may develop within a few hours or days. Examples of injuries which can lead to disease problems are loss of mucus, loss of scales, damage to the integument, and internal damage. Internal injuries occur when fish are not properly restrained or sedated during handling. It is common for a fish to jump out of the worker's hand and fall onto a hard surface. Internal injuries can easily result, especially in females with developing ovaries. The use of wet wool gloves when handling fish is a great aid because they make it possible to keep the fish still without squeezing it.

In instances where tissue samples are taken or fish are handled roughly, it is common practice to provide prophylactic disease treatment. This may involve dipping the fish in an antibiotic, bactericidal, or parasiticidal solution. Another technique is to inject an antibiotic at the time of handling. These techniques and others are described thoroughly in the fish hatchery manual by Piper et al. (1982).

5.2.2 Water Quality Management

Additional stress may be related to declining water quality which precedes or accompanies the actual collection process. For example, when ponds are harvested, they are often partially drained before fish are collected. Fish that have not been captured before the water level is reduced are forced into a small volume of water which has high turbidity, low levels of dissolved oxygen, and rapidly changing temperature. Rapid and dangerous changes in temperature most commonly occur in winter and summer when air and water temperatures can be quite different. In warm climates, the handling of fish during the summer should be limited to the early morning, and sufficient water of normal temperature should be run into the pond, raceway, or tank to keep the water from warming excessively. This flow of water, provided it is properly aerated, will also reduce low oxygen stress.

Once fish have been placed into a holding facility, water quality should be monitored frequently and maintained within the optimum range for the species involved. Optimum fish holding conditions occur when an adequate supply of well-aerated water is available. Dissolved oxygen should always be at or above a level of 5 mg/l. While some species can tolerate lower levels, it is best to maintain the above value. It may be necessary to aerate the water to maintain this level. Methods for aerating holding facilities include agitators, air stones supplied by air blowers or compressors, spraying inflow water in a narrow stream to inject atmospheric oxygen, and bubbling in bottled or liquid oxygen. Pure oxygen is rarely used in holding facilities, but it should be available as a back-up source of aeration if the primary supply fails.

Temperature in holding water should be as similar to temperature at the collection site as possible so that thermal shock is avoided. If a temperature change of more than two to three degrees C cannot be avoided, temper the fish into the new environment. Place a container of fish at one temperature into water of the second temperature and allow sufficient time for the two temperatures to equilibrate. Alternatively, introduce receiving water slowly into the container of fish until equilibration is achieved. Limit tempering to about two degrees C hourly, and maintain dissolved oxygen at or above the level of 5 mg/l while tempering is taking place. Remember that fish are often being moved to a temperature which is less suitable for the species than the one in which they were living. Although there is little experimental evidence to indicate that this causes higher mortality than rapid temperature changes toward the thermal optimum of the species, observational evidence seems to support the idea.

Although dissolved oxygen and temperature are perhaps the most important water quality variables, care should be taken to avoid any drastic change in water quality. Among the variables which should be checked are salinity, pH, hardness, and alkalinity. Salinity shock is usually only a problem with stenohaline species (those with narrow salinity tolerance), though it may be necessary to slowly temper even euryhaline species (those with a broad range of salinity tolerance) if change is great. Accomplish salinity tempering by diluting or concentrating the water from which the fish were captured until it is of the same salinity as the holding water. When differences in pH, alkalinity, or hardness are great between the water of capture and the holding water, tempering again may be required. No standards have been set for hardness and alkalinity, but most species can withstand relatively large changes without difficulty. Exceptions exist for some euryhaline marine fish, for example, the red drum *(Sciaenops ocellatus)*, which can survive transfer from saltwater to hard freshwater but may undergo severe osmostic stress when placed into soft freshwater. With respect to pH, holding facility water should be kept within the range of 6.5 to 8.5. In static and recirculating systems, place small amounts of calcium carbonate (e.g., in the form of crushed limestone or crushed oyster shells) into the water system. Flow-through systems also can be buffered by adding chemicals, but the cost is generally prohibitive.

When fish are crowded, metabolites become concentrated in the water, and the fish can become acutely stressed. Fish can usually be held or hauled for several hours before carbon dioxide, ammonia, and nitrite levels become problems, but at some point the accumulated chemicals will cause stress. In a flow-through system, metabolite concentrations can be controlled by maintaining a high enough exchange rate to flush them out. Static systems are the most subject to metabolite build-up and for that reason should not be used for long-term holding unless fish densities are low and metabolite levels are routinely monitored. Zeolite is becoming increasingly popular as an inexpensive ammonia absorbant and may be used in both holding and hauling activities.

Be certain that the holding water is free of toxic chemicals. Municipal water should not be used unless it is first de-chlorinated. The easiest method is to aerate the water for 24 hours prior to use. Commercially available de-chlorinating chemicals are also available. Ground-water should be free of hydrogen sulfide, supersaturated gases, and other toxic conditions. In cases where contamination is suspected, test surface or ground-water supplies for excessive levels of toxic substances.

Gentle handling and constant concern for the health of the fish is the primary responsibility of persons who handle live fish. Proper procedures can be taught, but perfection of them comes only through experience. Keep the environment as near optimum as possible given the constraints which always exist. Never handle fish until all facilities and equipment to be utilized in conjunction with their handling have been checked out and verified as being ready for use. Identify in advance problems which can lead to mortalities and take precautions to reduce risk.

5.3 GENERAL HANDLING AND HAULING PROCEDURES

5.3.1 Plastic Bag Technique

Plastic bags are utilized for transporting fish virtually throughout the world. A large percentage of goldfish and tropicals are shipped in polyethylene bags that are partially filled with water. Oxygen from compressed gas cylinders is then used to saturate the water and displace atmospheric gases from the bag. The bag is then inflated with oxygen, closed, and secured with rubber bands. Johnson (1979) determined normal carrying capacity for fish in plastic bags as a function of time required in transit (Table 5.1). The table assumes that the bags are properly filled and secured against oxygen loss. To reduce temperature changes during transport and to prevent punctures, plastic bags are commonly placed in insulated cardboard boxes.

Table 5.1 Fish biomass (g/l) which can safely be placed in plastic bags for various lengths of time and for fish of various sizes at 18 C (after Johnson 1979)

Kind of fish (and size)	Duration of Transport (hrs)			
	1	12	24	48
Food fish fingerlings (8 cm)	100	75	50	25
Food fish fry (0.6 cm)	50	40	30	—
Bait fish (2.5 cm)	100	75	50	25
Bait fish (8 cm)	200	150	100	75
Pet fish (2.5-5.0 cm)	100	75	50	25

5.3.2 Live Tanks and Live Wells

Small boats often come equipped with live wells fitted with a water pump or aerator to help maintain good water quality. Those with water pumps move water from outside the boat through the live tank on a continuous basis and are generally useful to the fishery scientist who wishes to keep fish alive until they can be released again. If the live tank does not have a pump, oxygen depletions can occur fairly rapidly, particularly when fish density is high. In this case, an aerator is necessary, and the fish should be worked up and released as quickly as possible after capture. Having enough workers to collect both fish and data quickly is of critical importance.

5.3.3 Live Hauling

Basic designs for hauling tanks and trucks for all types of fish are similar (Fig. 5.2), though size varies considerably. Hauling tanks may be small enough to fit on the

Figure 5.2. Large numbers of fish can be moved long distances in trucks equipped with holding tanks that have appropriate methods of aeration.

bed of a pickup truck or large enough to require a tractor-trailer rig. The actual tanks may have one large unit fitted with baffles or several smaller compartments in which fish are retained without access to other compartments within the unit (Copeland 1947; Henegar and Duerre 1964). All tanks larger than a few liters capacity should be fitted with baffles to retard sloshing of water.

Circulating pumps are often used with transportation tanks to ensure that aerated water is distributed evenly throughout the tank. Gasoline pumps or those driven by a power takeoff or hydraulic system on a truck or tractor are often used for this purpose. The effectiveness of water circulation, as indicated by Norris et al. (1960), depends upon such factors as location and efficiency of aeration devices, the shape of the hauling tank, the pattern of water circulation, and the rate of water circulation.

Hauling tanks are generally fitted with mechanical agitators, liquid oxygen, bottled oxygen, an air compressor, an air blower, or some combination of these aerating devices. Oxygen is generally used only for supplementation, but in case the primary system fails, oxygen may be utilized as the basic source of aeration.

The biomass of fish which can be hauled in a given volume of water depends on the type and size of fish, water temperature, concentration of dissolved oxygen, carbon dioxide, and metabolites, and other factors. Smith (1978) discussed the hauling of salmonids and cited unpublished data from various sources which indicated that fish loading should be reduced 5.6% for each degree F rise in temperature. Conversely, the same rate of increase in fish carrying capacity could be realized for each degree that temperature is lowered. Smith (1978) also reported that under ideal conditions the maximum load of catchable-size rainbow trout (4.0 to 10.6 fish/kg) is about 0.3 to 0.4 kg/l of hauling water if the hauling period does not exceed 10 hours. Similar loading capacities have been reported for other trout species of the same size. Smith (1978) provided information on calculating loading capacities for fish of various sizes. For warmwater fish, on the other hand, carrying capacities in hauling tanks are more standardized (Table 5.2).

5.4   REDUCING MORTALITY

5.4.1   Maintenance of Holding and Hauling Conditions

For short-duration holding of live fish, providing proper aeration may be all that is required. Agitators, bottled oxygen, and other means of aeration described earlier also can be utilized in holding facilities. For most short-term holding, agitators are commonly utilized, although blowers and air compressors are also used.

Temperature control in holding facilities may be required, particularly when the

Table 5.2.   Estimated weight (kg/l) of various kinds of fishes which can be hauled at 18 C for various lengths of time (after Johnson 1979)

Type of Fish (and Size)	Duration of Transport (hrs)			
	1	12	24	48
Fingerling food fish (5 cm)	0.24	0.18	0.12	0.12
Fingerling food fish (20 cm)	0.36	0.36	0.24	0.18
Adult food fish (36 cm)	0.48	0.48	0.36	0.24
Bait fish (5 cm)	0.24	0.24	0.24	0.24
Bait fish (8 cm)	0.36	0.24	0.12	0.12

temperature of the water differs greatly from ambient air temperature. While hauling tanks on large trucks are usually insulated, small hauling tanks and many holding tanks are not insulated, and, therefore, the temperature of the water in them fluctuates. Immersion heaters and chillers can be utilized with both hauling and holding tanks, or if a holding facility is indoors, regulation of room air can be used to control water temperature. Ice can be used to keep water from warming during holding or hauling.

In closed systems, even those with biofiltration, pH control will be necessary if the fish are to be held for long periods of time. The actual time required for pH to change drastically in response to the addition of carbon dioxide and other metabolites will depend, in part, on stocking density, fish species, temperature, and feed addition. Because of the buffering capacity of the water in the system, prediction of pH changes is difficult. In water of low alkalinity, pH will begin to decrease more rapidly than in highly alkaline water, though the buffering capacity of any water system can be increased if some source of carbonate ions is available. Limestone or oyster shell will supply calcium carbonate and should be incorporated into a closed recirculating water system. Tris buffer (tris-hydroxymethyl-amino-methane) at the level of 1.5 to 2.5 g/l has been used to buffer both fresh and saltwater (McCraren and Millard 1978).

5.4.2 Limitation of Feeding

Fish being hauled should not be fed. The increased metabolic rate and production of feces which results from feeding can lead to rapid deterioration of water quality. Do not feed fish for 72 hours prior to hauling. Even if fish are to be held only for brief periods of time, feeding is not recommended. Long-term holding and conducting experiments under controlled laboratory conditions do, of course, require daily feeding, and care should be taken to maintain water quality under those circumstances.

Fish are generally not fed on days that they are handled. For example, fish on an experiment might be weighed at intervals of two or three weeks to evaluate growth rate and recalculate feeding levels. Fish that are to be handled should never be fed before such handling because additional stress will occur when the fish have feed in them and weight determinations will be biased by the additional weight of the feed in the stomachs and intestines. Some species, such as catfish, stop feeding for at least several hours after handling. Others, such as tilapia, will often feed immediately after handling. Each species, therefore, requires evaluation independently. If fish do not feed actively after handling, they should not be fed because the deteriorating feed only increases water quality problems and provides a substrate for the growth of undesirable fouling organisms.

5.4.3 Anesthesia

Anesthesia is often utilized to calm fish during surgical procedures, stripping of eggs and sperm, and other handling operations. The two most popular anesthetics for fish are quinaldine and MS-222 (tricaine methanesulfonate). Quinaldine was first described as a fish anesthetic by Muench (1958) who found that concentrations of 2.5 to 20 ppm effectively anesthetized various species of fish within a few minutes. MS-222 came into routine use in the handling of both marine and freshwater fishes during the 1950s (Schoettger and Julin 1967).

According to Johnson (1979), quinaldine and MS-222 are best used to quiet fish at the time of loading. Sedated fish do not become as excited as those that are not sedated and, consequently, do not place as much oxygen demand on hauling tank water. Anesthetics are frequently utilized during routine handling for data collection to keep the fish from moving around and to help protect them from injury.

Long-term use of anesthetics is not advised, and overdosage can lead to direct mortality. In most cases, the fish are placed into an anesthetic solution for a few minutes until the desired level of sedation is reached. Thereafter, the fish may be handled as required and then returned to anesthetic-free water before they are released or hauled. Always test a few fish in any anesthetic solution before exposing an entire population to it.

Anesthesia with MS-222 requires varying concentrations of the drug, depending upon water quality and the species of fish being treated. In most instances, levels of 100 mg/1 or less have been effective. For example, Wedemeyer (1970) determined that rainbow trout yearlings held in 10 C water of 20 ppm hardness were readily anesthetized with 80 ppm of MS-222.

Other anesthetics which have been used include benzocaine (Wedemeyer 1970; Dawson and Gilderhus 1979), sodium bicarbonate, and carbonic acid. Concentrations of sodium bicarbonate of 142 to 642 mg/1 with pH in the range 6.5 to 7.5 were effective on brook trout and common carp (Booke et al. 1978). Post (1979) found that baths containing 150 to 600 mg/1 of carbonic acid would anesthetize fish.

A fish that has not been allowed to recover its equilibrium sufficiently following

Table 5.3 Some chemicals used in handling and hauling fish and their registration status as of 1979 (after Schnick et al. 1979)

Chemical	Use and Registration Status
Calcium hypochlorite	Disinfectant at 5-10 mg/1 to control algae and bacteria; 200 mg/1 for 1 hour to sanitize. Cleared for food fish.
Furanace	Antibacterial for myxobacteria at 0.05-0.1 mg/1 indefinitely; 1.0 mg/1 for 5-10 minutes. Non-food fish use.
MS-222	Anesthetic at 15-66 mg/1 for 6-48 hours sedation; 50-330 mg/1 for 1-40 minutes anesthesia. Food fish use after 21-day withdrawal period.
Salt (NaCl)	Osmoregulatory enhancer at 5-10 ppt for indefinite period. Cleared for food fish.
Sulfamerazine	Antibacterial against furunculosis at 22 g/100 kg of fish daily for 14 days in feed. Cleared for use in salmonids.
Oxytetracycline	Antibacterial at 5.5-8.25 kg/100 kg of fish daily for 10 days in feed. Food fish use approved.
Carbon dioxide	Anesthetic; declared to be food additive. Not labeled for fishery use.
Sodium bicarbonate	Anesthetic; declared to be food additive. Not labeled for fishery use.

Table 5.4. Chemicals tested by Johnson (1979) and found to be ineffective in retarding bacterial growth under simulated hauling tank conditions

Chemical	Concentration (ppm)
Acriflavin	1-2
Oxytetracycline	5-50
Sodium chloride	2,000
Methylene blue	0.1-5
Potassium permanganate	2-5

anesthesia should not be released into any environment where predators could take advantage of the helplessness of the sedated fish. Allow fish to recover until they behave normally before releasing them.

The use of anesthetics on fish that are to be sold for human consumption is not currently permitted (Table 5.3), except in the case of MS-222. Fish anesthetized with MS-222 can be sold for human consumption after 21 days since they were last exposed to the chemical (Schnick et al. 1979).

5.4.4 Chemical Additives

Holding units should be cleaned as frequently as possible to help control bacterial growth. The addition of various bactericidal compounds to retard bacterial growth has been practiced for many years, although few chemicals are currently registered for this use (Table 5.3). Among the chemicals used in bacterial control are acriflavin, combiotic, nitrofurazone, and oxytetracycline. According to a survey conducted by McCraren and Millard (1978), acriflavin is the most widely used, at typical concentrations of 1-2 ppm. Combiotic has been used at 15 ppm. Other chemicals thought to retard bacterial growth are sodium chloride, methylene blue, and potassium permanganate. Johnson (1979) tested several of these and found little retardation in bacterial numbers when test conditions simulated those found in hauling tanks (Table 5.4). Those chemicals, therefore, are not recommended for use as bacteriostatic agents.

Although sodium chloride or calcium chloride may have no significant bacteriostatic effect, they have been added to transportation water to reduce the effects of injury incurred during hauling. The detrimental effects of scale, skin, and mucus losses are reduced by the addition of 2 ppt (0.2%) sodium chloride (Johnson 1979). Striped bass are generally hauled in 10 ppt saltwater (McCraren and Millard 1978), and transporting most saltwater species in an isotonic or nearly isotonic salt solution (10-12 ppt) probably is beneficial (Norris et al. 1960).

Box 5.1.   Registering Chemicals

Chemicals utilized in fisheries must be registered with the appropriate federal governmental agency (e.g., United States Food and Drug Administration or United States Environmental Protection Agency). The United States Fish and Wildlife Service assigned primary responsibility of facilitating registration to the Fish Control Laboratory in La Crosse, Wisconsin. Testing is conducted by the latter agency and private industry. Costs involved in having a chemical registered may run into several millions of dollars.

Excessive foam production is a common problem when hauling fish or when holding fish in crowded conditions with closed recirculating or static water systems. Causes include excessive mucus production, the use of certain drugs in the water, and the build-up of excreted proteins in the water. Commercial anti-foaming agents are available.

5.4.5 Sanitation

Sanitary conditions should always be maintained when fish are handled—especially when fish are removed from tanks, ponds, or other water bodies, handled, and returned to their original locations. In such instances, nets and other equipment utilized in the handling process should be sanitized between uses. This reduces the possibility that diseases will be transmitted from one group of fish to another. Chlorine solutions and solutions of other disinfectants, especially iodophores (Table 5.3), are commonly used to maintain sanitation.

Large seines and other collection devices which cannot be readily disinfected by dipping them in chemical solutions should be dried thoroughly between uses. Temporary holding and hauling tanks should be disinfected between uses to reduce the possibility of disease transmission.

5.5 REFERENCES

Booke, H. E., B. Hollender, and G. Lutterbie. 1978. Sodium bicarbonate, an inexpensive fish anesthetic for field use. *Progressive Fish-Culturist* 40:11-13.

Copeland, T. H. 1947. Fish distribution units. *Progressive Fish-Culturist* 9: 192-202.

Dawson, V. K., and P. A. Gilderhus. 1979. *Ethyl-p-aminobenzoate (benzocaine): efficacy as an anesthetic for five species of freshwater fish.* United States Fish and Wildlife Service Investigations in Fish Control No. 87, Washington, District of Columbia, USA.

Feast, C. N., and C. E. Hagie. 1948. Colorado's glass fish tank. *Progressive Fish-Culturist* 10: 29-30.

Henegar, D. L., and D. C. Duerre. 1964. Modified California fish distribution units for North Dakota. *Progressive Fish-Culturist* 26: 188-190.

Johnson, S. K. 1979. *Transport of live fish.* Texas Agricultural Extension Service, Fish Disease Diagnostic Laboratory Publication FDDL-F14, College Station, Texas, USA.

McCraren, J. P., and J. L. Millard. 1978. Transportation of warmwater fishes. Pages 43-88 *in Manual of fish culture,* Section G: fish transportation. United States Fish and Wildlife Service, Washington, District of Columbia, USA.

Muench, B. 1958. Quinaldine, a new anesthetic for fish. *Progressive Fish-Culturist* 20: 42-44.

Norris, K. S., F. Borcato, F. Calandrino, and W.N. McFarland. 1960. A survey of fish transportation methods and equipment. *California Fish and Game* 46: 6-33.

Piper, R. G., I. B. McElwain, L. E. Orme, J. P. McCraren, L. G. Fowler, and J. R. Leonard. 1982. *Fish hatchery management.* United States Department of the Interior, Fish and Wildlife Service, Washington, District of Columbia, USA.

Post, G. 1979. Carbonic acid anesthesia for aquatic organisms. *Progressive Fish-Culturist* 41: 142-144.

Schnick, R. A., F. P. Meyer, and H. D. Van Meter. 1979. Announcement of compounds registered for fishery use. *Progressive Fish-Culturist* 41: 36-37.

Schoettger, R. A., and A. M. Julin. 1967. *Efficacy of MS-222 as an anesthetic on four salmonids.* United States Bureau of Sport Fisheries and Wildlife Investigations in Fish Control, No. 13, Washington, District of Columbia, USA.

Smith, C. E. 1978. Transportation of salmonid fishes. Pages 9-41 *in Manual of fish culture,* Section G: fish transportation. United Stated Fish and Wildlife Service, Washington, District of Columbia, USA.

Wedemeyer, G. 1970. Stress of anesthesia with M.S. 222 and benzocaine in rainbow trout *(Salmo gairdneri). Journal of the Fisheries Research Board of Canada* 27:909-914.

Chapter 6
Passive Capture Techniques

WAYNE A. HUBERT

6.1 INTRODUCTION

Passive capture gears involve the capture of fish or other aquatic animals by entanglement or entrapment in devices that are not actively moved by man or machine (Lagler 1978). The gears used in passive sampling of fishery stocks are similar to those derived for food gathering over the centuries. The use of nets and traps was widespread among primitive peoples, and many currently applied techniques were used by the ancient Egyptians, Greeks, and Romans (Alverson 1963).

Passive sampling gears are divided into two groups on the basis of their mode of capture, entanglement versus entrapment. Entanglement devices capture fish by holding them ensnared in a fabric mesh. Gill nets and trammel nets are both entanglement gears. Entrapment gears capture organisms which enter an enclosed area through one or more funnel or V-shaped openings and cannot find a means of escape. Hoop nets, trap nets, and pot devices are examples of entrapment gears.

6.1.1 Advantages of Passive Gears

Passive gears are simple in their design and construction. They are generally handled without mechanized assistance other than a boat, and they require little specialized training to operate. Passive gears can be used to sample fish for many purposes and can yield fairly precise data on relative abundance for many species and water bodies. Nets fished in a similar manner and time each year can give reasonable estimates of changes in stock density. Pre-sampling can be employed to determine the variability of experimental netting and to determine needed sample sizes in management and research projects.

6.1.2 Disadvantages of Passive Gears

All passive gears are selective to some extent for certain species, sizes, or sexes of animals. Commercial fishermen recognize the selectivity of gear types and design and use gears in particular habitats to capture specific species or sizes of fish (Carter 1954; Starrett and Barnickol 1955). The act of capturing a fish involves several stages: the fish and gear must overlap in time and space, the fish must encounter the gear, the fish must be caught by the gear, and finally the fish must be retained by the gear until it is retrieved. Selectivity occurs at each stage of the capture sequence. An understanding of the selectivity of a gear is needed to interpret data, but, in general, there is little information available.

Theoretically, the catch-per-unit-effort (CPUE) of passive sampling gears should be directly proportional to the abundance of fish in the stock, but many other variables also influence CPUE. Some of the more important variables influencing capture efficiency are season, water temperature, water level, turbidity, and currents.

Changes in fish behavior result in a great degree of variability in CPUE among species and among year classes within a species because capture efficiency with passive gears is a function of fish movement. Many movements are unpredictable as a result of our poor understanding of the ways in which environmental factors influence movement tendencies. The efficiency of entanglement gears is further influenced by fish morphology. Gill nets and trammel nets tend to be more efficient in the capture of species with external protrusions and less efficient in the capture of species with compressed bodies.

6.2   ENTANGLEMENT GEARS

6.2.1 Gill nets

Gill nets are vertical walls of netting normally set out in a straight line (Fig. 6.1). Capture is based on fortuitous encounter with the net. There are three ways in which fish are caught by gill nets: (1) wedged—held by the mesh around the body, (2) gilled—held by mesh slipping behind the opercula, or (3) tangled—held by teeth, spines, maxillaries, or other protrusions without penetration of the mesh. Most often fish are gilled. Fish of a size for which the net is designed swim into the net and pass only part way through a single mesh. When the fish struggles to free itself, the twine slips back of the gill cover and prevents escape.

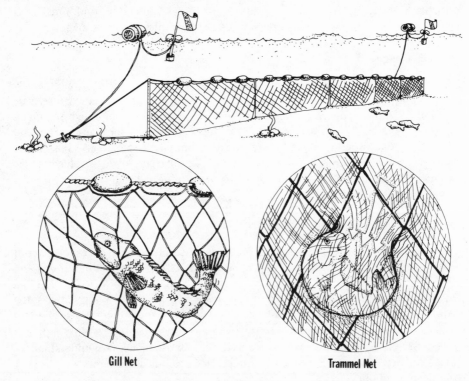

Gill Net   **Trammel Net**

Figure 6.1   Gill net and trammel net (from Dumont and Sundstrom 1961).

Construction. A gill net is made of a single wall of webbing held vertically in the water by weights and floats. The manner in which the webbing is hung, the "hanging ratio," determines the shape of the mesh (Gebhards 1966). The hanging ratio is the ratio between the length of the float line and the length of the stretched mesh hanging from the float line. For example, if a net of 100 m stretched length were hung on 40 m of float line, the hanging ratio would be 40/100 or 0.40. For most experimental nets a series of diamond-shaped meshes is desired, but the effectiveness of the mesh shape varies with the morphology of the fish species being sought (Welcomme 1975).

The mesh size of gill nets is generally expressed in either of two ways, bar measure or stretch measure. Bar measure is the distance between knots. Stretch measure is the length of a single mesh when the net is stretched taut. In general, the units of bar measure are half that of stretch measure. For example, a 5-cm bar measure mesh is the same as a 10-cm stretch measure mesh.

Gill net webbing has been made of cotton, linen, nylon, and monofilament twine. Cotton and linen nets have been replaced by multistrand nylon and monofilament because these new materials make more efficient nets, do not deteriorate as rapidly, and require less maintenance (Lagler 1978). Monofilament nets are more difficult to repair and to handle in cold weather than nets of other materials, but they are less visible to fish, easy to clean, and more durable. The choice of webbing material can influence catch due to its breaking strength, visibility, stiffness, elasticity, and smell (Welcomme 1975; Jester 1977).

Wooden, cork, or plastic floats strung on the float line have been used to float gill nets. The lead line has incorporated lead weights strung or crimped onto the line to weight the net. Both floats and leads tend to tangle with the meshes and to catch on obstructions. Because of this, foam-core float lines and lead-core lead line are becoming popular.

Experimental work is often conducted using a net of several mesh sizes to reduce the effects of size selectivity. A common design for an experimental net is 2 m deep with 8-m panels of 38-, 51-, 64-, 76-, 87-, and 102-mm stretch mesh webbings (Lagler 1978). Regier and Robson (1966) suggested that for general sampling purposes it would be more efficient for the mesh sizes to increase in a geometric progression rather than the usual arithmetic progression.

Deployment and retrieval. Gill nets can be set in many different ways, depending on the species desired and the habitats involved (Von Brandt 1964; Nedelec 1975). The most common manner of deployment is the stationary bottom set. The net, anchored at both ends, is set as an upright fence of netting along the bottom. Prior to setting, rig the net with appropriate anchors, lines, and buoys. When setting the net, drop the anchor over the bow, and back the boat as the net is played out by handling the float line and shaking out tangles in the mesh. Nets set perpendicular to the wind in water less than 5 m deep tend to roll up and tangle due to wave action (Skidmore 1970). To fish a gill net in shallow water, attach the net to stakes driven in the bottom. A rope can be strung for setting gill nets or other sampling gear under the ice using the willow stick, Murphy stick, or "jigger" methods (Hamley 1980) (*See* Box 6.1.).

Start retrieving gill nets at the downwind end and pull the net over the side of the boat. Stack the gill net into a wash tub or similar container in coils or figure eights. Remove fish as they come out of the water. Removal of fish is easier if you use a short stick with a nail driven into the end, head cut off, and flattened. The end can be bent into a hook or used as a spatula to lift meshes over the opercula and slide them off the body (Lagler 1978).

Box 6.1   Methods for Stringing a Rope under Ice

1. Willow stick method
    A long, pliable "willow" stick is used to pass a rope down through one hole and up through a second. The process can be repeated several times to extend the rope as far as desired.

2. Murphy stick method
    The Murphy stick is a modern adaptation of the willow stick named for "Murphy's Law," i.e., what can go wrong will go wrong (James Reynolds, personal communication). The method utilizes a 3-m section of 2.5 - 3.0-cm diameter aluminum pipe handle hinged to a second piece of pipe (0.5 - 2.0 m long) which extends as a probe under the ice. The second piece is fashioned with a sponge rubber sleeve for floatation and a snap at the far end. The probe can vary in length to accommodate different water depths. To operate, the far end of the second pipe is pushed through one hole and maneuvered toward a second hole where the attached rope is hooked and pulled up through the hole. This technique is especially useful in currents or under thick ice.

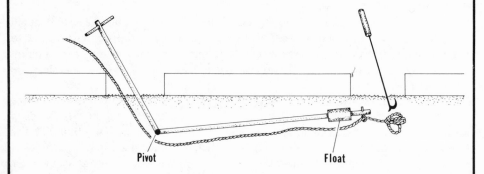

Pivot   Float

3. Jigger
    The jigger is a device which maneuvers a rope under the ice by manipulating a metal claw with the attached rope. When the rope is pulled, the metal claw moves the jigger away from the hole. [Adapted from Hamley (1980), with permission.]

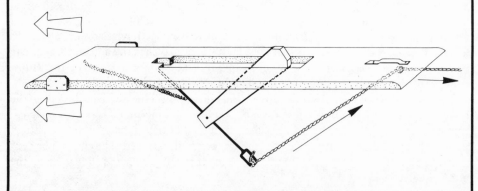

A bottom set can be adapted to sample fish off the bottom at midwater depths. The net may be either suspended by droplines from large buoys or held below the surface by lines attached to anchors (Von Brandt 1964).

Gill nets can be floated at the surface. In a riverine situation the net may be stretched and pulled downstream by the attachment of floating hydrofoils or "mules." Floating gill nets have also been pulled through pools in an active manner similar to seines (Biggins and Cressey 1973).

Gill nets are set with the long axis perpendicular to the water surface in order to determine the vertical distribution of fish in water up to 50 m deep. A variety of methods have been used, but they generally involve a mechanism similar to a window shade. The nets are wound around a cylindrical column which may also suffice as a float. The nets are unwound to the bottom with periodic placement of lightweight "spreader bars" to hold the net open (Horak and Tanner 1964; Lackey 1968; Bartoo et al. 1973; Kohler et al. 1979).

Gill nets can be fished by setting them around concentrations of fish or areas suspected to harbor fish. Following placement of the net, movement of fish into the net can be stimulated by use of noise, light, electrical fields, chemicals, or air screens. White (1959) described several net sets using encircling tactics employed by commercial fishermen on the Tennessee River.

Habitat applications. Gill nets are used to sample fish in a wide range of habitats. They are generally considered a shallow water gear, although bottom sets may be made at depths exceeding 50 m. The use of gill nets is limited to areas free of obstructions, snags, and floating debris, as well as locations with little or no current. Generally gill nets are not considered applicable in lotic habitats, but they have been drifted, anchored in eddies, and used like seines in rivers. Gill nets have found wide use in the monitoring of lake and reservoir fish populations. They are useful in remote areas where access can be attained only by small plane or boat.

Target organisms. Many fish species are caught in gill nets, but gill nets are especially selective for species that move substantial distances in their daily routines. The species selectivity of experimental gill nets has been documented in some waters (Berst 1961; Heard 1962; Trent and Pristas 1977; Yeh 1977). Because many fish caught by gill nets die in the net or are injured upon removal (for example, percids and clupeids), the net is less appropriate for use when live fish must be obtained or released.

Biases. The size selectivity of various mesh sizes is a problem in gill net sampling. For a particular mesh size, fish of some optimum size are held most securely; smaller or larger fish are less likely to be caught. Very small fish can swim right through, and very large fish cannot penetrate deep enough into a mesh to become stuck. A typical gill net size selectivity curve is bell-shaped, falling to zero at both sides of a maximum (Pope et al. 1975). Size selectivity curves for various mesh sizes have been computed for many species (Hamley 1975). Two generalizations regarding size selectivity are (1) the optimum girth for capture is about 1.25 times the mesh perimeter and, (2) fish more than 20% longer or shorter than the optimum length are seldom caught (Hamley 1980). Substantial variation in the shape and magnitude of the curves has been observed. Hamley and Regier (1973) described a biomodal curve for walleye. The shape of the curve can vary within a species by sex and season (Hamley 1975).

In general, larger meshes are more efficient. Young fish are not the most abundant age groups in samples with experimental nets because small fish are not likely to push through the mesh. Size selectivity influences estimates of growth rate, mortality, and length-weight relations because larger meshes select larger fish of each age (Hamley 1980). The most important factors influencing the size selectivity of a gill net for a particular species are mesh size, elasticity of the twine, hanging ratio, strength and flexibility of the twine, and visibility of the twine, as well as the time and manner in which the net is fished.

The mesh material can have a substantial influence on gill net efficiency. In general, nylon mesh is more efficient than linen or cotton, while monofilament is more efficient than all other materials (McCombie and Fry 1960; Berst 1961; Pristas and Trent 1977). In addition, the diameter, flexibility, and color of a particular mesh influences efficiency (Jester 1977).

The hanging ratio also can influence efficiency. The influence of hanging ratio is greatest among species generally caught by tangling. The lower the ratio, the more efficient the gill net becomes in entangling deep-bodied fish (Welcomme 1975).

Species selectivity, size selectivity, and efficiency of gill nets are governed to a great degree by the construction of the net. Substantial control of variability in the catch can be achieved by using standardized nets. Monitoring or assessment projects should utilize gill nets of identical design, material, and construction. Lack of concern for variables as "insignificant" as the diameter of the net twine or the hanging ratio of a net can greatly influence the selectivity and efficiency of a gill net and quality of subsequent sampling data.

Capture of fish in gill nets depends on fish swimming into direct physical contact with the net and then being held by gilling or entangling. Fish activity is governed by many factors. The activity of many species is related to the degree of light under water which leads to diurnal movement patterns. Most species exhibit nocturnal or crepuscular activity peaks that are consistent from day to day within a particular season. Seasonal patterns in movement and distribution occur as a result of spawning activity and habitat requirements.

Many physical and chemical variables have been shown to influence the movement and distribution of fish (Berst 1961; Welcomme 1975; May et al. 1976; Pristas and Trent 1977). The variables include turbidity, movement of fronts, currents, water temperature, water depth, water level fluctuations, and thermocline location. Many of these variables can be measured, and predictive relations to fish activity and subsequent gill net catch can be defined.

The duration of a gill net set also influences sampling results. Catches do not accumulate in a gill net at a uniform rate (Kennedy 1951; Austin 1977). The efficiency of a gill net decreases as fish accumulate in it. Eventually the number of captured fish reaches a saturation point and does not increase further. Catch is generally not related to the duration of the set or "soak time" in a linear fashion, and "saturation" generally occurs when only a small percentage of the meshes are occupied.

The sampling regime (i.e., season, time of day, location, duration of sets) also can influence catch. A standardized sampling scheme can be employed from year to year, as well as between water bodies, to minimize the variability that can be generated by physical and chemical variables. A rigidly defined design identifying the season, time, location, and duration of sets, coupled with precise gear and deployment specifications can reduce much of the random variability among gill net samples.

Gill nets are most applicable in lakes or rivers with low current velocities where the target species are highly mobile fishes. They are most effective in waters where fish visibility is reduced and where relatively level, snag-free bottoms are present. Gill nets generally are most effective when set overnight and emptied each day.

6.2.2 Trammel Nets

A trammel net is composed of three panels of netting suspended from a float line and attached to a single lead line (Fig. 6.1). Two outer panels or walls are of a large mesh netting while the inner panel is of a small mesh. The inside small mesh panel has greater depth and hangs loosely between the two other panels.

Fish generally are captured in a "bag" or "pocket" of netting. A fish striking from either side can pass through the large-mesh outer panel and hit the small-mesh inner panel which it carries through one of the large openings of the opposite large-mesh outer wall. This action forms a bag or pocket in which the fish is entangled. Small fish or very large fish may be wedged, gilled, or tangled in the netting.

Construction. Trammel nets are generally constructed of cotton or nylon webbing. Numerous designs have been employed by commercial fishermen (Dumont and Sundstrom 1961; Nedelec 1975). Following is an example of a trammel net which is used in reservoir and riverine sampling. A 2-m deep net with 250-mm bar mesh outer walls and a 25-mm bar mesh inner wall would be hung with two-thirds greater depth or 3.3-m deep netting. A float line with floats or foam core and a lead line with either lead weights attached or a lead core could be utilized.

Deployment. Trammel nets are set in the same ways as gill nets: setting stationary bottom sets, drifting, floating, or encircling. Drifted trammel nets have been used to sample channel habitats of a large river (Hubert and Schmitt 1982a).

Trammel nets are most effective when set around an aggregation of fish; the fish are then frightened or driven into the nets. A typical set of this type involves surrounding an area of aquatic vegetation with trammel netting (White 1959).

Habitat applications. Trammel nets can be used in many types of habitats, including large rivers. Stationary bottom sets can be made in lakes, backwaters, and quiet sections of rivers. The nets can be floated or drifted in channel habitats. In spite of their versatility, trammel nets are not often employed in fishery stock assessment.

Target organisms. Trammel nets are selective for fish species with rough surfaces and protrusions, such as sturgeon, catfish, or temperate basses. They are particularly useful in sampling species that occur in shallow water and can be frightened into hitting the net. A major advantage of trammel nets over gill nets is that fish pocketed in a trammel net generally are in much better condition upon removal than fish gilled in a gill net.

Biases. The same sampling biases exist with trammel nets as with gill nets. The only exception is that trammel nets are much less size selective than simple gill nets.

6.3 ENTRAPMENT DEVICES

Aquatic animals enter an entrapment device by their own movement. They are captured because the entrance happens to stand in the exact path of their movement, when they attempt to move around a barrier, or they are attracted to the enclosure by the "cover" it may provide or the presence of bait or other fish within the trap. Once in the entrapment device, a fish may move until it escapes, or it may pass through one or more additional funnel-shaped openings into additional compartments in which the chances of escape are reduced.

A wide variety of entrapment gears exist (Dumont and Sundstrom 1961; Von Brandt 1964; Nedelec 1975; Everhart and Youngs 1981). Knowledge of migration, cover-seeking habits, escape reactions, and diet of aquatic animals have influenced the design of entrapment devices used for various organisms. Entrapment devices used in fishery sampling are often small, portable versions of commercial fishing devices.

6.3.1   Hoop Nets

A hoop net is a cylindrical or conical net distended by a series of hoops or frames, covered by web netting. The net has one or more internal funnel-shaped throats whose tapered ends are directed inward from the mouth (Nedelec 1975) (Fig. 6.2). Local terminology for hoop nets can be confusing because the name often refers to the species selectivity of a particular design. Some names given to hoop nets include buffalo nets, bait nets, fiddler nets, and even fyke nets (Starrett and Barnickol 1955).

Construction. Hoops are made of wood or steel. Hoop diameter varies from 0.5 m to over 3 m with four to eight hoops in a net. The cotton or nylon webbing tied around the hoops can range from 10-mm to over 100-mm bar mesh. Generally, two funnel-shaped throats are attached, one to the first hoop and a second to the third hoop from the mouth. The throats may be of two basic designs, square or finger throats. A square throat is simply a square to circular opening in the constricted end of the funnel. A finger throat is composed of two half-cones of twine on each side of the mouth secured to a back hoop (Hansen 1944). A finger throat is often placed in the mouth of the second throat to lessen the chances of escape. The closed end of the net, where fish accumulate, is called the cod end or the "pot." A drawstring is attached to the cod end for removing captured fish. Hoop nets are protected from deterioration by periodic dipping in tar-based net-coating materials.

Deployment and retrieval. In riverine habitats, hoop nets are set with the mouth opening downstream, at depths that entirely cover the hoops of the net, by attaching a rope to an anchor or stake at the stream bottom. Stakes can be set in water up to 5 m deep using a "driver pole". A driver pole is a long pole with a 2 to 4-cm diameter sleeve at one end. The pole is used to drive a wood or steel stake with a rope attached into the stream bottom.

Commercial fishermen generally set hoop nets without buoys. Landmarks are used to identify the location. Retrieval is made by dragging the bottom with a grappling hook until the net or rope is hooked.

Current keeps the hoops separated and the net stretched. Hoop nets are often baited with cheese scraps or pressed soybean cake. Generally hoop nets are set for 24-hour sampling periods.

Habitat applications. Hoop nets most often are employed in channel habitats of rivers because they can be set and fished effectively in strong currents without being washed away or becoming clogged with debris. Hoop nets have been used in several reservoir habitats from tailwaters to the lower ends of impoundments. They are relatively simple to set and can be deployed from a small open boat.

Target organisms. Hoop nets are highly selective for fish species attracted by "cover," bait, or other fish. In the Mississippi River basin, hoop nets are especially selective for catfish, buffalo, and carp (Starrett and Barnickol 1955). Many species, such as smallmouth bass or gizzard shad, seldom enter hoop nets despite high abundance in the habitat (Lagler 1978). Hoop netted fish are generally captured unharmed and can be released with little or no injury to the fish.

Biases. Net construction (hoop diameter, mesh dimensions, and mouth size) has dramatic influence on the species and size selectivity of hoop nets. For example, in a study on the Mississippi River with two sizes of hoop nets, the larger diameter nets captured twice as many fish and half again as many species with equal sampling effort

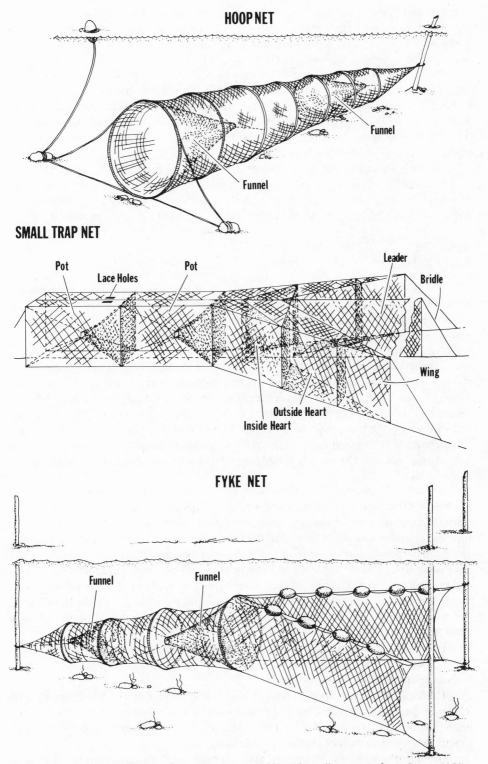

Figure 6.2 Hoop net, fyke net (from Sundstrom 1957) and small trap net (from Crowe 1950).

(Hubert and Schmitt 1982b). The species selectivity and CPUE can also be influenced by baiting hoop nets (Pierce et al. 1981). Escapement rates of different species from hoop nets also influence sampling results. It has been shown that some species are more adept at escape than others (Hansen 1944).

Physical and chemical variables can have a significant influence on CPUE. Variation in water temperature, current velocity, turbidity, dissolved oxygen, and habitat type within a river all affect CPUE of individual species (Hubert and Schmitt 1982b). The most important variables appear to be water temperature and turbidity. Their influence varies by species, but in most cases as water temperature declines or turbidity increases, the CPUE declines.

As with entanglement gears, the design and construction of hoop nets can be standardized, along with sampling time and location, to reduce sampling variability. However, because of the highly dynamic nature of riverine habitats, little can be done to control variability of physical factors such as turbidity or current velocity.

6.3.2 Fyke Nets

Fyke nets are modified hoop nets with one or two wings or a leader of webbing attached to the mouth to guide fish into the enclosure (Fig. 6.2). The net is set so that the wings and leader intercept the movement of fish. When fish follow the wing or leader in an attempt to get around the netting, they swim into the hoop net. Fyke nets are also known as wing nets, frame nets, trap nets, and hoop nets.

Construction. Fyke nets are often constructed identically to hoop nets insofar as the net itself is concerned. Attached to the first hoop may be one or two pieces of netting called "wings." The wings are set at 45° angles to the axis of the hoop net and secured in position with poles or anchors. A "leader," a piece of netting extending outward along the axis, may be attached. The first one or two hoops of the net are often changed to rectangular or semi-circular frames to keep the net from rolling. In nets of this type, the throat is recessed from the mouth and attached to the second or third frame.

Deployment and retrieval. Generally, fyke nets are set in shallow water as deep as the height of the wings or leader and the height of the first hoop or frame. The net and leader are set taut by anchors or poles driven into the bottom. A single-pot set involves one leader and hoop net. The end of the leader is set on the shore, and the leader is extended perpendicular to shore so that fish cannot swim around it. When setting the net from a small boat, it is placed on the bow with the pot down and leader on top. The end of the leader is staked or anchored and played out as the boat is moved in reverse. When the leader is fully extended, the hoop net is put overboard with the float line of the leader upright. The hoop net is played out, stretched, and staked or anchored in position.

Fyke nets are also deployed in pairs with a single leader between them (Nedelec 1975). This type of set is generally made parallel to shore along the outer edge of weed beds or along shallow off-shore reefs.

The location of fyke nets in lakes, as well as the manner in which they are set, can influence catch (Bernhardt 1960). Seasonal variation in fyke net catches is a general phenomenon (Hansen 1953; Kelley 1953). Standardized nets, sampling locations, and times are needed to reduce sampling variability.

Habitat applications. Fyke nets are generally used in shallow areas of ponds, lakes, and reservoirs. They have been used to sample fish in streams, sloughs, and sluggish sections of rivers with moderate current velocity (Swales 1981). Fyke nets also can be utilized in heavily-vegetated, marsh-type habitats by cutting paths in submerged or emergent vegetation, but damage by furbearing animals can be substantial in these habitats (Kelley 1953).

Target organisms. As with hoop nets, fyke nets are selective for certain species and sizes of fish. Cover seeking, mobile species, such as esocids and centrarchids, seem to be the most susceptable to capture.

6.3.3   Trap Nets

A variety of large entrapment devices have been used in coastal marine and large lake fisheries (Dumont and Sundstrom 1961; Alverson 1963; Grinstead 1968). These include the pound net fishery of the Atlantic Coast (Reid 1955) and the deep trap net fishery of the Great Lakes (Van Oosten et al. 1946). Modified, scaled-down versions of commercial trap nets have been employed in fishery research (Crowe 1950; Beamish 1972).

Construction. A portable trap net was described by Crowe (1950) which weighed 45 kg including the anchors (Fig. 6.2). This net was constructed of 3-cm stretched mesh netting with a 1.3-m deep pot that was 2.6 m long and 2 m wide. The net had two wings leading to an "outside heart" of 100-mm stretched mesh and to an "inner heart" of 62-mm mesh. Both hearts were covered with similar netting. A 30-m long, 1.3-m deep leader was attached.

Deployment and retrieval. Crowe's (1950) portable unit was usually set below the water surface with the leader perpendicular to shore, but it sometimes was deployed with a single leader between two traps. Fish were removed from the net by lifting only the pot into a small boat.

Habitat applications. The portable trap net can be used in water up to 12 m deep over a clean, firm bottom. It was not applicable over soft bottoms because of failure of the anchors to hold. Crowe's portable net and modifications of it have been used in a variety of lake and reservoir habitats.

Target organisms. Trap nets are very effective in the capture of migratory species that tend to follow shorelines. Trap nets are less size and species selective than gill nets in a particular habitat. They have been found to be selective for larger fish of size classes above the minimum imposed by the physical dimensions of the net (Latta 1959). Some selectivity occurs because of variable escape rates relative to season, species, and size of fish (Hansen 1944; Partriarche 1968).

6.3.4   Pot Gears

Pot gears are portable traps that fish and shellfish enter through small openings. They are rigid devices of various designs and dimensions for specific kinds of animals (Sundstrom 1951; Dumont and Sundstrom 1961; Alverson 1963; Rounsefell 1975; Nedelec 1975). Pot gears are generally small enough that large numbers can be put on a boat.

Pot gears are most effective in the capture of bottom dwelling species seeking food or shelter (Everhart and Youngs 1981). In order to reach a receptacle containing bait, the fish or crustacean must pass through a more or less conical-shaped funnel. In some traps they pass through more than one funnel, making escape more difficult.

An example of a typical pot gear is the American lobster pot. A traditional lobster trap is the "half-round pot" (Dumont and Sundstrom 1961; Alverson 1963; Everhart and Youngs 1981) (Fig. 6.3). It is constructed with a rectangular base and three half-round bows, one at each end and one near the center. The bows are covered with lathe, and a door is constructed on one side for removal of lobsters. The pot contains two inside compartments. A smaller "chamber" has an opening on each side with a funnel of netting. From the chamber a funnel attached to the middle bow leads to the larger "parlor." Bait is placed on a hook or in a mesh bag attached to the center bow.

The lobster pot is deployed by weighting it with bricks or stones and attaching a buoy line to a lower corner of the "chamber" end. Commercial fishermen fish the pots in strings of 10 to 15, spaced 10 to 20 m apart.

Other pot gears include crab pots, eel pots, minnow traps, slat traps (Perry 1979), and a variety of wire traps used to capture both fish and shellfish (Carter 1954; Beall and Wahl 1959) (*See* Fig. 6.3). One of the most studied pot gears is the Windermere perch trap (Worthington 1950; Bagenal 1972). The trap is constructed with three semi-circular wire hoops on a 67- x 76-cm base and covered with 1.3-cm hexagonal wire netting. One end has a funnel with a 8.5-cm diameter opening, and the other end has a door for removal of the catch. The traps are set unbaited. The trap is cheap, easy to make, and does not require significant manpower to fish. The trap is best suited for estimating age structure and condition of fish. It is not as efficient for estimating relative abundance because of the variability in catch between traps and sets.

6.3.5 Weirs

Weirs are barriers, built across an entire stream, which divert fish into a trap. They are most suited for capturing migratory fish as they move up or down streams. A wide variety of weirs have been used; they can be permanent or temporary structures (Von Brandt 1964; Welcomme 1975; Craig 1980). The Wolf-type weir (Wolf 1951) is an efficient design used to capture salmon and trout moving downstream. The basic design has been applied in large salmon streams with extremely variable runoff (Hunter 1954). Two-way fish traps have been designed for use in small trout streams (Whalls et al. 1955; Twedt and Bernard 1976). Fish passage facilities in dams can serve the same function as weirs (Welcomme 1975).

Weirs have been used to gather data on age structure, condition, sex ratios, spawning escapement, smolt production, and the abundance of sexually mature adults. The use of weirs is restricted to small rivers because of the construction expense, formation of navigation obstacles, and tendency to clog with ice or debris, which can cause flooding or collapsing of the structure.

6.4 USES OF PASSIVE SAMPLING DATA

Appraisal of sport or commercial management or assessment of environmental impacts are often possible with the use of passive sampling gears (Allen et al. 1960; Hocutt and Stauffer 1980). However, problems with sampling variability and gear selectivity are generally universal.

Relative abundance is an index of population density. Estimation of relative abundance assumes that CPUE is proportional to stock density. Variability in fish behavior can cause large fluctuations in CPUE and hamper interpretation of CPUE data concerning relative abundance. In order to utilize CPUE, attention must be given to reducing variability by standardizing gear, methods, and design (Welcomme 1975).

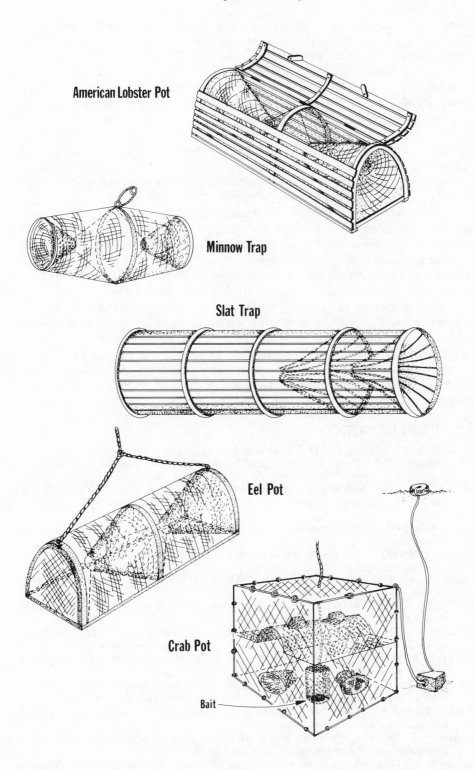

American Lobster Pot

Minnow Trap

Slat Trap

Eel Pot

Crab Pot

Bait

Figure 6.3 Pot gears. American lobster, crab pot, and slat trap from Sundstrom (1957).

With such controls, passive capture gears have been fished in a similar fashion and time each year to give reasonable estimates of the relative abundance of species in lakes and reservoirs (Carlander 1953; Walburg 1969; Le Cren et al. 1977).

Several statistics have been derived to describe the composition of the catch of a particular species. Length frequency distributions are commonly described. Length frequency information can be summarized into statistics such as proportional stock density (*see* chapter 15). Age composition of the catch can also be used to describe samples and to assess differences between stocks or over time. Data on age composition can be summarized into estimates of survival using catch curves or the method of successive ages. When computing survival, adjustments must be made for size selectivity of the gear (Hamley 1975), as well as variation in catchability related to size (Latta 1959) and growth rate (Welcomme 1975).

It is generally not possible to determine the species composition of a community using passive gears because of species and size selectivity (Allen et al. 1960). The species compostion of samples can be used, however, to assess differences between communities and changes in a community over time. Some statistics that can be computed to describe the species composition of samples include species richness, species diversity, and similarity indices (Hocutt and Stauffer 1980).

Passive capture gears have been used to estimate a variety of life history parameters such as growth, reproductive cycles, migratory patterns, distribution within a water body, diurnal activity trends, and diet. Fish are collected with passive gears to monitor contaminant levels, such as chlorinated hydrocarbon pesticide residues, polychlorinated biphenyls, and heavy metals. Entrapment devices have been used for mark and recapture experiments to estimate abundance (Latta 1959; Welcomme 1975; Craig 1980) and exploitation.

6.5 REFERENCES

Allen, G. H., A. D. Delacy, and D. W. Gotshall. 1960. Quantitive sampling of marine fishes—a problem in fish behavior and fishing gear. Pages 448-511 *in* E. A. Pearson, editor. *Waste disposal in the marine environment*. Pergammon Press, New York, New York, USA.

Alverson, D. L. 1963. Fishing gear and methods. Pages 45-64 *in* M. E. Stansby, editor. *Industrial fishery technology*. Robert E. Krieger Publishing Company, New York, New York, USA.

Austin, C. B. 1977. Incorporating soak time into measurement of fishing effort in trap fisheries. *Fishery Bulletin* 75:213-218.

Bagenal, T. B. 1972. The variability in the number of perch, *Perca fluviatilis* L. caught in traps. *Freshwater Biology* 2:27-36.

Bartoo, N. W., R. G. Hansen, and R. S. Wydoski. 1973. A portable vertical gill net system. *Progressive Fish-Culturist* 35:231-233.

Beall, H. B., and R. W. Wahl. 1959. Trapping bluegill sunfish in West Virginia ponds. *Progressive Fish-Culturist* 21:138-141.

Beamish, R. J. 1972. *Design of a trap net for sampling shallow water habitats*. Fisheries Research Board of Canada Technical Report 305, Winnipeg, Canada.

Bernhardt, R. W. 1960. Effect of fyke-net position on fish catch. *New York Fish and Game Journal* 7:83-84.

Berst, A. H. 1961. Selectivity and efficiency of experimental gill nets in South Bay and Georgian Bay of Lake Huron. *Transactions of the American Fisheries Society* 90:413-418.

Biggins, R., and S. Cressey. 1973. A technique for capturing trout with gill nets in deep streams. *Progressive Fish-Culturist* 35:106.

Carlander, K. D. 1953. Use of gill nets in studying fish populations, Clear Lake, Iowa. *Proceedings of the Iowa Academy of Science* 60:623-625.

Carter, E. R. 1954. An evaluation of nine types of commercial fishing gear in Kentucky Lake. *Transactions of the Kentucky Academy of Science* 15:56-80.

Craig, J. F. 1980. Sampling with traps. Pages 55-70 *in* T. Bachiel and R. L. Welcomme, editors. *Guidelines for sampling fish in inland waters.* Food and Agriculture Organization of the United Nations, European Inland Fisheries Advisory Commission Technical Paper 33, Rome, Italy.

Crowe, W. R. 1950. Construction and use of small trap nets. *Progressive Fish-Culturist* 12:185-192.

Dumont, W. H., and G. T. Sundstrom. 1961. *Commercial fishing gear of the United States.* United States Government Printing Office, United States Fish and Wildlife Circular No. 109, Washington, District of Columbia, USA.

Everhart, W. H., and W. D. Youngs. 1981. *Principles of fisheries science,* 2nd edition. Comstock Publishing Associates, Ithaca, New York, USA.

Gebhards, S. V. 1966. Repairing nets. Pages 110-125 *in* A. Calhoun, editor. *Inland fisheries management.* State of California, Department of Fish and Game, Sacramento, California, USA.

Grinstead, B. S. 1968. Comparison of various designs of Wisconsin-type trap nets in TVA reservoirs. *Proceedings of the Annual Conference of the Southeastern Association of Game and Fish Commissioners* 22:444-457.

Hamley, J. M. 1975. Review of gill net selectivity. *Journal of the Fisheries Research Board of Canada* 32:1943-1969.

Hamley, J. M. 1980. Sampling with gill nets. Pages 37-53 *in* T. Bachiel and R. L. Welcomme, editors. *Guidelines for sampling fish in inland waters.* Food and Agriculture Organization of the United Nations, European Inland Fisheries Advisory Commission Technical Paper 33, Rome, Italy.

Hamley, J. M., and H. A. Regier. 1973. Direct estimates of gill net selectivity to walleye (*Stizostedion vitreum vitreum*). *Journal of the Fisheries Research Board of Canada* 30:817-830.

Hansen, D. F. 1944. Rate of escape of fishes from hoop nets. *Transactions of the Illinois Academy of Science* 37:115-122.

Hansen, D. F. 1953. Seasonal variation in hoop net catches at Lake Glendale. *Transactions of the Illinois Academy of Science* 46:216-266.

Heard, W. R. 1962. The use and selectivity of small-meshed gill nets at Brooks Lake, Alaska. *Transactions of the American Fisheries Society* 91:263-268.

Hocutt, D. C., and J. R. Stauffer. 1980. *Biological monitoring of fish.* D. C. Heath and Company, Lexington, Massachusetts, USA.

Horak, D. L., and H. A. Tanner. 1964. The use of vertical gill nets in studying fish depth distribution, Horsetooth Reservoir, Colorado. *Transactions of the American Fisheries Society* 93:137-145.

Hubert, W. A., and D. N. Schmitt. 1982a. Factors influencing catches of drifted trammel nets in a pool of the Upper Mississippi River. *Proceedings of the Iowa Academy of Science* 88:121-122.

Hubert, W. A., and D. N. Schmitt. 1982b. Factors influencing hoop net catches in channel habitats of Pool 9, Upper Mississippi River. *Proceedings of the Iowa Academy of Science* 88:84-91.

Hunter, J. G. 1954. A weir for adult and fry salmon effective under conditions of extremely variable runoff. *Canadian Fish-Culturist* 16:27-33.

Jester, D. B. 1977. Effects of color, mesh size, fishing in seasonal concentrations, and baiting on catch rates of fishes in gill nets. *Transactions of the American Fisheries Society* 106:43-56.

Kelley, D. W. 1953. *Fluctuation in trap-net catches in the upper Mississippi River.* United States Fish and Wildlife Service, Special Scientific Report—Fisheries 101, Washington, District of Columbia, USA.

Kennedy, W. A. 1951. The relationship of fishing effort by gill nets to the interval between lifts. *Journal of the Fisheries Research Board of Canada* 8:264-274.

Kohler, C. C., J. J Ney, and A. A. Nigro. 1979. Compact, portable vertical gill net system. *Progressive Fish-Culturist* 41:34-35.

Lackey, R. T. 1968. Vertical gill nets for studying depth distribution of small fish. *Transactions of the American Fisheries Society* 97:296-299.

Lagler, K. F. 1978. Capture, sampling and examination of fishes. Pages 7-47 *in* T. Bagenal, editor. *Methods for assessment of fish production in fresh waters.* Blackwell Scientific Publications, Oxford, England.

Latta, W. C. 1959. Significance of trap-net selectivity in estimating fish population statistics. *Papers of the Michigan Academy of Science* 44:123-138.

LeCren, E. D., C. Kipling, and J. C. McCormack. 1977. A study of the numbers, biomass and year-class strengths of perch *(Perca fluviatilis* L.) in Windermere from 1941 to 1966. *Journal of Animal Ecology* 46:281-307.

May, N., L. Trent, and P. J. Pristas. 1976. Relation of fish catches in gill nets to frontal periods. *Fishery Bulletin* 74:449-453.

McCombie, A. M., and F. E. J. Fry. 1960. Selectivity of gill nets for lake whitefish, *Coregonus clupeaformis. Transactions of the American Fisheries Society* 89:176-184.

Moyle, J. B., and R. Lound. 1960. Confidence limits associated with means and medians of series of net catches. *Transactions of the American Fisheries Society* 89:53-58.

Nedelec, C., editor. 1975. *Catalogue of small-scale fishing gear.* Fishing News (Books) Ltd., Surrey, England.

Patriarche, M. H. 1968. Rate of escape of fish from trap nets. *Transactions of the American Fisheries Society* 97:59-61.

Perry, W. G. 1979. Slat trap efficiency as affected by design. *Proceedings of the Annual Conference of the Southeastern Association of Fish and Wildlife Agencies* 32:666-671.

Pierce, R. B., D. W. Coble, and S. Corley. 1981. Fish catches in baited and unbaited hoop nets in the Upper Mississippi River. *North American Journal of Fisheries Management* 1:204-206.

Pope, J. A., A. R. Margetts, J. M. Hamley, and E. F. Okyuz. 1975. *Manual of methods for fish stock assessment. Part III. Selectivity of fishing gear.* Food and Agriculture Organization of the United Nations Fisheries Technical Paper No. 41, Revision 1, Rome, Italy.

Pristas, P. J., and L. Trent. 1977. Comparisons of catches of fishes in gill nets in relation to webbing materials, time of day, and water depth in St. Andrew Bay, Florida. *Fishery Bulletin* 75:102-108.

Reid, G. K., Jr. 1955. The pound-net fishery in Virginia. Part 1. History, gear description, and catch. *Commercial Fisheries Review* 17:1-15.

Regier, H. A., and D. S. Robson. 1966. Selectivity of gill nets, especially to lake whitefish. *Journal of the Fisheries Research Board of Canada* 23:423-454.

Rounsefell, G. A. 1975. *Ecology, utilization, and management of marine fisheries.* C. V. Mosley Company, St. Louis, Missouri, USA.

Skidmore, W. J. 1970. *Manual of instructions for lake survey.* Minnesota Department of Natural Resources, Division of Fish and Wildlife, Section of Fisheries. Special Publication No. 1, Minneapolis, Minnesota, USA.

Starrett, W. C., and P. G. Barnickol. 1955. Efficiency and selectivity of commercial fishing devices used on the Mississippi River. *Illinois Natural History Survey Bulletin* 26:325-366.

Sundstrom, G. T. 1957. *Commercial fishing vessels and gear.* United States Fish and Wildlife Service, Circular 48:1-48. Washington, District of Columbia, USA.

Swales, S. 1981. A lightweight, portable fish-trap for use in small lowland rivers. *Fisheries Management* 12:83-88.

Trent, L., and P. J. Pristas. 1977. Selectivity of gill nets on estuarine and coastal fishes from St. Andrew Bay, Florida. *Fishery Bulletin* 75:185-198.

Twedt, T. M., and D. R. Bernard. 1976. An all-weather, two-way fish trap for small streams. *California Fish and Game* 62:21-27.

Van Oosten, J., R. Hile, and F. Jobes. 1946. The whitefish fishery of Lakes Huron and Michigan with special reference to the deep-trap-net fishery. *Fishery Bulletin* 50:297-394.

Von Brandt, A. 1964. *Fish-catching methods of the world.* Fishing News (Books), Ltd. London, England.

Walburg, C. H. 1969. *Fish sampling and estimation of relative abundance in Lewis and Clark Lake.* United States Fish and Wildlife Service. Technical Paper 18. Washington, District of Columbia, USA.

Welcomme, R. L., editor. *Symposium on the methodology for the survey, monitoring, and appraisal of fishery resources in lakes and large rivers.* Food and Agriculture

Organization of the United Nations. European Inland Fisheries Advisory Commission Technical Paper 23 (Supplement 1), Rome, Italy.

Whalls, M. J., K. E. Proshiek, and D. S. Shetter. 1955. A new two-way fish trap for streams. *Progressive Fish-Culturist* 17:103-109.

White, C. E., Jr. 1959. Selectivity and effectiveness of certain types of commercial nets in the T.V.A. lakes of Alabama. *Transactions of the American Fisheries Society* 88:81-87.

Wolf, P. 1951. A trap for the capture of fish and other organisms moving downstream. *Transactions of the American Fisheries Society* 80:41-45.

Worthington, E. B. 1950. An experiment with populations of fish in Windermere, 1938-48. *Proceedings of the Zoological Society of London* 120:113-149.

Yeh, C. F. 1977. Relative selectivity of fishing gear used in a large reservoir in Texas. *Transactions of the American Fisheries Society* 106:309-313.

6.6 APPENDIX - REPAIRING NETS

STACY V. GEBHARDS

(reprinted from Inland Fisheries Management,
California Fish and Game, with permission)

A properly mended net can mean the difference between catching many fish or none at all. This discussion is intended to acquaint beginners with mending procedures. The techniques described are those used by commercial fishermen in the Illinois River Valley, with some modifications by the author.

6.6.1 Trimming

The first step is to trim the hole (Fig. 6.4) so it can be rewoven in one continuous operation. Each knot has 4 unbroken strands (quarter meshes) leading from it. Around the edges of a tear you will find knots with 1, 2, or 3 unbroken strands. Trim as follows:

1. Start at the top of the hole and leave one knot with 3 unbroken strands. This will be the starting point for the reweaving.
2. Work down one side of the hole, knot by knot. Hereafter, when finding a knot with 3 unbroken strands, cut out the lower strand. Leave knots with 2 or 4 unbroken strands as they are.
3. Trim down one side to the bottom and then trim the other side. Leave one knot at the bottom with 3 unbroken strands. This will be the last tie in the weaving.
4. The hole is now ready for weaving (Fig. 6.4). Each knot around the edge of the hole should have 2 or 4 unbroken strands, except the starting point at the top and the finishing point at the bottom.

6.6.2 Weaving

The twine used for weaving is wound on a shuttle filled by passing the twine beneath the tongue, around the notch at the bottom, up, and beneath the tongue from the opposite side.

The basic knot is the sheet bend. Gill nets which utilize synthetic threads in construction (nylon, orlon, dacron, etc.) require special knots to prevent slippage. Carrothers (1957) describes some of these special knots. Two nonslip knots for nylon are shown in Fig. 6.10.

Figure 6.5 illustrates the sequence of knots in weaving. Details are shown in Fig. 6.6 through 6.10. Knake (1947) describes variations practiced in New England.

When weaving from left to right, bring the shuttle up through the mesh. Do the same for sider knots. When weaving from right to left, pass the shuttle down through the mesh.

6.6.3 Section Replacement

Sometimes a section of net must be replaced (Fig. 6.11). The starting knot and finishing knot begin and end at a 3-strand knot. The remaining knots along the edges are all 2- and 4-strand knots (Fig. 6.12). The seaming procedure follows:

1. Trim each edge of the hole straight, with one continuous row of meshes (Fig. 6.11B).
2. Cut the new section (Fig. 6.11 C) to the same depth as the hole and 2 meshes narrower than the original section.

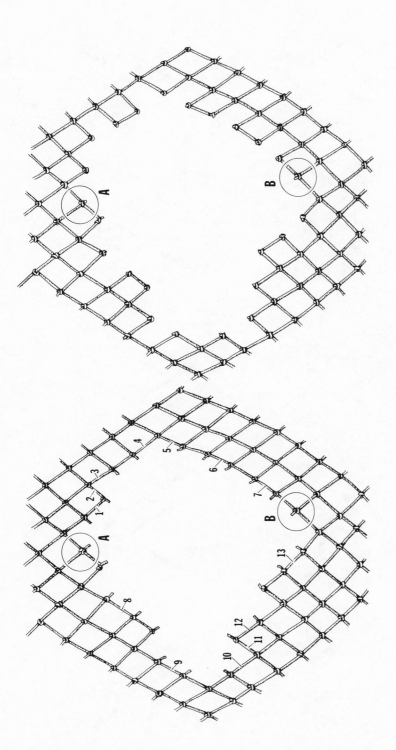

Figure 6.4 Trimming a hole for repair. Left, numbers *1-13* indicate the sequence in cutting. Knots *A* and *B* are the two knots with 3 unbroken strands which are not cut. Knot *A* is the starting point, and knot *B* is the finishing point. Right, the hole trimmed and ready for weaving. Note that all knots other than *A* and *B* have either 2 or 4 unbroken strands.

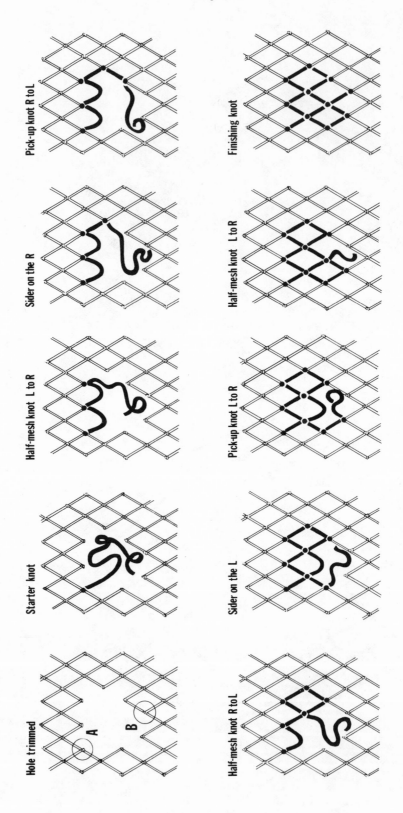

Figure 6.5 Sequence of knots in weaving. Knot-tying steps shown in later figures.

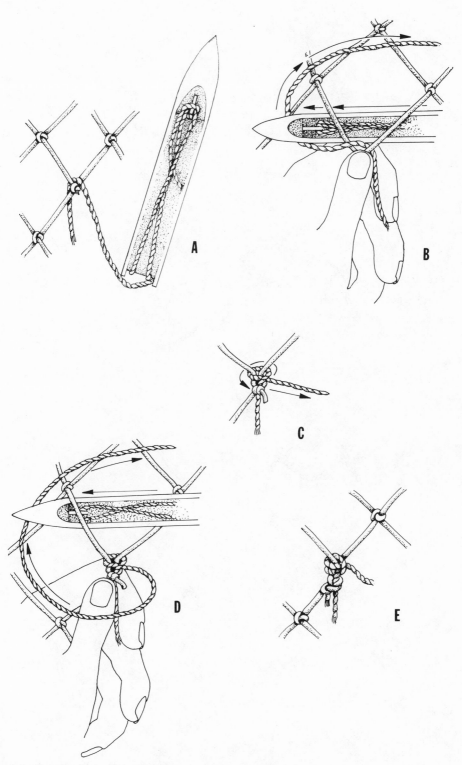

Figure 6.6 Starting knot. Steps in tying proceed from *A* to *E*.

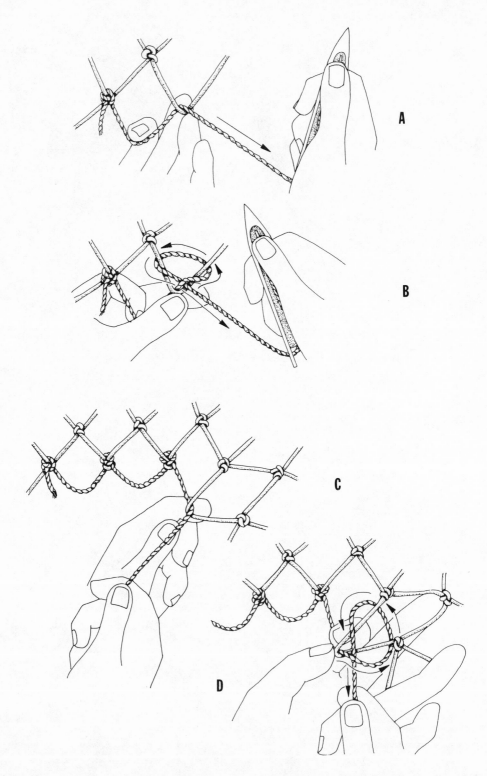

Figure 6.7   Half-mesh knot, left to right *(A* and *B)*; sider knot on the right *(C* and *D)*.

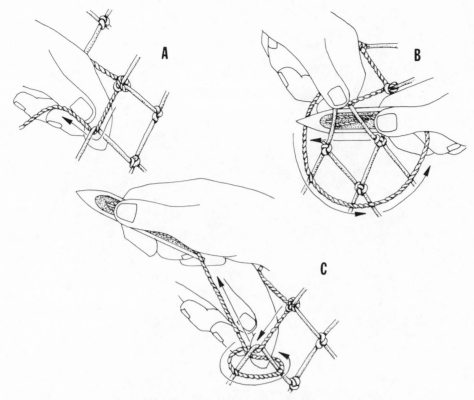

Figure 6.8   Pick-up knot, right to left. Steps in tying proceed from *A* to *C*.

3.   Fig. 6.12 shows the sequence of knots used in seaming. As in mending, the strands are gaged with the fingers. The beginning and finishing ties are half-meshes; the others are all quarter-meshes.

6.6.4   Maintenance of Nets

Synthetic fibers have greatly reduced the problem of rotting associated with cotton and linen nets. However, some care is still necessary to insure maximum life from synthetic-fiber nets. Copper and creosote-base preservatives for natural fibers are discussed in various sources listed in the bibliography.

Nylon breaks down in sunlight twice as fast as linen or cotton (Carrothers 1957). Hence, it should be protected from direct sunlight. Dips for synthetics which contain compounds that screen out sunlight can be purchased in heavy and light viscosities. The heavy grades reduce abrasion.

Fish slimes may create acidic conditions which will damage nylon. Carrothers (1957) recommends dipping nylon gill nets in a 2 percent copper sulphate solution long enough to remove the slime and rinsing in clean water before drying. Strong solutions or residues can reduce wet mesh strength.

Chlorine, oxidizing bleaches, and drying oils, such as linseed oil, also damage nylon.

Sunlight damages many other synthetics less seriously than nylon. Orlon is highly resistant. Dacron is resistant to acids but not to alkaline conditions. Excessive heat reduces the tensile strength and elasticity of synthetic fibers.

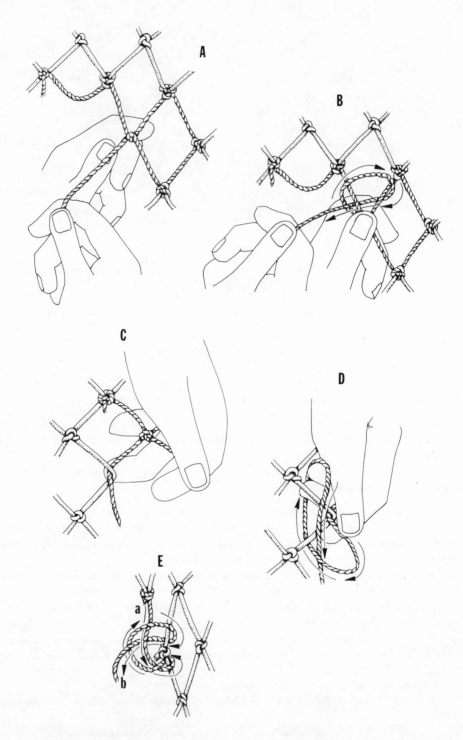

Figure 6.9   Half-mesh knot, right to left *(A* and *B);* sider knot on the left *(C* and *D).* An alternate sider knot is shown at *E;* it is tied the same on right and left sides. Pull the hitch tight below the knot first, then weave and tighten the hitch above the knot.

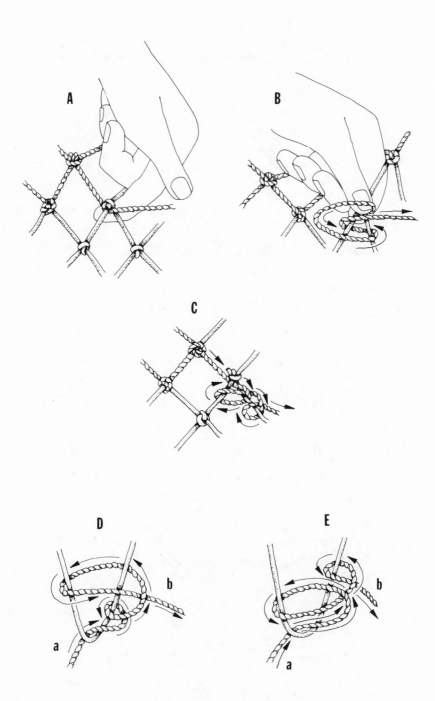

Figure 6.10 Pick-up knot, left to right *(A* and *B)*; finishing knot *(C)*. Two variations of the "knot-and-a-half" used in hand tying nylon nets are shown at *D* and *E*. The hitch at *a* is pulled tight before making the second hitch at *b*.

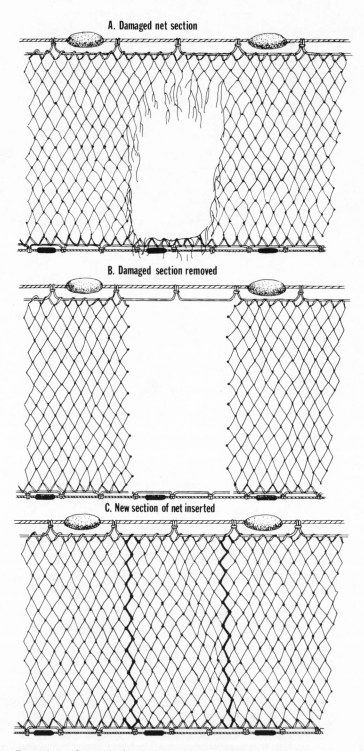

A. Damaged net section

B. Damaged section removed

C. New section of net inserted

Figure 6.11   Procedures for replacing a damaged section of net. The darkened meshes in *C* indicate meshes woven with twine to join the new and old sections of net.

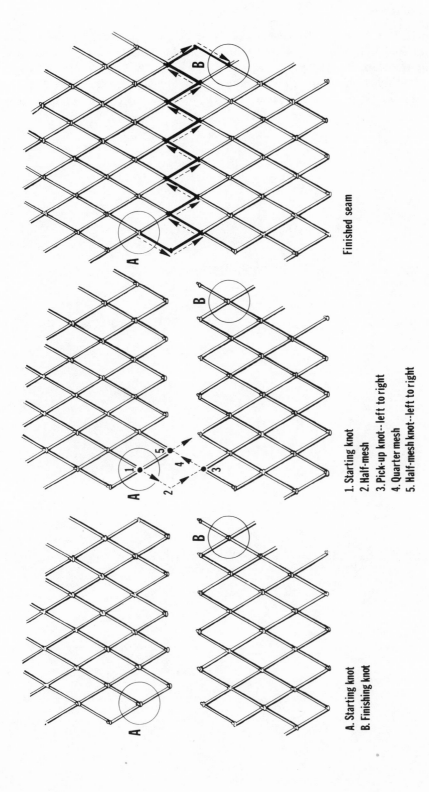

Finished seam

1. Starting knot
2. Half-mesh
3. Pick-up knot--left to right
4. Quarter mesh
5. Half-mesh knot--left to right

A. Starting knot
B. Finishing knot

Figure 6.12 Procedure for seaming two net sections together, after trimming as shown in Fig. 16.11.

6.6.5 References

Carrothers, P. J. G. 1957. *The selection and care of nylon gill nets for salmon.* Fisheries Research Board of Canada, Industrial Memorandum no. 19, Ottawa, Ontario, Canada.

Knake, B. O. 1947. *Methods of net mending - New England.* United States Fish and Wildlife Service, Fishery Leaflet no. 241, Washington, District of Columbia, USA.

Chapter 7
Active Fish Capture Methods

MURRAY L. HAYES

7.1 INTRODUCTION

Active fish capture techniques include those methods that capture (take) fish and macro-invertebrates by sieving them from the water medium by means of mesh panels or bags. These include trawls, seines, dredges, and their variations. Such methods were used by primitive men in gathering food, and major gear components—netting, lines, hooks, weights, and floats—date back to ancient times. Modern techniques have evolved from ancient concepts through application of man's ingenuity, engineering, and scientific talents, producing a wide variety of gear and techniques to meet different fishing conditions and sampling requirements. There are many excellent manuals and catalogs that describe active fish capture gear and fishing techniques. While the majority of these references are aimed at commercial applications, they provide a vast fund of ideas and designs for scientific fish sampling gear. (*See* annotated list of reference manuals on gear designs in Box 7.1.) This chapter provides information needed to select, construct, and operate active fishing gear to sample fish and macro-invertebrates in a wide range of habitats. My approach is to describe active fish capture techniques in the context of their use in scientific sampling problems as opposed to their use in commercial fishing applications.

7.2 SELECTION OF METHODS

Variations in the kinds, sizes, behavior, and distribution of fish precludes the use of any single gear or technique for all scientific sampling. Each fishing system has its limitations in a particular habitat for a particular target organism. As discussed in chapter 1 of this book, the method selected depends on the scientific objectives for sampling.

7.2.1 Habitat as It Affects Gear Selection

The bottom type, water depth, transparency, and current may affect the selective properties and type of gear applicable in any particular habitat. In littoral habitats, gradient, substrate, vegetation, wave action, and currents limit the operation of various types of active fish sampling gear. In most cases, vessel support for sampling in the littoral zone is limited to small boats and skiffs which have limited space and little power for mechanization for gear handling and towing. In some cases, tractors, winches, and other powered equipment ashore may substitute for vessels, but, in general, littoral sampling uses gear that can be handled by manpower. Cliffs or steep shorelines present very different sampling problems from low gradient gravel, sand, or mud beaches. Boulders, rock outcrops, debris, stumps, and aquatic vegetation may

Box 7.1   General Information on Fishing Gear and Its Construction

Kristjonsson, H., editor. 1959. *Modern Fishing Gear of the World,* volume 1. Fishing News Books, Ltd., Farnham, Surrey, England.

Kristjonsson, H., editor. 1964. *Modern Fishing Gear of the World,* volume 2. Fishing News Books, Ltd., Farnham, Surrey, England.

Kristjonsson, H., editor. 1971. *Modern Fishing Gear of the World,* volume 3. Fishing News Books, Ltd., Farnham, Surrey, England.

These three volumes are compendia of papers presented at three International Fishing Gear Congresses sponsored by the Food and Agriculture Organization of the United Nations. They constitute a comprehensive source for gear design and application.

Backiel, T., and R. L. Welcomme. 1980. *Guidelines for sampling fish in inland waters.* EIFAC (European Inland Fisheries Advisory Commission) Technical Paper No. 33. United Nations Food and Agriculture Organization, Rome, Italy.

von Brandt, A. 1972. *Fish catching methods of the world.* Fishing News Book, Ltd., Farnham, Surrey, England.

Sainsbury, J. C. 1971. *Commercial fishing methods: an introduction to vessels and gear.* Fishing News Books, Ltd., Farnham, Surrey, England.

These three volumes describe a large number of methods that may provide ideas to solve many fish-sampling problems.

Libert, L., and A. Maucorps. 1973. *Mending of fishing nets.* Fishing News Books, Ltd., Farnham, Surrey, England.

Klust, G. 1982. *Netting materials for fishing gear.* Fishing News Books, Ltd., Farnham, Surrey, England.

Garner, J. 1962. *How to make and set nets.* Fishing News Books, Ltd., Farnham, Surrey, England.

These three volumes provide practical instructions on how to make and repair fishing gear.

Ne'de'lec, C., editor. 1975. *FAO Catalogue of small scale fishing gear designs.* Fishing News Books, Ltd., Farnham, Surrey, England.

Scharfe, J., editor. 1978. *FAO catalogue of fishing gear designs.* Fishing News Books, Ltd., Farnham, Surrey, England.

These two volumes contain diagrams for construction of nets and rigging. The first volume is particularly valuable for small-scale designs that may find application in fish sampling.

Knake, B. O. 1947. *Methods of net mending—New England.* United States Department of the Interior, Fish and Wildlife Service, Fishery Leaflet 241, Washington, District of Columbia, USA.

Continued

Knake, B. O. 1956. *Assembly methods for otter-trawl nets.* United States Department of the Interior, Fish and Wildlife Service, Fishery Leaflet 437, Washington, District of Columbia, USA.

Sundstrom, G. T. 1957. *Commercial fishing vessels and gear.* United States Department of the Interior, Fish and Wildlife Service, Circular 48, Washington, District of Columbia, USA.

Dumont, W. H., and G. T. Sundstrom. 1961. *Commercial fishing gear of the United States.* United States Department of the Interior, Fish and Wildlife Service, Circular 109, Washington, District of Columbia, USA.

These are older references that are out of print but are available in most libraries that hold United States government fish and wildlife publications.

obstruct passage of gear along the bottom. Even when the bottom is smooth and unobstructed, currents and wave action may limit operation of gear. Fish in the littoral zone are commonly sampled using beach or haul seines, push nets, small scale otter or beam trawls, and, off cliffs or steep shorelines, small purse seines.

In pelagic habitats, the physical factors affecting sampling are quite simple in comparison to the littoral and benthic zones. Water depth, visibility, stratification, and currents are the principal factors to be considered. Gear size and configuration are virtually unlimited, and many options are available. Surface pelagics, especially schooling fishes, are commonly sampled using encircling gear such as purse seines and lampara nets. Tow nets with rigid frames extending above the surface are used, often at night, to sample more widely dispersed species that remain near the surface. Sampling midwater fish with trawls is a problem in three dimensions. Echo sounders on the ship and transducers mounted on the net (netsonde) are commonly used to position the net to sample schooled or stratified fish at depth, i.e., aimed trawling. Midwater sampling gears also may be towed in such a way that they sample different portions of the water column—vertical haul, oblique haul, or tow at depth—and, thus, sample ichthyoplankton and small fish.

Benthic or demersal fish sampling is affected by gradient, substrate, depth, and currents. As in the littoral zone, rough, shelved, or obstructed substrates severely limit the areas that can be sampled. Since the gear cannot be seen while in operation, good echo-sounding and local experience are needed to sample successfully. Bottom trawls are weighted to settle on the substrate, and the spatial problem is reduced to two dimensions. For quantitative bottom sampling, adequate locating or navigation equipment is required to determine distance towed.

7.2.2 Behavior of Target Organism as It Affects Gear Selection

The availability of fish and their reaction to the gear are important criteria in gear selection (Laevastu and Hayes 1981). Fish distribution for any given species may vary with life history; for example, juveniles are often found in shallow water or estuaries at the same time that adults are found offshore. Many demersal fishes have seasonal depth migrations, and, at any given time, individual species may be separated in depth. Consequently, bottomfish sampling is most often conducted at selected depths along the bottom contour. Pelagic fish have diurnal patterns of depth distribution and concentration. Therefore, it may be very important to sample at specific times in the day. Seasonal migrations for spawning or feeding may result in fish concentrations that simplify or complicate sampling. Temperature, salinity, and currents affect the distribution of fish. For example, stratification may restrict the depth range of a particular species or group. Thermal or salinity fronts may restrict horizontal range. Finally, fish respond to the lunar cycle (tides and light at night) and weather. Consequently, the fish available at any given time and place will vary with fish behavior.

Gear selection and deployment also depend upon the response of the target organism to the gear (Merdinyan et al. 1979). Very important in this regard is the swimming speed and endurance of the fish. Many pelagic fish swim continuously, and larger fish swim faster than smaller fish. The structure and behavior of fish schools is complex, and aggregations of fish often respond differently than individuals. Many demersal fishes, especially flatfish, will attempt to swim away ahead or in front of a sweepline advancing over the bottom. Many semi-pelagic midwater fish will swim upward and keep pace with an advancing wall of webbing. Surface schooling pelagic fish will often swim along a net wall or circle continuously in an enclosure. In general, the fish will continue these routine actions and responses until they tire or until the net becomes too narrow or too close, at which time the fish may attempt to escape.

Differential behavior of fish in response to the gear also results in stratification of the catch by species and size within the net (Hughes, 1976). In subsampling, it is important to take the sample in a manner that considers this bias. Consequently, gear design may incorporate sweeplines, wings, leads, overhangs, shelves, funnel shapes, and diminishing mesh sizes to take advantage of the behavioral characteristics of the target fish. This means that the gear design must exceed some minimum size and move or be set at some minimum speed in order to encompass or cover the fish before they react to the gear. Further, the mesh size in the part of the net where the fish react to its presence must be small enough to retain the fish.

7.2.3 Monitoring Gear Performance

In scientific sampling, it is important to understand the ability of the gear to capture fish—what it can do—and to know that it is performing as expected—what it does do. In new sampling situations, it is particularly important to examine these factors. The design and performance of active gear types is a field of research and technology in itself.

The performance of the gear may be judged from the catch or from the condition of the gear. Species, size, and sex composition of the catch may be compared among different gear types or with data from other sources. Catch of bottom fish or substrate materials may indicate the bottom-tending characteristics of a trawl or drag net. The polish or shine of metal parts or the abrasion on soft materials from bottom contact may indicate whether a given gear component is performing as expected. Interpretation of these indirect factors is largely a matter of experience.

Instruments mounted on or accessory to the gear are commonly used to monitor its performance. For example, modern research trawlers measure the amount of towing cable out and its tension on shipboard instruments. Successful midwater trawling requires information on depth of the net. Modern midwater trawlers are equipped with acoustic transducers mounted on the headropes of the net. These netsonde units provide continuous records on the ship's echo sounder display of the distance that the net is above the bottom or below the surface, the mouth-opening height, and, often, temperature at the net depth.

Experiments using arrays of acoustic instruments to measure the mouth-opening configuration of trawls have been used to investigate the variability to be expected in certain standard research trawls. Results demonstrate that trawl behavior may be erratic and suggest that continuous monitoring of performance and some form of positive on-bottom indicator would be desirable additions to research trawls.

7.2.4 Limitations and Biases of Data

Every type of sampling gear has its limitations, and the data must be interpreted with these in mind. Active capture techniques have the distinct advantage of enclosing or encompassing geometrically definable sampling spaces in which the target organisms may be separated from the water by means of the sieving action of the gear (Fig. 7.1). For example, the sampling volume of a midwater trawl or plankton net is approximately equal to the cylindrical volume defined by the area of the net mouth times the distance towed. In the case of benthic trawls, the volume would be rectangular or a flattened oval in cross section. However, it is common practice in trawl surveys to ignore the vertical sampling dimensions and calculate benthic biomass on an area-swept basis. For beach seines, the sample volume is more difficult to measure, and, like bottom trawls, area-swept may be used to expand samples to the total area. For purse seine gear, the sample volume is essentially a cylinder, but seine gear may be held open in a current or across the migration path of a fish school, making the volumetric calculation less useful.

Of those fish that encounter the sampling gear, some may react by swimming over or under the net. Others may be herded into the net or be repelled from the net by the noise, vibration, or pressure waves. Still others may be small enough to pass through the mesh opening. Many other factors, including skill of the operator, affect the performance of the gear. These factors enter into the selective properties of the particular gear and may result in biased samples. To control variation, it is important to measure factors known to influence gear operation (times, rates, dimensions, tensions, bottom contact, etc.), to specify experience or skill level of the operator, and to use standard protocols or operating procedures.

The choice of sampling gear must then consider the habitat limitations on gear operation; the behavior of the fish, both as it affects their distribution and their response to the gear (availability); and the selective properties of the gear for that proportion of the target population encountered (selection).

7.3 NET MATERIAL AND CONSTRUCTION

Some familiarity with materials and methods used in construction of fishing nets is needed to select, procure, and apply active fish capture techniques. Fishing nets are constructed using panels of mesh or web that are cut to shape, sewn together, and attached to ropes or wire cables to form the net shape or enclosure desired. Floats, weights, frames, hardware, and specialized accessories are added to complete the fishing system. (*See* Box 7.1 for list of manuals on materials and construction.)

7.3.1 Fibers, Twine, and Rope

Natural fibers, such as hemp, cotton, and linen, were the historic materials used to manufacture twines and fabricate fishing nets. In the early 1950s, synthetic fibers began to replace natural fibers because of their greater strength, resistance to decay,

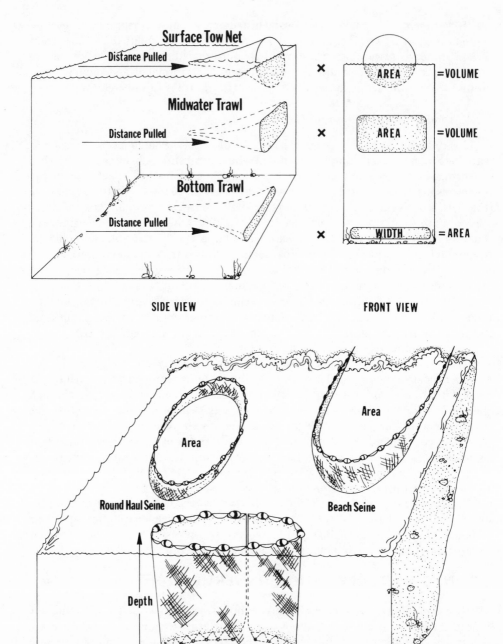

Figure 7.1 Geometric space sampled by active capture techniques. Catches are usually reported in terms of volumes sampled for surface tow nets, midwater trawls, and purse seines, and in terms of areas swept for bottom trawls, round haul seines, and beach seines.

and more reasonable price. Natural fibers are rarely found in modern fishing gear.

Modern netting is made from a large number of synthetic materials. The most common materials used in net fibers are polyamides (nylon-type), polyesters (dacron-type), polyethylenes, and polypropylenes. (Nylon and dacron are trade names for synthetics manufactured by the Dupont Corporation.) Because polyamides have a specific gravity near 1.14 and polyesters near 1.4, they sink; polyethylene has a specific gravity near .91 and it floats. Polypropylene is widely used in rope and for chafing gear in heavy trawls, but because it maintains knots poorly, it is seldom used for netting.

Twines are made from single fiber types or blends of fiber types using twisting, folding, braiding, and cabling techniques. There are many trade names for different twines, and the type selected will depend upon the preference of the user. The size of the net twine is often specified according to the number of threads that occurred in cotton twine of similar diameter. The number of threads in the cotton twine multiplied by two was the approximate tensile strength of the cotton twine. Cotton net twine was ordered by thread count—for instance, 12-thread, 15-thread, 18-thread—whereas the synthetic counterparts are ordered as #12, #15, #18, based on the cross section diameter of the original cotton net twines.

The proliferation of modern synthetic materials and fiber blends has led toward specification of twine size in the "tex" system, and "R-tex" has been adopted as a reference unit. The "R-tex" number indicates the size of the finished twine by its weight in grams per 1000 meters of length. The strength of modern net twines varies with fiber composition: polyamides are strongest, followed by polyesters and polyethylenes.

Net mesh or web is usually manufactured on machines or looms that form netting over a range of sizes. The most common netting is of knotted construction using single or double reef knots or sheet bends. Single knot mesh is used in most trawls, purse seines, and large beach seines which are made of relatively heavy twines. Double knots are common in gill net fabrics. Various forms of woven, braided, or crocheted net fabrics in both rhombic and hexagonal mesh patterns are coming into wider use. Such "knotless" webbing is commonly used in seines, dip nets, trawl liners, and other fish-handling applications to reduce abrasion and scale loss caused by knots. Hexagonal webbing also provides significant savings in material. Knotless webbing is more difficult to sew and repair, however, than conventional knotted netting.

7.3.2 Gear Construction, Mesh Size, and Nomenclature

Although the web in fishing nets is normally fabricated by machines, the process of making a finished unit of gear is still done by hand. Many of the sewing procedures and methods in use originated in ancient times and have been passed down from generation to generation. There are many publications on construction and repair of fishing nets (*See* Box 7.1 and section 6.6.).

The most important factor in net design is size and direction of the mesh. Mesh size is usually specified by the length of one bar side of the diamond-shaped mesh opening (bar measure) or by the length of the mesh stretched to bring the "side" knots together (stretch measure). Note that one row is one-half mesh and that a bar is one side of a mesh. Usually these measurements include the length of one end knot and should be taken under standard conditions and tensions. For regulatory and research purposes, the size and shape of the mesh opening may be the most important measurement. Special mesh calipers or pass-through gauges have been used for such measurements.

The path of a single yarn or twine in a conventional knotted netting determines the direction of the mesh. The direction which is perpendicular to the general direction of the twine is termed the "run" of the net. Because knotted netting is stronger in the direction of the run, it is very important that the tension or strain on the net when it is fished be in the direction of the run. Also, repairs are made with the run, that is, the general direction of the twine used in the repair must be in the same direction as that in the net.

7.3.3 Attaching Net Panels to Ropes

Attaching net panels to ropes is fundamental in construction of fishing gear, and the way it is done determines the shape of both the mesh opening and the net itself. At least three methods are used to express the amount of web hung on a length of rope (Table 7.1). The hanging ratio (E) is defined as the ratio between the length of the rope divided by the stretched length of the netting attached to it (Fig. 7.2). Note that the shape of the mesh opening and the angles of the sides vary with the hanging ratio. The hanging percentage (P) expresses the "looseness" of the hanging net and may be defined as the percentage that the length of the stretched mesh exceeds the length of the rope ($P = 100 (1-E)\%$). In a third method, "hang in" may be expressed in terms of the fraction or percentage of excess mesh beyond the stretched length hung on a unit length of rope. For example, a haul seine may be constructed by hanging 15 meters of stretched web on a 10-meter length of rope. In this case, $E = .66$, $P = 33\%$, and "hang-in" = $1/2$ or 50%. Common hanging ratios (E) for beach seines, haul seines, and trawls are 0.6 to 0.8; for purse seines are 0.8 to 0.9; and for gill nets are 0.4 to 0.6.

7.4 DRAGGED OR TOWED GEAR—TRAWLS

A trawl is a bag-shaped net which is dragged along the bottom or through the water column to collect fish or other biological samples by straining them from the water. It is normally towed by one or two powered vessels and may be designed as a bottom, midwater, or surface sampler. Trawls vary in size from small hand-operated nets towed from small boats to very large mechanically handled trawls towed from commercial fishing vessels.

Table 7.1 Hanging of webbing in net construction [a]

Hang-in or looseness percentage	Ratio of line to webbing E	Angle of mesh	Height of mesh	Width of mesh	Filtering coefficient (area of mesh)
10	.90	128 20'	0.436	0.90	0.785
20	.80	106 20'	0.599	0.80	0.958
30	.70	89—	0.713	0.70	0.998
40	.60	73 40'	0.801	0.60	0.961
50	.50	60 00'	0.866	0.50	0.866
60	.40	47 10'	0.916	0.40	0.733
70	.30	34 50'	0.954	0.30	0.572

[a] Adapted from Lusyne (1959).

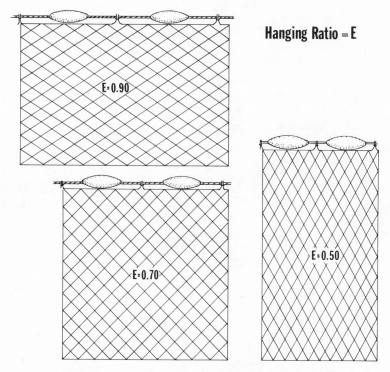

Hanging Ratio = E

Figure 7.2 Appearance of net panels of same mesh size and lengths hung at three different hanging ratios. Note difference in shape of the openings.

7.4.1   General Characteristics

Trawls may be classified by their sampling function, bag construction, or method of maintaining the mouth opening. Function may be defined by the part of the water column sampled—surface, midwater, or bottom trawls—or by the target group sampled—flounder, shrimp, or pollock trawls.

There are two principal methods of construction. Figure 7.3 illustrates a generalized otter trawl, its nomenclature, and the two principal methods of construction. Two-seam trawls are constructed from two panels of netting. The mouth of such a net opens like an envelope and is generally much wider than it is high. Four-seam (multi-seamed) trawls are constructed from four or more panels of netting. The mouth of such a net opens like a grocery bag and is generally rectangular.

The mouth of the trawl may be held open by means of frames or beams, doors, and/or tension in the towing cables. In pair trawling, the mouth opening is maintained by the spacing between the towing vessels and the outward forces generated by the angle of the tow cables.

7.4.2   Variations

Bottom trawls. Beam trawls use rigid frames or poles to hold the mouth of the net open (Fig. 7.4). There are many variations. The most frequent design consists of two D-shaped, sled-like runners held apart by a pole or beam to which the open end of the net is attached. The bags of such nets are usually of simple two-seam construction and relatively short in length. The mesh and twine size depend upon the size of the fish to be

TRAWL NET

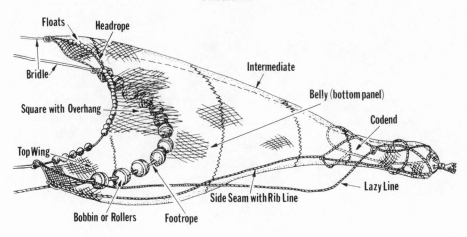

2 SEAM TRAWL NET

4 SEAM TRAWL NET

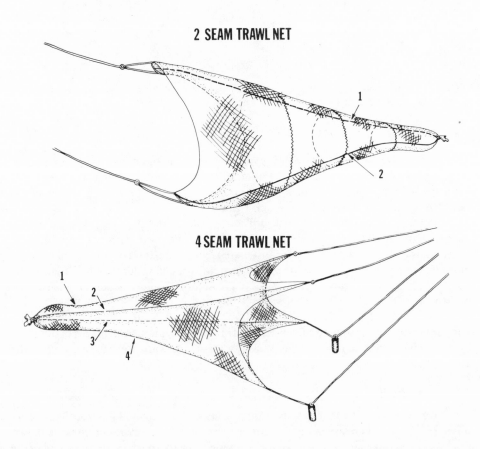

Figure 7.3 A generalized trawl, showing names of parts found on most trawls. Lower panels illustrate difference between 2- and 4-seam construction.

2 BOAT MIDWATER TRAWL

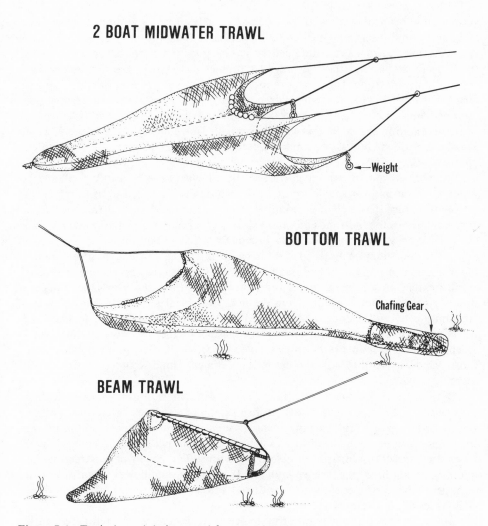

Figure 7.4   Typical trawl designs used for sampling fish.

sampled. The headrope is often attached directly to the beam. The footrope is somewhat longer than the headrope and often weighted with chain. The net is usually towed from a single cable or trawl warp attached to the beam ends with a short bridle. Such nets are easy to construct and operate in convenient sizes for many fish sampling applications. They can be maneuvered easily to avoid obstructions. Large beam trawls, however, are clumsy to handle. The plumb staff beam trawl is a variation which allows the net to be detached from the beam to improve handling of large beam trawls on small vessels.

In otter trawls the mouth opening is maintained by outward forces generated by water pressure and bottom friction against door-shaped boards (otter boards) towed at an angle to the net direction. There are many door designs—flat, V-type, oval, polyvalent—and those used in fish sampling are normally chosen according to local experience.

The two-seam type trawl is the traditional bag design for bottomfish trawls; however, side panel inserts and multi-panel designs may be used to increase the vertical mouth opening for semi-pelagic fish. The headrope is equipped with floats, and the groundline is weighted, often with chain. To prevent fish from escaping upwards, bottomfish nets are designed with overhang—that is, the headrope is shorter than the footrope, and the top panel of the net passes over the fish before they encounter the footrope. The mesh size for bottom trawls depends upon the size of the target organisms, but, to reduce drag, it is generally larger in the wings and intermediate sections than in the codend. Bottomfish nets are normally of rugged construction, and the codend is usually protected by chafing gear to absorb wear. Small mesh codend liners are commonly used to retain small fish in scientific sampling applications.

Otter trawls may be equipped with rollers attached to or below the footrope to improve operation over rough bottoms. Shrimp trawls are rigged to fish slightly off bottom to reduce the catch of bottomfish, invertebrates, and trash. A tickler chain is rigged in front of the footrope and causes shrimp to react upward into the net. Flatfish trawls have heavy footropes that stay in contact with the bottom.

Trawl doors may be attached to the net with different arrangements of sweeplines and bridles. In shrimp trawls and small fish trawls, the door may be connected directly to short extensions of the headrope and footrope. In larger otter trawls the doors are usually rigged at some distance from the net, using bridles from the wing tips and sweeplines between the doors and bridles. These arrangements improve net handling procedures but also affect the selective properties of the net. For example, long sweeplines tend to funnel certain fish species, especially flatfish, into the net mouth. Short sweeplines tend to hide the net in the sediment cloud set up by action of the doors.

Bottom trawls may be towed from a vessel with single or double towing cables (trawl warps). Shrimp trawls and small fish sampling trawls are often towed with a single trawl warp and bridle arrangement. In such cases the bridles should be two or three times as long as the headrope to achieve a proper net mouth opening. Larger trawls are usually towed from double trawl warps. The length of the trawl warps should be about three times the water depth. In shallow water, however, warps should be longer than three times the water depth and, in very deep water, shorter than three times water depth (Pereyra 1963).

Midwater trawls. Four-seam or multi-seam net designs are the most frequently used type for offbottom and midwater fishing. These nets have rectangular or polygon-shaped mouth openings. Generally the net is designed symmetrically without overhang. Sometimes a moderate extension of the bottom panels is made to improve capture of fish that tend to dive under the net.

There has been a rapid evolution in design of midwater trawls. Early midwater trawls were very large, slow-moving nets. This was a consequence of low-powered vessels coupled with lack of modern fishing electronics. Indeed early midwater trawling efforts relied on pressure sensors at the doors or on the net and electrical conductors in the trawl cables to transmit net depth to the vessel. Early experiences demonstrated that aimed midwater trawling could be very productive and led to rapid development of netsonde instrumentation and new net designs to reduce drag.

The mesh size of midwater trawls depends upon the size and behavior of the target organisms. Because many pelagic fish orient to visual cues, they tend to move along rather than through mesh panels. Consequently, the mesh may be extremely large in the forward part of the net. This graduated mesh size principle is used in the design of modern super-mesh trawls and in "rope" trawls in which the forward net panels are replaced with ropes extending from the net mouth perimeter to an intermediate part. Such graduated-mesh nets have lower towing resistance than smaller mesh nets, allowing the use of much larger and faster moving nets for a given power application.

The mouths of midwater trawls are held open using a number of techniques. Hydrofoil-shaped otter boards may be used to spread the net laterally. The net may be depressed by a weighted footrope and large weights hung on the towing bridles near the lower net corners. The "inflation pressure" of the water passing through the moving net is more important in midwater trawls than in bottomfish nets and decreases the relative importance of floats on the headrope and weights on the footrope. Various forms of planing floats and "kites" on the headrope, as well as hydrofoil depressors on the footropes, have been used. The net is maneuvered to depth by changing the length of the towing cables and/or changing vessel speed.

Two-boat trawling (pair trawling) is commonly used in commercial fishing to enable lower powered vessels to fish large nets (Fig. 7.4). In these applications, the net mouth is held open by the outward force of the towing cables. Pair trawling also eliminates the need for cumbersome doors with their hydraulic and friction resistance. Such techniques are simple to use in small scale fish sampling applications.

Small midwater trawls. A number of specialized, small-scale midwater trawls have been developed by biological oceanographers for sampling pelagic fish and zooplankton (Gehringer and Aron 1968). In smaller mesh nets towed at higher speeds, the volume strained may be significantly different than the geometric volume and should be calibrated using flowmeters (*See* chapter 9). The Isaacs-Kidd midwater trawl is a small mesh, rectangular-opening net (1-3 square meters) with a special fixed angle depressor attached to the footrope. In the Isaacs-Kidd net the bridle arrangement, coupled with use of the depressor, tends to keep the towing cable out of path of the net. This principle is used in other midwater samplers which combine framed nets with other depressors or weight to control depth. A number of sampling procedures are used. The net may be lowered to depth rapidly and recovered slowly during the tow. This oblique tow collects a single sample integrated over depth. To sample a deep layer, the net may be rapidly lowered, towed for the time period necessary to collect a sample, and retrieved. In this process, the sample is contaminated by organisms collected during setting and recovery. To sample discrete depths without contamination, opening and closing devices may be added to the net. Recently, a number of multiple net samplers with opening and closing devices have developed. These nets are designed to take samples at up to 10 discrete depths on command during a single tow (Sameoto et al. 1980).

Surface tow nets. Various forms of surface tow nets have been developed to sample neuston and small fish. Neuston nets are simple framed nets with flotation designed to sample the surface and top portion of the water column. Larger surface tow nets with rigid frames may be towed from the side of a sampling vessel or towed between two small boats. These techniques have been used to sample juvenile salmonids in lakes or the sea and are most successful at night.

7.4.3 Applications

Trawling and trawling-acoustic surveys are the primary tool used to assess the composition, distribution, abundance, and biological characteristics of marine fish populations. These surveys represent very extensive geographical investigations and apply complex statistical design and quantitative sampling plans. Applications cited in this section were selected to exemplify current methodology.

In the northwest Atlantic Ocean on the continental shelf from Cape Hatteras to Nova Scotia, bottom trawling surveys have been conducted since 1963 using a stratified random sampling plan (Clark 1979). These surveys cover about 187,000 square kilometers and sample about 300 stations twice each year to provide data on recruitment, growth, mortality, and stock abundance. Two different survey trawls have been used: a "36 Yankee" and a modified high opening "41 Yankee." These are standard design, two-seam, commercial fishing bottom trawls. To compare results, fishing power coefficients have been calculated to convert data from either trawl to a common basis (Sissenwine and Bowman 1978). Results are presented in terms of stratified mean catch per tow and provide a basis to compare abundance over time.

In the eastern North Pacific, trawling surveys have been used for exploratory fishing since 1948 (Alverson et al. 1964). Such surveys were designed to help fishermen locate fishable concentrations in new areas. Results were presented in terms of catch rates and areas with densities sufficient to support commercial exploitation.

Trawling surveys for resource assessment in the Pacific began in 1961 and are designed to support fisheries management and conservation by estimating population characteristics. In the eastern Bering Sea, for example, bottom trawling surveys have been conducted since 1955. These surveys now use 950 stations which extend from the Alaska Peninsula to the U.S.-USSR convention line and north to the latitude of Nome, an area of about 650,000 square kilometers (Smith and Bakkala 1982). In contrast to the surveys in the northwest Atlantic, the trawling survey in the eastern Bering Sea is a stratified systematic design using stations located on 20-nautical mile centers, except for the area near the continental slope where additional stations are added to sample the different depths.

Over the years the sampling gear used in the North Pacific has evolved from the simple application of the commonly used commercial gear to a specialized suite of standard trawls developed to sample different biological communities over different substrates and throughout the water column. The performance of this suite of gear has been studied (Wathne 1977) and is routinely monitored by netsonde and special instrumentation. Bottom trawling surveys target crabs and demersal fish, especially flounders. Again, two standard survey trawls have been used—a "400 Eastern" and a somewhat larger and higher opening version of the same net designated by its headrope-footrope lengths, the "83/112 bottom trawl." These nets are traditional two-seam design bottom trawls that have proven successful in commercial fishing. They open 14-17 meters in width and 1.5-3.7 meters in height.

For fish distributed off the bottom, a high-opening, four-seam design is used. The "Nor'eastern net" is a design developed for commercial fishermen. The research version opens about 13.4 meters in width and about 8.8 meters in height (Gunderson and Sample 1980). This net, equipped with roller gear, has proved particularly useful for Pacific rockfish and other "round" bottomfish species that occur over hard bottoms near the edge of the continental shelf.

For semipelagic fish, such as Pacific hake and walleye pollock, a combination of bottom trawling surveys for the near bottom portion of the stock and hydroacoustic-midwater trawling surveys for the off-bottom portion of the stock is used (Dark et al. 1980). The midwater trawl is used to identify the acoustic targets detected and to collect biological data. The net used is a "Canadian diamond" design which is a near symmetric, six-panel, cone-shaped net with graduated mesh to reduce towing resistance.

In more localized studies, a wide variety of trawls have been used for life history and environmental studies. These studies are geographically intensive and usually apply comparative designs and semi-quantitative sampling methods. While many of these nets are local designs based on small scale commercial fishing gear used in the same area, the small "try nets" used in commercial shrimp fisheries have been used widely. These try net designs are usually small two-seam nets that are selective for juvenile and small fish. Mearns (1970) collected and tested a number of small scale otter trawls used by different investigators in California and found significant differences in their performance. As a result, he has recommended that biologists use uniform designs and operating procedures for environmental monitoring.

Small beam trawls have been used frequently to sample the benthic macrofauna (Carey and Heyamoto 1972). These nets have constant width and height and, when coupled with distance towed, provide samples of the epifauna over a defined area. The distance towed may be measured with metering wheels attached to the net (Carney and Carey 1980) or determined from high precision navigation equipment, such as Loran C or radar.

Trawling to sample fish in fresh water is much less common than in marine habitats. In the Great Lakes and certain large reservoirs, commercial scale bottom trawling has been used in resource investigations (Greenwood and Boussu 1967). Small bottom trawls have been used to sample certain age-0 fishes with only modest success. Midwater trawls (Lewis 1967) and tow nets have been used to sample pelagic fish in lakes with more success. Two-boat tow netting at night is a routine method used to sample juvenile sockeye salmon and the open water community in large lakes.

7.5 DREDGES

7.5.1 Concept

Dredges are rugged, frame-supported box- or bag-shaped devices used to sample benthic organisms when dragged over the bottom. They are often equipped with blades, rakelike teeth, or hydraulic jets to scrape or dig into the substrate. They range in size from less than a meter to more than eight meters in width.

7.5.2 Variations

While dredges used in certain large scale commercial fisheries have evolved to near-standard design, the vast majority of dredges used in biological sampling are of local and special purpose design. The principles used in commercial gear for sea scallops and surf clams may suggest variations that can be applied to local sampling problems (Fig. 7.5). Commercial scallop dredges sample the larger epibenthic fauna. A scallop dredge consists of a sledlike metal frame with a scraper bar across the front and a bag constructed of metal rings attached to the rear to collect the epibenthic organisms. The scraper bar is a flat blade designed to skim the substrate and deflect epibenthic organisms upward into the bag. While scallop dredges operate in close contact with the bottom, they do not dig or cut into the substrate in normal operation. The bag is fabricated of metal rings held together by "hog rings" that are pressed into position with large, special-purpose pliers. Such dredges are heavy, very rugged, and easily repairable with spare rings.

Epibenthic dredges may be adapted to other sampling situations by various modifications. Rakelike teeth may be added to the scraper blade to dislodge oysters or mussels. The metal ring bag may be replaced in part or whole by a wire mesh box or by

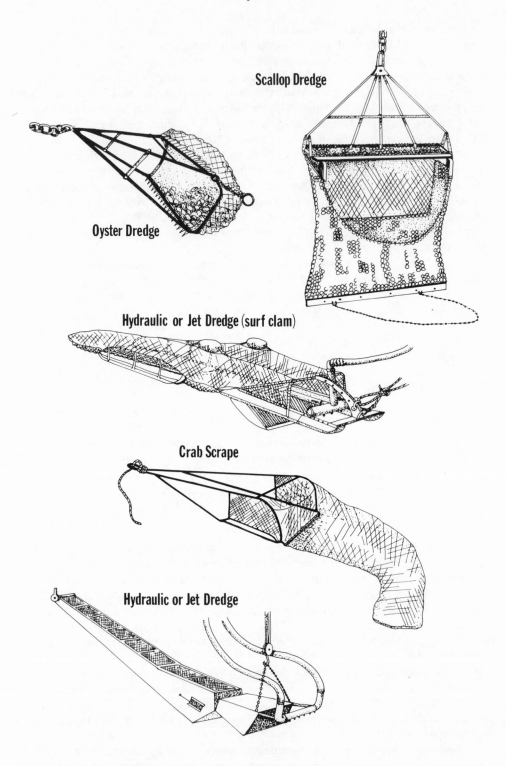

Figure 7.5 Typical dredge designs used for sampling invertebrates such as molluscs, shrimp, and crabs.

a net fabric bag. A crab or shrimp scrape is a lightweight epibenthic dredge usually operated in shallow water. A "pipe dredge" is essentially an open-ended box or basket, sometimes constructed of pipe, used to sample epibenthic organisms.

Many forms of mechanical clam harvesters have been developed for clams and may be adapted to sample the larger infauna. The hydraulic escalator shellfish harvester was developed in Chesapeake Bay. It consists of a bank of water jets located in a cutting assembly in front of an escalator-type conveyor belt. The device is attached to the side of a vessel, lowered to the bottom, and pushed forward. A high volume pump on the vessel supplies water to the jets which erode the sediment and wash the clams onto the upward-moving conveyor. At the surface, the catch is sorted, and unuseable material is returned to the water.

In deeper water, surf clams are harvested using hydraulic-assisted cutting or digging dredges. These dredges are similar in principle to the hydraulic cutting head of the escalator-type dredge, but the conveyor belt is replaced by a collecting bag. The hydraulic clam dredge is a rigid boxlike structure with an adjustable cutting blade and hydraulic manifold at the forward end. As the dredge is dragged forward, the knife cuts a trench several inches deep. The substrate cuttings are plowed upward into the boxlike portion of the dredge. Water is pumped through a hose to the hydraulic manifold and through nozzles which are arranged to wash the sediment from the clams and other macrofauna. Other nozzles and the hydraulic force of water passing through the dredge move the clams into a collecting bag made of steel rings.

While the hydraulic clam dredge is a complex and cumbersome piece of equipment, its principles have been incorporated into small scale portable systems. Such equipment may be adapted to sample the larger infaunal benthos that are poorly sampled using clamshell or boxcore dredges. These range from small versions of the commercial-scale equipment to diver-operated hydraulic jets used to wash specimens into hand-held collecting bags.

7.5.3 Dredge Applications

Dredges and scrapes are used to sample the larger forms of the benthic epifauna and infauna that may not be adequately sampled by means of conventional small scale bottom samplers (*see* chapter 4). Dredges or scrapes have proved superior to small otter trawls or push nets for quantitative sampling of blue crabs in shallow waters with vegetation (Miller et al. 1980). Conventional scallop dredges and hydraulic clam dredges are used to assess the commercial abundance of these species. In the eastern Bering Sea, large populations of surf clams were identified using commercial-type hydraulic clam dredges (Hughes 1981).

7.6 SURROUNDING OR ENCIRCLING NETS

A seine is an active fishing system that traps fish by enclosing or encircling them with a long fencelike wall of webbing. The top edge of the net is buoyed upwards by a series of plastic, wood, or cork floats (the corkline), and the bottom edge is weighted downward by lead weights, chain, or lead-cored groundlines (the leadline). They often have a specially constructed bunt or bag in which the fish are concentrated as the net is closed.

7.6.1 Variations

Beach or haul seines are used in shallow water situations where the net wall extends from the surface to the bottom (Fig. 7.6). Purse seines and lampara nets are used in open water or over smooth bottom. The bottom edge of the net is drawn together to enclose the fish. "Danish" and "Scottish" seining are commercial methods used to operate a submerged bag seine to catch groundfish.

Beach seine. Beach or haul seines are constructed of mesh panels hung from a float line with a weighted leadline attached to its lower edge. Mesh size varies with intended use, but it is generally small relative to the circumference of the fish to avoid gilling. Modern seines are usually constructed of knotless synthetic mesh to avoid abrasion of the captured fish. They are available from net suppliers as complete units, or they may be ordered by length.

The simplest type of haul seine consists of uniform panels of mesh. Lengths may vary from a few meters to a hundred meters or more, and depth is usually proportional to length. Small seines are often handled with sticks or poles ("two-stick seines"). Larger haul seines are often equipped with spreaders and/or bridles at the ends for ease in towing and hauling.

Large haul or beach seines are usually modified or specially constructed to meet the sampling situation. Bags or bunts are commonly added to the net in the part where fish collect as the net is hauled through the water or onto the beach. The bag is usually located in the center panel of the net, but in situations where the net sweep is made by hauling or towing one end of the net, it may be located near one end. Larger beach seines or ones used on steeper beaches may be tapered toward the ends to reduce the amount of mesh to be handled. Larger sized mesh may be used in the end panels to reduce drag, and this graduated mesh principle may be combined with tapered panels to reduce total resistance (Sims and Johnsen 1974).

Small haul seines or simple two-stick seines may be operated from shore by manpower and wading. Larger nets and operations in deeper water require powered boats or seine skiffs. Often winches, reels, tractors, or other power equipment are used to mechanize the handling and hauling of the net.

Purse seines. A purse seine is a net with flotation on the corkline adequate to support the net in open water without touching bottom. A rope or wire cable is strung through rings attached along the bottom edge of the net and is used to pull the bottom of the seine together like a drawstring. The length, depth, and mesh size of purse seines depend upon the application. Purse seines are generally large scale gear made up of long individual strips of mesh sewn together horizontally. The selvage strips nearest the leadline, and sometimes the corkline, are often of heavier construction to absorb wear and improve handling. The portion of the purse seine in which the fish are gathered at the end of a haul is termed the bunt and is constructed of smaller mesh and/or knotless webbing to improve fish handling.

Two methods of purse seining are common in North America. Two-boat seining is common in the menhaden fishery on the east coast. In this technique the seine is divided between two boats. To set the seine, the two boats are detached, and the net is played out from each as they surround the target fish school. The ends of the net are towed together to close the circle and trap the fish. The net is pursed and each boat recovers opposite ends of the net. The fish are trapped in the bunt located at the center of the net.

The second seining method, the single boat technique, is common in west coast

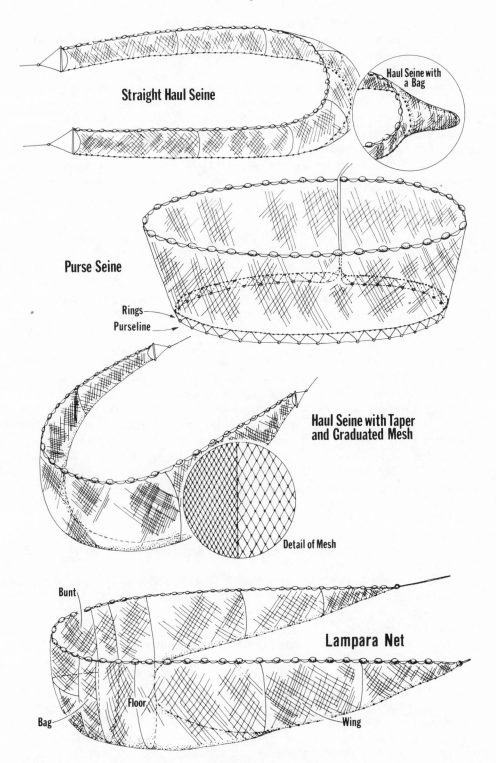

Figure 7.6 Typical seine designs used for sampling fish in shallow water or along beaches (haul seines) and in deeper water (purse and lampara seines).

herring, tuna, and salmon fisheries. In this method, the seine is set and recovered from a single boat. A net or seine skiff is used to hold the free end of the seine in place while fishing. It is also used to tow the seine boat or net during hauling to aid in the net recovery process.

Lampara seine. A lampara net is a type of open water seine with tapered ends and a relatively deep, loosely hung center section or bunt. The leadline is shorter than the corkline. The net is set in a circle to surround the target fish school, the two ends are brought together, and the net is towed in the direction away from the center bunt section. The towing tension on the wings tends to bring the two portions of the leadline parallel and to close the bottom of the net like a clam shell. The net is hauled aboard the boat by the leadlines, leaving the fish enclosed in the pocket-shaped bunt.

7.6.2 Applications

Beach or haul seines are the most commonly used active fish capture methods in hatcheries, lakes, along shorelines, and in streams with slow currents. Beach seines are generally selective for small, slow-moving, or schooling fish whose normal habitat is shallow water. Seining is most effective in shallow water over smooth bottoms. To prevent fish from escaping, the leadline must remain in contact with the bottom and not become caught in or ride up over obstructions. As the net is pulled onto the beach, special care must be taken in handling the leadline to prevent fish from escaping under the net.

Special seining techniques are used in rivers or where there are strong along-shore currents. If downstream migrating fish are to be captured, the net may be held open in an arc by a powered boat and closed upstream against the current. If upstream migrating fish are to be captured, the net may be attached to a stake or anchored at some intermediate point on its length, with the upstream end attached to shore and the downstream end trailing in the current to form a trap. To close, the midpoint anchor is released and the free end of the net is rapidly drawn to the beach. Similar "seine trap" configurations may be used to determine direction of fish movement along a shoreline by setting the trap alternately open in one direction and then in the other.

In shallow water areas with no appropriate beach, such as tidal flats, the seine may be set in a circular pattern and closed by drawing the moving wall of webbing past the end held in place or attached to a stake or post. In this "stake seining" technique, the fish are herded into the bunt or bag, located at the fixed end of the net, as the circle of webbing becomes smaller.

To sample very large areas, seines may be attached end to end or constructed in modular sections that may be extended to cover large areas. Larger scale applications are specialized and require mechanized equipment to handle the net and haul it through the water.

Where bottoms are obstructed or in streams, seines may be set in an opportune area, and the fish may be herded into the net using electro-fishing or other means.

Other specialized seining techniques have been developed to operate haul seines under ice and on the bottom with the corkline submerged. Bottom seining in deep water is a commercial technique (Scottish or Danish seining) used to catch groundfish that do not swim upward to avoid the net, primarily flounders and soles. The seine has long wings and a deep bag or codend. Fishermen set the net by anchoring one end of the seine warp with a float attached. They move the vessel in a triangular pattern, laying the anchored warp in a straight line, turning to set the net along a second side, turning again to set the second warp along the third side, and returning to the float.

They move the vessel away from the apex of the triangle, closing the net by a combination of towing and bringing the ends of the seine together.

Purse seines are used in both open water and shallow water situations in contact with the bottom. Purse seining is normally a large scale and complex fishing operation. Most research applications have used commercial scale nets operated by professional fishermen. Purse seines have been used to sample salmon abundance on the high seas, for tagging, and to determine the direction of their migration. For the latter purpose, the purse seine would be set and held open in different directions on succeeding sets. Small mesh purse seines have been used to sample juvenile salmon in the ocean to determine distribution, growth, and mortality. Small scale and miniature purse seines have been developed for research applications (Hunter et al. 1966; Johnsen and Sims 1973; and Levi 1981). Purse seines also have been used in sampling inland lakes and reservoirs to assess fish stocks (game and non-game species). Seines designed to sample less than one hectare of water have been used successfully (Whaley and Fowden 1982).

7.7 OTHER ACTIVE SAMPLING GEAR

Fisheries scientists have adapted active sampling techniques to many specialized situations. These techniques encompass or surround a sampling space and capture any organisms present by physically separating them from the water medium.

Various forms and sizes of nets attached to frames may be pushed or dropped through the water column to sample fish where a net pulled by a wire or rope might become fouled or scare the fish. Push nets with rollers may be pushed through and over submerged vegetation to sample fish and invertebrates in a manner similar to the operation of a lawn mower. Such nets have long handles and triangular or D-shaped frames and may be pushed by wading or from a boat in shallow water. Framed nets pushed in front of a small boat to catch small fish in the surface layer may be more effective than towed nets (Kriete and Loesch 1980).

Lift nets with rectangular frames, hoops, or crossed spreaders are effective in capture of many small schooling fish. Such nets may be set below the water surface or on the bottom and lifted to capture fish as they pass over the net. Fish may be attracted over such nets using bait or light attraction at night.

Cast nets are conical-shaped nets which can be thrown or cast to drop over an area to be sampled. The net is constructed so that the circumference of the open end of the mesh cone is greater than the circumference of the leadline. As the net is retrieved, the extra mesh forms pockets that trap the fish. Drop or gravity nets may be released from boats or stationary frames built over shallow water (Kjelson 1975). They may be equipped with purse seine-type closing systems. Special precautions must be taken in the use of drop nets to keep the leadline on the bottom as the net is closed.

7.8 REFERENCES

Aron, W. 1962. Some aspects of sampling the macroplankton. *Rapports et Proces-Verbaux des Reu'ions* 153:29-38.

Alverson, D. L., and W. T. Pereyra. 1969. Demersal trawl explorations in the northeastern Pacific Ocean—An evaluation of exploratory fishing and analytical approaches to stock size and yield forecasts. *Journal of the Fisheries Research Board of Canada* 26:1985-2001.

Alverson, D. L., A. T. Pruter, and L. L. Ronholt. 1964. *A study of demersal fishes and fisheries of the northeastern Pacific Ocean.* H. R. MacMillen Lectures in Fisheries, Institute of Fisheries, University of British Columbia, Vancouver, British Columbia, Canada.

Bardach, J. E., J. J. Magnuson, R. C. May, and J. M. Reinhart, editors. 1980. *Fish behavior and its use in the capture and culture of fishes.* International Center for Living Aquatic Resource Management, ICLARM Conference Proceedings 5, Manila, Philippine Islands.

Beene, E. L., and A. M. Landry, Jr. 1978. The roller-net: a new marine sampling gear. *Proceedings of the Annual Conference Southeastern Association of Fish and Wildlife Agencies* 32: 517-519.

Bowman, R. E., and E. W. Bowman. 1980. Diurnal variation in the feeding intensity and catchability of silver hake *(Merluccius bilinearis). Canadian Journal of Fisheries and Aquatic Sciences* 37:1565-1572.

Carey, A. G., Jr., and H. Heyamoto. 1972. Techniques and equipment for sampling benthic organisms. Pages 378-408 *in* A. T. Pruter and D. L. Alverson, editors. *The Columbia River estuary and adjacent ocean waters.* University of Washington Press, Seattle, Washington, USA.

Carney, R. S., and A. G. Carey, Jr. 1980. Effectiveness of metering wheels for measurement of area sampled by beam trawls. *Fishery Bulletin* 78:791-796.

Clark, S. H. 1979. Application of bottom-trawl survey data to fish stock assessment. *Fisheries* 4 (3):9-15.

Dark, T., M. O. Nelson, J. Traynor, and E. P. Nunnallee. 1980. The distribution, abundance and biological characteristics of Pacific hake *(Merluccius productus)* in the California-British Columbia region during July-September 1977. *Marine Fisheries Review* 42 (3/4):17-33.

Durkin, J. T., and D. L. Park. 1967. A purse seine for sampling juvenile salmonids. *Progressive Fish-Culturist* 29:56-59.

Dyer, M. F., W. G. Fry, P. D. Frey, and G. J. Cranmer. 1982. A series of North Sea benthos surveys with trawl and headline camera. *Journal of the Marine Biological Association of the United Kingdom* 62:297-313.

Gehringer, J. W., and W. Aron. 1968. Field techniques. *Monograph on Oceanographic Methodology 2.* UNESCO (United Nations Educational, Scientific, and Cultural Organization). Place de Fontenoy, 75 Paris-7e, France.

Gjernes, T. 1979. *A portable midwater trawling system for use in remote lakes.* Pacific Biological Station, Department of Fisheries and Oceans, Fisheries and Marine Service, Technical Report 888, Nanaimo, British Columbia, Canada.

Greenwood, M. R., and M. F. Boussu. 1967. Efficient fishing systems—potential for increasing effectiveness of commercial fisheries in reservoir management and utilization. Pages 467-476 *in Reservoir Fishery Resources Symposium.* Southern Division, American Fisheries Society, Bethesda, Maryland, USA.

Gunderson, D. R., and T. M. Sample. 1980. Distribution and abundance of rockfish off Washington, Oregon, and California during 1977. *Marine Fisheries Review* 42 (3/4): 2-16.

Hanks, R. W. 1963. *The soft-shell clam.* United States Bureau of Commercial Fisheries, Circular 162:1-16.

Houser, A., and H. E. Bryant. 1967. Sampling reservoir fish populations using midwater trawls. Pages 391-404 *in Reservoir Fishery Resources Symposium.* Southern Division, American Fisheries Society, Bethesda, Maryland, USA.

Hughes, S. E. 1976. System for sampling large trawl catches of research vessels. *Journal of the Fisheries Research Board of Canada* 33:833-839.

Hughes, S. E., and N. Bourne. 1981. Stock assessment and life history of a newly discovered Alaska surf calm *(Spisula polynyma)* resource in the southeastern Bering Sea. *Canadian Journal of Fisheries and Aquatic Sciences* 38:1173-1181.

Hunter, J. R., D. C. Aasted, and C. T. Mitchell. 1966. Design and use of a minature purse seine. *Progressive Fish-Culturist* 28:175-179.

Johnsen, R. C., and C. W. Sims. 1973. Purse seining for juvenile salmon and trout in the Columbia River estuary. *Transactions of the American Fisheries Society* 102:341-345.

Jurkovich, J. E. 1981. Principles and innovations in commercial fishing gear. Pages 23-32 *in Science, Politics, and Fishing: a series of lectures developed by Robert W. Schoning.* Oregon Sea Grant Program publication ORESU-W-81-001, Corvallis, Oregon, USA.

Kimura, D. K. 1978. Logistic model for estimating selection ogives from catches of codend whose ogives overlap. *Journal du Conseil, Conseil International pour l'Exploration de la Mer* 38:116-119.

Kjelson, M. A., W. R. Turner, and G. N. Johnson. 1975. Description of a stationary drop-net for estimating nekton abundance in shallow water. *Transactions of the American Fisheries Society* 104:46-49.

Kriete, W. H., Jr., and J. G. Loesch. 1980. Design and relative efficiency of a bow-mounted pushnet for sampling juvenile pelagic fishes. *Transactions of the American Fisheries Society* 109:649-652.

Laevastu, T., and M. L. Hayes. 1981. *Fisheries oceanography and ecology.* Fishing News Books, Ltd., Farnham, Surrey, England.

Levi, E. J. 1981. Design and operation of a small two-boat purse seine. *Estuaries* 4:385-387.

Lusyne, P. A. 1959. Some considerations on net making. Pages 102-106 *in* Hilmer Kristjonsson, editor. *Modern fishing gear of the world,* volume 1. Fishing News Books, Ltd., Farnham, Surrey, England.

Magnuson, J. J., S. B. Brandt, and D. J. Stewart. 1980. Habitat preferences and fishery oceanography. Pages 371-382 *in* J. E. Bardach et al., editors. *Fish behavior and its use in the capture and culture of fishes.* International Center for Living Aquatic Resource Management, Manila, Philippine Islands.

McClenegham, K., and J. L. Houk. 1978. A diver-operated net for catching large numbers of juvenile marine fish. *California Fish and Game* 64:305-307.

McNeely, R. L. 1968. Mid-water trawling: where we are now ... where we are going. Pages 160-164 *in* D. W. Gilbert, editor. *The future of the fishing industry of the United States.* University of Washington, Publications in Fisheries, New series, volume 4, Seattle, Washington, USA.

Mearns, A. J., and M. J. Allen. 1978. *Use of small otter trawls in coastal biological surveys.* United States Environmental Protection Agency Ecological Research Series EPA-600-3-78-083, Corvallis, Oregon, USA.

Merdinyan, M. E., C. D. Mortiner, and L. Melbye. 1979. *Bibliography: the relationship between the development of fishing gear and the study of fish behavior.* University of Rhode Island, Marine Memorandum 59, Kingston, Rhode Island, USA.

Miller, R. E., D. W. Campbell, and P. J. Lunsford. 1980. Comparison of sampling devices for the juvenile blue crab, *Callinectes sapidus. Fishery Bulletin* 78:196-198.

Pearcy, W. G. 1980. A large opening-closing midwater trawl for sampling oceanic nekton, and comparison of catches with an Isaacs-Kidd midwater trawl. *Fishery Bulletin* 78:529-534.

Pereyra, W. T. 1963. Scope ratio-depth relationships for beam trawl, shrimp trawl, and otter trawl. *Commercial Fisheries Review* 25 (12):7-10.

Sameoto, D. D., L. O. Jaroszynski, and W. B. Fraser. 1980. BIONESS, a new design in multiple net zooplankton samplers. *Journal of the Fisheries Research Board of Canada* 37:722-724.

Schott, J. W. 1975. Otter trawl cod-end escapement experiments for California halibut. *California Fish and Game* 61:82-94.

Sims, C. W., and R. H. Johnsen. 1974. Variable-mesh beach seine for sampling juvenile salmon in Columbia River estuary. *Marine Fisheries Review* 36 (2):23-26.

Sissenwine, M. P., and E. W. Bowman. 1978. Fishing power of two bottom trawls towed by research vessels off the northeast coast of the U.S.A. during day and night. *International Commission for the Northwest Atlantic Fisheries, Research Bulletin* 13:81-87.

Smith, G. B., and R. G. Bakkala. 1982. *Demersal fish resources of the eastern Bering Sea: spring 1976.* NOAA (National Oceanic and Atmospheric Administration) Technical Report NMFS SSRF-745, Washington, District of Columbia, USA.

Wathne, F. 1977. Performance of trawls used in resource assessment. *Marine Fisheries Review* 39 (6):16-23.

Waley, R. A., and M. Fowden. 1982. *Progress report for projects in the 1981 work schedule reservoir research.* Wyoming Game and Fish Department, Cheyenne, Wyoming.

Chapter 8
Electrofishing

JAMES B. REYNOLDS

8.1 INTRODUCTION

Electrofishing, in the strictest sense, is the use of electricity to capture fish. However, electricity has also been used to guide or block the movements of fish and to anesthetize or quickly kill them after capture. In fact, the earliest attempts to use electricity on fish, in the late 19th and early 20th centuries, were often aimed at manipulating fish movements and not at capturing fish. The serious development of electrofishing as a technique for fishery science began after World War II. Research in the 1940s and 1950s was aimed primarily at development of field equipment for capturing fish and studying the reaction of fish in controlled electrical fields. In the 1960s and 1970s, much attention was given to the effects of electrofishing on the physiology of fish and other aquatic organisms (Friedman 1974). In recent years, particularly the past decade, significant development and evaluation of field equipment and techniques for capture have occurred.

This chapter offers the reader a primer on electrofishing. It provides some general principles and guidelines for the understanding, construction, and safe, efficient use of electrical fishing devices. Although it does not give detail on procedures for a given situation, references are provided for additional direction. In writing this chapter, I have relied heavily on information from Aronson and Kocher (1970) on basic electricity, Novotny and Priegel (1974) on electrical fields and electrofishing equipment, Vibert (1967) on fish response to electroshock, and Simpson (1978) on factors affecting electrofishing efficiency. It is very important to talk to as many knowledgeable people as possible before beginning or modifying an electrofishing operation and investing much time and money. Experienced biologists, commercial suppliers, and electricians or electronic technicians represent an information source that can steer a project away from costly mistakes.

8.2 PRINCIPLES

8.2.1 Basic Electricity

Terms and relations. All matter consists of particles which attract and repel each other because of the positive or negative charges the particles bear. Electricity is the form of energy that results from this attraction or repelling of particles. A circuit is a closed path along which an electric charge moves. The rate of flow or intensity of the charge is the current and is measured in amperes. The electromotive force that moves the charge is voltage and is measured in volts. The size and quality of the circuit limits the current it can carry and is referred to as resistance measured in ohms. Electrical power is the rate at which electrical work is done and is measured in watts. One watt of power results when a current of one ampere flows through a resistance of one ohm

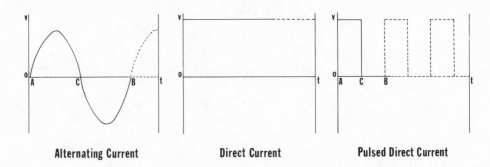

Figure 8.1 Characteristics of alternating, direct, and pulsed direct current. Solid lines represent one cycle, *A-B;* dashed lines are subsequent cycles. Duty cycle is *AC/AB.* Other letters designate voltage *(v)* and time *(t)*, both beginning at zero (adapted from Vibert (1967), with permission).

under the force of one volt. The flow of current in a circuit is like the flow of water in a pipe. The pressure (voltage) drives a flow (current) through the pipe (circuit). The amount of flow the pipe can handle depends on its size and material (resistance). As the flow reaches the end of the pipe, it releases energy to do work at some rate (power). The intensity of an electrical current is directly proportional to the electromotive force and inversely proportional to the resistance of the circuit (Ohm's Law):

$$\text{Current (amperes)} = \text{Voltage (volts)} / \text{Resistance (ohms)} \qquad (8.1)$$

Electrical currents. There are two basic types of electrical current (Fig. 8.1). Direct current (dc) flows only in one direction because the negative and positive ends (electrodes) of the circuit are always the same. Current leaves the negative electrode (cathode) and enters the positive electrode (anode). The storage battery is a common source of dc. Alternating current (ac) flows alternatingly in both directions because the anode and cathode switch positions, back and forth, on the two electrodes. In ac, voltage follows a sine wave, increasing from zero to a maximum and back to zero as current flows in one direction, and then repeats the same pattern in the opposite direction. A cycle of ac is one complete revolution of current that begins and ends at zero voltage and includes one flow in each direction. Frequency of ac is measured as the number of cycles per second or hertz (Hz). One hertz equals one cycle per second. The portable generator is a common source of ac, usually at frequencies of 60, 180, or 400 Hz.

Both ac and dc can be modified to produce various current shapes that have different effects on fish. Pulsed dc gives a unidirectional current with periodic interruptions that result in square waves (pulses) of voltage (Fig. 8.1). The ratio of time on to total time within one cycle (i.e., from the beginning of one pulse to the beginning of the next pulse) is called duty-cycle and often is expressed as a percentage; a 50% duty-cycle means that dc flows during half of each cycle.

A greater variety of modifications are possible with alternating current than with direct current. Half-wave rectified (pulsed) ac results when ac is passed through a rectifier, leaving a sequence of half-sine waves in the same direction which are separated by pauses of equal duration. Fully rectified ac is composed of an uninterrupted sequence of half-sine waves in the same direction. Smooth rectified current is dc derived from rectified and "filtered" ac; when insufficiently modified, the current has weak sinusoidal variations and is called ripple or undulating current. Other variations of ac are possible but not commonly used in electrofishing.

8.2.2   Electrical Fields and Fish

Current distribution. The current and voltage that leaves the circuit distributes itself about the electrode in a complex way; this distribution is critical to electrofishing effectiveness. The three parameters which applied to the entire circuit—voltage, current, and resistance—now differ at different points in the water. They must be expressed in terms of linear distance, using Ohm's Law (eq. 8.1), because the flow weakens as it gets farther from the electrode (Fig. 8.2):

$$\text{Current density (amperes/cm}^2) = \frac{\text{Voltage gradient (volts/cm)}}{\text{Resistivity (ohm-cm)}} \tag{8.2}$$

Current density is the current that flows through a 1-cm^2 plane. Voltage gradient is the change in voltage over a 1-cm distance. Resistivity is the resistance of 1 cm^3 of water and is equal to the inverse of water conductivity if the latter is measured as mho/cm^3. (Appropriately, mho is ohm spelled backwards.) Since conductivity is usually measured as micromhos/cm^3, its inverse value must be multiplied by 10^6 to calculate resistivity. Conductivity is a convenient measure of the quality of water for electrical transmission. Distilled water has very low conductivity (0.5 - 4.0 micromhos/cm). The conductivity of most freshwater bodies is between 50 and 1,500 micromhos/cm. On

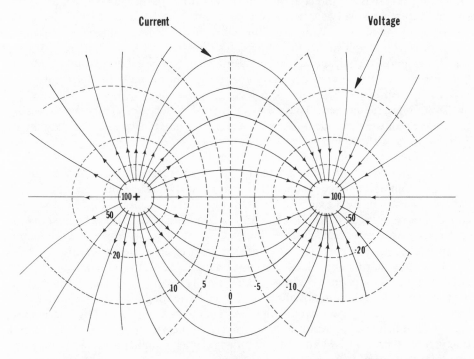

Figure 8.2   An electrical field produced by two electrodes. The solid lines connecting the two fields represent constant current. Dashed lines surrounding each electrode represent constant voltage. Voltage gradient and current density are highest near the electrodes and decrease rapidly with linear distance from the electrodes (adapted from Novotny and Priegel (1974), with permission).

the average, sea water is 500 times more conductive than freshwater.

Electroshock effectiveness. The results of controlled studies indicate that current density is the electrical field parameter most directly related to the effects of electricity on fish. Current density is highest near the electrode and decreases rapidly with linear distance from the electrode (Fig. 8.2). High current densities will kill fish, moderate densities will stun them, and low densities will allow them to escape. Effective electrofishing will maximize the "stun" field while minimizing the "kill" and "escape" fields. The actual values, or limits, of these fields will vary with the fish.

Because a fish has resistance, a given current density at one end will result in a lowered density at the other, producing a voltage gradient in the fish. Voltage gradients of 0.1 to 1.0 volts/cm are most effective for stunning fish; these gradients can be maintained in freshwaters of normal conductivity (100-500 micromhos/cm) by adjusting circuit voltage to produce a current of 3-6 amperes. At high conductivities, water becomes less resistive than fish and the current tends to flow around them, resulting in little or no voltage effect; this is the reason that electrofishing is not widely used in brackish or salt water. At low conductivities (less than 100 micromhos/cm), the water is more resistant than fish, but the electrical field is limited to the immediate area of the electrode. When this occurs, a fish may not be affected until it touches the electrode—then it suddenly receives a high voltage gradient and dies.

Electrofishing effectiveness is usually limited by inadequate current density, either because of too little voltage gradient, too much resistivity, or both (eq. 8.2). Decreasing resistivity by direct application of salt (ions) to the field has met with little success except in small, restricted waters (Lennon and Parker 1958). Increasing voltage gradient can be accomplished by increasing the circuit voltage or increasing the electrode size, which, in turn, increases circuit current or electrical "drain". Increasing voltage requires a large generator which may be expensive and unsafe for both fish and biologist; consider using a larger generator, but consider it carefully. Increasing electrode size is safer and less expensive, but this option is limited by available voltage and practical physical size. Increased electrode size means increasing the *diameter* of the electrode. Lowering an electrode into the water only creates more current drain without increasing voltage gradient horizontally. Larger electrodes result in larger stun fields and smaller kill fields (Fig. 8.3).

Electroshock effects. Fish react to electroshock in several ways, depending on the type of current and voltage gradient they encounter. Under the influence of ac or dc, fish in the "escape" field exhibit a "fright" response, and those in the "kill" field, of course, are electrocuted. In the "stun" field, however, fish exhibit a variety of responses. Any electrical current can cause electrotaxis (forced swimming), electrotetanus (muscle contractions), and electronarcosis (muscle relaxation) in fish.

In responding to ac, a fish tends to assume a position perpendicular to the electrical current to minimize voltage gradient in the body. It may undulate in rhythm with the ac cycle, exhibiting oscillotaxis (forced swimming without orientation). At higher current, tetany occurs, and the fish is immobilized. Unmodified ac is most damaging to fish and can result in hemorrhaged tissue, ruptured swim bladders, and fractured vertebrae. Pulsed and fully-rectified ac ellicit a similar response but are not as potentially harmful as unmodified ac.

In a dc field, a fish typically turns toward the anode and exhibits galvanotaxis (forced swimming with orientation) followed by galvanonarcosis (muscle relaxation caused by dc). As a result, fish tend to move toward the anode where they roll over and are easily captured. This can be helpful in turbid or turbulent water because fish may

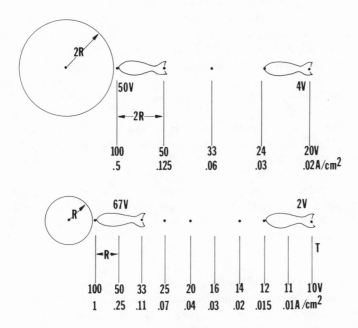

Figure 8.3 Advantage of larger electrode size on effectiveness of an electrical field. In this example, 200 volts *(V)* are applied to two electrodes, one twice the radius *(R)* of the other. At any given distance, the larger electrode produces twice the voltage and current density *(A/cm²)* as the smaller one, maximizing the "stun" field. At electrode surface, current density of the larger electrode is half that of the smaller, minimizing the "kill" field (adapted from Novotny and Priegel (1974), with permission).

be "pulled" to the surface near the anode. Pulsed dc is better than unmodified dc because it sustains forced swimming with less damage to the fish. In addition, pulsed dc requires less voltage than does ac and thereby requires a smaller, less expensive electrical source. Whether pulsed or not, dc tends to do less damage to fish than ac.

8.3 EQUIPMENT

8.3.1 Electrofishing Boats

Mechanical system. Flat-bottom boats at least 4 m long and 1.5 m wide that are made of heavy-gauge aluminum have proven effective and safe for electrofishing (Fig. 8.4). Grounding of electrical equipment is easily accomplished by direct attachment to the hull. Wood or fiberglass boats do not provide grounding and should not be used. Outboard motors of moderate horsepower with a trolling speed are best for electrofishing boats. Seats should not be removed because they contain flotation material and provide rigidity to the boat. The boat should have a forward deck large enough to accommodate two standing adults as dip-netters; its minimum load rating should be 300 kg.

One or two booms, extending forward from the bow for electrode support, should be made of sturdy, non-conductive material—fiberglass is best, but wood is an adequate, cheaper alternative. The booms should be retractable or removable. They should be easy to repair or replace because they occasionally break during use.

Electrical system. A portable generator is the best power source; it consists of an alternator to produce ac and a gasoline-powered engine to drive the alternator. A three-phase, 230-volt generator with 3500-watt capacity gives the greatest flexibility in electrofishing. Although generators with a 60-Hz frequency are commonly used, higher frequencies, such as 180 or 400 Hz, are also effective. The higher the frequency, the smaller and lighter the generator. New generators normally have the neutral or ground wire of the 230-volt outlet attached to the generator frame; this must be disconnected to avoid the possibility of a lethal shock when touching the frame of a running generator. The frame should then be connected to the metal hull to eliminate shock hazard from the generator. It is important to buy only generators in which the circuit ground wire can be easily disconnected. A three-phase generator can be damaged by unequal loading, that is, unequal current in the three circuits. This can result from unequal electrodes, blown fuses, or broken wires; such conditions should be remedied to avoid costly repairs. Any generator used must have a fuse or circuit breaker, preferably the latter, to protect it when overloaded.

A good quality of stranded, insulated, gasoline-resistant wire rated for the maximum current must be used; number 10, 12, or 14 wire is desirable for electrofishing boats. The choice of wire depends upon the generator size. All wiring should be encased in metal conduit which is grounded to the hull. Wire connections must never be made by splicing and taping. Where connections are necessary, use a plastic wire nut with a rating greater than or equal to the wire rating and protect it in a metal, watertight junction box grounded to the hull. Switches should be weatherproof

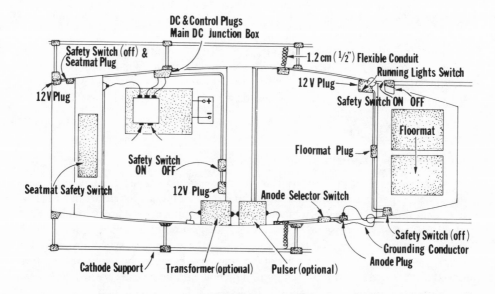

Figure 8.4 Configuration of a boat modified for electrofishing with alternating, direct, or pulsed direct current (adapted from Novotny and Priegel (1974), with permission).

and operate on low voltage circuits (less than 24 volts) which are carried in a conduit separate from that carrying high voltage.

Operation of lights, aeration units, and other electrical accessories should be done with a 12-volt dc system and not with the electrofishing circuit. A 12-volt, "deep charge" battery (aviation or marine type) connected through a separate circuit to watertight outlets will accommodate these accessories. The battery can be charged continuously through a battery charger from a 110-volt ac outlet on the generator during electrofishing. Of course, the more accessories powered on the 12-volt dc circuit, the less voltage available for the electrofishing circuit. When the generator is shut down and the bow lights turned off, the battery can operate essential accessories, such as work lights and aerators, for periods sufficient to meet most fish handling needs. Bow lights should be mounted well above water line for reduced construction costs, easy replacement, and safety.

If modifications of the generator-produced ac are needed for electrofishing, a control box between the power source and electrodes is necessary. These units can be made locally if experts are available or purchased from commercial suppliers. The unit should be sturdy and weatherproof for field use and should provide ac or dc, pulsed and unpulsed, for the electrofishing circuit. Output voltage selections should be possible, as well as frequency and duty cycle selections for pulsed dc. The unit should be protected by both manual and automatic circuit breakers. Meters to monitor all voltages and output currents are essential.

Electrode system. In addition to providing an effective electrical field for fish capture, an electrode system should minimize regions of high voltage gradients to avoid power wastage and fish damage, be adjustable to changes in water conductivity, be capable of negotiating weeds and obstructions, be easy to assemble and disassemble, and minimize disturbance of the water for better viewing. Cylinder and ring electrodes best meet these requirements, particularly the latter. A cylinder electrode is pipe-shaped, usually about 1 m long. A ring electrode consists of about 12 "dropper" electrodes suspended at equal intervals from a horizontal aluminum ring about 1 m in diameter. Each "dropper" electrode is a 15-cm long stainless steel tube attached to a length of insulated wire 30-45 cm long (Novotny and Priegel 1974). Electrodes of other shapes may be easy to use, but they do not produce as effective an electrical field as do ring and cylinder electrodes. When using ac, allow electrodes to extend no more than 1-2 m into the water; given a normal generator, they are not effective at greater depths. When using dc, keep anodes within 0.5 m of the surface to take advantage of galvanotaxis. Best electrode materials are stainless steel, flexible conduit, and copper tubing, in that order. Diameters of individual electrodes usually should be 0.5-2.5 cm. Adjustment for varying conductivities may be accomplished by interchanging electrodes—smaller diameter electrodes in more conductive water—or by exposing more or less of each electrode with a movable rubber sleeve.

In an ac operation, each electrode produces its own field. A three-phase generator needs three electrodes to avoid unequal loading, but two electrodes will not cause excessive imbalance. Current density can be increased in waters of low conductivity by either increasing the number of electrodes or their size; increasing their size is preferable because lower, less harmful voltage gradients are produced near the electrodes, compared to the voltage gradients caused by more electrodes of a small diameter.

In a dc operation, the anodes are in front of the boat, and the cathodes along either side. The distance between cathode and anode should be as great as the combined length of boat and booms will permit. For safety, the boat hull should never be used as a cathode. Anodes and cathodes must have similar resistances to avoid power wastage at the cathodes; if anode size is increased, cathode size should be increased accordingly. Cathodes should be well spaced to distribute the current and reduce resistance. Flexible conduit makes excellent cathodes for electrofishing boats. Anodes should affect a large area near the water surface to enhance fish capture; ring electrodes do this best.

Safety system. The boat should have ample room for safe movement from bow to stern. Passageways must be skid proof. Railings, insulated and padded, must surround the entire bow area to protect dip-netters. Side railings are a very desirable feature. Positively-activated foot switches should be placed in the bow for dip-netter(s) and in the stern for the boat operator. The boat operator should also be able to operate the electrofishing unit from a seated position while guiding the boat. The face of the control box should be lit during operation in darkness. All hazardous surfaces, electrical or otherwise, should be clearly marked in words or color codes (e.g., red-danger, yellow-caution). Generator exhaust should be piped to the side of the boat; hot exhaust pipes must be encased in screening or insulation. Generators should be loosely hooded, if at all—they will overheat without ample air circulation, and explosion of accumulated fumes may result. Batteries should be held in an acid-proof, non-metallic container and placed on the deck in a low traffic area away from the generator.

8.3.2 Backpack Shockers

Mechanical system. Except for leads and electrodes, the entire backpack shocker unit should be housed in a weatherproof metallic container that is fastened securely to a comfortable pack frame (Fig. 8.5). Backpack shockers that are safe and effective may be made locally with expert direction or purchased from commercial suppliers.

Electrical system. The power source is either a 12-volt, deep charge battery or a 115-volt ac generator. If a battery is used, the unit should allow selection of output voltage, frequency, and duty-cycle. Lightweight generators may be carried on a pack frame or used as a remote power source through additional circuitry in the backpack unit. (Personally, I think it is too dangerous to carry a generator on my back and avoid doing so.) Number 18 wire is best for most backpack units.

Electrode system. The electrodes are hand-held and must be insulated from the operator by handles 1.5-2.0 m long, preferably made of fiberglass. A horizontal ring electrode attached to the end of the handle is easiest and most effective to use. A spatula electrode, made of metal mesh at the end of an insulated handle, is also effective but uses more voltage.

Safety system. Positively activated switches on each electrode handle are essential safety features. The power source should have both automatic and manual circuit breakers. Backpack generators, like those in boats, must be modified to eliminate the grounding to the generator frame.

Alternative method. Shore-based electrofishing is frequently used in larger streams and rivers where a backpack shocker produces a field too small and weak to be effective. The main difference between this method and a backpack operation with a remote power source is that, in the latter, one person carries about current control equipment in the stream; in shore-based electrofishing all equipment except long leads and electrodes are on shore, or less often, in a towed raft. Large generators are used for

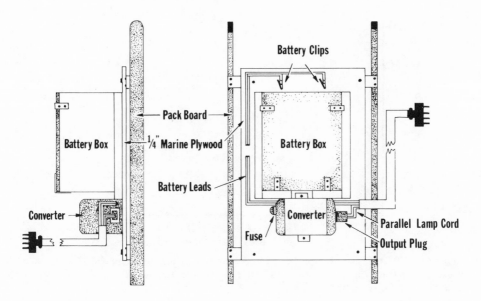

Figure 8.5 Configuration of a simplified backpack electroshocker (adapted from Braem and Ebel (1961), with permission).

more effective electrofishing, but this results in additional hazards (*see* section 8.5.1). In ac operations, both electrodes are hand-held within 2-3 m of each other. In dc operations, both anode and cathode can be hand-held, or the cathode can be placed as a metal grid on the stream bottom, usually near the generator, and one or more anodes can be hand-held for fishing in a wide area; this, too, presents a hazard (*see* section 8.5.1).

8.4 EFFICIENCY

Many factors influence electrofishing efficiency. For most factors, efficiency is highest in a given range and drops off at extreme values of the variable. If one understands these effects, data from electrofishing samples can be useful; one can either accept the biases that exist and state their extent and direction or reduce the biases by standardizing sampling conditions. Each determinant of electrofishing efficiency can be placed in one of three categories: fish characteristics, habitat characteristics, and operating conditions; many are inter-related and their effects are difficult to isolate.

8.4.1 Fish Characteristics

Vulnerability to electrofishing varies among species due to innate differences in anatomy and behavior. Bony fishes conduct current more readily than do cartilaginous fishes. Among the bony freshwater fishes, some species vary widely in their tissue resistance (Edwards and Higgins 1973). Because electrofishing is a sampling technique for shallow waters, habitat preference among species will affect their vulnerability (Larimore 1961). Species which inhabit shoreline habitat, such as centrarchids, cyprinids, and percids, are generally more vulnerable than pelagic or benthic species, such as clupeids, osmerids, and ictalurids.

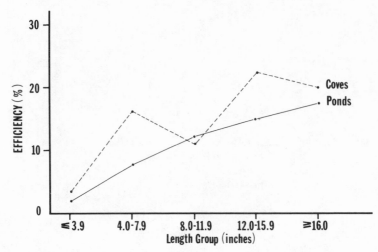

Figure 8.6 Effect of size of largemouth bass on electrofishing efficiency in 35 ponds and 16 reservoir coves in the midwestern United States (adapted from Simpson (1978), with permission).

Fish size is an important determinant of electrofishing efficiency (Fig. 8.6). At a given voltage gradient, total body voltage increases with length, resulting in greater electroshock to larger fish. Also, large fish are more visible to dip-netters than small ones and may be inadvertently selected for. Size selectivity of electrofishing has been amply demonstrated in the field (Sullivan 1956; Junge and Libosvarsky 1965; Reynolds and Simpson 1978).

Other fish characteristics are more subtle and have not been adequately documented. Predators and spawners may be more vulnerable than other fish to electrofishing, not only because of their large size, but also because their territorial behavior makes them less likely to avoid an oncoming electrofishing operation. On the other hand, schooling fish may more readily escape capture by group fright response. Cross and Stott (1975) demonstrated that repeated electrofishing at intervals of two hours reduced the catchability of fish; at intervals of one day or more this did not happen. They felt that previously shocked fish had not learned avoidance but, rather, were often too weak to be drawn to the anode another time. Finally, the abundance of fish may affect electrofishing efficiency. In a study of factors affecting electrofishing efficiency, Simpson (1978) found that population density of largemouth bass was inversely related to efficiency; he postulated that dense populations may result in offshore distributions, group fright response, or less effective dip-netting, and, thus, reduced efficiency.

8.4.2 Habitat Characteristics

Water temperature affects efficiency both directly and indirectly. High temperature increases fish metabolism, resulting in an increased ability to perceive and escape an electrical field; it also increases conductivity which is advantageous in waters of low conductivity. Low temperature decreases flotation of stunned fish, making capture less likely. Indirectly, temperature affects efficiency by changing fish distribution. Adults typically move shoreward or upstream to spawn in conjunction with a temperature increase; however, very high temperatures force fish into deeper water, away from electrofishing areas.

Water transparency, like temperature, has a dome-shaped relation to electrofishing efficiency. High transparency may shift fish distribution toward deeper water and may allow fish to see the boat and avoid capture, but low transparency also reduces capture rate by making stunned fish less visible to dip-netters. Kirkland (1962) observed an increase in catch rate of largemouth and spotted bass when normally clear water became turbid; he concluded that the bass were less able to detect and avoid capture.

Dissolved oxygen concentration in the hypolimnion of eutrophic waters can be very low in late summer. When this occurred in small Midwest impoundments, Simpson (1978) found that electrofishing efficiency for largemouth bass and bluegill increased. He concluded that the fish had moved shoreward out of deeper, oxygen-poor water.

Morphometry of the water body, including surface area, shoreline slope, and shoreline development, has important influences on electrofishing. Fish are less vulnerable to capture in larger water bodies than in smaller ones, perhaps because a greater proportion of the latter is suitable for electrofishing. Fish are also more vulnerable in lakes and impoundments with high shoreline development. Very steep shoreline slopes make electrofishing less effective because of limited habitat for sampling.

Substrate affects electrofishing efficiency. Bottoms of fine particles and organic debris are more conductive than coarse bottoms with little organic content. Thus, mud and silt will reduce the current density of an electrical field and will reduce electroshock efficiency more than gravel and rubble reduce it.

Cover, such as submerged brush, trees, and rooted macrophytes, has a concentrating effect on fish distribution. Its influence on electrofishing is difficult to quantify because cover itself is difficult to describe accurately. Electrofishing in experimental ponds containing known numbers of largemouth bass was always more effective in sections containing artificial structure than in sections without such structure (Simpson 1978).

8.4.3 Operating Conditions

Although most of the operating conditions that exist while electrofishing (for example, electrofishing equipment) can be controlled, weather cannot. Rain and wind have the most immediate effects on electrofishing by disrupting surface visibility for dip-netters, but they are less important on small water bodies. Springtime frontal systems may cause adult fish tending nests to abandon temporarily their role as parents (Summerfelt 1975); such cool fronts may cut catch rates by half during electrofishing operations on large reservoirs (Simpson 1978).

The decision to electrofish at night or during the day may be very important. At night, larger, predatory fish are inshore feeding, and all fish seem less apt to avoid capture. As a result, several studies have shown that electrofishing at night catches more species, larger individuals, and more of them than electrofishing during the day (Loeb 1958; Witt and Campbell 1959; Sanderson 1960; Kirkland 1962). Sampling dates can also be controlled; they should be selected so as to increase electrofishing efficiency, particularly as efficiency relates to temperature and seasonal events such as spawning.

The organizational aspects of an electrofishing operation will influence its efficiency. The number of crew members, their experience, and the proven reliability of their equipment and procedure can make the difference between success and failure;

only trial and error over time can provide solutions to these problems of electrofishing. Well-defined sampling objectives are essential to operational effectiveness. If survey and inventory are needed, all stunned fish should be collected. If, however, single stock assessment has priority, capturing only the appropriate fish will improve efficiency— in time and money, if not in catch rate.

8.5 PROCEDURES

8.5.1 Operational Guidelines

Safety aspects. Electrofishing is hazardous work. The batteries and generators used in electrofishing provide more than enough current to electrocute a person. It is the current that passes through the body and vital organs that is particularly dangerous. Currents of 20-500 Hz and as low as 0.0002 amperes can cause serious injury or death. Most people who grasp a wire carrying as little as 0.01-0.02 amperes cannot let go because of tetany. Death is usually a result of respiratory arrest, asphyxia (caused by contracted chest muscles), or ventricular fibrillation (uncontrolled, irregular heart beats). Cardiopulmonary resuscitation (CPR) can restore breathing, but for ventricular fibrillation, CPR is only a stalling tactic—a defibrillator must be used to restore normal heart rhythm.

All members of an electrofishing crew should understand the system they are using and the risks involved. Before a field operation, new crew members should receive orientation on equipment and procedure; a practice run with generator on and electrodes off can improve efficiency and safety. At least two and preferably all crew members should have CPR and first aid training. An electrofishing system should receive a thorough safety inspection before a field operation.

Lightweight, Coast Guard-approved life vests and hip boots or waders should be worn by all personnel in or on the water. All types of electrofishing operations need a large first aid kit and fire extinguisher (ABC type), preferably placed near the generator. Gasoline should be stored in spill-proof, vented containers of approved design and not exceeding 19-liter capacity. For people in electrofishing boats, continuous generator noise at close range may cause eardrum damage. Ear plugs or mufflers are effective protection against noise. Another solution is the battery-powered, communication headset; this device is easy and inexpensive to install, cuts out operational noise, and gives clear voice contact between dip-netter and pilot. Voice-activated headsets, without batteries, are available but are more expensive.

As electrofishing is about to begin, several rules should be observed. NEVER ELECTROFISH ALONE. Conversely, avoid excessively large crews; they cause confusion and lead to mistakes. Every crew should have at least one biologist experienced at electrofishing who is in charge during an operation; that person should also have control of the power source. The person in charge should make all initial settings on the equipment, start the generator if one is used, make sure everyone is ready and in place, and then start with a pre-arranged signal.

While under way, some other rules are important. Avoid operating near other people, pets, or livestock that are in or on the water or shore. Although light rain is not a serious problem, thundershowers or rough water are; operations should cease when the person in charge says so. Electrofishing should proceed slowly and deliberately— avoid fish-chasing and other acrobatics. Resist the urge to hand-capture a stunned fish; a missed fish is better than electroshock. Always shut down the electrical field when equipment changes, repairs, or refueling become necessary. Rest often enough to avoid fatigue.

Finally, the members of an electrofishing crew must stay close together during an operation. With boat and backpack shockers this is easy to do because the power source moves with the crew; this is not the case in shore-based operations. Shore-based operations are dangerous because they combine the riskiest elements of the other methods: high voltages of boat shockers and instream work of backpack shockers. As crews get farther from the power source, more spread apart, or both, the probability increases of someone falling unseen; sustained electroshock is likely if that person has no safety switch (e.g., a dip-netter). The safest procedure is to use the "buddy system" all the time that a shore-based operation is in effect.

Technical aspects. Certain accessories enhance sampling efficiency. Dip nets with long, fiberglass handles are essential to any electrofishing operation; carry extras. They should not serve as electrodes because they are constantly being pulled out of the water and waved about. Do not use excessively deep net bags because they splash water onto electrical and mechanical surfaces and increase the hazard. A large holding tank should be placed within reach behind the dip-netters with no equipment between them and the tank. The tank should be equipped with an aerator (closed system) or circulating pump (open system) to oxygenate the water. There should be enough space on either side of the tank for safe passage. Small stock-watering tanks with corrugated walls are readily available, inexpensive, sturdy, and of sufficient capacity. In backpack or shore-based operations, a holding tank must be kept near the dip-netters on a towed raft. Wire baskets, made of framed hardware cloth, make excellent in-stream live cars; they permit water flow while shielding the fish they contain from electroshock. A 12-volt dc searchlight permits the boat operator to see fish missed by the dip-netters at night. He may either make note of them or retrieve them with a short-handled net; if the latter is done, an additional holding tank or small tub must be within reach. The search light also has obvious safety advantages not inherent in fixed working lights. A conductivity meter is essential, if specific conductance of the water is unknown, in order to select the electrodes, initial voltage, and current type that are most effective. The meter should be checked periodically for accuracy.

Before an operation begins, the amount of effort either by distance, time, or desired sample size should be at least tentatively agreed upon. Duration of electrofishing time is generally a reliable index of effort; it can be measured either as the difference between time of start and stop or the amount of time that the electrodes are energized as indicated by a circuit clock. Hours or minutes are the best units for effort by time. Distance of shoreline sampled is also a good effort index, but this requires shoreline measurements for each sample unless permanent stations are repeatedly used. The best unit for effort by distance is a minimum that is always sampled, such as 100 m or 1 km. A useful variation of the distance index is one complete lap of the shoreline of a small lake or blocked cove of a lake (Simpson 1978); effort is roughly proportional to surface area of the water body sampled. In the early development of a sampling program, both distance and time for each sample should be measured. The variable most predictive of catch, both in terms of linearity and confidence limits, is the one to use consistently as an effort index. With either measure of effort, it is preferable to report catch-per-unit-effort by interpolation, not extrapolation (e.g., if 15 fish are captured in 0.25 hour and 500 m, express adjusted catch as 1 fish/minute or 3 fish/100 m and not as 60 fish/hour or 30 fish/km).

Proper care of fish after capture will reduce handling mortality (*see* chapter 5). Fish held in a dip net while others are gathered will be subjected to prolonged electroshock. Likewise, the longer fish are held in the tank with other fish, the more likely that stress-induced mortality will occur; the warmer the water, the greater the likelihood of mortality. Therefore, fish should be processed at frequent intervals. One way is to stop often enough to handle and release fish before they weaken; a collapsible pen attached to the side of the boat allows a final check on recovery of fish before their release. Another way is to process fish while under way. The processor sits amidship within reach of the tank; a hooded light and foot- or voice-activated tape recorder with microphone permits the measuring, weighing, and marking of fish as they are caught. If the catch rate is more than the processor can handle, stops become necessary. Fish are released to the side and rear; if electroshock is not severe, they recover quickly and seldom swim into the electrical field again. By processing fish while under way, sample sizes can be reduced through sequential sampling (*see* section 8.5.3).

8.5.2 Sampling Guidelines

Like any other sampling method, electrofishing is best deployed with consistency and objectivity. Unless the sampling objective is species or stock-specific, dip-netters should collect all individual fish possible; otherwise selectivity may occur while dip-netting. These guidelines, therefore, are starting points only; they are based on the assumption that all stunned fish, irrespective of species and size, are to be collected.

The generator should be set for just enough voltage to obtain the desired current, generally 3-6 amperes. Lower amperage may still electroshock fish but is selective for larger ones. Higher amperage will often injure fish, particularly if ac is used. If higher amperage is necessary, use pulsed dc which has greater range than unmodified dc.

Low pulse rate (5-40 pulses per second) seems effective for spiny-rayed fish such as walleye, white bass, yellow perch, and bluegill (Novotny and Priegel 1974); very low pulse rate (3-5) is effective for large catfish. Higher pulse rates (40-120) have a greater effect on young fish and soft-rayed fishes, e.g., trout, salmon, carp, bullheads, and minnows. Higher pulse rates are also necessary for strong swimmers such as adult salmon, pike, and bass. Duty cycle does not seem as important as pulse rate; 25% is as effective as 50% and conserves power, but 10% is not effective (Novotny and Priegel 1974).

Standing waters. Electrofishing is best done from boats in lakes and impoundments. Spring, while adults and young are mixed in the warmer inshore waters, is the best season for sampling; autumn may be the next best time because adult fish move inshore as the lake's waters cool. Electrofishing is most productive at night or twilight, particularly just after sunset as predators move inshore to feed. One lap of the shoreline of a small lake or blocked cove is the most complete unit; record both distance and time of the lap. In large lakes, or when a full lap is not possible, sample a fixed length of shoreline and record the time. In waters where Secchi disk readings are generally 0.1-1.0 m, use either ac or pulsed dc, whichever works best; where readings are generally less than 0.1 m or more than 1.0 m, pulsed dc is usually more effective than ac. To ensure consistency, avoid changing current type after it is selected. A slow, deliberate capture style is not only safer than fishchasing, but it also yields more representative samples. Catches can be increased in areas of submerged cover by moving in with power on but circuit off and then energizing electrodes for an element of surprise; or the cover can be surrounded with a block net and then sampled inside the net. In areas with extensive cover, use pulsed dc if possible.

Flowing waters. Boat shockers can be used in large streams, backpack shockers in small ones, and shore-based units in either. During night operations, use a boat or restrict wading to shallow, low flow areas. Avoid electrofishing during high water. For safety, effectiveness, and consistency, restrict sampling to normal or low water stages. Adults often can be sampled effectively in spring, soon after cessation of high waters, while spawning in shallow mainstream areas or in tributaries. In boat shockers, float downstream with the current, using the motor mainly for maneuvering near shore and in rapid water; this keeps the boat next to the stunned fish. When wading, upstream movement is preferable for sampling, with dip-netters behind or beside electrode handlers to minimize fish avoidance. As in lakes, sample stream cover methodically, using pulsed dc if possible. One side of a stream should be electrofished for a distance sufficient to ensure that all habitats associated with the inside and outside of bends are sampled. Record effort as time fished over a fixed distance.

8.5.3 Data Analysis

Species detection. For a given type of electrofishing operation (and other gear types, for that matter), species detection is a useful statistic for deciding whether to use electrofishing in surveys and inventories. Species detection is the percentage of sampled water bodies known to contain a given species in which at least one specimen was captured. Generally, if detection success is high for a given species (75-100%), the sampling method is useful for population assessments wherever the species occurs. Species detection of centrarchids by boat electrofishing was high in small Midwest impoundments (Reynolds and Simpson 1978). However, because of the aforementioned species selectivity, electrofishing samples alone should not be used to assess fish community structure.

Population abundance. Catch-per-unit-effort by electrofishing is a useful, easily obtained index to the abundance of the populations of many fish species. This relative measure is more useful when applied to adult fish because of size selectivity, but trends in the abundance of young fish can be monitored if electrofishing equipment and procedures are evaluated with this objective in mind.

Estimates of absolute abundance based on electrofishing samples may be seriously biased, but it is often difficult to ascertain in which direction the bias occurs. Mark-recapture estimates may be too high because marked (i.e., previously electroshocked) fish are more vulnerable than unmarked ones (Grinstead and Wright 1973) or the reverse may be true (Cross and Stott 1975). The practice of "importing" marked fish into a blocked cove from other areas of a reservoir is also questionable; for example, Simpson (1978) found that electrofishing efficiency, on the average, was about 18% for these fish, compared with 12% for marked residents of a cove (*see* also chapter 11). It is probably better to capture fish by another method for marking and release and then to use electrofishing for a single census estimate; multiple census estimates by electrofishing at frequent intervals (e.g., every few hours) are probably quite biased (Cross and Stott 1975).

Catch depletion estimates may be less seriously affected by electrofishing samples because recaptured fish are not used in the estimate. Cross and Stott (1975) recommended elimination of the first catch as a data point and the use of subsequent catches to estimate catchability by regression; the slope of the regression is then used to draw a line from the first data point to obtain population size by extrapolation.

Population structure. Because of size selectivity, length-frequency data from electrofishing samples should be regarded with caution. This is especially true of the data regarding the relative abundance of small fish; it is probably biased too low. Length-frequencies of the stock-size fish (harvestable fish) based on electrofishing data, however, are reasonably accurate and can provide the basis for indices of stock structure (*see* chapter 15).

The fish-by-fish nature of capture in electrofishing makes the method suitable for sequential sampling (Weithman et al. 1980). Measurement of stock-size fish continues while electrofishing proceeds; when a category or index of stock structure can be assigned at an acceptable level of confidence, sampling stops. In this way, sample size can often be reduced by almost 50%. If electrofishing continues until 100 stock-size fish have been measured and no decision about stock structure can be reached, sampling stops and an endpoint estimate is made using the binomial distribution.

Population dynamics. The size-selectivity of electrofishing is the greatest hinderance to its use for estimating the dynamic rates of a population—reproduction, growth, and mortality. Rates of reproduction or recruitment are severely compromised because sampling efficiency is low for juvenile, particularly age-0, fish. Estimates of growth rate are less suspect, but they may be biased too high if electrofishing tends to take larger individuals of a year class, resulting in samples of faster-growing fish. Mortality rates based either on catch curves or mark-recapture data, may be biased; in the former, smaller, younger fish will be disproportionately low in electrofishing samples, causing an underestimate of mortality. Recapture data can give biases in either direction, depending upon whether marked (previously shocked) fish are more or less vulnerable to electrofishing than unmarked ones. Studies of population dynamics should not depend on electrofishing as the sole source of data.

8.6 REFERENCES

Aronson, M. H., and C. W. Kocher. 1970. *Voltage/current relations.* Measurements and Data Home Study Course 24. (Available from Measurements and Data Corporation, 2994 West Liberty Avenue, Pittsburgh, Pennsylvania 15216).

Braem, R. A., and W. J. Ebel. 1961. A back-pack shocker for collecting lamprey ammocoetes. *Progressive Fish-Culturist* 23:87-91.

Cross, D. G., and B. Stott. 1975. The effect of electrical fishing on the subsequent capture of fish. *Journal of Fish Biology* 7:349-357.

Edwards, J. L., and J. D. Higgins. 1973. *The effects of electric currents on fish.* Georgia Institute of Technology, Engineering Experiment Station, Final Technical Report, Projects B-397, B-400 and E-200-301, Atlanta, Georgia, USA.

Friedman, R. 1974. *Electrofishing for population sampling: A selected bibliography.* United States Department of the Interior, Office of Library Services, Bibliography Series 31.

Grinstead, B. G., and G. L. Wright. 1973. Estimation of black bass, *Micropterus* spp., population in Eufaula Reservoir, Oklahoma, with discussion of techniques. *Proceedings of the Oklahoma Academy of Science* 53:48-52.

Junge, C. O., and J. Libosvarsky. 1965. Effects of size selectivity on population estimates based on successive removals with electrical fishing gear. *Zoologicke Listy* 14:171-178.

Kirkland, L. 1962. A tagging experiment on spotted and largemouth bass using an electric shocker and the Petersen disc tag. *Proceedings of the Southeastern Association of Game and Fish Commissioners* 16:424-432.

Larimore, R. W. 1961. Fish population and electrofishing success in a warm-water stream. *The Journal of Wildlife Management* 25:1-12.

Lennon, R. E., and P. S. Parker. 1958. *Applications of salt in electrofishing.* United States Fish and Wildlife Service, Special Scientific Report - Fisheries 280.

Loeb, H. A. 1958. A comparison of estimates of fish populations in lakes. *New York Fish and Game Journal* 5:66-76.

Novotny, D. W., and G. R. Priegel. 1974. *Electrofishing boats: Improved designs and operational guidelines to increase the effectiveness of boom shockers.* Wisconsin Department of Natural Resources, Technical Bulletin 73.

Reynolds, J. B., and D. E. Simpson. 1978. Evaluation of fish sampling methods and rotenone census. Page 11-24 *in* G. D. Novinger and J. G. Dillard, editors. *New approaches to the management of small impoundments.* North Central Division, American Fisheries Society, Special Publication 5.

Sanderson, A. E. 1960. Results of sampling the fish population of an 88-acre pond by electrical, chemical and mechanical methods. *Proceedings of the Southeastern Association of Game and Fish Commissioners* 14:185-198.

Simpson, D. E. 1978. *Evaluation of electrofishing efficiency for largemouth bass and bluegill populations.* Master's thesis. University of Missouri, Columbia, Missouri, USA.

Sullivan, C. 1956. The importance of size grouping in population estimates employing electric shockers. *Progressive Fish-Culturist* 18:188-190.

Summerfelt, R. C. 1975. Relationship between weather and year-class strength of largemouth bass. Pages 166-174 *in* H. Clepper, editor. *Black bass biology and management.* Sport Fishing Institute, Washington, District of Columbia, USA.

Vibert, R., editor. 1967. *Fishing with electricity: Its application to biology and management.* Fishing News Books Limited, Surrey, England.

Weithman, A. S., J. B. Reynolds, and D. E. Simpson. 1980. Assessment of structure of largemouth bass stocks by sequential sampling. *Proceedings of the Annual Conference of the Southeastern Association of Fish and Wildlife Agencies* 33:415-424.

Witt, A., Jr., and R. S. Campbell. 1959. Refinements of equipment and procedures in electrofishing. *Transactions of the American Fisheries Society* 88:33-35.

Chapter 9
Fish Eggs and Larvae

DARREL E. SNYDER

9.1 INTRODUCTION

Interest in the study of fish eggs (embryos) and larvae in North America has grown substantially since the mid-1960s, largely in response to increased public concern for the environment and the need to better assess and monitor fishery resources. Such studies are utilized to appraise and monitor fish populations; to predict and monitor environmental impacts; to develop culture techniques; to increase our knowledge of fish reproduction, ontogeny, and early life history; and to increase taxonomic and systematic knowledge.

Collections of early life history stages of fish are frequently used to identify nursery areas and delimit spawning grounds and seasons. Data on the distribution and abundance of fish eggs and larvae are used to evaluate the current status, document year-by-year fluctuations, and detect long-term trends of a particular community, population, or stock. These data also are used sometimes to differentiate various stocks of the same species. Unfortunately, successful prediction of the future strength of cohorts based on the abundance of eggs and larvae is still difficult, if not untenable. Such predictions depend on an understanding of the various factors affecting early survival, the ability to predict or determine seasonal and year-by-year variations in those factors, and the ability to estimate mortality of larvae. All of these requirements are problematic (Smith 1981).

The ecological roles and habitat requirements of fish eggs, larvae, and early juveniles can be quite distinct from those of later juveniles and adults, and they usually change significantly as fish grow and develop. *See* Box 9.1 for commonly used subdivisions of the larval period. Accordingly, information on the habits and habitat requirements of early life history stages can be critical not only to effective management of fishery resources, but also to evaluation of the impacts of man's activities on specific fish populations or entire ecosystems. Entrainment of fish eggs and larvae through the water intakes of power and manufacturing plants and of agricultural and domestic water supplies is particularly important. The early stages of fish are also generally more sensitive to environmental changes than older fish and can serve as useful biological indicators of pollution. Indeed, fish eggs, larvae, and early juveniles are often used in toxicity tests and related studies.

This chapter is an introduction to collecting and processing fish eggs and larvae. More specific, marine-oriented recommendations are provided by Smith and Richardson (1977). While no comparable manual exists for nearshore and inland waters, Nellen and Schnack (1975) and Bagenal and Nellen(1980) discuss sampling early life history stages in such waters, and Bowles et al. (1978) describe

Box 9.1.   Terminology for Phases of Larval Fish Development

There are three commonly accepted terminologies for phases of the larval period. *See* Snyder (1976a) for a discussion of these and others.

(1)   HARDY ET AL. (1978): This system is based on traditional terms and criteria dating back to at least the early 1950s and is essentially a modification of terms given by Mansueti and Hardy (1967). The primary function is to distinguish between endogenously and exogenously feeding larvae based on the presence of yolk material.

Yolk-sac larva - Phase between hatching and absorption of yolk.

Larva - Phase between absorption of yolk and acquisition of minimum complement of adult fin rays.

Prejuvenile - Phase between acquisition of minimum adult fin ray complement and assumption of adult body form; used only where strikingly different from juvenile (cf. Hubbs 1958; *Tholichthys* stages of butterfly fishes, querimana stages of mullets, etc.). (Sometimes referred to as the *Transitional* phase).

(2)   AHLSTROM ET AL. (1976): This system is based on the morphogenesis of the homocercal tail. The primary functions are to facilitate the presentation and consideration of morphometric data based on standard and notochord lengths and the preparation of comparative descriptions.

Preflexion larva - Phase between hatching and the upward flexing of the tip of the notochord (or the appearance of the first caudal rays).

Flexion larva - Phase during upward flexion of the tip of the notochord (usually considered to terminate with formation of all principal caudal rays and the appearance of the first dorsal secondary or procurrent caudal rays).

Postflexion larva - Phase following upward flexion of the tip of the notochord (and ending with complete formation of rays in all fins and sometimes other criteria depending on the species group; a *transitional* or *prejuvenile* phase follows for some species).

(3)   SNYDER (1976a, 1981). This system is based largely on the morphogenesis of the median finfold and fins. The primary functions are to facilitate comparative descriptions of fish larvae and the production of keys based on fish in similar states of development (particularly allowing use of median fin soft-ray counts and relative fin positions in keys and descriptive accounts for the last phase).

Protolarva - Phase between hatching or parturition and the appearance of the first median fin ray or spine (dorsal, anal, or caudal fins).

Mesolarva - Phase from the appearance of the first median fin ray or spine to acquisition of the pelvic fins or fin buds and the full complement of principal soft rays in the median fins.

Metalarva - Phase from the acquisition of the pelvic fins or fin buds and the full complement of principal soft rays in the median fins to (1) the acquisition of the adult complement of spines and rays in all fins, including secondary rays, (2) the appearance of segmentation in at least a few rays in each fin characterized by segmented rays, and (3) the loss beyond recognition of all finfolds and atrophying fins, if any (very rare).

Continued

Comments: At times it might be desirable to denote the presence of yolk material while using either the Ahlstrom or Snyder terminology. This can be accomplished by modifying the name with either the adjective "yolk-sac" or prepositional phrase "with yolk" (e.g., yolk-sac flexion or metalarva with yolk). For most fish, preflexion and proto larva are nearly synonymous. Some fish hatch or are born as postflexion larvae or as mesolarvae or metalarvae.

ichthyoplankton sampling gear for power plant impact assessment. Marcy (1975) and Graser (1977) also discuss entrainment sampling.

9.2 SPECIAL SAMPLING CONSIDERATIONS

9.2.1 Where to Sample

Within an investigator's physical and economic limitations, design of a sampling program and choice of collecting gear depend on specific study objectives and the species, life history stages, and habitats to be sampled. Generally, more diverse habitats or fish communities require more sampling stations and a greater variety of gear and sampling methodologies. The distribution of eggs or larvae depends on the mode of reproduction, the specific spawning and nursery grounds, and the very nature of the eggs or larvae themselves. Balon (1975) proposed a useful ecological classification of reproductive strategies that can aid in the design of a sampling program if the target species are known. Much species-specific information on reproduction and some on early life history have been summarized by Breder and Rosen (1966), Carlander (1969, 1977) and the authors of many regional guides.

9.2.2 When to Sample

Most sampling programs begin prior to the expected initiation of spawning activity and terminate after the stages of interest are no longer present or vulnerable to the gear being used. The eggs and larvae of most fish are present for only a short period of time each year, depending on their specific spawning seasons, incubation periods, and rates of larval growth and development in the waters of concern. Because these factors differ substantially from one species to another, however, fish eggs and larvae are likely to be present in some habitat of most aquatic systems throughout at least the spring and summer months in temperate regions and year-round in more tropical waters and the oceans. For a particular species, the time when eggs and larvae are present varies among waters and from year to year within the same water, depending on climatic conditions and other differences in the environment.

Most larval fish investigations require sampling from twice a week to twice a month depending on objectives. Except for exploratory purposes, monthly or less-frequent sampling is of very limited value—entire species or cohorts may be missed, even when sampling is somewhat more frequent.

The distribution of fish larvae and their vulnerability to certain gear can differ significantly with the time of day. Like many invertebrates, some fish larvae rise toward the surface at night and retreat to lower levels or the bottom during daylight hours. Some larvae exhibit daily movement to and from the littoral zone. Others in freshwater streams exhibit diel drift. Many of these, including at least some of the catostomids, drift predominately at dusk or during the night. (Perhaps they become disoriented as light levels decline.) During daylight, some of these night drifters reside in the quieter and shallower waters along the shore. The larvae of most fishes are able to evade collecting gear in ever increasing numbers as they grow, especially in the photic zone during daylight hours (Cada et al. 1980; Smith 1981). Evasion is at least in part dependent upon visual cues. Intensive studies of fish eggs and larvae typically involve sampling during both day and night hours, at least initially, and sometimes include stations sampled continuously or periodically during 24-hour periods. More limited programs may be designed for sampling only at night or, usually for reasons of convenience and safety, only during daylight. In any of these cases, sampling biases must be considered in conclusions drawn from the data.

9.2.3 How to Sample

Some questions can be answered with qualitative collections alone (e.g., what species or developmental stages are present?), but most surveys of fish eggs and larvae require quantitative data. The ease and accuracy with which volume or area sampled can be determined is an important factor in the selection of gear and sampling methodology. For comparability among samples, it is essential to standardize units of effort for samples taken with the same gear even when area or volume is measured.

The number and type of samples, number and distribution of sampling stations, and frequency of sampling needed to adequately represent a study area increase with increasing unevenness or heterogeneity in the distribution of target organisms. In part because of variable environmental conditions over area and time (e.g., light, wind, currents, temperature, salinity), the distribution of targeted fish eggs and larvae is rarely uniform. Even in relatively homogeneous environments, distribution is usually patchy. While increased sampling effort (either larger or more numerous units) can more accurately represent the station sampled, sampling effort its limited by biological, physical, economic, and environmental constraints. Patchy distribution and limited sampling effort, as well as the biases and inefficiencies of the sampling gear and techniques employed, all contribute to variability in the data. Replicate samples are needed to measure and assess that variability.

9.3 COLLECTION GEAR AND TECHNIQUES

Almost any gear or sampling technique that filters water through a small-mesh material can be used to collect fish eggs and larvae. The water can be filtered *in situ* or transported to a remote filtering device. Eggs or larvae buried in or adhering to various substrates can be dislodged and filtered from the water or removed with the substrate and subsequently separated from it. Much of the gear discussed is not available commercially and must be custom built. Table 9.1 summarizes some of the advantages and disadvantages of various gears for collecting planktonic fish eggs and larvae.

9.3.1 Plankton Nets

Coarse-mesh plankton nets are the most commonly used gear for collecting the smaller, planktonic stages of fish, often referred to as ichthyoplankton. A plankton net is essentially a funnel of fine-mesh netting designed to filter water passing through its mouth and concentrate the remaining organisms and debris in a collecting bag or bucket at the other end. In pelagic waters, this gear is used almost exclusively. With

Table 9.1 Some general advantages and disadvantages of ichthyoplankton sampling gear. From Bowles et al. (1978).

Design	Examples	Advantages	Disadvantages
Low speed nets	Bongo nets Meter nets Benthic sleds Tucker trawl Neuston nets Henson nets	Sample large water volume in short time. Require only small vessel to deploy. Can be fished fixed or towed Inexpensive equipment	Clogging of meshes is major problem. Fishing characteristics highly variable in turbulence. Avoidance by larger larvae. Manpower needs are relatively high. Limited by water body morphometry.
High speed nets	Jet nets Gulf nets Miller nets	Reduced avoidance by larger larvae Sample large volume of water in short time. Moderately expensive equipment.	Requires larger vessels and usually winches to deploy. May be some extrusion of smaller organisms through mesh. Manpower needs relatively high.
Plankton recorders	Hardy plankton recorders	Can be fished from high speed commercial vessels. Can provide sample from a narrow band over a long distance.	Samples frequently mutilated. Extrusion of sample through mesh. Limited by bottom morphometry.
Mid-water nets	Isaacs-Kidd mid-water trawl British Columbia trawl	True pelagic samples. Large sample volume.	Manpower needs are relatively high. Must maintain constant boat speed.
Pumps	Centrifugal pumps Trash pumps Sewage pumps	Can sample turbulent areas or areas inaccessible to nets. Reduced manpower needs. Can replicate samples easily.	Small sample volumes may be inaccurate. Large size of pumps makes them difficult to handle in field.

appropriate hardware, plankton nets can be towed, pushed, pulled or buoyed up, dropped down, or fastened in a stationary position in the current. Relatively comprehensive comparisons of various plankton net gear and discussions of design, selection, and use are provided by Tranter and Fraser (1968), Jossi (1970), Smith and Richardson (1977), Bowles et al. (1978), and Marcy and Dahlberg (1980). Problems, causes, and factors associated with the use of plankton nets for surveying or monitoring fish eggs and larvae are summarized in Table 9.2.

Plankton nets can vary immensely in size with mouth diameters or widths ranging from about 10 cm to a meter or more. Nets used at low velocities are typically larger than those employed at higher velocities. A net is typically much longer than the

Table 9.2   Problems, causes, and factors associated with the use of plankton nets for surveying or monitoring fish eggs and larvae. Retitled from Marcy and Dahlberg (1980).

Problems	Causes	Factors
Mechanical Problems		
Passive avoidance	Clogging	Density of plankton and debris, duration of tow, small mesh
Extrusion	Mesh openings too large, high velocity through net	Shape and size of mesh openings, size and shape of organisms, tow speed
Damage to ichthyoplankton	Long tow, high velocity through net, predation in net	Fragility of organisms, stage of development, water temperature
Gear failure	Gear design, user's experience	Working conditions
Meter inaccuracy	Inconsistent calibration	Clogging
Contamination	Inconsistent cleaning, net open during retrieval	Speed of retrieval, depth of tow
Biotic Problems		
Active avoidance	Warning signals (vibration, turbulence, gear visibility)	Net size, tow speed, boat wash, temperature, water clarity, swimming speed of larvae, bridles, hydraulics, visual cues
Uneven Distribution (Horizontal)		
Patchiness	Variability of spawning, development rates, and mortality	Water temperature, sample size, predation, currents
Mass movements	Currents (river, tidal, seiche)	Meteorological conditions
Uneven Distribution (Vertical)		
Stratification	Preference for light intensity, food, temperature	Weather, currents
Diel variation	Response to light and food	Meterological conditions, currents
Nonvulnerable Stages		
Demersal	Species-specific habits	Distribution of spawning grounds
Shelter-seekers	Species-specific habits	Availability of shelter, distribution of spawning grounds

diameter or width of the mouth to increase filtration efficiency and reduce clogging and damage to the specimens collected. For maximum sampling efficiency (up to 95% without a mouth reduction cone) and minimal losses in efficiency because of clogging in circular-mouth nets, Tranter and Smith (1968) recommended combination cylindrical-conical nets with at least a 5:1 open mesh area to mouth area ratio (9:1 for meshes less than 1/3 mm). The mesh size of a plankton net must be small enough to retain with minimum damage the smallest organisms of interest but as large as possible to minimize clogging and sample sorting time because of retention of smaller non-target organisms and debris. Mesh sizes most commonly used to sample fish eggs and larvae range between 0.3 and 0.8 mm; however, especially in inland waters, some researchers extend the range to 1.0 mm or more. Monofilament netting with relatively square aperatures is generally preferred over other material, and nets are sometimes dyed a dark color to minimize visibility.

The cod end (bag end) of a net usually consists of a removable bag, jar, or bucket with screened windows. Generally the latter is preferred to a solid unit for more effective concentration of the filtrate and ease in removing the sample. When the collected eggs, larvae, or early juveniles are to be maintained alive, specially designed live-boxes can be attached to the end of the net.

To quantify data obtained from plankton net samples, the volume of water filtered must be determined. Properly functioning, calibrated, and positioned flow meters in the mouth of a net provide relatively accurate measures of the amount of water passing through the net. Tranter and Smith (1968) showed that for bridled, circular-mouth nets, flow meters should be positioned about one third the radial distance to the center where velocity is about average. Flow meter position in the mouth of unbridled circular nets is less critical since flow across the mouth is more nearly uniform. For certain purposes, such as measuring filtration efficiency, a flow meter can be mounted outside as well as inside the mouth of the net.

Low-speed samplers. Plankton nets with mouth diameters or widths of 0.5 m or larger (Fig. 9.1) are typically sampled at velocities less than 2 m/sec (often near 1 m/sec) for time intervals of less than 30 minutes (usually between 5 and 20 minutes). In open waters these samplers are generally towed obliquely through all depths or horizontally at discrete depths (e.g., surface, mid-depth, and near bottom). Depressors or weights attached to the tow-line or sampler frame are usually required to keep the nets submerged during towing. Depth is controlled by tow-line length and speed of tow and usually is monitored by measuring the angle of the tow-line. To assure that all depths are sampled evenly in oblique tows, a specified tow-line angle is maintained by adjusting vessel speed while the tow-line is released and retrieved at constant rates (Smith and Richardson 1977). Some samplers have opening and closing mechanisms to avoid contamination of collections from discrete depth as the net is deployed. For sampling as close to the bottom as possible without grounding the net, low-speed samplers are mounted on sleds or runners. To sample consistently at or just below the surface, flotation devices are often fastened to the net frame. Surface and near-surface tows should be made well clear of the wake of the vessel. Surface samples are also taken alongside or in front of vessels by fixing nets to side-arms or other structures that can be extended or lowered into the water (Fig. 9.2). The latter approach is particularly useful on smaller boats for sampling in shallow waters and close to shore. For neuston samples, the top of the net must ride steadily above the surface.

Some plankton nets can be sunk to the bottom or other specified depth and hauled vertically. The volume filtered depends on the depth and is usually much less than that for an oblique or horizontal tow. Under ideal conditions, this technique can provide a very uniform sample of all depth strata under a surface area equivalent to that of the mouth of the net. When deployed from a vessel in rough or rolling waters, however, the net tends to ascend in a series of rapid rises followed by sharp decelerations and possibly short reversals in direction (Smith and Richardson 1977).

Bowles et al. (1978) concluded that nets with a 1-m diameter mouth sample pelagic eggs better than any other type of gear and sample older larvae better than half-meter nets with a 0.5-m diameter mouth, but that both bridleless bongo nets and high-speed samplers (discussed below) are more effective for sampling larger larvae than meter nets. A bongo net is a pair of nets held side-by-side in a frame and is towed by a cable that attaches to the frame between the two nets. Smith and Richardson (1977)

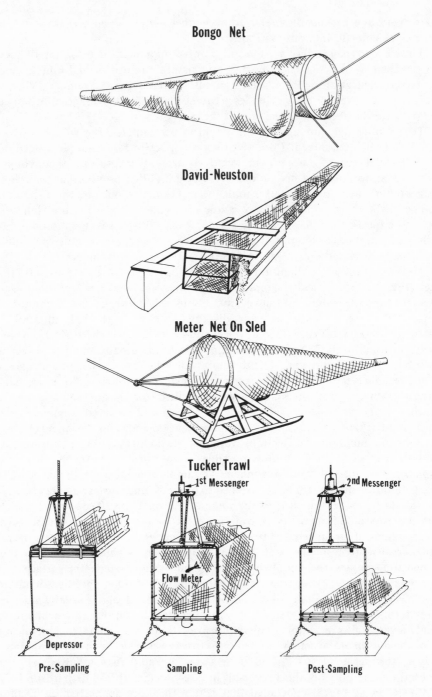

Figure 9.1. Examples of towable, low-speed, plankton-net samplers: a set of 60-cm bongo nets, without opening and closing mechanism (Smith and Richardson 1977); David-neuston sampler modified with a second net under the surface net. (Hempel and Weikert 1972); a 1-m square modified Tucker trawl with opening and closing mechanism, (left to right, net in pre-sampling, sampling, and post-sampling position) (Sameoto and Jaroszynski 1976); and a 1-m net on sled (Dovel 1964). Illustrations adapted, with permission, from indicated sources.

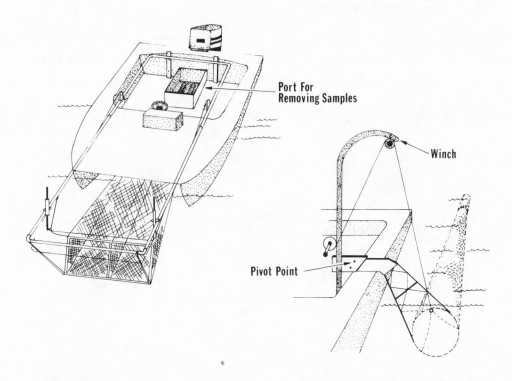

Port For
Removing Samples

Winch

Pivot Point

Figure 9.2. Examples of vessel-mounted, low speed, plankton-net samplers: a side-arm sampling frame for a 1-m plankton net (Dovel 1964); an under bow, dual, push-net sampler with 60-cm square-mouth nets (Bagenal and Nellen 1980, after Miller 1973). Vessels can be anchored for stationary sampling. Illustrations adapted, with permission, from indicated sources.

recommended the bongo sampler (McGowan and Brown 1966) with 60-cm cylindrical-conical nets as standard gear for quantitative sampling of pelagic ichthyoplankton. Its primary advantage, aside from lack of bridles in front of the net, is the provision for two nets in the same frame — ideal for collecting replicate samples.

Among the more technologically advanced low-speed plankton samplers is the MOCNESS (multiple opening and closing net and environmental sensing system) sampler and its subsequent modifications (Sameoto et al. 1977). The sampler is essentially a multiple net version of the single net opening and closing Tucker trawl illustrated in Figure 9.1. In addition to providing for the collection of several discrete samples in one tow and using electronics to more reliably control the opening and closing of nets, this sampler provides and records data on net depth, pitch and roll, and water temperature.

Drift samplers. Where water is moving with sufficient velocity (e.g. ocean or lake currents, tidal zones, streams and rivers, and power plant or industrial water intakes and discharges), many of the previously discussed low-speed samplers can be fastened or suspended in a stationary position for passive sampling (Fig. 9.3) Rectangular-mouth nets are often anchored or staked in position when sampling the drifting biota of shallow streams. Simultaneous collection of drifting fish eggs and larvae and

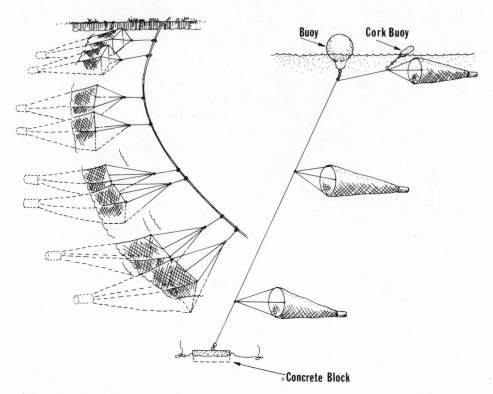

Figure 9.3. Examples of stationary deployment of plankton nets: Left, a series of 30 x 50-cm drift nets suspended across a shallow stream; right, a vertical series of 1-m conical plankton nets suspended from an anchored line. Illustration on right adapted, with permission, from Dovel (1964).

macroinvertebrates is often desirable and can be done with the same gear. However, since fish eggs and larvae are more fragile and in many cases less abundant than macroinvertebrates, nets for fish collection are usually larger than those intended for macroinvertebrates only and are typically sampled for shorter periods of time, usually less than 30 minutes.

High-speed samplers. High-speed samplers are generally smaller than low-speed samplers, are bridleless, are encased, have a reduced mouth opening, and are towed at speeds over 2 m/sec, sometimes much faster (Fig. 9.4). One of the serious disadvantages of most high-speed samplers is a high filtration pressure across the meshes of the net or screen which results in the extrusion of smaller eggs and larvae and mutilation of many specimens retained. This problem is reduced in the jet net by a series of expansions within the sampler body (Clark 1964). Another form of high speed sampler, the continuous plankton recorder, depends on the impingement of filtered specimens on a continuous band of net material that advances progressively during the tow and thereby records the density and frequency of plankton patches during long tows. Unfortunately, plankton recorders have many inherent biases and difficulties (Haury et al. 1976).

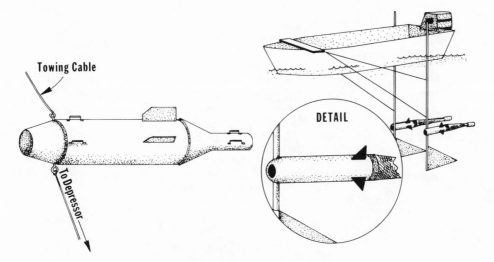

Figure 9.4. Examples of high-speed, plankton-net samplers: a 20-cm Gulf III sampler modified with a restricted tail unit, net suspended from a frame inside (Fraser 1968); a 10-cm Miller high-speed samplers mounted by pipes to a boat (Noble 1970). Illustrations adapted, with permission, from indicated sources.

9.3.2   Micronekton or Macroplankton Trawls

Larger pelagic larvae and early juveniles are more effectively captured with larger, coarser-meshed gear than plankton nets. These gear are often scaled down versions of trawls used to capture larger fish (*see* chapter 7), scaled up versions of plankton nets used to capture fish eggs and smaller larvae, or a combination of both. Some trawls are specifically designed to sample surface waters (Chapoton 1964); others are used at various depths (Houser 1976). Some pelagic or midwater trawls are not much larger than the larger plankton nets; the distinction is mesh size. A few trawls are truly immense, one being 42 m long with a multiple plankton net sampler serving as a cod end device (Pearcy 1980). Many utilize more than one mesh size, often between 5 and 15 mm in the body with a finer mesh in the cod end.

Aron (1962) compared various gear for sampling macroplankton and concluded that, in spite of limitations, Isaacs-Kidd midwater trawls (Fig. 9.5), because of their simplicity and capability of being towed at higher speeds, are among the most useful gear for sampling large plankton and small nekton. Smith and Richardson (1977) described the use of one version consisting of 2-mm knotless mesh and a mouth area of 2.9 m^2. Factors affecting the catch of midwater trawls, particularly the Isaacs-Kidd trawl, are discussed by Atsatt and Seapy (1974).

9.3.3   Seines and Dip Nets

Seines and other types of nets discussed in chapter 7 are often scaled down, at least in mesh size, for the collection of larval and early juvenile fish. Larger seines with sufficiently fine mesh can be very effective in collecting larvae but may be difficult to use, especially in currents or over areas with potential snags or soft mud bottoms. Specimen concentration and removal without excessive damage also can be a problem, especially without a sewn-in collection bag. Dip nets or scoop nets and small

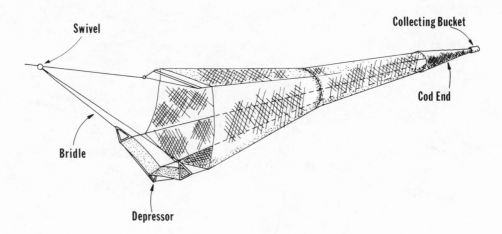

Figure 9.5. Modified Isaacs-Kidd midwater trawl for sampling larger larvae and early juveniles. Adapted, with permission, from Aron (1962).

one-man, fine-mesh seines (e.g., 1 m square units) are particularly useful when working irregular shorelines, tight coves or inlets, around submerged structures, in vegetated areas, or in shallow riffles ("kick" samples). Pushnets, which are essentially framed seines or very much enlarged dip nets with beveled boards or rollers on the leading edges (Strawn 1954), can be very effective in well-vegetated shallows.

Data acquired with seines, dip nets, and related gear are usually considered qualitative. However, special care and procedures for sampling uniformity and minimization of biases will allow quantification, at least on a per unit effort basis. Plankton purse seines have been successfully employed to provide more realistic estimates of the densities of larger larvae in near surface waters than those that can be expected from towed plankton nets (Hunter et al. 1966; Murphy and Clutter 1972).

9.3.4 Traps

The larvae of fish that bury their eggs, scatter them over specific substrates, or utilize nests can be quantitatively captured as they rise out of or off of those substrates by enclosing the surface of the nest or area of concern, often with specially designed nets (Fig. 9.6). Most such gears channel the emerging larvae into special collection bags, jars, or boxes and are used most often for the collection of salmonid larvae in flowing water. After larvae emergence from nests, drift nets or appropriately modified weirs, fences, or similar structures can be used in many streams to collect the larvae or early juveniles as they drift or migrate with the current (Forester 1930; Wolf 1951, chapter 6).

In more quiet, shallow waters, nests or spawning substrates can be enclosed with fine-mesh cylinders with open tops extending above the surface (Werner 1968). Similar but larger enclosure structures such as block-off nets, drop traps, and related gear (*see* chapter 7) are useful for quantitative sampling of later larvae and larger fish in shallow but well-vegetated areas. Once the target area is enclosed, the vegetation and debris is usually removed in a systematic fashion, collecting any adhering eggs or larvae in the process and removing the remaining fish with fine mesh seines or dip nets.

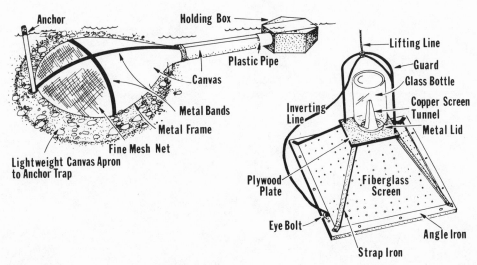

Figure 9.6. Examples of larval fish emergence traps. Illustrations adapted, with permission, from Porter (1973) (left) and Collins (1975) (right).

In deeper waters, drop nets (Hoagman 1977) and buoyant nets (Bagenal and Nellen 1980) can provide vertical samples from surface to bottom or vice versa.

Specially designed passive traps of clear plastic or netting can be used to collect larvae and early juveniles as they move about the littoral zones or range through deeper or pelagic regions. Some of these are scaled down modifications of adult trap nets with wings and leaders to direct the fish into the sampler (Hamilton et al. 1970; Beard and Priegel 1975; Trippel and Crossman 1981). Others are essentially "minnow" traps with funnels leading to a holding box or container. Lighted modifications (Nagiec 1975; Faber 1982) sampled at night are far more effective for capturing larval and early juvenile fish than similar unlighted traps. These collections, however, are biased in favor of fish attracted to light. Captures by passive traps are also affected by local disturbances that influence fish behavior.

9.3.5 Pumps

Pumps sample by sucking or air-lifting water with organisms, debris, and, in the case of benthic samples, substrate materials through an intake structure and hose or pipe to a remote site where the water is filtered through screens or nets (Fig. 9.7). The pumps themselves are usually positioned on-shore or mounted on a boat or raft.

Pumps can be used to sample in nearly any habitat, including those not accessible by plankton nets or most other gear. However, while the pump sampling efficiency for pelagic fish eggs and larvae is comparable to that of meter and half-meter plankton nets (Aron 1958; Portner and Rohde 1977; Gale and Mohr 1978), even the largest pumps (over 8 m³/minute) sample only about 18% of the volume filtered per unit of time by meter nets towed at 1 m/sec. Indeed, most pumps sample less than 5% of that volume.

Pump samplers are of greatest value for sampling benthic substrates and around submerged structures. To this end, the amount of suction or velocity of water and other materials and specimens drawn into the intake is more important than the volumetric capacity of the pump. With just the open end of the intake hose or pipe or

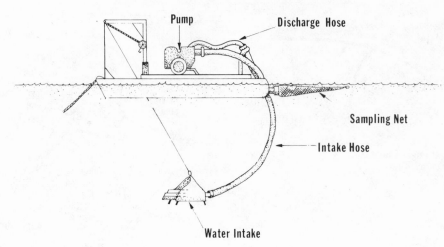

Figure 9.7. A towable pump sampler and intake sled. Adapted, with permission, from Gale and Mohr (1978).

with special sweeper-like intake attachments, the bottom and submerged structures can be vacuum cleared of the looser materials or organisms. The data for these samples can be quantified on the basis of the actual area swept by the intake structure.

Specimen damage is often much greater in pump samples than in plankton net samples. Bowles et al. (1978) suggested that bladeless impeller units cause less specimen damage than other pumps if the sample must pass through the pump before filtering. Damage to the larvae can be reduced by filtering the water before it passes through the pump (Manz 1961) or by using a venturi-type sampler in which the water sampled by-passes the pump entirely (Allen and Hudson 1977). Additional specimen damage results from abrasion with other materials and from high pressures across the meshes of the filtering screen or nets. Ecological Analysts Inc. (1976) described a special "larval fish table" to reduce filtration pressure by expanding the area occupied by the sample as it passes through a net.

9.3.6   Other Substrate Collection Techniques

Other gears and techniques for collecting benthos also can be used to collect fish eggs or larvae in, adhering to, or lying on bottom substrates. Dredges, grabs, and corers (Viljanen 1980) can be used in waters of various depths and actually remove the substrate to be sampled. For other techniques the substrate is stirred-up *in situ,* either manually or hydraulically, and the dislodged organisms are collected in a net or other container. Most of these samplers can be used only in shallow waters, but the dome sampler also can be used by divers in deeper waters (Gale and Thompson 1975). The adhesive, clumped, or nest-spawned eggs of some fish can be sampled directly by hand or by using small suction devices such as pipettes, basters, or slurp guns (Werner 1968; Bowles et al. 1978) or by collecting the actual rocks, debris, or vegetation upon which the eggs were deposited. Artifical substrate samplers can be effective for quantitatively sampling the eggs of fish that broadcast demersal eggs or spawn on specific substrates (Gammon 1965; Stauffer 1981).

9.3.7 Electrofishing

A few researchers have applied electrofishing technology (chapter 8) to the collection of fish larvae and early juveniles. In flowing water, the fish can be either driven into set nets or stunned upstream of the nets and allowed to drift into the nets. Noble (1970) found an electric grid positioned in front of a Miller high-speed sampler (Fig. 9.4) improved the catch of larger larvae and juveniles but had little effect on the catch of smaller larvae.

9.3.8 Direct Observation

Data about fish eggs and larvae sometimes can be effectively obtained without collection of specimens by using underwater cameras or by direct observation through viewing ports or while diving (Shealy 1971; Vogele 1975; Elliot 1976). Such techniques are discussed in chapter 19 and by Aggus et al. (1980), who noted that underwater observations are particularly useful for studying fish reproduction.

9.4 SAMPLE HANDLING AND PRESERVATION

Fish eggs and larvae are fragile and easily damaged. Care must be taken during collection, concentration, fixation, and subsequent manipulation and curation to minimize specimen damage and deterioration. Ahlstrom (1976), Richards and Berry (1973), Scotton et al. (1973), and Smith and Richardson (1977) describe procedures for handling and preserving samples.

9.4.1 Fixation and Preservation

Fixatives and preservatives. Fixation is the process of rapidly killing and chemically stabilizing tissues to maintain anatomical form and structure. Buffered or unbuffered solutions (depending on purposes) of 5 to 10% formalin (2 to 4% formaldehyde) in water are recommended for fixing samples of fish eggs and larvae. Weaker solutions can result in inadequate fixation; stronger solutions can produce excessively rigid specimens.

Preservation is the means by which fixed tissues are maintained in that condition. While the formalin solutions recommended for fixation are also suitable as preservatives, more dilute solutions are less hazardous to human workers. Buffered or unbuffered solutions of 3 to 5% formalin (1 to 2% formaldehyde) in water are recommended as preservatives for fish eggs and larvae. The use of alcohol solutions for preservation, while still standard practice in many museums, is strongly discouraged. Dehydration effects can result in substantial specimen distortion and shrinkage. Also, when working with specimens in alcohol solutions, the high volatility of the preservatives can result in rapid and detrimental fluctuations in concentration and osmotic pressures.

Buffers. While slightly basic fixatives and preservatives are optimal for maintaining ossified structures, neutral or acidic solutions are preferred to prevent eventual clearing (lysis) of some soft tissues and long-term fading of melanophore pigmentation (Taylor 1977). Unbuffered formalin solutions are acidic and the process of fixation can result in progressively lower pH values. To minimize decalcification while providing a sufficiently low pH for good fixation and preservation of most specimens, formalin solutions should be maintained at near neutral pH, about 6.5 to 7.5 (Steedman 1976a). A much lower pH during initial fixation is necessary for leptocephalus-type larvae (e.g., eels and tarpons; Steedman 1976b).

Commonly used buffers for samples of fish eggs and larvae are borax and calcium carbonate (usually as marble chips or limestone powder). Use of borax requires care to avoid undesirably high pH values (Steedman 1976a; Smith and Richardson 1977). Observing lysis of some soft tissues even in less alkaline solutions with borax, Taylor (1977) concluded that borax should not be used for purposes of long-term storage in formaldehyde solutions. Calcium carbonate is generally recommended as a safer buffer for long-term storage. Unless solid calcium carbonate is included with the sample, buffered fixatives should be replaced within several days after initial fixation to assure a proper pH level.

Markle (in press) reported the successful use of phosphate-buffered formalin for fixation and preservation of fish larvae but he recommended that fish eggs be maintained in unbuffered solutions to avoid osmotic collapse of thin chorions. Phosphate buffers can be tailored for very specific pH values. Markle recommended a pH of 6.8, maintained by adding 1.8 g of sodium phosphate monobasic and 1.8 g of anhydrous sodium phosphate dibasic to each liter of 5% formalin in distilled water.

Color Preservation. While the relatively stable black and brown pigments, the melanins, are maintained well in acidic to neutral formalin solutions and no additives are required to preserve them, most other natural pigments (e.g., reds in *Thunnus* larvae) are readily oxidized or dissolved and are lost or altered shortly after initial fixation. To slow loss of the latter pigments, Berry and Richards (1973) and Scotton et al. (1973) recommended 0.2 to 0.4% solutions of IONOL CP-40 (a 40% emulsified concentrate of butylated hydroxytoluene, BHT) in formalin fixatives and preservatives. Based on experiments with larger fish, Gerrick (1968) recommended a 1% solution of another antioxidant, erythorbic acid. Taylor (1981) discussed many of the problems in using color preservatives, and Ahlstrom (1976) suggested that, considering the variable results with antioxidants, the best method for study and utilization of evanescant pigments is to examine specimens immediately after collection and initial fixation.

9.4.2 Sample Handling

Most mechanical damage to specimens is a result of sampling technique and handling prior to fixation. In plankton net samples, larvae may cling to the netting and must be carefully rinsed down and concentrated for transfer to a sample container. Deterioration of tissues in dead specimens can be significant within several minutes of collection, especially at warmer temperatures. It is imperative that samples be fixed as soon as possible.

For proper fixation, Steedman (1976a) recommended that the specimen fraction be no more than 10% of the total volume occupied by the specimens and fixative and that the specimens remain in fixative for at least 10 days. Smith and Richardson (1977) extended the maximum volume for the sample to 25% of the container when the sample is "wet" (partially drained) plankton. Rinse large debris such as sticks and leaves and remove it prior to fixation. Such plant material reduces the amount of formaldehyde available for animal tissues, and, in some instances, it can eventually release substances, such as tannin, which darkly stain eggs and larvae. To avoid burning specimens, predilute concentrated formalin or add it to the sample only after a substantial amount of water has already been added, keeping in mind the effect of the sample itself on fluid volume and ultimate fixative concentration. Fill sample containers almost completely to minimize specimen damage during transit and to prevent specimens from drying on the sides of the container or undersurface of the lid above fluid level. Once a sample jar is topped-up and sealed, invert and gently shake it a few times to distribute fixative uniformly throughout the sample.

Protect samples from heat and sunlight, which can significantly affect the quality of the specimens being fixed or preserved. Heat beyond normal room temperature can be particularly detrimental if sustained; cooler storage temperatures are preferred. Taylor (1981) suggested that normal room lighting will not affect preservation of fixed tissues or malanophore pigmentation. If properly fixed, preserved, and stored, specimens should remain in good condition indefinitely.

9.5 SAMPLE PROCESSING

With the exception of samples requiring immediate on-site processing, as in entrainment mortality studies, most samples of fish eggs and larvae are fixed and returned to the laboratory for sorting and subsequent identification, enumeration, and determination of other specimen-specific information (e.g., size, developmental state, age, and gut contents). After samples have been sorted and processed, they should be archived systematically for future reference, verification, or subsequent study (*see* chapter 14). Fish eggs and larvae in good condition and rare specimens, even if in poor condition, should never be discarded. When no further use is anticipated, specimens should be donated to an appropriate depository—centers of larval fish study, museums, or universities with programs for curating and utilizing the material.

9.5.1 Subsampling

Subsampling prior to sorting or identification is not recommended for the objectives of most studies. The fish components of plankton samples often account for less than one percent of the sample biomass. Even in samples that are largely fish, a solitary specimen of one species commonly occurs among hundreds of another species. Rarer developmental stages or species might be missed entirely if collections are subsampled. If the macroinvertebrate or invertebrate zooplankton portion of a collection is to be subsampled, the fish should be removed before splitting. Even when subsampling is consistent with the objectives of a study (e.g., determining densities of one or more target species with abundant eggs or larvae) or necessary because of excessive volume or time required to sort the samples, consideration should still be given to first removing the larger larvae and early juveniles. The Folsom plankton splitter is typically recommended for subsampling purposes (Smith and Richardson 1977).

9.5.2 Sorting

The goal of sorting is to segregate 100% of the fish eggs and larvae from the other organisms and debris in a sample. The optimal approach and time required to sort plankton or benthos samples vary considerably with (1) the size of the sample; (2) the number, size, and form of fish eggs and larvae present; and (3) the nature of other organisms and debris in the sample. Some relatively clean samples with few eggs and larvae can be sorted in a fraction of an hour while others with many small specimens in an abundance of plankton, filamentous algae, or debris can require a day or longer.

Sort samples with very small eggs and larvae a fraction at a time in small dishes under low-power stereo-microscopes. Sort other samples in larger dishes or trays with or without the aid of a magnifying lens. It is often desirable to re-sort samples two or more times. Scotton et al. (1973) reported that in one study over 20% of the larvae were missed on the initial sort. Re-sorts of randomly selected samples are often required for quality control.

Prior to sorting a portion of a sample, drain the alliquot of fixative or preservative and rinse it in water to reduce formaldehyde fumes. To avoid significant reversal of the fixation process while sorting in water, return the specimens removed and the remaining sample to preservative as soon as possible. To minimize handling damage, manipulate and transfer specimens carefully with eyedroppers or small pipettes, wire loop probes, and flexible forceps. If sharp pointed or rigid forceps are needed, use them very gently.

For taxonomic purposes, biologists generally prefer not to stain fish eggs and larvae for sorting. However, in many situations, use of rose bengal (or certain other stains; Williams 1974) can dramatically reduce sorting time and increase the efficiency with which smaller specimens are located and removed (Mitterer and Pearson 1977). Dry rose bengal powder on the wetted end of a probe or a few milliliters of a 1%-solution usually produce the desired moderate to deep rose color in a liter of sample and preservative. Invert the sample and gently agitate it several times during the staining period, which can last from several minutes to several hours depending upon the nature of the sample.

When sorting unstained material with reflected light, a black or dark background is typically recommended for best contrast with eggs and larvae. However, transmitted light (via scope base or light table) under a clear glass sorting container is preferred by many sorters and is more effective for stained specimens than is a white background (Williams 1974). Dorr (1974) described a side-lighted sorting chamber which provides an evenly lit field of view with a smooth non-reflecting background and little glare; organisms appear well defined in this lighting.

9.5.3 Identification

Accurate specimen identification is an especially critical aspect of sample processing; the validity and usefulness of all subsequent taxon related data analyses, results, and conclusions depend on it. The specific objectives of a study determine whether all fish specimens or just the target species are to be identified and to what taxonomic level. Identification of larvae to family is seldom a problem for the experienced biologist. At the species level, however, many larvae can not be positively identified because of lack of adequate keys, guides, descriptions, or taxonomic experts familiar with the candidate species. Identification of eggs can be difficult even to the family level. In spite of positive evidence for identifying a specimen as a particular species, the potential existence of another very similar but undescribed species in the study area usually requires that the identification be considered tentative. Intraspecific variation and differences in condition and pigmentation brought about by differing environmental conditions also can complicate identification. Tentative identifications should always be so indicated (usually with a question mark or footnote) or left at a higher taxon level. Specimens that cannot be identified to family or order should be designated as "unidentified"; specimens so mutilated or deteriorated as to preclude any likely possibility of confident identification should be designated as "unidentifiable".

The identification of fish eggs and larvae is in part a process of elimination. Even before examination of a single specimen, the range of possibilities can be narrowed by knowledge of the fishes which occur within or near the study area. Knowledge of their spawning seasons, temperatures, habitats and behavior, coupled with information about egg deposition and larval nursery grounds and behavior, also are useful in reducing the possibilities.

Expert assistance. It is often advantageous to seek training or taxonomic assistance from individuals with more experience in identification of fish eggs and larvae. Assistance, or references to others who can provide such assistance, can be obtained by contacting centers of larval fish identification (e.g., Canadian and United States National Marine Fisheries Service ichthyoplankton sorting centers; National Museum of Natural Sciences, Ottawa, Ontario; Great Lakes Regional Fish Larvae Collection, University of Michigan, Ann Arbor; Tennessee Valley Authority Regional Larval Fish Identification and Information Center, Norris, Tennessee; Colorado State University Larval Fish Laboratory, Fort Collins), the curators of various other museum, university, government laboratory, and consulting firm collections (e.g., the Lower Mississippi River Basin collection of Louisiana State University, Baton Rouge), and the Early Life History Section of the American Fisheries Society.

Literary aids. Snyder (1976b) reported that only about 15% of freshwater and anadromous fishes in North America had been described and illustrated in their early life history stages, and Smith and Richardson (1977) suggested the same for perhaps less than 10% of the world's marine fishes. Fortunately, the eggs and larvae of a majority of the commercial species in North America and others commonly taken in ichthyoplankton samples have been described. The major bibliographies included in the reference list at the end of this chapter are invaluable for locating much of the descriptive literature. Regional identification manuals are listed in Table 9.3.

Reference collections. Comparison of an unknown specimen with known reference specimens can be one of the most useful and certain approaches to identification. Reference series can be assembled from the specimens collected during the course of a project by tracing development backward from identifiable juveniles to the earliest stages collected, but identities should be verified by other means, if possible. Such series can serve the additional purpose of a voucher collection. Known reference specimens frequently are obtained by raising fish in captivity and sometimes can be obtained by loan or exchange with other researchers or various depositories of early life history stages.

Diagnostic characters. The morphological characteristics of fish eggs and larvae that are useful for identification vary with developmental stage and species. These characters are particularly well discussed by Berry and Richards (1973), Hempel (1979), Ahlstrom and Moser (1980), and Snyder (1981). To reduce identification errors it is important to use as many characters as possible.

Among the characteristics commonly used to identify fish eggs are egg diameter and shape, nature of the chorion surface, size of the perivitelline space, nature of the yolk, and (at specific stages) the size, number, and position of oil globules in the yolk. The presence and nature of attachment threads, filaments or stalks, an obvious micropyle, multiple chorionic membranes, and gelatinous or adhesive coats, and the type of cleavage also can be diagnostic. The latest embryonic stages are sometimes identified using characteristics for recently hatched larvae.

Diagnostic characters for fish larvae can be grouped into five major categories: meristics, morphometrics, structural form and position, developmental state relative

Table 9.3. Illustrated manuals for the identification of the early life history stages of North American fishes. Most manuals include regional notes on distribution, ecology and spawning.

Author (s) and Publ. Date	Region	Comments
Fish, 1932	Lake Erie	62 species accounts, several based on misidentifications.
Mansueti and Hardy; 1967	Chesapeake Bay Region	45 species accounts, Acipenseridae through Ictaluridae; updated as Volume 1 of Hardy et al., 1978.
May and Gasaway, 1967	Oklahoma	Key with descriptive notes and photographs of 18 species.
Colton and Marak, 1969	Northeast Coast, Black Island to Cape Sable	27 species accounts.
Scotton et al., 1973	Delaware Bay Region	56 species accounts.
Lippson and Moran, 1974	Potomac River Estuary	88 species accounts with some keys.
Hogue et al. 1976	Tennessee River	Keys with brief descriptions; includes photographs of over 32 species.
Hardy et al., 1978 (6 volumes, each individually authored).	Mid-Atlantic Bight, including tidal freshwater zones	278 species accounts; includes very comprehensive coverage of published literature.
Drewry, 1979	Great Lakes Region	Punch-card key to families of yolk-sac larvae, line drawings.
Wang and Kernehan, 1979	Delaware Estuaries	113 species accounts with some keys.
Elliott and Jimanes, 1981	Beverly-Salem Harbor Area, Massachusetts	47 species accounts.
Snyder, 1981	Upper Colorado River System, Colorado	19 species accounts, Cyprinidae and Catostomidae, with key to metalarvae.
Wang, 1981	Sacramento-San Joaquin Estuary and Moss Landing Harbor Elkhorn Slough, California	74 species accounts with keys or comparison tables.
Auer, 1982	Great Lakes Basin, emphasis on Lake Michigan	148 species accounts with keys.

to size, and pigmentation. Metalarvae often can be identified, at least in part, on the basis of adult characteristics such as fin meristics and position.

The criteria by which counts and measures are made are not well standardized and must be considered when reporting or using such values in the literature. Figure

9.8 illustrates many of the counts and measures utilized and recommended for taxonomic analysis. Depending on the species, other counts and measures or other criteria than those illustrated may have greater diagnostic value.

Because myomeres, the serial segments of body musculature, are obvious morphological features and correspond directly with the number and positions of vertebrae, total and partial counts of these structures are among the most useful characters available for identifying protolarvae and mesolarvae at higher taxonomic levels and sometimes at the species level. Myomeres are usually present in their full complement prior to or shortly after hatching and are chevron-shaped throughout the protolarval and much of the mesolarval phase.

Polarizing filters can be used effectively by placing them above and below thin and translucent specimens and intensely lighting the field from below; this increases the contrast between myomeres and mysepta and thereby facilitates more accurate counts. Rotate specimens against the filter-blackened background until the best effect is attained. Reflected light set at very oblique angles will often create highlights and shadows on the myomeres of specimens for which polarizing filters are not effective. Immersion in glycerin or various stains sometimes makes myomeres more prominent.

Total myomere counts are usually given as the number between the most anterior and posterior myosepta. Unfortunately, the most anterior and posterior myosepta are often difficult to discern, especially on larvae with high myomere counts. The most anterior myomeres are often observed only in the epaxial (dorsal) half of the body with the first (sometimes deltoid in shape) located immediately behind the occiput. Contrary to practice by many researchers, some biologists (e.g., Scotton et al. 1973; Fuiman 1982) recommend that the urostylar myomere, which lies posterior to the last myoseptum, be included in total myomere counts.

Partial myomere counts are used frequently to reference the location of various structures. The most frequently used counts are those anterior to and following the posterior margin of the vent (preanal and postanal myomere counts, respectively). Following Siefert (1969), most biologists include all myomeres transected by a vertical line from the point of interest as part of the count of myomeres anterior to that point.

Morphometrics are the dimensional measures of the body and associated structures. Like myomere counts, they also are used to reference the location of specific structures. Such measures are usually made under a low-powered stereomicroscope with dial calipers or an ocular micrometer. Body lengths or portions thereof are measured along lines parallel to the horizontal axis of the fish while depths and widths are measured in planes perpendicular to the horizontal axis. For comparative purposes, such measures usually are reported as a percentage or proportion of another measure such as total, standard, or notochord lengths. Many measures are allometric, changing in proportion as the fish grows; accordingly, the measures must be related to size or developmental stage.

The general shape of a fish larva and the form and position of specific and sometimes very unusual (Moser 1981) anatomical structures provide some of the most obvious characters for identification. Leptocephalus larvae have large, thin, leaf or ribbonlike bodies and huge fanglike teeth. Other larvae are characterized by eye, fin, or vent migration or by fin development remote from the body in voluminous finfolds. Still others have long trailing guts or pendulous abdomens, precocious pelvic fins with elongate spines or rays, dorsal rays longer than the body, tentaclelike structures, heads or bodies armed with enormous spines, photophores, balloonlike skins, eyes on very long movable stalks, or other unusual structures. More common, diagnostically

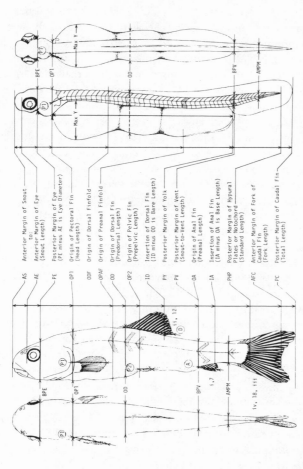

Figure 9.8. Selected counts and measures used by Snyder (1981) for taxonomic analysis of larval fishes. Yolk sac and pterygiophores are included in width and depth measures but fins and finfolds are not. *B* in *BPE* and *BPV* means immediately behind, i.e. not including the eye or vent, respectively. *AMPM* is the anterior margin of the most posterior myomere. Location of width and depth measures at *OD* prior to *D* formation is approximated to that of later larvae. Once the dorsal fin is formed, length to *ODF* is measured only as long as finfold remains anterior to it. *PHP* is measured to the end of the notochord until the adult complement of principal caudal rays are observed. For fish in which *PV* and *OA* approximate each other, *OA* is oftern deleted and assumed to be the same. Fin lengths (*D, A, P1,* and *P2* encircled) are measured along the plane of the fin from the origin to the most distal margin. Principal rays of the median fins are darkened and the number given in Arabic numerals, while secondary rays are outlined and given in lower case Roman numerals; when counted, the rays of paired fins are not so dividied. Fin spines, if present, would be given in upper case Roman numerals. The first and last myomeres, as well as the last myomeres in counts to specific points of reference, are shaded.

valuable characters include the folds or loops of the gut and the size, shape, and position of the air bladder, yolk sac, oil globules, mouth, finfolds, and fins.

The sizes at which certain ontogenetic events occur also are useful in determining the identity of larvae. Such events include hatching, yolk absorption, notochord flexion, and the attainment of and change in previously mentioned structures. Among the more useful size-related events are development of the median and paired fins, particularly the size at which the first and later the full complement of fin rays in the individual fins are formed. In most fish, the caudal fin is the first fin to differentiate from the median finfold. Although the pectoral buds usually form prior to or shortly after hatching, they often remain rayless until late in the mesolarval phase and are sometimes, along with the rudimentary or secondary-ray portion of the caudal fin, the last fins to acquire the adult complement of rays. Pelvic fin buds often appear after principal soft rays of the dorsal and anal fins have already formed, but in some fish they are evident prior to hatching and acquire ray or spinelike structure during the protolarval phase. While most fish hatch as protolarvae, some fish, such as trout and salmon, ictalurid catfishes, killifishes, and certain darters, hatch as mesolarvae with median fin development already underway.

The basic patterns of pigment distribution and changes in those patterns as fish grow can be particularly useful as characters for identification at the species level. The predominant pigmentation in most fish embryos and larvae is that of the melanin bearing chromatophores (melanophores). The presence, absence, or manner in which the pigment is displayed (from tightly contracted spots to widely expanded reticular networks) varies with environmental conditions, particularly background coloration.

9.5.4 Other Specimen-specific Data

It is often useful or necessary to derive additional information from the specimens collected to better understand their early life history in the waters being studied and evaluate the effects of modifications of the environment. Food habits can be studied using techniques similar to those applied to adults (chapter 17). Much literature has been devoted to the food habits, nutritional state, and survival of fish larvae (e.g., Blaxter 1974; Lasker and Sherman 1981). Relatively precise data on age and growth can be determined by examining the daily growth rings of otoliths (Brothers et al. 1976; Taubert and Coble 1977). Finally, for observation of internal structures and particularly study of larval fish osteology, techniques are available for clearing whole specimens and staining cartilage, bone, and other tissues (Taylor 1967; Galat 1972; Fritzsche and Johnson 1979; Ono 1980). For skeletal study, "soft" x-ray photographs of larvae also have proven useful (Miller and Tucker 1969).

9.6 GONAD MATURATION AND FECUNDITY

9.6.1 Gonad Maturation

The state of gonad maturation can be used to determine current reproductive status, size or age at first spawning, proportion of a stock that is reproductively mature or active, and nature of the reproductive cycle for a particular population or species. Hoar (1969), Hempel (1979), and Woynarovich and Horvath (1980) provide good descriptions of gonad and gamete development.

Phases of maturation. In some studies it is sufficient simply to distinguish immature or juvenile fish from the adult. Near or during the spawning season, this is often easily accomplished without significant harm to the fish by simple examination. The conclusions are most obvious if the fish is running-ripe, if the gametes can be extruded with gentle pressure on the abdomen, if the body cavity is strongly distended with eggs, or if secondary sexual characteristics are strongly displayed (e.g., brilliant breeding coloration, elongated fins, hooked snouts, or tubercles, depending on the species).

In many situations, however, it may be necessary or desirable to document more precisely the state of the gonads or gametes. For species with short annual spawning seasons, two types of classifications have been developed—one based largely on the state and size of the gametes themselves and requiring microscopic examination and the other based mostly on the gross appearance and size of the gonads and visible ova (Box 9.2).

Gonadosomatic index (GSI). Another common approach to documenting the development or maturity of the gonads and their products is to calculate the gonadosomatic index (GSI), also known as the maturity coefficient. This is simply the weight of the gonads expressed as a percentage of the body weight prior to gonad removal or after complete evisceration (Nikolsky 1963). If based on eviscerated fish, the value will be quite different; therefore, the method should be duly noted. Also, if possible, the fish and gonad specimens should be weighed in fresh condition; weights can change considerably when specimens are preserved (Laevastu 1965). For most seasonal spawners, the size of the gonads changes dramatically as they pass through the various phases of gamete maturation. For tropical fishes, however, which reproduce almost year-round, the character of the GSI can be quite different— spawning is usually partial and variation in weight of the gonads is comparatively small (Nikolsky 1963).

9.6.2 Fecundity

Fecundity is the primary link between quantitative estimates of spawning intensity (based on collections of eggs and larvae) and the size of the spawning stock (Saville 1964; Smith and Richardson 1977). Fecundity data also are used to estimate survival, to determine the number of fish required for brood stock, and to help characterize specific races, populations, or stocks of fishes. The topic is particularly well covered by Bagenal (1968, 1978). Lagler (1956), Laevastu (1965), and Nikolsky (1963) discuss various methods for determining fecundity.

Fecundity is often referred to as total, absolute, or individual fecundity and is generally defined as the number of ripening oocytes and mature ova or eggs just prior to spawning. Representing the spawning potential, usually for the whole season, fecundity is specifically distinguished by some authors from fertility, which they define as the number of eggs actually shed (Bagenal 1978). The primary difference lies in the fact that in most fish, some, and occasionally many, of the developing and mature eggs remain in the ovary after spawning is completed and are eventually resorbed.

Indices of fecundity. Fecundity data often are related to length, weight, and age. To compare fecundities among individuals, populations, races, or species, many authors have used an index referred to as relative fecundity, the number of eggs per unit weight. However, since body weight is significantly affected by the increase in ovary weight and other changes in condition as fish approach the spawning season, Bagenal (1968, 1978) suggested that length is a much better index of fecundity than is weight. For many fish, fecundity is nearly proportional to length cubed, but in some

Box 9.2. Phases of Gonad Maturation

This summary of phases of gonad maturation in fish is based on accounts by
Nikolsky (1963), Laevastu (1965), Bagenal (1968), Harden-Jones (1968), and
Woynarovich and Horvath (1980).

IMMATURE PHASES: The gonads of juvenile fish are generally very small and
threadlike or stringlike, and they lie close to and under the vertebral column.
They appear transparent, translucent, colorless or grey; in some species, the
ovaries are wine-red and the testes are whitish to grey-brown. The oocytes in
the ovaries are very small, transparent, and not visible to the unaided eye.

DEVELOPING, RIPENING, OR MATURING PHASES: The gonads enlarge
until they finally fill the body cavity, reach a maximum weight, and have
produced gametes in a mature, ripe, or gravid state. Testes generally turn from
a reddish color to white. Ovaries become opaque turning from red to yellow or
orange color because of the process of vitellogenesis, during which yolk is
accumulated in the enlarging oocytes. Because of numerous blood capillaries,
the ovaries may appear tinged with red. The developing eggs themselves
change from very small, transparent oocytes to large, opaque, clearly
discernible ova filled with yolk material, most becoming round and
transparent just prior to spawning. Towards the end of this phase the testes
may yield small amounts of milt when pressure or a milking force is applied to
the abdomen, but they will not be in a running-ripe condition. These phases
are frequently subdivided according to the color and size of the gonads,
amount of body cavity filled, and size and appearance of the developing eggs.

MATURE, RUNNING-RIPE, OR SPAWNING PHASES: The gametes of fish in
this condition are in a mature form with eggs typically round and transparent.
Both eggs and milt flow freely from the body with little or no pressure applied
to the abdomen. Gonad size and weight decline rapidly during spawning
activities.

SPENT OR RECOVERY PHASES: The testes and ovaries are emptied of most
mature gametes and are considerably reduced in size and weight with organ
walls striated. The gonads are red with prominent blood vessels; the ovaries
may be bloody at the beginning of this phase. Residual sperm and advanced or
mature eggs are broken down and resorbed. If the fish is to begin another
cycle, small transparent oocytes may already be developing but remain
invisible to the unaided eye.

cases it increases at a much faster rate, making it proportional to length at higher
powers up to about seven (Hempel 1979). Nikolsky (1963) discussed two additional
indexes referred to as specific fecundity. To better characterize the reproductive
capacity of a population, these indices variously include individual fecundity, onset of
sexual maturation, sex ratio, and periodicity and frequency of spawning throughout
the life of the individual.

   Sampling considerations. Fecundity should be determined from gravid or pre-
running-ripe fish. Capture and handling of running-ripe fish sometimes can result in
the inadvertent extrusion of eggs, and similar losses could have occurred naturally
prior to capture. In fish with less mature ovaries, it can be difficult to distinguish the
currently maturing eggs from younger oocytes, sometimes referred to as recruitment

stock (usually distinguished by their lack of vacuoles or yolk). Also, egg counts from such fish would likely include many eggs destined to become atretic (eggs that will cease development at various stages in the maturation process and will eventually be resorbed). Fecundity can be particularly difficult to determine for multiple or batch spawners in which batches of maturing oocytes are likely to be in various states of development, from eggs barely distinguishable from reserve oocytes to those approaching running-ripe condition. Many fish of broad geographic distribution spawn during a very limited season in the northern latitudes but become partial or extended season spawners in the southern portions of their range. Presumably a count of all yolk-containing eggs gives the fecundity for the season, assuming spawning is not a year-round activity.

Preservation of ovaries. Although formaldehyde solutions are sometimes used to fix and preserve ovaries for egg counts, often as part of an entire specimen, a modified Gilson's fluid (100 ml 60% ethanol or methanol, 880 ml water, 15 ml 80% nitric acid, 18 ml glacial acetic acid, and 20 g mercuric chloride) is generally recommended. This fixative hardens eggs and helps break down ovarian connective tissue. Bagenal (1968) suggested that the ovaries be split longitudinally and turned inside-out to assist penetration by the preservative (large ovaries should be cut into separate sections). The ovaries should be left in Gilson's fixative for at least 24 hours and up to 3 months and should be periodically shaken with vigor. Prior to counting or subsampling the eggs for estimates, it is necessary to clean the eggs of ovarian tissue and fixative. It may be necessary to tease some eggs loose from connective tissue that has not broken down.

Enumeration. While direct and total counts are usually most accurate, such counts require considerable time and patience, especially if the number of eggs is large and an automatic counter (Parrish et al. 1960) is not available. Estimates based on subsamples are made using gravimetric, volumetric, or volume displacement methods. By these methods, the total number of eggs is calculated by multiplying total weight, settled volume, or displaced volume for the entire lot of eggs by the mean of the number of eggs per unit measure in several subsamples. If maturing eggs are distributed relatively homogeneously in the ovaries, egg number can be estimated similarly from fresh or preserved ovaries. In this method, the total weight or volume of the ovaries is determined and the number of eggs per unit measure of ovary is counted in three or more subsamples from various portions of each ovary. The volume or weight of the ovaries and subsamples must be determined before the eggs in the subsamples are removed and counted. In all gravimetric and volumetric approaches, the eggs or ovaries should be reasonably free of excess moisture before weight or volume is determined.

H. von Bayer (1910) prepared a table (reprinted by Lagler 1956) relating the mean diameter of eggs (based on subsample) to the number of eggs per unit settled volume (in fluid). The latter value is multiplied by the settled volume for the entire lot of eggs to determine the number of eggs in that lot. An egg measuring trough is recommended for use in determining mean egg diameter (von Bayer 1910). Burrows (1951) found the method somewhat less accurate than the volumetric and gravimetric methods mentioned above. The following equations closely approximate the results given by von Bayer's table:

$$N = (1272 \text{ mm}^3/\text{ml}) (V/D^3), \tag{9.1}$$

where N is the calculated number of eggs in the lot, V is the settled volume of that lot in milliliters, and D is the mean egg diameter of a subsample in millimeters; or

$$N = (67.67 \text{ in.}^3/qt) (V/D^3) \tag{9.2}$$

for V in quarts and D in inches, the original units of the von Bayer table.

At least two other methods have been used to approximate the total number of maturing oocytes or mature ova in the ovaries. One is a variation of the volumetric methods above and is best used with small oocytes or ova. Water and eggs of a known combined volume are mixed well, providing a relatively homogeneous distribution of eggs throughout the volume of water. A subsample, also of known volume, is withdrawn quickly with a pipette or similar device. The eggs in the subsample are counted and the total number of eggs calculated by proportion. Counts have also been approximated by area. Eggs are spread one layer thick in a rectangle or square and the number of eggs along two adjacent sides are counted; the product of these figures approximates the total number of eggs.

9.7 REFERENCES

Aggus, L.R., J.P. Clugston, A. Houser, R. M. Jenkins, L. E. Vogele, and C. H. Walburg. 1980. Monitoring fish in reservoirs. Pages 149-175 *in* C. H. Hocutt and J. R. Stauffer, Jr., editors. *Biological monitoring of fish.* Lexington Books, D. C. Heath and Company, Lexington, Massachusetts USA.

Ahlstrom, E. H. 1976. Maintenance of quality in fish eggs and larvae collected during plankton hauls. Pages 313-318 *in* H. F. Steedman, editor. *Zooplankton fixation and preservation.* United Nations Educational, Scientific and Cultural Organization, Monographs on Oceanographic Methodology 4, Paris, France.

Ahlstrom, E. H., and H. G. Moser. 1980. Characters useful in identification of pelagic marine fish eggs. *California Cooperative Oceanic Fisheries Investigations Reports* 21:121-131.

Ahlstrom, E. H., J. L. Butler, and B. Y. Sumida. 1976. Pelagic stromateoid fishes (Pisces, Perciformes) of the eastern Pacific: kinds, distributions, and early life histories and observations on five of these from the northwest Atlantic. *Bulletin of Marine Science* 26:285-402.

Alderice, D. F., R. A. Bams, and F. P. J. Velsen. 1977. *Factors affecting deposition, development and survival of salmonid eggs and alevins. A bibliography, 1965-1975.* Canadian Fisheries and Marine Service, Nanaimo, British Columbia.

Allen, D. M., and J. H. Hudson. 1970. *A sled-mounted suction sampler for benthic organisms.* United States Fish and Wildlife Service, Special Scientific Report, Fisheries 614, Washington, District of Columbia, USA.

Aron, W. 1958. The use of a large capacity portable pump for plankton sampling, with notes on plankton patchiness. *Journal of Marine Research* 16:158-173.

Aron, W. 1962. Some aspects of sampling the macroplankton. *Conseil International pour l'Exploration de la Mer, Rapports et Proces-Verbaux des Reunions* 153:29-38.

Atsatt, L. H., and R. R. Seapy. 1974. An analysis of sampling variability in replicated midwater trawls off southern California. *Journal of Experimental Marine Biology and Ecology* 14:261-273.

Auer, N. A., editor. 1982. *Identification of larval fishes of the Great Lakes Basin with emphasis on the Lake Michigan Drainage.* Great Lakes Fishery Commission, Special Publication 82-3, Ann Arbor, Michigan, USA.

Bagenal, T. B. 1968. Fecundity. Pages 160-169 *in* W. E. Ricker, editor. *Methods for assessment of fish production in freshwater.* International Biological Programme Handbook 3, Blackwell Scientific Publications, Oxford, England.

Bagenal, T. B. 1978. Aspects of fish fecundity. Pages 75-101 *in* S. D. Gerking, editor. *Ecology of freshwater fish production.* Halsted Press, John Wiley and Sons, New York, New York, USA.

Bagenal, T. B., and W. Nellen. 1980. Sampling eggs, larvae and juvenile fish. Pages 13-36 *in* T. Backiel and R. L. Welcomme, editors. *Guidelines for sampling fish in inland waters.* Food and Agriculture Organization of the United Nations, European Inland Fisheries Advisory Commission, Technical Paper 33, Rome, Italy.

Balon, E. K. 1975. Reproductive guilds of fishes: a proposal and definition. *Journal of the Fisheries Research Board of Canada* 32:821-864.

Beard, T. D., and G. R. Priegel 1975. Construction and use of a 1-foot fyke net. *The Progressive Fish-Culturist* 37:43-46.

Berry, F. H., and W. J. Richards. 1973. Characters useful to the study of larval fishes. Pages 48-65 *in* A. L. Pacheco, editor. *Proceedings of a workshop on egg, larval, and juvenile stages of fish in Atlantic coast estuaries.* United States National Marine Fisheries Service, Middle Atlantic Coastal Fisheries Center, Technical Publication 1, Highlands, New Jersey, USA.

Blaxter, J. H. S. 1969. Development: eggs and larvae. Pages 177-252 *in* W. S. Hoar and D. J. Randall, editors. *Fish physiology, volume 3: reproduction and growth, bioluminescence, pigments and poisons.* Academic Press, New York, New York, USA.

Blaxter, J. H. S., editor. 1974. *The early life history of fish.* Springer-Verlag, New York, New York, USA.

Bowles, R. R., J. V. Merriner, and G. C. Grant. 1978. *Factors associated with accuracy in sampling fish eggs and larvae.* United States Fish and Wildlife Service, FWS/OBS-78/83, Ann Arbor, Michigan, USA.

Braum, E. 1978. Ecological aspects of the survival of fish eggs, embryos and larvae. Pages 102-131 *in* S. D. Gerking, editor. *Ecology of freshwater fish production.* Halsted Press, John Wiley and Sons, New York, New York, USA.

Breder, C. M., Jr. 1960. Design for a fry trap. *Zoologica* (New York Zoological Society) 45:155-167.

Breder, C. M., Jr., and D. E. Rosen 1966. *Modes of reproduction in fishes.* The Natural History Press, Garden City, New York, USA.

Brothers, E. B., C. P. Mathews, and R. Lasker. 1976. Daily growth increments in otoliths from larval and adult fishes. *Fishery Bulletin* 74:1-8.

Burrows, R. E. 1951. An evaluation of methods of egg enumeration. *The Progressive Fish-Culturist* 13:79-85.

Cada, G. F., J. M. Loar, and K. D. Kumar. 1980. Diel patterns of ichthyoplankton length-density relationships in Upper Watts Bar Reservoir, Tennessee. Pages 79-90 *in* L. A. Fuiman, editor. *Proceedings of the fourth annual larval fish conference.* United States Fish and Wildlife Service, FWS/OBS-80/43, Ann Arbor, Michigan, USA.

Carlander, K. D. 1969. *Handbook of freshwater fishery biology,* 3rd edition, volume 1. Iowa State University Press, Ames, Iowa, USA.

Carlander, K. D. 1977. *Handbook of freshwater fishery biology,* volume 2. Iowa State University Press, Ames, Iowa, USA.

Chapoton, R. B. 1964. Surface trawl for catching juvenile American shad. *The Progressive Fish-Culturist* 26:143-144.

Clarke, W. D. 1964. The jet net, a new high-speed plankton sampler. *Journal of Marine Research* 22:284-287.

Collins, J. J. 1975. An emergent fry trap for lake spawning salmonines and Coregonines. *The Progressive Fish-Culturist* 37:140-142.

Colton, J. B., Jr., and R. R. Marak. 1969. *Guide for identifying the common planktonic fish eggs and larvae of continental shelf waters, Cape Sable to Block Island.* United States Bureau of Commercial Fisheries Biological Laboratory, Laboratory Reference No. 69-9, Woods Hole, Massachusetts, USA.

Dorr, J. A. 1974. Construction of an inexpensive lighted sorting chamber. *The Progressive Fish-Culturist* 36:63-64.

Dovel, W. L. 1964. An approach to sampling estuarine macroplankton. *Chesapeake Science* 55:77-90.

Drewry, G. E. 1979. *A punch card key to the families of yolk sac larval fishes of the Great Lakes region.* Virginia I. Drewry Publishing Company, Waldorf, Maryland, USA.

Ecological Analysts, Incorporated. 1976. *Bowline Point Generating Station entrainment survival and abundance studies.* Ecological Analysts, Incorporated, 1975 Annual Interpretive Report 1, Melville, New York, USA.

Elliott, E. M., and D. Jimenez. 1981. *Laboratory manual for the identification of ichthyo-plankton from the Beverly-Salem Harbor area.* Massachusetts Division of Marine Fisheries, Boston, Massachusetts, USA.

Elliott, E. M., and I. M. Kushlan. 1980. *Annotated bibliography of ichthyoplankton in Massachusetts Bay.* Massachusetts Division of Marine Fisheries, Boston, Massachusetts, USA.

Elliott, G. V. 1976. Diel activity and feeding of schooled largemouth bass fry. *Transactions of the American Fisheries Society* 105:624-627.

Faber, D. J. 1982. Fish larvae caught by a lighttrap at littoral sites in Lac Heney, Quebec, 1979 and 1980. Pages 42-46 *in* C. F. Bryan, J. V. Connor, and F. M. Truesdale, editors. *The Fifth Annual Larval Fish Conference.* Louisiana Cooperative Fishery Research Unit and The School of Forestry and Wildlife Management, Louisiana State University, Baton Rouge, Louisiana, USA.

Fish, M. P. 1932. Contributions to the early life histories of sixty-two species of fishes from Lake Erie and its tributary waters. *Bulletin of the Buerau of Fisheries* (United States) 47:293-398.

Foerster, R. E. 1930. An investigation of the life history and propagation of the sockeye salmon *(Oncorhynchus nerka)* at Cultus Lake, British Columbia. No. 3. The downstream migration of the young in 1926 and 1927. *Contributions to Canadian Biology and Fisheries* 5:57-82.

Fritzsche, R. A., and G. D. Johnson. 1979. Striped bass vs. white perch: application of a new morphological approach to ichthyoplankton taxonomy. Pages 19-29 *in* R. Wallus and C. W. Voigtlander, editors. *Proceedings of a workshop on freshwater larval fishes.* Tennessee Valley Authority, Norris, Tennessee, USA.

Fuiman, L. A. 1982. Correspondence of myomeres and vertebrae and their natural variability during the first year of life in yellow perch. Pages 56-59 *in* C. F. Bryan, J. V. Connor, and F. M Truesdale, editors. *The Fifth Annual Larval Fish Conference.* Louisiana Cooperative Fishery Research Unit and The School of Forestry and Wildlife Management, Louisiana State University, Baton Rouge, Louisiana, USA.

Fraser, J. H. 1968. The history of plankton sampling. Pages 11-18 *in* D. J. Tranter and J. H. Fraser, editors. *Zooplankton sampling.* United Nations Educational, Scientific, and Cultural Organization, Monographs on Oceanographic Methodology 2, Paris, France.

Galat, D. L. 1972. Preparing teleost embryos for study. *The Progressive Fish-Culturist* 34:43-48.

Gale, W. F., and J. D. Thompson. 1975. A suction sampler for quantitatively sampling benthos rocky substrates in rivers. *Transactions of the American Fisheries Society* 104:398-405.

Gale, W. F., and H. W. Mohr, Jr. 1978. Larval fish drift in a large river with a comparison of sampling methods. *Transactions of the American Fisheries Society* 107:46-55.

Gammon, J. R. 1965. Device for collecting eggs of muskellunge, northern pike, and other scatter-spawning species. *The Progressive Fish-Culturist* 27:78.

Gennings, R. M. 1965. Description of a portable fish egg pump. Pages 65-72 *in Oklahoma Fishery Research Laboratory Semi-annual Report* (January-June 1965). University of Oklahoma, Norman, Oklahoma, USA.

Gerrick, D. J. 1968. A comparative study of antioxidants in color preservation of fish. *Ohio Journal of Science* 68:239-240.

Graser, L. F. 1977. Selectivity of larval fish gear and some new techniques for entrainment and open water larval fish sampling. Pages 56-71 *in* L. L. Olmsted, editor. *Proceedings of the First Symposium on Freshwater Larval Fish.* Southeastern Electric Exchange, Atlanta, Georgia, USA.

Hamilton, J. A. R., L. O. Rothfus, M. W. Erho, and J. D. Remington. 1970. *Use of a hydroelectric reservoir for rearing of coho salmon (Oncorhynchus kisutch).* Washington Department of Fisheries, Research Bulletin 9, Olympia, Washington, USA.

Harden-Jones, F. R. 1968. *Fish migration.* Edward Arnold Limited, London, England.

Hardy, J. D., Jr., G. E. Drewry, R. A. Fritzsche, G. D. Johnson, P. W. Jones, and F. D. Martin. 1978. *Development of fishes of the Mid-Atlantic Bight, an atlas of egg, larval and juvenile stages.* United States Fish and Wildlife Service FWS/OBS-78/12 (6 volumes), Washington, District of Columbia, USA.

Haury, L. R., P. H. Wiebe, and S. H. Boyd. 1976. Longhurst-Hardy plankton recorders: their design and use to minimize bias. *Deep-Sea Research* 23:1217-1229.

Heard, W. R. 1964. Phototactic behavior of emerging sockeye salmon fry. *Animal Behaviour* 12:328-388.

Hempel, G. 1979. *Early life history of marine fish: the egg stage.* Washington Sea Grant Program, Seattle, Washington, USA.

Hempel, G., and H. Weikert. 1972. The neuston of the subtropical and boreal north-eastern Atlantic Ocean—a review. *Marine Biology* 13:70-88.

Hettler, W. F. 1979. Modified neuston net for collecting live larval and juvenile fish. *The Progressive Fish-Culturist* 41:32-33.

Hoagman, W. J. 1977. A free-falling drop net for quantitatively sampling a water column. *Transactions of the American Fisheries Society* 106:140-145.

Hoar, W. S. 1969. Reproduction. Pages 1-72 *in* W. S. Hoar and D. J. Randall, editors. *Fish physiology, volume 3: reproduction and growth, bioluminescence, pigments and poisons.* Academic Press, New York, New York, USA.

Hogue, J. J., Jr., R. Wallus, and L. K. Kay. 1976. *Preliminary guide to the identification of larval fishes in the Tennessee River.* Tennessee Valley Authority, Technical Note B19, Norris, Tennessee, USA.

Houser, A. 1976. Trawling gear for sampling pelagic young-of-the-year fishes in large reservoirs. *Proceedings of the Annual Conference of the Western Association of State Game and Fish Commissioners* 56:265-272.

Hoyt, R. D. 1979. *Proceedings of the third symposium on larval fish.* Department of Biology, Western Kentucky University, Bowling Green, Kentucky, USA.

Hubbs, C. L. 1958. *Dikellorhynchus* and *Kanazawaichthys:* Nominal fish genera interpreted as based on prejuveniles of *Malacanthus* and *Antennarius,* respectively. *Copeia* 1958:282-285.

Hunter, J. R., D. C. Aasted, and C. T. Mitchell. 1966. Design and use of a miniature purse seine. *The Progressive Fish-Culturist* 28:175-179.

Jossi, J. W. 1970. *Annotated bibliography of zooplankton sampling devices.* United States Fish and Wildlife Service, Special Scientific Report-Fisheries 609, Washington, District of Columbia, USA.

Kernehan, R. J. 1976. *A bibliography of early life stages of fishes.* Ichthyological Associates, Incorporated, Bulletin 14, Ithaca, New York, USA.

Laevastu, T. 1965. *Manual of methods in fisheries biology, section 4-research on fish stocks.* Food and Agriculture Organization of the United Nations, Manuals in Fisheries Science 1, Fascicule 9, Rome, Italy.

Lagler, K. F. 1956. *Freshwater fishery biology.* W. C. Brown Company, Dubuque, Iowa, USA.

Lasker, R., and K. Sherman, editors. 1981. The early life history of fish: recent studies. *Conseil International pour l'Exploration de la Mer, Rapports et Proces - Verbaux des Reunions,* 178.

Lippson, A. J., and R. L. Moran. 1974. *Manual for identification of early developmental stages of fishes of the Potomac River estuary.* Martin Marietta Corporation, Environmental Technology Center, PPSP—MP-13, Baltimore, Maryland, USA.

Mansueti, A. J., and J. D. Hardy, Jr. 1967. *Development of fishes of the Chesapeake Bay region; an atlas of egg, larval, and juvenile stages; part I.* Natural Resources Institute, University of Maryland, Baltimore, Maryland, USA.

Manz, J. V. 1961. A pumping device used to collect walleye eggs from offshore spawning areas in western Lake Erie. *Transactions of the American Fisheries Society* 93:201-206.

Marcy, B. C. 1975. Entrainment of organisms at power plants, with emphasis on fishes - an overview. Pages 89-106 *in* S. B. Saila, editor. *Fisheries and energy production, a symposium.* Lexington Books, D. C. Heath and Company, Lexington, Massachusetts, USA.

Marcy, B. C., Jr., and M. D. Dahlberg. 1980. Sampling problems associated with ichthyoplankton field-monitoring studies with emphasis on entrainment. Pages 233-252 *in* C. H. Hocutt and J. R. Stauffer, Jr., editors. *Biological monitoring of fish.* Lexington Books, D. C. Heath and Company, Lexington, Massachusetts, USA.

Marcy, B. C., Jr., V. R. Kranz, and R. P. Barr. 1980. Ecological and behavioral characteristics of fish eggs and young influencing their entrainment. Pages 29-74 *in* C. H. Hocutt and J. R. Stauffer, Jr., editors. *Power plants: effects on fish and shellfish behavior.* Academic Press, New York, New York, USA.

Markle, D. F. (In press). Phosphate buffered formalin for long-term preservation of formalin fixed ichthyoplankton. *Copeia.*

May, E. B., and C. R. Gasaway. 1967. *A preliminary key to the identification of larval fishes of Oklahoma, with particular reference to Canton Reservoir, including a selected bibliography.* Oklahoma Department of Wildlife Conservation, Oklahoma Fishery Research Laboratory Bulletin 5, Norman, Oklahoma, USA.

McGowan, J.A., and D. M. Brown. 1966. *A new opening-closing paired zooplankton net.* University of California, Scripps Institute of Oceanography, Reference 66-23, La Jolla, California, USA.

Miller, J. M. 1973. A quantitative push-net system for transect studies of larval fish and macro-zooplankton. *Limnology and Oceanography* 18:175-178.

Miller, J. M., and J. W. Tucker, Jr. 1979. X-radiography of larval and juvenile fishes. *Copeia* 1979:535-538.

Mitterer, L. G., and W. D. Pearson. 1977. Rose bengal stain as an aid in sorting larval fish samples. *The Progressive Fish-Culturist* 39:119-120.

Moser, H. G. 1971. Morphological and functional aspects of marine fish larvae. Pages 89-131 *in* R. Lasker, editor. *Marine fish larvae—morphology, ecology, and relation to fisheries.* Washington Sea Grant Program, Seattle, Washington, USA.

Murphy, G. I., and R. I. Clutter. 1972. Sampling anchovy larvae with a plankton purse seine. *Fishery Bulletin* 70:789-798.

Nagiec, M. 1975. Description of a ring net and a trap net for the sampling of pike perch fry. Pages 566-570 *in* R. L. Welcomme, editor. *Symposium on the methodology for the survey, monitoring and appraisal of fishery resources in lakes and large rivers-panel review and relevant papers.* Food and Agriculture Organization of the United Nations, European Inland Fisheries Advisory Commission Technical Paper 23 (Supplement 1), Volume 1, Rome, Italy.

Nellen, W., and D. Schnack. 1975. Sampling problems and methods of fish eggs and larvae investigations with special reference to inland waters. Pages 538-551 *in* R. L. Welcomme, editor. *Symposium on the methodology for the survey, monitoring and appraisal of fishery resources in lakes and large rivers - panel reviews and relevant papers.* Food and Agriculture Organization of the United Nations, European Inland Fisheries Advisory Commission Technical Paper 21, Rome, Italy.

Nikolsky, G. V. 1963. *The ecology of fishes.* Academic Press, New York, New York, USA. (Translated from Russian by L. Birkett).

Noble, R. L. 1970. Evaluation of the Miller high-speed sampler for sampling yellow perch and walleye fry. *Journal of the Fisheries Research Board of Canada* 27:1033-1044.

Ono, R. D. 1980. A silver impregnation technique to demonstrate muscle-bone-cartilage relationships in fishes. *Stain Technology* 55:67-70.

Parrish, B. B., I. G. Baxter, and M. I. D. Mowat. 1960. An automatic fish egg counter. *Nature* (London) 185:777.

Pearcy, W. G. 1980. A large, opening-closing midwater trawl for sampling oceanic nekton, and comparison of catches with an Isaacs-Kidd midwater trawl. *Fishery Bulletin* 78:529-534.

Phillips, R. W. 1966. A trap for capture of emerging salmonid fry. *The Progressive Fish-Culturist* 28:107.

Porter, T. R. 1973. Fry emergence trap and holding box. *The Progressive Fish-Culturist* 35:104-106.

Portner, E. M., and C. A. Rohde. 1977. *Tests of high volume pump for ichthyoplankton in the Chesapeake and Delaware Canal.* Johns Hopkins University, PPSE-T-3, Baltimore, Maryland, USA.

Richards, W. J., and F. H. Berry. 1973. Preserving and preparing larval fishes for study. Pages 12-19 *in* A. L. Pacheco, editor. *Proceedings of a workshop on egg, larval and juvenile stages of fish in Atlantic coast estuaries.* United States National Marine Fisheries Service, Middle Atlantic Coastal Fisheries Center, Technical Publication 1, Highlands, New Jersey, USA.

Sameoto, D. D., and L. O. Jaroszynski. 1976. *Some zooplankton net modifications and developments.* Canadian Fisheries and Marine Service, Technical Report 679, Research and Development Directorate, Dartmouth, Nova Scotia.

Sameoto, D. D., L. O. Jaroszynski, and W. B. Fraser. 1977. A multiple opening and closing plankton sampler based on the MOCNESS and N.I.O. nets. *Journal of the Fisheries Research Board of Canada* 34:1230-1235.

Saville, A. 1964. Estimation of the abundance of a fish stock from egg and larval surveys. *Conseil International pour l'Exploration de la Mer, Rapports et Proces - Verbaux des Reunions* 155:164-170.

Scotton, L. N., R. E. Smith, N. S. Smith, K. S. Price, and D. P. de Sylva. 1973. *Pictorial guide to fish larvae of the Delaware Bay, with information useful for the study of fish larvae.* University of Delaware, College of Marine Studies, Delaware Bay Report Series 7, Newark, Delaware, USA.

Shaw, R. F. 1980. *A bibliography of the eggs, larval and juvenile stages of fishes: including other pertinent references.* Maine Sea Grant Program, Technical Report 61, Orono, Maine, USA.

Shealy, M. H. 1971. Nesting bass observed with underwater television. *New York's Food and Life Sciences Quarterly* 4 (4):18-20.

Shih, G. T., A. J. G. Figueira, and E. H. Grainger. 1971. *A synopsis of Canadian marine zooplankton.* Fisheries Research Board of Canada, Bulletin 176, Ottawa, Ontario, Canada.

Siefert, R. E. 1969. Characteristics for separation of white and black crappie larvae. *Transactions of the American Fisheries Society* 98:326-328.

Smith, P. E. 1981. Fisheries on coastal pelagic schooling fish. Pages 1-31 *in* R. Lasker, editor. *Marine fish larvae - morphology, ecology, and relation to fisheries.* Washington Sea Grant Program, Seattle, Washington, USA.

Smith, P. E., and S. L. Richardson. 1977. *Standard techniques for pelagic fish egg and larva studies.* Food and Agriculture Organization of the United Nations, Fisheries Technical Paper 175, Rome, Italy.

Smith, P. E., and S. L. Richardson. 1979. *Selected bibliography on pelagic fish egg and larva surveys.* Food and Agriculture Organization of the United Nations, Fisheries Circular 706, Rome, Italy.

Snyder, D. E. 1976a. Terminologies for intervals of larval fish development. Pages 41-60 *in* J. Boreman, editor. *Great Lakes fish egg and larvae identification; proceedings of a workshop.* United States Fish and Wildlife Service, FWS/OBS-76/23, Ann Arbor, Michigan, USA.

Snyder, D. E. 1976b. Identification tools: what's available and what could be developed. Report of working group 2. Pages 88-96 *in* J. Boreman, editor. *Great Lakes fish egg and larvae identification; proceedings of a workshop.* United States Fish and Wildlife Service, FWS/OBS-76/23, Ann Arbor, Michigan, USA.

Snyder, D. E. 1981. *Contributions to a guide to the cypriniform fish larvae of the Upper Colorado River System in Colorado.* United States Bureau of Land Management, Biological Sciences Series 3, Denver, Colorado, USA.

Stauffer, T. W. 1981. Collecting gear for lake trout eggs and fry. *The Progressive Fish-Culturist* 43:186-193.

Steedman, H. F. 1976a. General and applied data on formaldehyde fixation and preservation of marine zooplankton. Pages 103-154 *in* H. F. Steedman, editor. *Zooplankton fixation and preservation.* United Nations Educational, Scientific, and Cultural Organization, Monographs on Oceanographic Methodology 4, Paris, France.

Steedman, H. F. 1976b. Some technical aspects of leptocephalus fixation and preservation. Pages 321-322 *in* H. F. Steedman, editor. *Zooplankton fixation and preservation.* United Nations Educational, Scientific and Cultural Organization, Monographs on Oceanographic Methodology 4, Paris, France.

Strawn, K. 1954. The pushnet, a one-man net for collecting in attached vegetation. *Copeia* 1954: 195-197.

Taubert, B. D., and D. W. Coble. 1977. Daily rings in otoliths of three species of *Lepomis* and *Tilapia mossambica. Journal of the Fisheries Research Board of Canada* 34:332-340.

Taylor, W. R. 1967. An enzyme method of clearing and staining small vertebrates. *Proceedings of the United States National Museum* 122:1-17.

Taylor, W. R. 1977. Observations on specimen fixation. *Proceedings of the Biological Society of Washington* 90:753-763.

Taylor, W.R. 1981. On preservation of color and color patterns. American Society of Ichthyologists and Herpetologists, Subcommittee on Curatorial Supplies and Practices. *Curation Newsletter* 3:2-9.

Todd, I. S. 1966. A technique for the enumeration of chum salmon fry in the Frazer River, British Columbia. *Canadian Fish Culturist* 38:3-36.

Tranter, D. J., and J. H. Fraser, editors. 1968. *Zooplankton sampling.* United Nations Educational, Scientific and Cultural Organization, Monographs on Oceanographic Methodology 2, Paris, France.

Tranter, D. J., and P. E. Smith. 1968. Filtration performance. Pages 27-56 *in* D. J. Tranter and J. H. Fraser, editors. *Zooplankton sampling.* United Nations Educational, Scientific, and Cultural Organization, Monographs on Oceanographic Methodology 2, Paris, France.

Trippel, E. A., and E. J. Crossman. 1981. Collapsible fish trap of plexiglass and netting. *The Progressive Fish-Culturist* 43:157-158.

Viljanen, M. 1980. A comparison of a large diameter corer and a new hydraulic suction sampler in sampling eggs of *Coregonus albula. Annales Zoologici Fennici* 17:269-273.

Vogele, L. E. 1975. *Reproduction of spotted bass, Micropterus punctulatus, in Bull Shoals Reservoir, Arkansas.* United States Fish and Wildlife Service, Technical Paper 84, Washington, District of Columbia, USA.

von Bayer, H. 1910. A method of measuring fish eggs. *Bulletin of the Bureau of Fisheries* (United States) 28:1011-1014.

Wang, J. C. S. 1981. *Taxonomy of the early life stages of fishes—fishes of the Sacramento-San Joaquin Estuary and Moss Landing-Elkhorn Slough, California.* Ecological Analysts Incorporated, Concord, California, USA.

Wang, J. C. S., and R. J. Kernehan. 1979. *Fishes of the Delaware estuaries - a guide to the early life histories.* Ecological Analysts Incorporated, Towson, Maryland, USA.

Werner, R. G. 1968. Addendum: Methods of sampling fish larvae in nature. Pages 178-181 *in* W. E. Ricker, editor. *Methods for assessment of fish production in freshwater.* International Biological Programme Handbook 3, Blackwell Scientific Publications, Oxford, England.

Werner, R. G. 1976. A preliminary annotated bibliography of the literature relevant to descriptions of eggs and larval stages of fishes of the Great Lakes. Pages 107-200 *in* J. Boreman, editor. *Great Lakes fish egg and larvae identification; proceedings of a workshop.* United States Fish and Wildlife Service, FWS/OBS-76/23, Ann Arbor, Michigan, USA.

Williams, G. E. 1974. New technique to facilitate hand-picking macrobenthos. *Transactions of the American Microscopical Society* 93:220-226.

Wolf, P. 1951. A trap for the capture of fish and other organisms moving downstream. *Transactions of the American Fisheries Society* 80:41-45.

Woynarovich, E., and L. Horvath. 1980. *The artificial propagation of warm-water finfishes - a manual for extension.* Food and Agriculture Organization of the United Nations, Fisheries Technical Paper 201, Rome, Italy.

Yocum, W. L., and F. J. Tesar. 1980. Sled for sampling benthic fish larvae. *The Progressive Fish-Culturist* 42:118-119.

Chapter 10
Sampling with Toxicants

WILLIAM D. DAVIES AND WILLIAM L. SHELTON

10.1 INTRODUCTION

The use of fish toxicants for sampling fish populations is a common practice, particularly in impounded waters and streams in the southeastern United States. While the use of fish toxicants for reclamation is thoroughly reviewed by Lennon et al. (1970), reports addressing their use in sampling are scattered throughout the literature. The objective of this chapter is to bring into focus the varied aspects of sampling fish populations with approved toxicants. More specifically the chapter will (1) examine sources of bias associated with the techniques, (2) review fish toxicants presently registered for use, (3) assess the types of biological information that are normally gathered, and (4) discuss methodology for employing fish toxicants in lentic and lotic conditions.

10.2 FISH TOXICANTS

Lennon et al. (1970) discuss 30 toxicants used as piscicides while Cumming (1975) relates the history of their use in the United States. Two of these, rotenone (Derris, Cube) and antimycin (Fintrol), are commonly used to sample fish populations and have formulations currently registered for fishery use by the United States Environmental Protection Agency according to the Federal Environmental Pesticide Control Act (Schnick and Meyer 1978). Other toxicants also are commonly used for sampling outside the United States. Toxaphene, for example, while not registered for fishery use in the United States since 1963, continues to be used in other countries for sampling and reclamation of fish populations. Where certain toxicants are not restricted, as in many developing countries, common sense should prevail in their choice and application. A decision should be based not only on the efficacy of a toxicant, but also on its persistence in the environment, toxicity to other animals, and danger to man.

Rotenone is an extract from rotenone-bearing plants in the family Leguminosae. It is applied as a powder, a wettable powder, or a liquid containing from 2.5 to 5% rotenone. A powder form (Cube) looses its toxicity when exposed to air and is more difficult to apply; generally, the wettable powders and liquid formulations are easier and safer to use. Liquid formulations can be stored in sealed containers for periods up to one year without loss of efficacy. Regardless of the form, take care to avoid prolonged contact with the skin and especially the eyes, nose, and throat area.

Rotenone kills fish by blocking oxygen uptake, and the fish suffocates. Rotenone toxicity is primarily a function of the species, size of fish, and water temperature, although pH, oxygen concentration, and the presence of suspended matter also affect toxicity. For most species toxicity is greatest between 50 and 75°F; 0.5 mg/l of formulation (0.025 mg/l of rotenone) kills most fish species. However, 1.0 to 2.0 mg/l is usually applied to insure a complete kill. Depending on the temperature, stressed fish will surface within a few minutes to several hours. The suffocating action is reversible; Bouck and Ball (1965) showed that a 5-10 mg/l concentration of methylene blue can be used to revive stressed fish. The toxic effects can be eliminated almost immediately with potassium permanganate (KMnO4) at 1 mg/l for each 0.05 mg/l of rotenone. Potassium permanganate, however, is also toxic to some fish at concentrations of 3 to 4 mg/l and is hazardous to apply. Nose, throat, and eye protection should be used by anyone working with potassium permaganate. Rotenone loses its toxicity in several days under natural conditions. Post (1958) determined that duration of toxicity in days (d) could adequately be expressed as a function of temperature $(T$, in °F) where

$$d = 93 - 1.33 \ T \qquad (10.1)$$

for temperatures between 45 and 60°F, and

$$d = 38 - 0.43 \ T \qquad (10.2)$$

for temperatures between 60 and 80°F. Detoxification may take longer in very soft water or in deeper, stratified bodies of water.

While rotenone is generally nontoxic to most mammals and birds at concentrations used to sample fish populations, swine may be adversely affected. Application of 1.0 and 2.0 mg/l of rotenone is lethal to zooplankton and many aquatic invertebrates, but effects are usually short term.

Antimycin is an antibiotic, produced in cultures of *Streptomyces,* that is toxic to fish (Derse and Strong 1963). It is sold under the trade name "Fintrol" and is available in various formulations. Fintrol-5 and Fintrol-15 are formulations coated on sand grains that release the toxicant evenly in the first 5 and 15 feet (1.5 and 4.6 m) of depth, respectively. There is also a liquid formulation which was developed for use in flowing water.

Antimycin kills fish by inhibiting respiration, but at a different site than rotenone. Unlike rotenone, the effects are irreversible at lethal concentrations. The toxicity of antimycin is diminished by high alkalinity, high temperature, sunlight, and the metabolic activity of aquatic organisms. Marking and Dawson (1972) concluded that antimycin has a half-life of biological activity of 5-8 days in soft, acid waters, but only a few hours at pHs of 8.5 and above. As with rotenone, antimycin can be detoxified with potassium permanganate.

Fish species have been divided into three groups with high, moderate, and low sensitivity to antimycin (Walker et al. 1964). In general, scaled fish are more sensitive than bullheads or catfish. Sunfish are moderately sensitive while such species as gizzard shad and trout are highly sensitive.

Antimycin can be used to remove scaled fish (except goldfish) from a catfish culture pond or to selectively sample portions of the population. A concentration of 15 ppb of antimycin will eliminate all cyprinids, catostomids, percids, and centrarchids (Gilderhus et al. 1969). This concentration is less harmful than the recommended fish-killing concentration of rotenone to aquatic animals other than fish. Because of its selective qualities, antimycin may be more useful in the management of fish populations than as a sampling tool.

10.3 SAMPLING OBJECTIVES

Sampling with toxicants is beset by a host of problems, i.e., associated biases and sampling variations (chapter 1). Bias (or accuracy) deals with how close the best estimate obtained from the sampling is to the true or population value. Variance (or precision) relates to how much estimates differ among repeated sampling efforts. Keeping these two concepts in mind when developing any sampling scheme will aid in efficient planning of the study. The units of measurement commonly associated with this type of sampling (Hayne et al. 1968) are discussed in regard to their bias and expected variability.

10.3.1 Total Standing Crop Estimation

Sampling fish populations with toxicants may be one of the better methods for obtaining an estimate of total standing crop, or fish biomass, present at some specific time. The estimate may be used as a measure of stock density or expanded to the total area of the lake, reservoir, or stream and compared to standing crop data from similar bodies of water.

Ideally, a random sampling design would eliminate bias caused by site selection or area sampled (i.e., the mean standing crop from a series of randomly selected samples will approximate the true mean value.) However, if the choice of sampling sites in a reservoir is restricted to shoreline areas, then the sample estimate represents only that part of the reservoir from which the sample was drawn (i.e., the total shoreline area), and extending the estimate to the entire water body will cause bias. An alternate approach (Grinstead et al. 1978) is to select subjectively a sampling area deemed representative of the body of water.

Several other possible sources of bias exist. Even though block nets are commonly used, fish move past the nets. In this respect, researchers recovered 75% of fish that were tagged and released in the area to be sampled but only 32% of fish that were collected elsewhere and released in to the study area (Axon et al. 1980).

Another source of bias is incomplete recovery of all fish. This is especially troublesome if certain species or size groups within species are selected against systematically. Some insight into the influence of certain environmental factors that affect surfacing of dead fish following application of toxicants is provided by Parker (1970). Where water temperature was above 60°F, water depth was less than 3 meters, and rooted plants were absent, practically all rotenone-killed fish surfaced within a week. Fish from a sampling area, however, are normally collected for periods of 1-3 days; some biomass, especially that comprised of small fish, will be missed. Henley (1967), using SCUBA gear, determined that approximately 75% of the total number and 95% of the biomass of fish would normally surface during a three-day period. The percentages declined to 67 and 85%, respectively, for a two-day period. Adjustment factors for non-recovery of fish from the sampling area are presented in Table 10.1.

Distribution patterns among fish species and sizes within species affect to what extent shoreline standing crop estimates are biased when expanded to open water. Based on fish population studies in Lakes Barkley (Aggus et al. 1980) and Douglas (Hayne et al. 1968), where shoreline standing crop estimates were compared to estimates from larger areas of the reservoir including open water, there does not appear to be an appropriate single adjustment factor. In both studies, the relationship between estimates from shoreline and open water areas varied depending on the species present, their sizes, and the size of cove sampled. Aggus et al. (1980) suggest that adjustment factors derived from the Douglas Lake study could be applied to thermally stratified storage impoundments with high shoreline development, whereas those derived from the Barkley Lake study could be used on large, shallow,

Table 10.1 Adjustments to cove rotenone data for various species. To adjust data for nonrecovery of some individuals, multiply the observed data by the value in column *(1)*. To adjust data to represent open-water areas, multiply observed data by the value in column *(2)* or *(3)*, depending on the type of reservoir (*see* text.)[a]

Species	Length interval (in)	Adjustments for Non-recovery (1)	Adjustments for open-water	
			Douglas Lake study (2)	Barkley Lake study (3)
Gars	1-40	1.78	1.00	NA [b]
Bowfin	1-40	5.00	1.00	NA
Gizzard shad	1-16	1.33	1.00	4.20
Threadfin shad	1-12	1.33	1.00	21.09
Pikes	1-40	1.59	1.00	NA
Carp	1-8,	1.66	0.74	1.00
	9-12,	1.66	1.22	0.74
	13-40	1.66	1.21	1.82
Minnows	1-15	2.00	1.00	NA
Carpsuckers,	1-8,	1.52	0.40	NA
Hogsuckers,	9-12,	1.52	0.83	NA
and Redhorses	13-40,	1.52	2.52	NA
Bullheads and	1-5,	1.89	0.64	2.74
Other catfish	6-9,	1.89	1.62	3.29
	9-40	1.89	1.04	1.87
Flathead catfish	1-5,	1.27	0.64	2.74
	6-11,	1.27	1.62	3.29
	12-40	1.27	1.04	1.87
Madtoms	1-5,	1.89	0.64	NA
	6-9,	1.89	1.62	NA
Sunfishes	1-3,	1.85	0.50	0.21
	4-5,	1.85	1.00	0.59
	6-13	1.85	1.22	0.60
Black basses	1-4,	1.67	0.50	0.08
	5-9,	1.67	0.90	0.16
	10-28	1.67	1.34	0.24
Crappie	1-3,	1.64	0.97	5.84
	4-7,	1.64	1.38	26.02
	8-20	1.64	2.42	3.28
Yellow Perch	1-20	2.33	1.00	NA
Freshwater drum	1-5,	1.67	1.26	1.40
	6-9,	1.67	1.38	1.95
	10-40	1.67	2.56	1.68
Walleye and	1-8,	1.59	0.64	NA
Sauger	9-11	1.59	0.54	NA
Darters	1-7	2.00	1.00	NA

[a]Larry Aggus, National Reservoir Research Program, United States Fish and Wildlife Service, Fayetteville, AR, USA. (personal communication)
[b]Not available.

mainstream reservoirs in which thermal stratification is weak or nonexistent. Table 10.1 summarizes biomass adjustment factors presently being considered from each study.

Estimates of total standing crop from shoreline areas may change because of seasonal differences in the distribution of some adult fishes and possibly their reproduction, growth, and mortality during the year. Cove rotenone sampling of fish populations in April, June, and August on West Point Lake, Alabama-Georgia, provided estimates of largemouth bass standing crop of 42.8, 9.9, and 5.7 kg/ha, respectively. There appeared to be an off-shore movement of adult largemouth bass before June. While the June and August samples may have provided a more realistic density estimate for the reservoir as a whole, the size (and age) structure may be better represented in the April sample. Choosing which season to sample depends on the uses of the data.

Variability associated with estimates of standing crop has been commonly reported (Lambou and Stearn 1958; Hayne et al. 1968; Shelton et al. 1982) as coefficient of variation (CV), which is the standard deviation divided by the mean. Based on these data, the number of samples needed for a confidence interval of a chosen width can be estimated for various error probabilities. The sample size required for reasonably precise estimates becomes large for a CV of 1.0 or more, which is typical of standing crop data. Within practical limits of manpower and cost (sampling four to six sites per lake or reservoir annually), the confidence interval could only be within $\pm 50\%$ of the mean standing crop at a probability level of $\alpha = 0.05$. In West Point Lake, Alabama-Georgia, stratification by location and repeated sampling of the same area each year has not appreciably lowered variability (Shelton et al. 1982).

10.3.2 Other Uses of Data

Data collected with toxicants also can describe (1) relative abundance of year classes and size distribution within a species, (2) the proportional balance of groups of species, and (3) distribution of total biomass among species. The first can be used as a measure of reproductive success and total annual mortality in populations if bias related to size is constant within a species. The second is an index of the trophic-dynamic state of the community if bias is constant within groups of species. The third is a measure of the contribution of a species to total fish biomass and has been used frequently in conjunction with the proportional balance among groups of species to differentiate between balanced and unbalanced systems. Here sampling bias must be constant over all species concerned.

In most sampling situations the conditions for obtaining unbiased estimates are not satisfied. As a result, estimates may only have value in a relative sense (for example, among years for a single species in a single reservoir) and only then if the bias remains constant from year to year.

10.4 SAMPLING PROCEDURES

Sampling fish populations in littoral areas, either in coves or along straight shorelines, will be used as an example for detailing equipment needs and field procedures. A discussion of other applications (i.e., open water, stream sampling, small area sampling) will emphasize special considerations.

10.4.1 Cove Rotenone Sampling

Several publications (Chance 1958; Swingle 1958; Hall 1974) describe procedures for conducting fish population studies using rotenone. A summary of recommended procedures follows.

Large impoundments contain a wide range of habitats from which fish populations may be sampled. Fishery studies have typically used either random sampling or fixed sites that are repeatedly sampled. King et al. (1981) surveyed state fishery agencies seeking information about their sampling procedures; of 46 that responded, 21 used fixed sites exclusively, 5 used random sites exclusively, and 20 used both.

There are advantages if at least two fixed sites and two random sites each year (or season) are incorporated into the sampling design. A comparison of results from fixed versus randomly selected sites may provide a means for developing statistically unbiased estimates of population parameters in cases where only fixed sites are sampled.

Hayne et al. (1968) suggested that larger coves give better estimates (i.e., less bias, smaller variance) than do smaller coves. The coefficient of variation and bias determined for a variety of measurements for most species declined as the size of the cove was increased from 0.81 to 2.0 ha. Sizes of coves reported in the literature commonly range from 0.3 to 1.6 ha, although recently there appears to be an increase in the size of the area sampled. For example, the Tennessee Valley Authority (TVA) recently set 1.2 ha as a minimum size.

Since water temperature influences the effectiveness and detoxification of rotenone, sampling is usually limited to times when the water temperature is greater than 20°C. Within this time period, seasonal bias may be a consideration.

Usually an area of 0.4 to 1.5 ha can be efficiently sampled by a crew of 7-10 individuals. The procedure begins in the afternoon before the rotenone is applied. The block net is customarily set the afternoon before, and SCUBA-equipped divers (at least two for safety) ensure that the lead line is contiguous with the bottom. At this time, collect and mark and release fish, preferably from inside the cove. The subsequent rate of recapture by species and size can be used to correct for nonrecovery (or escapement). An alternative is to use the correction factors already established (Table 10.1). Measure and record water quality data (i.e., temperature and O_2 profiles) at this time. The volume of water in the cove must be determined, using methods described in chapter 4. Check the water level on the day the rotenone is to be applied.

If the area to be treated has been mapped in advance and the water level has changed appreciably since then, some adjustment may be necessary in the amount of rotenone to be applied. (A severe alteration in water level also will usually change the surface area to the extent that the standing stock calculation changes.) The appropriate concentration of toxicant (usually 1 mg/1) must be applied evenly, vertically and horizontally. For uniform distribution, dilute the rotenone formulation (1 part rotenone to 10 parts water) and pump it onto the deeper portions of the water body through a weighted hose from a boat. Cover the surface area in a boat in a systematic manner, by working back and forth in a zig-zag pattern. Reserve approximately 20% of the formulation for applying (spraying) in shallow water along the shore. See Box 10.1 for example computations.

Apply potassium permanganate outside the net to neutralize rotenone that drifts from the cove. Put an appropriate amount of permanganate into a weighted plastic

Box 10.1.   Information Needed for Applying Rotenone to Coves

Data required: *Example*
  Surface area   *(A)*—acres (square meters)   2.4 (10,100)
  Average depth   *(Z)*—feet (meters)   6.0 (1.83)
  Volume   *(V)*—acre-feet (cubic meters)   15.0 (18,483)
  Concentration   *(C)*—milligrams/liter (mg/1)   1.0 (1.0)

Volume:

$$V_{English} = ZA \qquad\qquad V_{Metric} = ZA$$
$$= 6 \,(2.4) \qquad\qquad\qquad = 1.83 \,(10,100)$$
$$= 15 \text{ acre-feet} \qquad\quad = 18,483 \text{ cubic meters (m}^3)$$

Rotenone (liquid, 5% formulation):

$$\text{Toxicant (gallons)} = V_E 0.33C \qquad \text{Toxicant (liters)} = V_M 0.001C$$
$$= 15 \,(0.33)\, 1.0 \qquad\qquad\qquad = 18,483 \,(0.001)\, 1.0$$
$$= 4.95 \text{ gallons} \qquad\qquad\qquad = 18.48 \text{ liters}$$

(if using powdered form, substitute 2.7 pounds for 0.33 gallons in the above formula)

Detoxification:

Apply potassium permanganate outside net to control drift; use formula for natural detoxification of rotenone to estimate amount remaining after second or third day pickup; apply 1 mg/1 to cove before removing block net to reduce subsequent kill of immigrant fish.

Source:

Two manufacturers currently supplying rotenone formulation for fishery use are

        S. B. Penick and Company   Blue Spruce Company
        1050 Wall Street West   50 Division Avenue
        Lyndhurst, NJ 07071   Millington, NJ 07946

Presently only one company markets antimycin as Fintrol:

        Aquabiotics Corp.
        3386 Commercial Avenue
        Northbrook, IL 60062

bag, and, after the rotenone has been distributed, perforate the bag to release the permanganate as the bag is pulled at various depths behind the boat. Wind direction will dictate whether the release should be concentrated at the surface or near the bottom. For example, wind blowing into the cove, will push water out along the bottom. Therefore, release permanganate near the bottom. As a rule-of-thumb, 10-20 pounds will reduce unwanted kill outside an average-sized cove. In some areas, it may be desirable to distribute additional potassium permanganate inside the cove before the block net is removed. This will reduce mortality of sensitive fish that enter the cove before all rotenone has dissipated.

At water temperatures above 21°C, fish (especially shad) begin to surface within a few minutes. Crews in boats and individuals wading the shoreline collect stressed fish. Continue this operation until stressed fish are no longer observed on the surface. An efficient pickup during the first day reduces handling the second day and probably reduces losses. On the second morning, continue collecting fish that have resurfaced overnight; a third day collection may be required, especially if water temperatures are below 20°C. *See* Box 10.2 for a summary of equipment and supplies needed.

Sort fish collected the first day by species. Then measure and weigh them by length intervals (i.e., inch - or cm-groups). Traditionally "inch-groups" (i.e., 1" $=\leq1.5$"; 2" = 1.5-2.5", etc.) have been used to summarize length-frequency distributions. However, for smaller species (e.g., gizzard shad, bluegill) centimeter grouping may provide additional information. Figure 10.1 depicts length-frequency distributions where centimeter-groups are superimposed on the traditional inch-groups. For both gizzard shad and bluegill, centimeter grouping shows modes that are not identified by the larger interval and, therefore, more clearly depict the size distribution in the populations. Record data on field forms (one page per species) in an indexed notebook.

Treat fish collected the second and third day the same as those collected the first, except do not record weights. Estimate weights of fish collected on the second and

---

Box 10.2.   Equipment Check List for a Rotenone Study

1) Appropriate vehicles (with 1⅞ and/or 2-inch trailer balls)
2) At least three boats (trailers/spare tires) with adequate freeboard/beam; motors in 15-25 hp range with gas cans
3) Paddles, boat cushions; life vest for each worker
4) Tool kit/spare parts for outboard motors, rotenone pump, and net repair items
5) Block nets (usually 210 x 30 ft of 3/8-inch mesh constructed of 43-lb knotless netting) with floats (4 x 3 inches) covered with sleeve of 1/4-inch mesh and a minimum of one lead (8 per lb) per two feet of net
6) Dip nets of various sizes (mesh and handle length)
7) Assorted pans and tubs
8) Rotenone formulation
9) Rotenone pump (medium pressure, 10 gal/min pump with 5 hp gasoline motor, including spray gun, suction hose, and discharge hose)
10) Potassium premanganate
11) Heavy duty plastic bags and nylon rope
12) Diving (SCUBA) gear (2 sets)
13) Survey equipment (plane table or transit/tripod and depth recorder)
14) Survey forms, scale envelopes, and calculator
15) Scales—hanging type of 60 pounds (20-30 kg), platform balance to nearest 0.01 lb (or 1 g) increments
16) Field table, sorting and measuring tables with inch and/or cm intervals, and measuring boards
17) Camping equipment
18) Miscellaneous: - shovel/mattocks
                   - first-aid kits
                   - paper towels
                   - hand cleaner
                   - water cooler
                   - locks/chains
                   - formalin
                   - jars (non-breakable)

third days from average weights from the first day's collection or from appropriate length-weight tables (*see* chapter 15).

Additional information such as weights and lengths of individual fish, reproductive condition, and scales can be obtained. Predators frequently gorge on stressed fish; thus, stomach contents taken from fish collected by rotenone may not be typical of their normal feeding habits.

Subsampling of fish often saves time when excessively large numbers of small fish are encountered. For each species, take a subsample of not less than 10% of the total weight of the species. Sort the subsample by size groups and count, measure, and weigh the fish. Determine the total weight and number by length interval by dividing by the fraction sampled (e.g., 0.10 if 10% of the total is sampled).

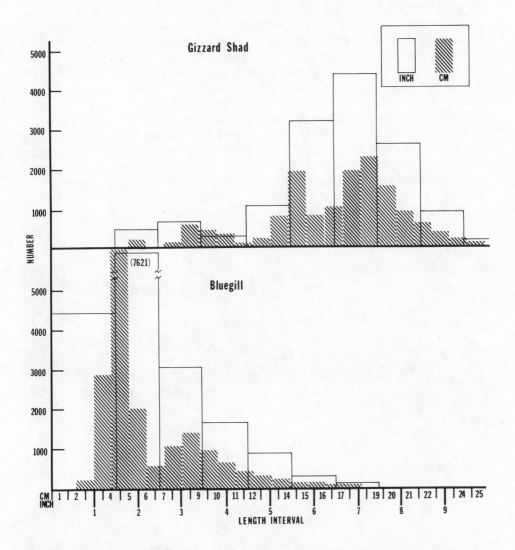

Figure 10.1. Gizzard shad and bluegill length-frequency plots from August cove rotenone sampling, West Point Lake, Alabama-Georgia, based on the same subsample that was grouped by inches and centimeters.

10.4.2 Other Sampling Conditions

Shallow water. Two methods for sampling fish populations in shallow water or along the shoreline have been evaluated. Wegener et al. (1974) describe the use of the "Wegener Ring" in shallow water (< 45 cm) typical of many Florida lakes. The ring consists of two hoops (2.25 m in diameter); one acts as a lead line while the other floats. The attached netting (0.3-cm, ace mesh) is 51 centimeters in depth. Two individuals throw the ring five to six meters and check to see that the lower ring is on the bottom; they then apply rotenone within the blocked-off area (0.0004 ha). The authors report that 12 to 15 areas can be sampled by two workers in an eight-hour day. Because of the variation expected from this procedure, the method seems most appropriate for determining trends in reproduction of prey throughout the season as well as for documenting growth and mortality of age-0 predators.

Timmons et al. (1979) sampled fish populations in shoreline areas of West Point Lake during May-October by blocking off a 0.015-ha area. The sampling procedure consists of enclosing a portion of the littoral zone with a net (30.5 m long, 2.7 m deep, 0.5-cm mesh) that has floats and adequate leads. Secure one end to the shore; feed the net out from the bow of a boat that is backed in an arc (Fig. 10.2) until the opposite end of the net can be secured to the shore. Check the lead line to determine that it is on the bottom. Apply a rotenone formulation to achieve a 1 mg/l concentration (100-150 ml of formulation needed). Collect stressed fish until no additional fish are noted; pull the net to shore in a seinelike fashion, and collect the remaining fish. This sampling procedure also allows a relatively large number of areas to be sampled. Two workers can easily sample 6-8 areas in an eight-hour day. This littoral rotenone technique has been used to estimate the relative abundance and size distribution of both age-0 predator and prey species throughout the growing season. Within-day coefficients of variation are relatively high (0.6-0.8) for abundance of age-0 largemouth bass. The

Figure 10.2.   Block net set from boat to enclose 0.015-ha area.

variability is highest in the spring but reduces to "acceptable" levels (i.e., CV = 0.5) in late summer and fall. Estimates of age-0 largemouth bass from 0.015-ha sample areas approximated those obtained from large cove samples taken at similar times (Shelton and Davies 1981). Intermediate-sized (0.08-ha) littoral rotenone samples also should be considered as an option (Shireman et al. 1981).

Open water. Sampling fish in open water usually involves a "standard" area delineated by blocking off 0.41 ha in a square (210 feet or 64 m on each side). Securely anchor corners and sides of the block net to maintain the squared configuration. Apply a rotenone formulation within the blocked area to achieve a concentration of 1 mg/1. A smaller area (0.08 ha) was evaluated by Shireman et al. (1981). They concluded that the smaller area provided data that were essentially the same as those estimated from standard sets, but required much less labor and equipment.

Aggus et al. (1980) estimated standing stock from open water areas in the Lake Barkley study. A cursory comparison of these to total population values gives the impression that estimates from open-water sets closely approximate the true values. Open-water sets, however, are more labor intensive than blocking off coves, and they are not feasible in deep bodies of open water; also in most situations it is difficult to avoid extensive fish kills outside the sampling area. This is particularly a problem where gizzard shad are encountered because it is difficult to counteract rotenone drift from an open area with potassium permanganate.

Sampling in streams. The problems associated with confining rotenone to the sampling area are magnified in streams. Field procedures for applying a rotenone formulation and its subsequent detoxification in large streams (3,500 cfs) have been discussed by Johnson and Pasch (1976) and Ober (1981). The same general procedures apply for smaller streams with lower flow rates, but there are fewer difficulties involving net placement and chemical application.

A concentration of 1 mg/1 of a 5% rotenone formulation over a 45-60 minute period at water temperature greater than 18°C is recommended for most warmwater streams. The rate that the liquid formulation is applied to achieve the desired concentration for the recommended time is a function of the size of the area and water discharge, usually measured in cubic feet per second (cfs). *See* Box 10.3 for examples of calculations. Determine current velocity with a mechanical current meter (taking measurements at several locations and depths along a transect across the stream) or less precisely with a float. Use transects (upstream, midstream, and downstream sections of the sample area) to measure stream width. Take a series of at least five equally spaced depth measurements at each transect (plus zero values at each end) to estimate average depth. Multiply the length of the stream segment by average widths to estimate the sampling area. *See* chapter 4 for detailed surveying techniques.

Apply rotenone in two stages. Apply a lower concentration (0.75 mg/1) for approximately half the recommended time and then increase the concentration to 1 mg/1 to immobilize more resistant species. Rotenone can be applied to a stream section by several methods. On larger streams, pump the toxicant through a perforated hose and supplement it with a sprayer in slack water areas. Calibrate the pumping system so that the correct amount is applied over the specified time. In some streams, current may be so slow as to require modification of these procedures. If a site is essentially lentic, apply rotenone as it is applied in coves.

Downstream detoxification of rotenone with potassium permanganate requires that sufficient quantities be applied to first overcome any demand resulting from organic matter in the water and then to remain in sufficient concentration to detoxify

Box 10.3. Computation for Rotenone Application to Streams

Data required: *Example*
   Surface area   *(A)*—acres (hectares)   1.5 (0.61)
   Average depth   *(Z)*—feet (meters)   3.8 (1.16)
   Average width   *(W)*—feet (meters)   55 (16.76)
   Length   *(L)*—feet (meters)   1188 (362.1)
   Current velocity   *(V)*—feet/second   1.2 (0.37)
                              (meters/second)
   Concentration   *(C)*—milligrams/liter   1.0 (1.0)
                              (ppm)

Surface Area:

$$A_{English} = WL \qquad\qquad A_{Metric} = WL$$
$$= 55(1188) \qquad\qquad\quad = 16.76(362.1)$$
$$= 63{,}340 \text{ ft}^2 \text{ (1.5 acre)} \qquad = 6{,}069 \text{ m}^2 \text{ (0.61 hectares)}$$

Discharge:

$$D_E = WZV \qquad\qquad\quad D_M = WZV$$
$$= 55(3.8)1.2 \qquad\qquad\quad = 16.76(1.16)0.37$$
$$= 250.8 \text{ ft}^3/\text{second} \qquad\quad = 7.19 \text{m}^3/\text{second}$$

Rotenone (liquid, 5% formulation):

   Dispensing Rate (ml/min) $= D_E 1.692C$ $= D_M 59.76C$
                                $= 250.8\,(1.692)(1.0)$ $= 7.19(59.76)(1.0)$
                                $= 424 \text{ ml/min}$ $= 430 \text{ ml/min}$

Detoxification (potassium permanganate, granular):

   $DR_{(oz/min)}$ $= D_E 0.0595C$ $DR_{(g/min)}$ $= D_M 60C$
                   $= 250.8(0.0595)2.0$ $= 7.19(60)2.0$
                   $= 29.8 \text{ oz/min}$ $= 863 \text{ g/min}$

rotenone. Holder (1975) describes a method for approximating the amount of potassium permanganate by observing the time required for a treated water sample to turn from pink to brown. The lowest concentration in which the color remains pink over a ten-minute period is the amount required for oxidizing the organic matter in the sample. An additional 3-4 mg/l is added for rotenone detoxification. A chemical test for determining the potassium permanganate demand is reported by Engstrum-Heg (1971).

   Drop an orange into the water at the upper end of the sampling area to determine its travel time to the lower end. This provides a reasonable estimate of time to begin detoxification of the rotenone at the downstream area.

   Always use a block net at the downstream limit; a net may also be employed at the upper end of the sampling area. A net described by Johnson and Pasch (1976) is illustrated in Fig. 10.3. The length, depth, and mesh size of the net depends on the

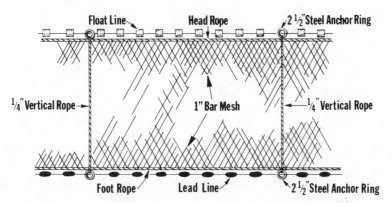

Figure 10.3. Detail of net for blocking a stream, as described by Johnson and Pasch (1979).

sampling conditions. If current velocity or stream size dictates use of larger mesh nets, attach subsampling bags constructed of smaller mesh (at random locations) along the net. In slower streams, use a small mesh size so that all sizes of fish are retained. Various involved procedures for setting the net on larger streams with substantial flow have been developed (Ober 1981).

Soon after rotenone is applied, fish will begin to surface; have enough people to collect the fish. Within an hour substantial numbers of fish will collect in the downstream net and subsampling bags. Collect these by carefully pursing the net while removing it from the stream. One major disadvantage of stream sampling is that no second day pick-up is possible; thus, fish that are snagged on the bottom will not be collected. As with cove sampling, marking fish and releasing them in the sampling area will provide estimates of nonrecovery rates.

10.5 CONCLUSIONS

Sampling with toxicants is one of the most effective means of obtaining population information, but this method is not without complications. The effort and associated cost limit the extent of sampling, thus affecting statistical precision. There are also potential problems with public relations, and some agencies are questioning the relative value of the information obtained. With reference to the data collected, perhaps standing crop information has been overemphasized, particularly since it is so variable, while other units of measurement provide more insight into the population dynamics. Data on size and age structure and estimates of recruitment are effectively collected with rotenone sampling. Even though food habit data are not useable, equally valuable characterizations of prey availability/predator relationships can be derived.

If rotenone were removed from our sampling options, we would lose considerable capability; therefore, it is important to consider ecological factors and public reaction to continued use and to utilize the data collected as efficiently as possible.

10.6 REFERENCES

Aggus, L. R., D. C. Carver, L. L. Olmsted, L. L. Rider, and G. L Summers. 1980. Evaluation of standing crops of fishes in Crooked Creek Bay, Barkley Lake, Kentucky. *Proceedings of the Annual Conference of the Southeastern Association Fish and Wildlife Agencies* 33:710-722.

Axon, J. R., L. Hart, and V. Nash. 1980. Recovery of tagged fish during the Crooked Creek Bay rotenone study at Barkley Lake, Kentucky. *Proceedings of the Annual Conference of the Southeastern Association Fish and Wildlife Agencies* 33:680-687.

Bouck, G. R., and R. C. Ball. 1965. The use of methylene blue to revive warm water fish poisoned by rotenone. *Progressive Fish-Culturist* 3:161-162.

Chance, C. J. 1958. How should population surveys be made? *Proceedings of the Annual Conference of the Southeastern Association Game and Fish Commissioners* 11:84-89.

Cochran, W. G. 1963. *Sampling techniques.* John Wiley and Sons, New York, New York, USA.

Cumming, K. B. 1975. History of fish toxicants in the United States. Pages 5-21 *in* P. H. Eschmeyer, editor. *Symposium on rehabilitation of fish populations with fish toxicants.* North Central Division, American Fisheries Society, Bethesda, Maryland, USA.

Derse, P. H., and F. M. Strong. 1963. Toxicity of Antimycin to fish. *Nature* (London) 200:600-601.

Engstrum-Heg, R. 1971. Direct measurement of potassium permanganate and residual potassium permanganate. *New York Fish and Game Journal* 18:118-122.

Gilderhus, P. A., B. L. Berger, and R. E. Lennon. 1969. *Field trials of Antimycin A as a fish toxicant.* United States Bureau of Sport Fisheries and Wildlife, Investigations in Fish Control, Number 27, Washington, District of Columbia, USA.

Grinstead, B. G., R. M. Gennings, G. R. Hooper, C. A. Schulty, and D. A. Whorton. 1978. Estimation of standing crop of fishes in the predator-stocking-evaluation reservoirs. *Proceedings of the Annual Conference of the Southeastern Association of Fish and Wildlife Agencies* 30:120-130.

Hall, G. E. 1974. Sampling reservoir fish populations with rotenone. Pages 249-259 *in* R. Welcomme, editor. *Symposium on the methodology for the survey, monitoring and appraisal of fishery resources in lakes and large rivers.* Food and Agriculture Organization of the United Nations, European Inland Fisheries Advisory Commission, Technical Paper No. 23, Rome, Italy.

Hayne, D. W., G. E. Hall, and H. M. Nichols. 1968. An evaluation of cove sampling of fish populations in Douglas Reservoir, Tennessee. Pages 244-297 *in Reservoir Fisheries Resources Symposium,* Southern Division, American Fisheries Society, Washington, District of Columbia, USA.

Henley, J. P. 1967. Evaluation of rotenone sampling with scuba gear. *Proceedings of the Annual Conference of the Southeastern Association of Game and Fish Commissioners* 20:439-446.

Holder, D. R. 1975. *A technique for using rotenone to sample fish populations in Georgia's warmwater streams.* Georgia Department of Natural Resources, Game and Fish Division, Technical Report No. 1, Atlanta, Georgia, USA.

Johnson, T. L., and R. W. Pasch. 1976. Improved rotenone sampling equipment for streams. *Proceedings of the Annual Conference of the Southeastern Association of Game and Fish Commissioners* 29:46-56.

King, T. A., J. C. Williams, W. D. Davies, and W. L. Shelton. 1981. Fixed versus random sampling of fishes in a large reservoir. *Transactions of the American Fisheries Society* 110:563-568.

Lambou, V. W., and H. Stern, Jr. 1958. An evaluation of some of the factors affecting the validity of rotenone sampling data. *Proceedings of the Annual Conference of the Southeastern Association of Game and Fish Commissioners* 11:91-98.

Lennon, R. E., J. P. Hunn, R. A. Schnick, and R. M. Burress. 1970. *Reclamation of ponds, lakes and streams with fish toxicants: a review.* Food and Agriculture Organization of the United Nations, Fisheries Technical Paper No. 100:1-99, Rome, Italy.

Marking, L. L., and V. K. Dawson. 1972. The half-life of biological activity of antimycin determined by fish bioassay. *Transactions of the American Fisheries Society* 101:100-105.

Ober, R. D. 1981. Operational improvements for sampling large streams with rotenone. Pages 364-369 *in* L. A. Krumholz, editor. *The warmwater streams symposium.* Southern Division, American Fisheries Society. Allen Press, Inc., Lawrence, Kansas, USA.

Parker, J. O., Jr. 1970. Surfacing of dead fish following application of rotenone. *Transactions of the American Fisheries Society* 99:805-807.

Post, G. 1958. Time versus water temperature in rotenone dissipation. *Annual Proceedings of the Western Association of Game and Fish Commissioners* 38:279-284.

Schnick, R. A., and F. P. Meyer. 1978. *Registration of thirty-three fishery chemicals: Status of research and estimated costs of required contract studies.* United States Fish and Wildlife Service, Investigations in Fish Control No. 86:1-19, Washington, District of Columbia, USA.

Shelton, W. L., and W. D. Davies. 1981. West Point Reservoir—A recreational demonstration project. Pages 1419-1431 *in* H. G. Stefan, editor. *Proceedings of the symposium on surface water impoundments,* Minneapolis, Minnesota, USA.

Shelton, W. L., W. D. Davies, D. R. Bayne, and J. M. Lawrence. 1982. (In press). *Fisheries and limnological studies on West Point Reservoir, Alabama-Georgia.* United States Army Corps of Engineers, District Mobile, Job Completion Report, Mobile, Alabama, USA.

Shireman, J. V., D. E. Colle, and D. F. DuRant. 1981. Efficiency of rotenone sampling with large and small block nets in vegetated and open-water habitats. *Transactions of the American Fisheries Society* 110:77-80.

Swingle, H. S. 1958. How fish population surveys should be reported. *Proceedings of the Annual Conference of the Southeastern Association of Game and Fish Commissioners* 11:103-104.

Timmons, T. J., W. L. Shelton, and W. D. Davies. 1979. Sampling reservoir fish populations in littoral areas with rotenone. *Proceedings of the Annual Conference of the Southeastern Association of Fish and Wildlife Agencies* 32:474-484.

Walker, C. R., R. E. Lennon, and B. L. Berger. 1964. *Preliminary observations on the toxicity of antimycin A to fish and other aquatic animals.* United States Bureau of Sport Fisheries and Wildlife, Investigations in Fish Control No. 2:1-18, Washington, District of Columbia, USA.

Wegener, W. D., D. Holcomb, and V. Williams. 1974. Sampling shallow water fish populations using the Wegener ring. *Proceedings of the Annual Conference of the Southeastern Association of Game and Fish Commissioners* 27:663-673.

Chapter 11
Tagging and Marking

RICHARD WYDOSKI AND LEE EMERY

11.1 INTRODUCTION

Tagging and marking are important techniques used to study individual aquatic animals or populations. It is uncertain when fish were first marked, but several centuries ago wealthy European landowners tagged the salmon and trout living in their streams (Jackobsson 1970). Fish were first tagged in the United States in 1873 when Atkins successfully tagged large numbers of Atlantic salmon in Maine (Rounsefell and Kask 1945). Since then, much has been written about marking fish. Emery and Wydoski's (1983) indexed bibliography on marking and tagging aquatic organisms is an excellent reference source of information about the application, retention, effects, and recovery of tagged aquatic animals. This chapter provides a synthesis of the different types of tags, marks, and techniques that have been used successfully on aquatic organisms. Radio and sonic tags used in underwater biotelemetry are discussed in chapter 20.

Aquatic biologists tag or mark aquatic animals to obtain information necessary for research or management. Tagging studies can provide information about (1) stock identification to determine whether stocks or subpopulations are utilized by sport or commercial fisheries; (2) migrations, including the path and distance of migration, rate of movement, and homing tendencies of a species; (3) behavior, including factors that limit abundance such as habitat selection and intra- and interspecies interactions; (4) age, including validation of other aging methods and determining growth rates; (5) mortality rates, to follow the effects of natural and fishing mortality on a population; (6) abundance, using mark and recapture experiments; and (7) stocking success of hatchery-reared fish.

11.2 SELECTING A MARKING METHOD

Various tags or marks have different capabilities and limitations and, therefore, are used for different purposes. Several factors affect which mark is best to use. Consider the: (1) length of time the mark must remain on the organism, (2) availability of personnel for tagging and recovery, (3) life history information about the species, (4) methods used to capture the organism and handling techniques which minimize stress, and (5) whether it will be necessary to coordinate the marking program with other scientists (Everhart et al. 1975; Stonehouse 1978; Jones 1979). Although no precise rules can be followed in selecting the most appropriate mark or tag, experience of various biologists has provided a list of criteria that should be considered. General characteristics of major types of marks are shown in Table 11.1.

Table 11.1 General criteria for selecting a marking technique.

CHARACTERISTIC	BIOLOGICAL (Natural)	Immersion	CHEMICAL Injecting and Tattooing	Feeding	PHYSICAL Mutilation	Tags
Duration of Mark						
Days		X	X		X	X
Weeks		X	X	X	X	X
Months	X		X	X	X	X
Years	X		X	X	X	X
Individuality of Mark						
Very Low		X	X	X		
Low	X				X	
High						X
Size of Organism at Marking						
Very Large	X		X	X	X	X
Large	X		X	X	X	X
Medium	X		X	X	X	X
Small	X	X	X	X	X	X
Very Small	X	X	X	X?		X
Numbers to be Marked						
Low (<50)			X	X	X	X
Medium (>50-<200)		X	X	X	X	X
High (>200-<1,000)		X	X	X	X	X
Very High (>1,000)	X	X	X	X	X	X
Recovery Method						
Visual	X[a]	X	X[c]	X[a]	X	X
Non Visual						X[a]
Type of Organism to be Marked						
Crustacean		X	X		X	X
Mollusk					X	X
Fish	X	X	X	X	X	X
Cost						
Low		X		X		
Medium			X		X	X
High	X[b]					

[a] May be necessary to use a microscope or other equipment to detect the mark.
[b] Personnel must have specialized skill to identify mark and look at large numbers of fish.
[c] Ultraviolet light (black light) necessary to detect fluorescent pigments.

11.2.1   Basic Considerations

Objectives of marking. The objectives of a study are the most important considerations in choosing a mark or method of application. Never undertake a marking program before deciding exactly what information is needed.

Effect on survival, behavior, and growth. Some tags or marks increase mortality and affect the results of the study (e.g., population estimates or mortality rates). Marks that change the behavior of animals after release can affect the rate of tag recovery and can bias observations about the animals' habits. Some physical tags (e.g., jaw tags) reduce growth rate and, thus, are inappropriate for studies of growth, condition, or production.

Permanency of the mark. Various marks differ in their longevity on animals, and the chosen mark must be retained by the organism for the duration of the study. Some biological marks are permanent (e.g., some metallic compounds and antibodies are deposited and retained for life in the bones of organisms as a result of feeding or injection), while others may change seasonally (such as parasitic marks which change with the life stage of the parasite and the host). Chemical marks, such as stains, offer good short-term marks that are easily observed. Physical tags vary considerably in their retention and visibility.

Information content of the mark. Sometimes it is only necessary to identify animals as belonging to a particular group (i.e., stock or subpopulation), and a simple mark, such as a colored stain, is sufficient. In other studies, it may be necessary to recognize individual animals, and the mark must include a number or a detailed code. Also, if anglers are expected to return a tag, it must be large enough to contain an address or other instructions.

Number of organisms to be marked. The numbers of organisms to be tagged will influence the selection of a tag or technique. If many individuals need to be marked, the mark must be inexpensive and easy to apply. Chemical stains, for example, work well for marking large numbers of organisms (e.g., fry, shrimp) for short periods of time. If fewer individuals need to be marked, a more expensive and time-consuming mark that provides more information (e.g., dart tags) might be appropriate. Another relevant consideration here is how many resources, in terms of money and people, are available for the program. Techniques vary greatly in the number of people needed to capture, handle, mark, and recover the organisms. Proper planning is needed to insure that these resources are available.

Stress of capture, marking, and handling. All capture and handling methods affect the osmoregulatory ability and physiological condition of fish (Wydoski and Wedemeyer 1976). Some species become stressed more easily than others. Muskellunge and striped bass, for example, go into a state of shock during long handling periods, and the saucer scallop is particularly stressed by exposure to air during marking. Mutilation of an animal or the protrusion of an external tag from the animal's body can have greater effects on an organism than a mark that does not cause such changes. Therefore, a compromise is usually necessary between a mark that provides the needed information and a marking method that reduces stress.

Skills of project personnel. Many marking procedures require special skills, and all require some practice. The same is true for detecting marks on recaptured fish. It may be necessary to train personnel for a tagging operation; the time needed should be incorporated into the overall plan.

Coordination among agencies, states, and countries. It is sometimes necessary to coordinate large-scale recovery efforts to accomplish the objectives of a study. For example, tagging and recovery of anadromous salmon on the open seas has involved a joint effort among many different countries. On a smaller scale, managers tagging fish on a lake bordered by two states need to coordinate their activities to accomplish the objectives of the study.

11.2.2   Mark Visibility

The visibility of a mark can greatly influence the interpretation of data recovered from marked animals. Two main factors—color and location of the mark—affect visibility, but shape of the mark is also important. A rainbow of colored tags or marks are available. The selection of color involves a compromise because bright colors that are easily seen and recognized at recapture may also be conspicuous to predators and increase mortality (Lawler and Smith 1963). Luckily, some colors which are easily detected by man out of water are not as conspicuous to predators under water. Red, international orange, and yellow are the most frequently selected colors, but these have been shown to cause higher predation rates than some less conspicuous colors. Be aware that visibility of the mark can have an important effect on the recovery results, either directly through mortality by predation or indirectly through failure to observe the mark on a recaptured animal.

If large numbers of animals must be examined in a short time, the mark should be bright in color (e.g., international orange or yellow), placed in a conspicuous location (e.g., along the dorsal surface on one side of the animal), and be noticeable (e.g., streamer or vinyl tubing). If small numbers of animals are to be examined and time is not a constraint, then the mark should be dull in color (e.g., olive or a color that matches the organism) and placed in a location where it will not interfere with the animal's locomotion or attract predators (e.g., brands or internal body tags).

11.2.3   Recovery of Marked Organisms

The recovery of marked animals depends upon four main factors—retention of marks, movement of animals, differential mortality of marked animals, and recognition of marks. The retention of various tags or marks is an extremely important factor. If marks fall off or fade, the effective number of marked fish will decrease over time. Choose a mark that has a high retention rate for the length of the study.

Movement of organisms also affects recovery rate. If marked fish leave an area, the recovery area must be expanded or changed to include the new area of residence. Frequently, low budgets restrict the expansion of the recovery area, and biased data may result. In addition, differential movement of marked and unmarked animals can bias population estimates.

Another factor that may affect the recovery of marked organisms is mortality caused by stress imposed during capture, handling, marking, and recovery. Handling organisms during marking procedures can produce wounds or cause behavioral changes that make the animal more susceptible to predation or mortality from diseases. *See* chapter 5 for proper handling methods.

Finally, recovery of marked animals depends on recognition of the marks. Color, location on the animal's body, and shape of the mark affect visibility and recognition. In some marking techniques, visibility is not critical for recovery (body cavity, subcutaneous, and coded wire tags), but trained personnel or sensitive detectors are needed for the effective recovery of these hidden tags.

11.2.4   Characteristics of the Animal

The effects of marking on aquatic animals depend on the physical condition of the fish at the time of release. Fish that have lost many scales are recovered at a lower rate than fish released in good condition. Small fish are more susceptible to injury than large fish, and return of small marked fish is generally lower. Moreover, some species are more susceptible to injury than others. Tagging of some species must be done quickly, or high mortalities will occur; tuna, for example, must be tagged within 20 seconds.

Fish species with closed swim-bladders that are captured and retrieved from deep water may die because their bladders become distended or burst upon reaching the surface. Fish, shrimp, and other aquatic organisms that are marked after capture from deep water may be lowered in cages to the appropriate depth before release to minimize physiological problems and to protect them from predators in surface waters (Jones 1979; Farmer and Al-Attar 1981).

Wounds that are caused by marking normally heal satisfactorily without the use of antibiotics, except in special situations such as in warm water or where other stresses lead to secondary infections. Tags that are firmly fixed are superior to dangler tags in reducing wound size and show better retention. Strength of anesthetics and duration of exposure to the anesthetic are also factors that affect stress. Care in capture and handling is important so that vital organs are not injured and skin abrasions are minimized. Water quality during tagging must be optimal. Hyperactive fish can be calmed with mild anesthetics, and osmoregulatory problems caused by marking can be reduced by adding table salt to the water (Piper et al. 1982). *See* chapter 5 for details of proper handling procedures.

Tagged organisms can also be held in tanks for hours or days before releasing to determine which organisms are in poor and good condition. If some unmarked fish are also held in the tanks, this type of experiment can provide an estimate of the increased mortality rate caused by marking. This will help in calculating return rates and interpreting data collected during the actual tagging study.

11.3   BIOLOGICAL OR NATURAL MARKS

Biological or natural marks are morphological differences in animals created by natural processes. These marks are either inherited or gained from the environment. Using these marks often depends on mathematical interpretation of the frequency of presence on animals of a stock or subpopulation. For example, the decision to use a parasite as a natural mark on fish is determined by calculating the ratio of incidence of that parasite on one fish population to its incidence on another. Using biological marks generally requires much knowledge about the life history of the organism. Selecting a biological marker (e.g., parasite) can require extensive preliminary work to determine if the mark is natural and what process causes it. In general, natural marks can be divided into parasitic, morphological, and genetic marks.

11.3.1   Parasitic Marks

The presence of parasites is used for identifying various groups or stocks of fish and for determining fish movement or migration patterns (Sindermann 1961). Most parasitic markers have been used to study anadromous fish species on the open seas and to identify individual cetaceans (Wursig and Wursig 1977; Killingley 1980).

There are several conditions for using parasites as biological marks (Kabata 1963): (1) the parasite should be common in one host population and absent in another, (2) the marker fish should be the only host in the life cycle of the parasite, (3) the parasitic infections should be of long duration, (4) the presence of the parasite in the host should remain stable throughout the life cycles of the fish and parasite, and (5) the environmental conditions throughout the study area should be within the physiological range of the parasite.

Advantages of parasitic marks are that they are natural, have low survey costs, and can be used on large bodies of water. The disadvantages of parasitic marks are that much time is needed to determine if the parasite can be a mark, mathematical errors are possible, identification of individual animals is not generally possible, and trained personnel are required to detect the marks.

11.3.2 Morphological Marks

Morphological marks include meristic counts; pigmentation; proportional body parts; shape, size, or age marks of otoliths and scales; chemical composition of scales or bony parts; and internal marks on scales. The shape, size, and age marks of scales are the most frequently used morphological characters. This method is used to differentiate salmon stocks on the high seas (Major et al. 1972). Marks caused by starvation have also been used to identify hatchery-reared fish (Major 1962).

The use of morphologic characters has two general limitations: (1) the characters (e.g., counts of vertebrae, fin rays, or gill rakers) are subject to environmental influences, and the results of these influences are not easily detected, and (2) characters that depend on the genotype of the fish, such as meristic counts, frequently overlap among populations (Fig. 11.1), making interpretation subjective. These problems have led researchers to develop methods that look directly at genetic marks.

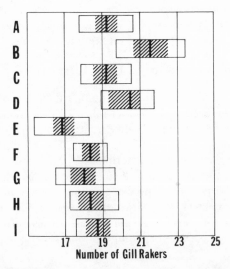

Figure 11.1 Meristic counts of fish gill rakers from different populations (A-I) showing overlapping characters. The dark line indicates one standard deviation, and the entire bar indicates the range for each fish population. Presentation of data in this fashion is commonly called a "Hubbsogram."

11.3.3 Genetic Marks

The use of genetic marks is more complex than the use of morphological or parasitic marks, but they have been shown to be effective for identifying fish stocks. The initial method developed was to look at the frequency of different blood groups among various stocks of fish. The serological method was used successfully in 1956 to differentiate sockeye salmon from Bristol Bay (Alaska) and the Columbia River (Washington). A refinement of serological techniques demonstrated that significant differences occur in blood groups of salmon and other fish species from different areas. Although serological methods have some advantages over the morphologic methods for studying fish populations, problems still remain in preserving blood samples and developing reagents.

The development of electrophoresis solved most of the problems associated with blood groups because this technique enabled biologists to differentiate discrete fish populations by examining many individual loci of genes (Allendorf et al. 1975). Electrophoresis is a process, similar to thin-layer chromatography, in which the genetically-controlled structure of proteins is displayed. Electrophoresis eliminates the need for developing specific reagents and is applicable to proteins from a variety of tissues. The electrophoretic technique appears to be the most promising genetic approach to fish stock identification.

11.4 CHEMICAL MARKS

11.4.1 Methods of Chemical Marking

Chemical marking of animals has been done with dyes, stains, inks, paints, liquid latex and plastics, metallic compounds, tetracycline antibiotics, and radioactive isotopes (Klima 1965; Arnold 1966). Animals of all sizes have been marked with chemicals by immersion, injection, tattooing, and feeding.

Immersion. Staining by immersion allows large numbers of organisms to be marked quickly, but the retention of the mark is usually less than one month. Staining shrimp with dyes has proven to be a useful short-term mark. The success of immersion as a short-term mark varies according to the species marked as well as its size, age, the water temperature and chemistry during marking, concentration of the stain, and duration of the immersion. Besides being toxic to the organism if not properly applied (Farmer 1981), the color (e.g., red) of the stain may attract predators and reduce the survival of the organism.

Injection. The injection of dyes or other substances into the body cavity or musculature can result in overall coloration of the animal. Injections into the body cavity should be made carefully to prevent penetration of internal organs. The injection-diffusion technique has been highly successful for marking crustaceans, where the material concentrates in the gills and is retained after molting. Injection of chemical substances to form spots is another technique which offers greater individuality of marks (Hart and Pitcher 1969). The success of the injection method varies with the diffusion tendencies of the substance, amount and concentration of the substance, and the anatomical or histological location of the injection. Injection of fluorescent pigments using compressed air has been successful in marking large numbers of salmonids in a short time (Phinney et al. 1967).

Tattooing. Tattooing with various substances allows the biologist to mark fish with letters and numbers for recognition of individuals. However, the technique is time-consuming. Tattoos or marks of this type are most useful in behavioral studies.

Feeding. Feeding dyes to fish has worked quite well under hatchery or laboratory conditions, but it is not applicable to most field situations. Dyes ingested by fish color the entire fish for a short time but eventually localize in specific areas. Various metallic compounds also have been fed to fish to produce marks, but results have been generally disappointing. McKee and Hublou (1963) found that bismuth was most promising because it was non-toxic and readily deposited in scales and bony tissues of the fish. Arnold (1966) reviewed specific information on various chemicals that have been used in marking fish and invertebrates and should be consulted if this marking method is contemplated.

11.4.2 Advantages and Disadvantages of Chemical Marks

The advantage of using chemicals for marks is ease of marking large numbers of organisms quickly. The time spent handling the organism is low, and chemicals can mark organisms which are too small for physical tagging methods. The main disadvantages of using chemicals as markers are that individual organisms cannot be recognized and that retention of the mark is generally brief because growth of the organism disperses the chemical or the chemical fades.

In addition, there are specific advantages and disadvantages for each of the chemical techniques. Feeding metallic compounds, radioactive isotopes, or antibiotics can produce marks in the bony parts of aquatic organisms that last a lifetime. Such marks have been useful in certain studies requiring a mark that will identify a particular time in the life of the animal (e.g., marking 10-cm and 15-cm salmon in a hatchery to determine which stocking size has the best survival). Such long-term marks require trained personnel to identify the marked organism. Injections of liquid latex under the skin or in the barbels of fish have produced mixed results, probably because of the reaction of different fish species to the latex and to location of the marks. For example, latex injected into the fins of lampreys was ejected from the body after two to four months (Hanson 1972); latex injected into the barbels of channel catfish was retained for one year (D. Kuntzelman, Fisheries Academy, Leetown, WV, personal communication).

Fluorescent pigments applied by injection under the skin with compressed air are good marks for stock identification and may last for years. Strange and Kennedy (1982) reported five-year retention of fluorescent particles on rainbow trout released into a reservoir. The main drawback of fluorescent marking is that individual handling under black light is necessary. Inexperienced observers in the field can easily miss the pigment marks or confuse them with natural fluorescence. A second mark (e.g., fin clip) may be needed to identify marked fish in an angler's creel.

11.5 PHYSICAL TAGS OR MARKS

Physical tags or marks offer the biologist the greatest variety of marking methods. Many of the physical tags employed today are the same design as those used many years ago in early tagging studies and differ only in the materials used for construction. The techniques range from simple, such as attaching paper clips to opercles or mutilating fins, to complex, such as branding with laser beams or identifying coded microtags with x-rays.

The advantages of using physical marks are that they allow recognition of individuals as well as groups, can be recovered by either visual or non-visual methods, and offer many choices. The general disadvantages of physical marks are variable retention rates, costly recovery methods, and time-consuming methods of application. All physical marks or tags can be divided into two basic categories, internal or external.

11.5.1 Internal Tags

Internal tags are advantageous because they do not require the removal of body parts and do not protrude from the body. Other advantages are that they are inexpensive, non-toxic, and do not stress fish as much as other tagging methods. Without trained personnel, however, an internal tag is difficult to recover unless an external mark is also used.

Body cavity tags. Internal tags were first used experimentally in flounders during the 1930s (Rounsefell and Kask 1945), when metal anchors were placed in the body cavities. Pieces of metal have also been placed in the body cavities of important commercial species such as herring. These internal tags are recovered by strong magnets which detect the tags during processing in canneries. However, there are recovery problems. The tags are often lost in the machinery before the fish come in contact with the magnets, the magnets fail to attract and retain all tags, and some tags are lost during the process of clearing tags and other debris from the magnets (Jones 1979). Body cavity tags have also been expelled from fish during spawning, and it is recommended that fish not be tagged until after spawning occurs (Winters 1977).

Subcutaneous tags. These tags are inexpensive plastic discs that contain printed serial numbers and legends. They are placed in the fish by making a slight incision through the outer layer of skin and inserting the tag between the skin and muscles. There is no need to sew the incision as it heals rather quickly. This type of tag usually requires another mark on the fish, such as a fin clip, to alert the person handling the fish to look for the subcutaneous tag. Both subcutaneous and body cavity tags are protected inside the fish and do not protrude to collect algae or other growths or attract predators.

Coded wire tags. Coded wire tags are small pieces of wire which are imbedded with a small applicator or by hand (Bergman et al. 1968) into the snouts of fish, necks or flippers of sea turtles, and bodies of prawns. These tags allow recognition of groups, either with color-coded markings or with notches that are read externally by x-rays. Most coded wire tags are inserted into fish by machine. The tagging machine has a head mold which holds the fish in the proper position for inserting the tag in the snout. Various shapes and sizes of head molds are available, and it is most important that the proper head mold be used because an improperly inserted tag will fall out. Tags are detected by a sensitive metal detector or by the presence of another external mark, such as a fin clip, which alerts the observer that an internal tag is present. In some recovery programs, anglers leave the heads of captured fish (those with adipose clips) at collection points where biologists can collect them. Recovery equipment is portable and can be used in the field with a common 12-volt battery. Fine adjustments can also be made to the detector while in the field to improve efficiency and detection. The coded wire tagging system (detector and applicator) has been very successful in large-scale tagging of small salmonids on the West Coast.

The advantages of coded wire tags are that large numbers of small fish can be tagged quickly, easily, and without altering the behavior of the fish. However, the marking equipment is expensive ($23,000 in 1982), which makes this method unaffordable for small-scale tagging studies. In large-scale marking operations, however, the cost of using wire tags can be competitive with traditional marking techniques (D. Ostergaard, Allegheny National Fish Hatchery, Warren, PA, personal communication).

Coded wire tags placed in the snout cartilage of salmonids are retained for long periods; Jefferts et al. (1963) found only a six percent loss after nearly two years in one study. Greater losses have occurred when tags were placed on other body parts. Retention of coded wire tags in other fish species has been mixed (Gibbard and Colura 1980; Kreiger 1982). Coded wire tags offer promise for use on other organisms, as indicated in studies of sea turtles (Schwartz 1981) and spot prawns (Prentice and Rensel 1977). The secret of good retention is careful application to insure the tag is properly positioned in a tissue (preferably cartilaginous) that will hold the tag in place. For example, if the organism is withdrawn from the needle applicator as the tag is inserted, the tag will be poorly positioned under the skin and will eventually fall out.

Microtags. Microtags are microscopic plastic chips (22,000 chips per gram) that function in much the same way as coded wire tags. Each chip contains seven layers of plastic that can be color-coded with ten different colors to give 300 million color combinations (available from 3M Center at $300 per pound, minimum 5 pound order). They can also be coded to contain a fluorescent or magnetic layer. Microtags were originally used to identify explosives. Studies are currently underway to determine if microtags used on wildlife can be effective on fish (Parker and Randolph 1981).

11.5.2   External Tags

The advantages of external physical marks are that they can be seen without dissection and most allow individual recognition of animals. The main disadvantages are that tags may attract predators and cause higher mortality, protrusion of tags can interfere with the locomotion of the organism, and wounds created by tagging make the organism susceptible to infections and disease. A prodigious variety of tags exist because there are many variables (e.g., objectives of the study, species differences, sampling problems, cost, personnel) to be considered when conducting a marking program.

Mutilation. Marking organisms through mutilation can be accomplished by clipping or punching fins or other body parts with scissors, nail clippers, side cutters, or other tools. Because mutilating a body part on an organism is a simple and quick process [experienced personnel have fin-clipped 900 fish per hour (Bailey 1965)], it has been a popular marking method employed on many different types of organisms. Although this technique can produce a permanent mark, certain body parts (e.g., toes, fins) can regenerate. Amphibians have always been difficult to mark satisfactorily because mutilated parts are regenerated after a short period of time. The number of marking combinations is low. Everhart et al. (1975) stated that there are only ten combinations of clips possible when using two of the paired fins in marking fish.

The effect on fish of partially or completely removing fins has been repeatedly tested, with highly variable results. A synthesis of the literature indicates that complete removal of the adipose fin probably has the least effect on fish. Complete removal of one pelvic or both pelvic fins is also a feasible marking method since the effects are not extremely severe on fish. Complete removal of pelvic and adipose fins in combination also allows a greater combination of marks. Fin-clipping does not appear to affect the sustained swimming ability of fish (Radcliffe 1950), but fins are important in maneuvering and controlling movements (Harris 1937). Therefore, it is not advisable

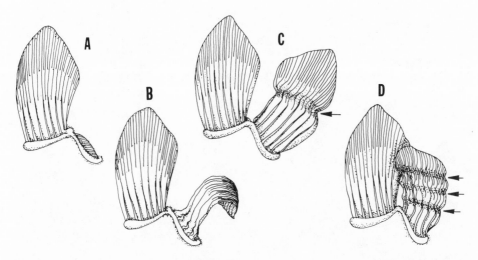

Figure 11.2 Appearance of fish fins after mutilation and regeneration. *(A)* Fins clipped close to the body usually do not regenerate. *(B)* Fins clipped away from the body will regenerate with greatly deformed rays. *(C)* Fins clipped farther away from the body will regenerate to normal size but will have some irregularity in the rays where the clip was made. *(D)* Fins with multiple clips show irregularities at the location of each clip.

to completely remove pectoral, dorsal, or caudal fins because these fins are used in maneuvering and their removal would modify behavior of the fish that, in turn, could affect growth (through altered metabolism) and survival (through competition and predation). Partial clips of either the dorsal or caudal fins can be used as short-term marks without adverse effects on survival or metabolism. Bergstedt's (1980) annotated bibliography provides a good reference for survival estimates for various studies involving fin clips. Because there are so many factors that can influence survival of marked fish in different studies, the survival values obtained from these studies should not be used as absolute values, but rather as relative differences in planning marking experiments (Cresswell 1981). *See* Fry (1961), Ricker (1949), Churchill (1963), Coble (1967), and Nicola and Cordone (1973) for specific studies.

Fin-clipping can permanently mark a fish if the rays are removed to the point of attachment to the bone (Stuart 1958; Eipper and Forney 1965; Jones 1979). If not, the fins will regenerate. The regenerated rays will appear distorted at the location of the clip (Fig. 11.2). Irregularities or distortion of regenerated fin-rays can also be used for repeated marking of the same fin. The adipose fin does not regenerate after removal by clipping, but partially clipped adipose fins may resemble short adipose fins (normal size for some species) after healing and pigmentation hides (masks) the clip and may make recognition of this mark difficult.

Partial fin-clips are often used in short-term mark-recapture experiments for estimating the numbers of fish in a particular body of water. The upper or lower lobe of the caudal fin is most often partially clipped for such studies. Some investigators have also made partial clips on the anterior or posterior portions of the dorsal fin, dorsal and anal fin spines (Rinne 1976), and maxillary bone. The maxillary and spinal clips are frequently used in conjunction with fin-clips to increase the number of marking combinations. Severed spines and maxillary bones regenerate quickly.

Other temporary marks include notches cut with scissors or holes punched with a

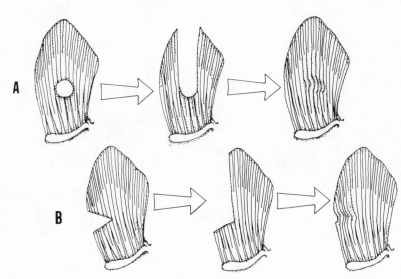

Figure 11.3 Punched *(A)* and notched *(B)* fins. Fins will shed the fin rays and have irregularities in the rays at the location of the mutilation.

paper punch (Fig. 11.3). Holes punched in the fins or opercles of fish are marks that quickly disappear. The ends of fin-rays that have been severed by notching or punching usually fall off shortly after marking, and the regenerated fin-rays show some irregularities. Retention of holes punched in species having bony opercles is better than in those species with cartilaginous ones. This technique is also useful for lobsters, shrimps, and prawns (Wilder and Murray 1956; Balazs 1973).

Mutilation is such an easy marking technique that it is often used in conjunction with other marking methods to alert the person recovering the organism to look for internal tags (e.g., capture of an adipose clipped coho salmon on the West Coast alerts the angler to the fact that this fish may also contain a coded wire tag or other internal tag).

Branding. Branding is the process of marking the surface of a fish by placing either a hot or cold instrument against the body of the fish for a few seconds. The brand displaces or concentrates the pigment at the branding site. The advantages of branding are that the body surface is not ruptured, there are no brightly colored tags or missing body parts, and the survival and growth of the fish are not affected. The primary disadvantage of branding is that the brands become illegible with time. Closed symbols (e.g., "B") lose identity faster than open symbols (e.g., "C"), and Arabic numerals are unsuitable for effective branding (Park and Ebel 1974; Joyce and El-Ibiary 1977). As the fish grows, the size of the original brand will become larger and the mark may become illegible.

The success of branding as a marking technique is influenced by several factors (Raymond 1974), the first of which concerns biological characteristics of the fish. A fish cannot be branded until it has scales or melanophores. Joyce and El-Ibiary (1977) suggested, for example, that hot branding channel catfish in warmwater, high-density culture should not be done when the fish are less than 30 weeks old and have a mean weight of 27 grams. Brands applied below the dorsal fin on most species produce the best marks. Cold branding techniques using liquid nitrogen work best on fish with fine scales (e.g. salmonids). Cold branding gives inconsistent results on coarsely scaled species (e.g., centrarchids). Hot brands are more successful on fish without scales (e.g., catfish) than on those with scales. Both cold and hot brands have been applied to crustaceans, amphibians, reptiles, and cetaceans with mixed success (Stonehouse 1978).

The techniques used to apply brands also affect success. Temperature of the branding tool and total time the branding tool is against the organism are two important variables which determine whether the brand will be retained. Cold branding with liquid nitrogen at -196°C for two seconds produced the best mark on salmonids and centrarchids (Raleigh et al. 1973). The brand became blurred if the application time exceeded two seconds, and an open sore developed if the branding iron was against the fish for five seconds. Coutant (1972) and Brock and Farrell (1977) reported similar results. A large variety of branding equipment has been used, including boiling water, propane torch, soldering iron, laser beam, hyfrecator, mixtures of dry ice and alcohol, mixtures of dry ice and acetone, and liquid nitrogen. Everest and Edmundson (1967) and others reported that during cold branding, frozen mucus and ice tend to accumulate on the branding tools, causing poor marks. This problem is eliminated if tools are cleaned regularly, but it also increases the time needed to mark animals.

The costs of branding equipment vary greatly depending on technique. An inexpensive cold branding iron can be made out of copper wire (e.g., 4-gauge; could be used for hot branding, too), and a coolant can be created by mixing dry ice and ethanol to -78°C. This technique produced marks that lasted ten months on trout.

Physical tags. Many types of physical tags are available for external application. Some of the principal tags and anatomical sites for attachment on (in) a fish are shown in Figure 11.4. A concise discussion of these prinicipal tags is provided below.

Petersen discs were widely used during the first 60 years of tagging. Although tag construction has changed from bone and metal to plastics and less expensive noncorrosive metal wires, the application and design of the tag have remained essentially unchanged. Apply the disc under the dorsal fin of the fish with a pin and pliers (Fig. 11.5). Slip one disc onto the pin, and push the pin through the muscle of the fish. Once the pin is in position, slip another disc onto the pin on the opposite side of the fish. Twist the pin into a loop to hold the tag in place, and cut off the excess wire with side cutters. Purchase pins of various lengths for use on fish of different thicknesses. The Petersen disc has also been used successfully on molluscs. The discs are glued to the shells with epoxy cement. Clams or conchs can remain out of water long enough to allow the adhesive to set and not suffer any physiological damage.

Petersen discs remain on an organism for life, especially if the fish is tough-skinned like sturgeon, sharks, or certain flounders. The biggest disadvantage is the long application time per tag. The Petersen disc sometimes causes open sores on some fish and injuries on prawns (Klima 1981).

Metal strap tags were adapted from cattle tags for use on fish. Later bird bands were applied to fish. Strap tags are made of light-weight, noncorrosive metal, are manufactured in all sizes, and are serially numbered. They are easily attached to fish with special pliers. Although metal strap tags have been traditionally applied to jaws of fish, jaw-tagging is now considered an unsatisfactory method because it affects growth. More recently, strap tags have been applied to other parts of the fish's body, including opercles, fins, and the caudal peduncle. Opercle tags have been the most successful although small (in relation to size of jaw) strap tags applied to the maxillaries have eliminated traditional growth problems. When tagging opercles, apply tags to the same side (e.g., left opercle) of all fish to increase efficiency during

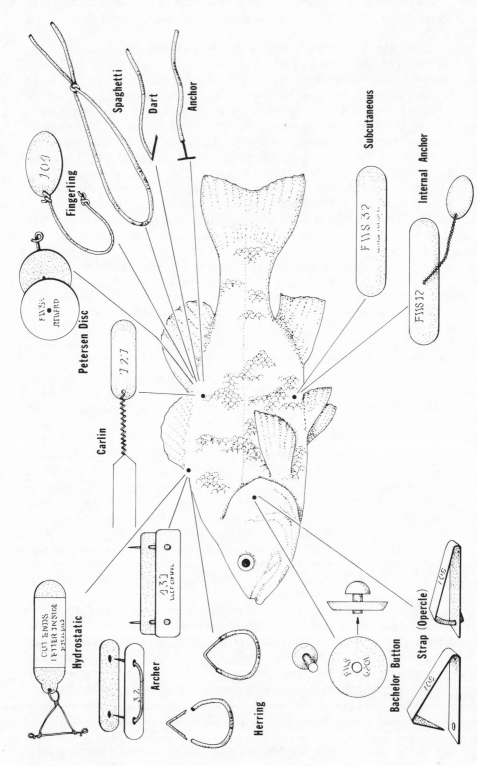

Figure 11.4 Principal tags and anatomical sites for attachment on (in) fish. (Note: The bridle for the hydrostatic tag trails behind the dorsal fin.)

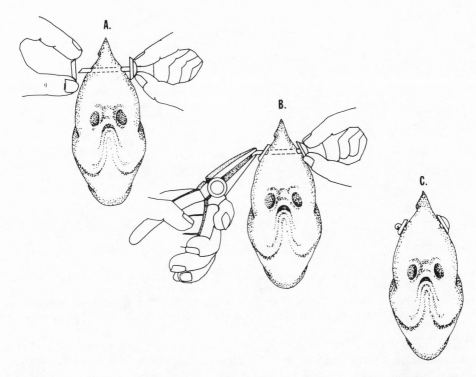

Figure 11.5 Steps in applying a Petersen disc tag to a fish. *(A)* One disc with pin is pushed through the fish. *(B)* A second disc is added on the pin on the opposite side of the fish and the pin is bent to hold the tag in place. *(C)* The excess part of the pin is removed and the tagging operation is complete. (Note: drawing has been enlarged to show detail; pinhead is actually much smaller.)

application and recovery.

Paper fasteners can be used like strap tags for short-term marks. These tags can be purchased at most general office supply stores and painted with fast-drying spray paints for easy detection. They are easy to apply (squeezed by hand until the metal barb perforates the opercle) and easy to remove. Paper fasteners work best on fish with bony opercles.

Strap tags are used on a variety of species, are permanent tags on some fish, and have been moderately successful on other animals such as turtles (Bacon 1971). The retention of strap tags on the opercles of fish depends on the strength of the opercle and the species. Strap tags are retained best on long-lived fishes such as sturgeon, flounder, halibut, and other fishes with bony opercles. In other species, the tags tend to tear free or erode through the opercles in a short period of time.

Dangler tags consist of a dangling tag (containing information) attached by wire to the fish. The most popular type is the Carlin tag, used in the United States for over 100 years. A smaller version, known as the finglering tag, has been developed for use on juvenile and small fishes. Another variation is the hydrostatic tag in which detailed instructions are placed inside a plastic capsule that is attached by wire to the fish.

The Carlin tag is relatively easy to apply and requires few tools (Fig. 11.6). The basic tool is a pair of hypodermic needles which have been soldered together near the handles. Push the two-needle unit through the fish (near and below the dorsal fin).

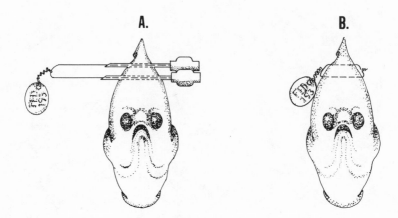

Figure 11.6   Steps in applying a Carlin tag to a fish using two hypodermic needles that have been soldered together. *(A)* The hypodermic needles are pushed horizontally through the fish and the Carlin tag wires are fed into the hollow needles. *(B)* The hypodermic needles are removed and the wires are tightened snugly against the body of the fish and excess wire removed.

Thread one of the two stainless steel tag wires into each of the hollow needles. Remove the hypodermic needles and the wires will protrude from the side of the fish opposite the tag. Pull the wires so that the tag rests snugly against the fish. Secure the lose ends to the skin by twisting them with hemostats. Remove excess wire.

For fingerling tags, the metal tag and stainless steel wire are replaced by a plastic disc and plastic or nylon thread, and a single thread holds the tag. Thread the tag through the fish with a regular sewing needle. Pull the thread through the fish until the tag is snug against the body. Secure the tag with a figure "8" knot. Tie the knot as close to the body as possible to prevent abrasions and open sores.

Dangler tags that are applied close to the body and are streamlined are retained well and can remain on fish for life. The loss rate has generally been low when tags are attached with stainless steel wire (Donahoo 1976), but losses may increase when extruded vinyl thread is used for attachment. Losses have also been higher in situations (hatchery raceways) where fish are handled frequently (K. Walewski, Aquaculture Production/Demonstration Station, Leetown, West Virginia, personal communication). Dangler tags are best suited for small studies involving few fish because applying the tags takes a great deal of time. Because the technique requires great dexterity for threading and tying the wires, it is less useful in cold weather.

Inexpensive vinyl tubing, in various colors, is now widely used for tagging fish and other animals. The three major types are spaghetti, dart, and anchor tags.

The spaghetti tag is a loop of vinyl tubing that passes through the fish. Information is printed directly on the tubing. Attach the spaghetti tag by threading tubing through the body of the fish with a hollow needle. Locate the tag below the dorsal fin so that the trailing knot is along the middle axis of the body. Secure the tag by tying the ends with a figure "8" knot or by stapling the ends together (Broadhead 1959). Two variations of the spaghetti tag are lock-on and cinch-up tags, which use special locking devices instead of a knot.

The spaghetti tag is retained well and is inexpensive, but application is time-consuming. The tag frequently creates wounds where the tubing erodes flesh, and vibration of the tubing as the fish moves prevents the wound from healing. These wounds may lead to disease and eventually may cause the tag to fall out. Although once quite popular, the spaghetti tag has been replaced in general use by dart and anchor tags.

Dart tags are made of a nylon shaft with a barbed end and a vinyl tube that fits over the upper end of the shaft. The barbed end of the shaft holds the tag in the fish, and the tube contains the tag information. Dart tags have become one of the most widely used physical marking methods. The tags are attached to the fish with a hollow needle in the same general location as spaghetti tags. The needle must penetrate past the bones supporting the dorsal fins so that the barb becomes interlocked with the skeleton of the fish and will not pull out.

The anchor tag is a modified dart tag in which a nylon T-bar replaces the harpoonlike head of the dart tag. Anchor tags are analogous to the tags used to attach prices to clothing. The tags are inserted with a gun (Dell 1968) which can be loaded with one or a clip of anchor tags, making the tagging of individuals or hundreds of organisms quick and easy. Like the dart tag, it is important that anchor tags penetrate deep enough into the fish that the T-bar interlocks with the skeleton. The needle should be inserted at a 45° angle so that the tag will be streamlined and flow with the body of the swimming fish.

The advantages and the disadvantages of the anchor tag and the dart tag are similar. Both types of tags are highly visible. Both tags are retained well if properly applied. Tag losses occur because the tag is inserted at the wrong angle or the barb or T-bar does not interlock with the skeleton. Information is lost if the tubing containing the legend separates from the shaft (Bruger 1981), but tag losses of this type have been greatly reduced. (This problem is also eliminated by using the anchor flag tag, in which the T-bar head ends in a flattened pendant that contains limited information. The flags can be numbered by a hand punch to identify individual fish.) Tags can now be chemically treated by the manufacturer to reduce problems caused by fouling (growths of algae and barnacles). Dart and anchor tags cost 15-20 cents each. Because anchor tags are attached with a tagging gun, tagging time is shorter than with dart tags. Anchor tags also work well on lobsters, crabs, and crayfish, provided they are attached in areas that allow the exoskeleton to be shed during molting.

11.6   COORDINATION OF LARGE MARKING OPERATIONS

Before beginning a large tagging operation, there are four main items to consider: (1) using standardized forms, (2) setting up a public relations program, (3) developing a reward system, and (4) securing intra- and inter-agency cooperation (Cogswell 1965).

11.6.1   Use of Standardized Forms

Standardized forms allow the collection of information in a systematic manner for easy compilation and analysis. Forms can be developed in the standard 80-column format for computer analysis. Various examples are available (e.g., Moberly et al. 1977) for use in designing forms that meet the objectives of the tagging operation

Box 11.1   Manufacturers and Distributors of Fish Marking Material, Equipment, and Related Supplies

Aero Brand Inks
18 Durton Avenue
Deer Park, NY 11729
Inks for all purposes

Carolina Biological Supply Company
2700 York Road
Burlington, NC 27215
Biological stains and chemicals

Fisher Scientific Company
1600 W. Glenlake Avenue
Itasco, IL 60143
Biological stains and dyes

Floy Tag & Manufacturing, Inc.
P.O. Box 5357
Seattle, VA 98105
Dart tags, spaghetti tags (tying
or locking), anchor tags, streamer tags,
Petersen tags, applicators

Guardsman Chemicals, Inc.
13535 Monster Road South
Seattle, WA 98178
Coded wire tags (magnetic, fluorescent,
color coded), injectors, detectors,
recovery systems

Howitt Plastics Co.
P.O. Box 162
Molalla, OR 97038
Plastic discs for Petersen tags,
subcutaneous plastic tags

Lee Mark Co.
635 Marina Vista
Martinez, CA 94553
Vinyl ink

National Band and Tag Co.
721 York Street
Newport, KY 41072

Monel strap tags, magnetic internal
tags, vinyl subcutaneous tags

Northwest Marine Technology
Shaw Island, WA 98286

Coded wire tags and related equipment

Phillips Process Co., Inc.
50 Commercial Street
Rochester, NY 14614
Industrial inks for all purposes

Pilgrim Plastic Products Co.
278 Babcock Street
Boston, MA
Internal tags, messages printed on fine
white plastic for capsule tags

Safety and Security Systems
Division/3M
223-3N 3M Center
St. Paul, MN 55144
Microtags

Salt Lake Stamp Co.
380 West Second South, Box 2399
Salt Lake City, UT 84101
Circular strap tags

Scientific Marking Materials
Box 24122
Seattle, WA 98121
Fluorescent pigments and equipment
for applying these to fish

Sigma Chemical Company
P.O. Box 14508
St. Louis, MO 63178
Biological stains and dyes

Teletronic Laboratories, Inc.
P.O. Box 971
El Monte, CA 91731

Vinyl tubing (for spaghetti tags; the firm
will print on the tubing)

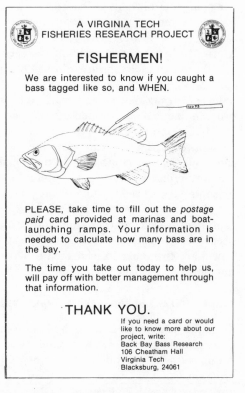

Figure 11.7 An example of a poster used to inform anglers about a fish tagging program.

(Eames and Hino 1981). Standardized forms assure that all persons involved in a tagging operation collect and tabulate information identically. This is absolutely necessary for proper interpretation of the results. Problems such as incomplete recording of tag information, nonstandard names for recovery locations, and failure to measure fish will be avoided with standardized forms and complete instructions.

11.6.2 Public Relations

Tag recoveries are often dependent upon sport or commercial fishermen (Youngs 1974; Osburn et al. 1980; Matlock 1981). Highest recovery rates from fishermen occur if large external or internal tags are used. Tagging experiments should be publicized through the press, radio, and television, through newsletters of fishing organizations, by distributing posters to fishermen and placing them in conspicuous places (Fig. 11.7), and by personal contacts (Eames and Hino 1981). Interest in a tagging program can be stimulated in many ways. Because most anglers are already interested in the movements of tagged fish, providing them with a letter about the fish they report (size of fish at release, where released, date, distance traveled, etc.) will improve the chances of a successful tagging program.

11.6.3 Reward Systems

Tagging programs using a reward system have greater return of tags than those that do not. A reward system can be used as an incentive to increase tag returns from fishermen. Rewards have been as large as $500,000 (Topp 1963) or as little as a personal letter of thanks. There are no hard and fast rules for implementing a reward system. A general rule, however, is that an appropriate reward equals the cost of tagging (price of tags, manpower, etc). As an example, for many years the National Marine Fisheries Service paid an average of $2.00 for return of a tagged shrimp; the actual value of each tag varied from $.05 to $5.00. Many shrimp fishermen, however, felt it was not worth their time to return tagged shrimp for $2.00. As an incentive, the payment for tag returns was changed to include rewards of $50.00 to $500.00. This program increased fishermen interest and tag returns.

It is important that tags contain adequate information in the legend so that persons recovering tags can return them to the right place to receive their rewards. The same information contained on the tag should also be clearly stated in all public relations programs. The reward and information about the fish should be sent to the person returning the tag immediately to motivate other anglers to return tags.

A reward system cannot eliminate all problems. Many anglers fail to report their capture of tagged fish even though they are aware of the tagging program. Fish have been double-tagged to increase detection. Paulik (1961) provides a table that shows how many fish must be recovered to have a probability of .50 to .99 of discovering nonreporting by fishermen.

11.6.4 Intra- and Interagency Cooperation

Intra- and interagency cooperation may be necessary for tagging studies which encompass large bodies of water and involve different states, provinces, and countries. Along a single state boundary, numerous agencies may be involved with a tagging program. Communications among various divisions or sections within a single agency must be open so that all persons involved with the tagging effort or others interested in the program can provide information to the program manager and the public.

11.7 REFERENCES

Allendorf, F. W., F. M. Utter, and B. P. May. 1975. Gene duplication within the family Salmonidae: Detection and determination of the genetic control of duplicate loci through inheritance studies and the examination of populations. Page 415-432 *in* C. L. Markert, editor. *Isozymes. IV. Genetics and evolution.* Academic Press, New York, New York, USA.

Arnold, D. E. 1966. *Marking fish with dyes and other chemicals.* United States Department of Interior, Bureau of Sport Fisheries and Wildlife, Technical Paper Number 10, Washington, District of Columbia, USA.

Bacon, P. R. 1971. Tagless turtles. *Journal of the International Turtle Tortoise Society* 5(3):26-27.

Bailey, M. M. 1965. Lake trout fin-clipping rates at two national fish hatcheries. *Progressive Fish-Culturist* 27:169-170.

Balazs, G. H. 1973. A simplified method for identifying experimental shrimp. *Progressive Fish-Culturist* 35:27.

Bergman, P., K. Jefferts, H. Fiscus, and R. Hager. 1968. *A preliminary evaluation of an implanted coded wire fish tag.* Washington Department of Fisheries Resource Paper 3(1):63-84, Seattle, Washington, USA.

Bergstedt, R. 1980. *Annotated bibliography on mortality associated with the marking of fish by fin clipping.* United States Fish and Wildlife Service, Great Lakes Fisheries Laboratory, Special Report, Ann Arbor, Michigan, USA.

Broadhead, G. C. 1959. Techniques used in the tagging of yellowfin and skipjack tunas in the eastern tropical Pacific Ocean during 1955-57. *Proceedings of the Gulf and Caribbean Fisheries Institute* 11:91-97.

Brock, J. A., and R. K. Farrell. 1977. Freeze and laser marking of channel catfish. *Progressive*

Fish-Culturist 39:138.

Bruger, G. E. 1981. Comparison of internal anchor tags and Floy FT-6B dart tags for tagging snook, *Centropomus undecimalis. Northeast Gulf Science* 4:119-122.

Churchill, W. 1963. The effect of fin removal on survival, growth, and vulnerability to capture of stocked walleye fingerlings. *Transactions of the American Fisheries Society* 92:298-300.

Coble, D.W. 1967. Effects of fin-clipping on mortality and growth of yellow perch with a review of similar investigations. *Journal of Wildlife Management* 31(1):173-180.

Cogswell, S. L. 1965. *Development and operation of a tagging unit in a marine biological laboratory.* United States Bureau of Commercial Fisheries, Laboratory Reference Number 65-18, Woods Hole, Massachusetts, USA.

Coutant, C. C. 1972. Successful cold branding of nonsalmonids. *Progressive Fish-Culturist* 34:131-132.

Cresswell, R. C. 1981. Post-stocking movements and recapture of hatchery-reared trout in flowing waters—a review. *Journal of Fish Biology* 18:429-442.

Dell, M. B. 1968. A new fish tag and rapid, cartridge-fed applicator. *Transactions of the American Fisheries Society* 97:57-59.

Donahoo, M. J. 1976. Modified Carlin-type tag for identifying experimental fish. *Progressive Fish-Culturist* 38:88-89.

Eames, M., and M. Hino. 1981. *An evaluation of the effects of four tags used for marking juvenile chinook salmon (Oncorhynchus tshawytscha).* Washington Department of Fisheries Technical Report 61, Seattle, Washington, USA.

Eipper, A., and J. Forney. 1965. Evaluation of partial fin-clips for marking largemouth bass, walleyes, and rainbow trout. *New York Fish and Game Journal* 12:233-240.

Emery, L., and R. Wydoski. 1983. (In press). *Marking and tagging of aquatic animals:* An indexed bibliography. United States Department of Interior, Fish and Wildlife Service, Resource Publication.

Everest, F., and E. Edmundson. 1967. Cold branding for field use in marking juvenile salmonids. *Progressive Fish-Culturist* 29:175-176.

Everhart, W., A. Eipper, and W. Youngs. 1975. *Principles of fishery science.* Cornell University Press, Ithaca, New York, USA.

Farmer, A. S. D. 1981. A review of crustacean marking methods with particular reference to penaeid shrimp. Pages 167-183 *in* A. S. D. Farmer, editor. *Proceedings of the international shrimp releasing, marking and recruitment workshop, Salmiya, State of Kuwait, 25-29 November 1978.* Kuwait Institute of Scientific Research, Kuwait Bulletin of Marine Science 2.

Farmer, A. S. D., and M. H. Al-Attar. 1981. Results of shrimp marking programmes in Kuwait. Pages 53-83 *in* A. S. D. Farmer, editor. *Proceedings of the international shrimp releasing, marking and recruitment workshop, Salmiya, State of Kuwait, 25-29 November 1978.* Kuwait Institute of Scientific Research, Kuwait Bulletin of Marine Science 2.

Fry, D. H., Jr. 1961. Some problems in the marking of salmonids. *Pacific Marine Fisheries Commission Bulletin* 5:77-83.

Gerking, S. D. 1963. *Non-mutilation marks for fish.* International Commission on North Atlantic Fisheries, Special Publication Number 4:248-254.

Gibbard, G. L., and R. L. Colura. 1980. Retention and movement of magnetic nose tags in juvenile red drum. *Annual Proceedings of the Texas Chapter of the American Fisheries Society* 3:22-29.

Hanson, L. H. 1972. An evaluation of selected marks and tags for marking recently metamorphosed sea lampreys. *Progressive Fish-Culturist* 34:70-75.

Harris, J. E. 1937. *The mechanical significance of the position and movements of the paired fins in the teleostei.* Papers of the Tortugas Laboratory, Carnegie Institute 31(7):173-189.

Hart, P. J. B., and T. J. Pitcher. 1969. Field trials of fish marking using a jet inoculator. *Journal of Fish Biology* 1:383-385.

Jakobsson, J. 1970. On fish tags and tagging. Pages 457-499 *in* H. Barnes, editor. *Oceanography and marine biology, an annual review,* volume 8. Hafner Publishing Company, New York, New York, USA.

Jefferts, K. B., P.K. Bergman, and H. F. Fiscus. 1963. A coded wire identification system for macro-organisms. *Nature* (London) 198:460-462.

Jones, R. 1979. *Materials and methods used in marking experiments in fishery research.* Food and Agriculture Organization of the United Nations, Fisheries Technical Paper Number

190 (FIRM/T190). Rome, Italy.

Joyce, J. A., and H. M. El-Ibiary. 1977. Persistency of hot brands and their effects on growth and survival of fingerling channel catfish. *Progressive Fish-Culturist* 39:112-114.

Kabata, Z. 1963. *Parasites as biological tags.* Pages 31-37 *in* International Commission on Northwest Atlantic Fisheries (ICNAF), Special Publication Number 4.

Killingley, J. S. 1980. Migrations of California gray whales tracked by oxygen-18 variations in their epizoic barnacles. *Science* 207:759-760.

Klima, E. F. 1965. *Evaluation of biological stains, ink, and fluorescent pigments as marks for shrimp.* United States Fish and Wildlife Service, Special Scientific Report-Fisheries Number 511.

Klima, E. F. 1981. The National Marine Fisheries Service shrimp research program in the Gulf of Mexico. Pages 185-207 *in* A. S. D. Farmer, editor. *Proceedings of the international shrimp releasing, marking and recruitment workshop, Salmiya, State of Kuwait, 25-29 November 1978.* Kuwait Institute of Scientific Research, Kuwait Bulletin of Marine Sciences 2.

Krieger, K. J. 1982. Tagging herring with coded wire microtags. *Marine Fisheries Review* 44(3):18-21.

Lagler, K. F. 1952. *Freshwater fishery biology.* William C. Brown Company, Dubuque, Iowa, USA.

Laird, L. M., and B. Scott. 1978. Marking and tagging. Pages 84-100 *in* T. Bagenal, editor. *Methods for assessment of fish production in fresh waters,* 3rd edition. IBP Handbook Number 3. Blackwell Scientific Publications, Oxford, England.

Lawler, G. H., and G. F. Smith. 1963. Use of colored tags in fish population estimates. *Journal of Fisheries Research Board of Canada* 20:1431-1434.

Major, R. 1962. *Marking sockeye salmon scales by short periods of starvation.* United States Fish and Wildlife Service, Special Scientific Report Number 416, Washington, District of Columbia, USA.

Major, R. L., K. H. Mosher, and J. E. Mason. 1972. Identifications of stocks of Pacific salmon by means of scale features. Pages 209-231 *in* R. C. Simon and P. A. Larkin, editors. *The stock concept in Pacific salmon.* H. R. MacMillan Lectures in Fisheries, 1970. University of British Columbia, Vancouver, British Columbia, Canada.

Matlock, G. C. 1981. Nonreporting of recaptured tagged fish by saltwater recreational boat anglers in Texas. *Transactions of the American Fisheries Society* 101:90-92.

McKee, T. B., and W. F. Hablou. 1963. An attempt to mark juvenile silver salmon by feeding selected metallic compounds. *Oregon Fish Commission, Research Briefs* 9(1):27-29.

Miles, H. M., S. M. Lochner, D. T. Michand, and A. L. Saliver. 1974. Physiological responses of hatchery reared muskellunge *(Esox masquinongy)* to handling. *Transactions of the American Fisheries Society* 103:336-342.

Moberly, S. A., R. Miller, K. Crandas, and S. Bates. 1977. *Mark-tag manual for salmon.* Alaska Department of Fish and Game, Division of Fisheries Rehabilitation, Enhancement, and Development, Juneau, Alaska, USA.

Nicola, S. J., and A. J. Cordone. 1973. Effects of fin removal on survival and growth of rainbow trout *(Salmo gairdneri)* in a natural environment. *Transactions of the American Fisheries Society* 102:753-758.

Osburn, H. R., G. C. Matlock, and H. E. Hegen. 1980. Description of a multiple census tagging program for marine fisheries management. *Annual Proceedings of the Texas Chapter of the American Fisheries Society* 2:9-25.

Park, D. L., and W. J. Ebel. 1974. Marking fishes and invertebrates. Part 2-Brand size and configuration in relation to long-term retention on steelhead trout and chinook salmon. *Marine Fisheries Review* 36(7):7-9.

Parker, N. C., and K. Randolph. 1981. Marking techniques. Pages 59-60 *in National Fisheries Center-Leetown annual report, October 1980-September 1981.* Kearneysville, West Virginia, USA.

Parker, R. R., E. C. Black, and P. A. Larkin. 1963. *Some aspects of fish-marking mortality.* International Commission of Northwest Atlantic Fisheries (ICNAF), Special Publication Number 4:117-122.

Paulik, G. J. 1961. Detection of incomplete reporting of tags. *Journal of the Fisheries Research Board of Canada* 18:817-832.

Phinney, D. E., D. M. Miller, and M. L. Dahlberg. 1967. Mass-marking young salmonids with fluorescent pigment. *Transactions of the American Fisheries Society* 96:157-162.

Piper, R. G., I. B. McElwain, L. E. Orme, J. P. McCraren, C. G. Fowler, and J. R. Leonard. 1982. *Fish hatchery management.* United States Fish and Wildlife Service, Washington, District of Columbia, USA.

Prentice, E. F., and J. E. Rensel. 1977. Tag retention of the spot prawn, *Pandalus plactyceros,* injected with coded wire tags. *Journal of the Fisheries Research Board of Canada* 34:2199-2203.

Radcliffe, R. W. 1950. The effect of fin-clipping on the cruising speed of goldfish and coho salmon fry. *Journal of the Fisheries Research Board of Canada* 8:67-73.

Raleigh, R. F., J. B. McLaren, and D. R. Graff. 1973. Effects of topical location, branding techniques, and changes in hue on recognition of cold brands in centrarchid and salmonid fish. *Transactions of the American Fisheries Society* 102:637-641.

Raymond, H. L. 1974. Marking fish and invertebrates. Part 1-State of the art of fish branding. *United States National Marine Fisheries Service Review* 36(7):1-6.

Ricker, W. 1949. Effects of removal of fins upon the growth and survival of spiny-rayed fishes. *Journal of Wildlife Management* 13:29-40.

Ricker, W. 1956. The marking of fish. *Ecology* 37:665-670.

Rinne, J. N. 1976. Coded spine clipping to identify individuals of the spiny-rayed fish *Tilapia. Journal of the Fisheries Research Board of Canada* 33:2626-2629.

Rounsefell, G. A. 1963. *Marking fish and invertebrates.* United States Department of Interior, Fish and Wildlife Sevice, Fisheries Leaflet 549.

Rounsefell, G. A., and J. L. Kask. 1945. How to mark fish. *Transactions of the American Fisheries Society* 73:320-363.

Schwartz, F. J. 1981. A long-term internal tag for sea turtles. *Northeast Gulf Science* 5:87-93.

Sindermann, C. J. 1961. Parasite tags for marine fish. *Journal of Wildlife Management* 25:41-47.

Stonehouse, B., editor. 1978. *Animal marking: Recognition marking of animals in research.* University Park Press, Baltimore, Maryland, USA.

Strange, C. D., and J. A. Kennedy. 1982. Evaluation of fluorescent pigment marking of brown trout (*Salmo trutta* L.) and Atlantic salmon (*Salmo salar* L.). *Fisheries Management* 13:89-95.

Stuart, T. A. 1958. *Marking and regeneration of fins.* Freshwater Salmon Fisheries Research, Edinburgh, Scotland 22.

Thoma, B., G. Swanson, and V. E. Dowell. 1959. A new method for marking fresh-water mussels for field study. *Proceedings of the Academy of Science* 66:455-457.

Topp, R. 1963. *The tagging of fishes in Florida, 1962 program.* Florida State Board of Conservation, Professional Papers Series 5.

Wilder, D. G., and R. C. Murray. 1956. Movements and growth of lobsters in Egmont Bay, Prince Edward Island. *Fisheries Research Board of Canada, Atlantic Progress Report* 64:3-9.

Winters, G. H. 1977. Estimates of tag extrusion and initial tagging mortality in Atlantic herring *(Clupea harengus harengus)* released with abdominally inserted magnetic tags. *Journal of Fisheries Research Board of Canada* 34:354-359.

Winters, G. H. 1977. Healing of wounds and location of tags in Atlantic herring *(Clupea harengus harengus)* released with abdominally inserted magnetic tags. *Journal of the Fisheries Research Board of Canada* 34:2402-2404.

Wursig, B., and M. Wursig. 1977. The photographic determination of group size, composition, and stability of coastal porpoises *(Tursiops truncatus). Science* 198:755-756.

Wydoski, R. S., and G. A. Wedemeyer. 1976. Problems in the physiological monitoring of wild fish populations. *Proceedings of the Annual Conference of the Western Association of Game and Fish Commissioners* 56:200-214.

Youngs, W. D. 1974. Estimation of the fraction of anglers returning tags. *Transactions of the American Fisheries Society* 103:616-618.

Youngs, W. D., and D. S. Robson. 1975. Estimating survival rate from tag returns: Model tests and sample size determination. *Journal of the Fisheries Research Board of Canada* 32:2365-2371.

Chapter 12
Hydroacoustics

RICHARD E. THORNE

12.1 INTRODUCTION

The term "fisheries acoustics" is generally applied to techniques which use sonars or depthsounders, that is, techniques in which sound is actively transmitted and information extracted from the returning echoes. Thus the term excludes techniques based on the sounds which fish make, usually termed "passive acoustics" or "bioacoustics" and also techniques using acoustic tags, usually termed "biotelemetry" (*see* Chapter 20).

Fisheries acoustics, a relatively new and rapidly developing field, has considerable potential for fisheries research and management. Because of its novelty as well as its technical complexity, it is one of the most misunderstood techniques in fisheries. Hydroacoustics has been shown to be an extremely powerful and very effective fishery assessment tool when properly applied under appropriate circumstances. The technique, however, is not a panacea for fisheries assessment problems and is not even applicable in some cases.

12.1.1 History

About 1930 it was discovered that the ultrasonic depthsounder could be used to detect fish, and since that time hydroacoustics has had a major impact on the fishing industry. The recent widespread application of such advances as sector scanning sonars combined with computer-based locational devices will challenge fisheries managers for some years to come. Developments in fisheries research and management have lagged behind commercial fishing technology. Substantial progress was not made until the late 1960s and early 1970s when the echo integrator and the supporting theory were developed (Cushing 1973; Margetts 1977). The first manual on hydroacoustics techniques was also published about this time (Forbes and Nakken 1972). Subsequent manuals dealing with fisheries acoustics are Burczynski (1979) and Saville (1978). The text, *Acoustical Oceanography* (Clay and Medwin 1977), devotes Chapter 6 to the scattering of sound by "things" in the sea and Chapter 7 to the use of acoustical data to make quantitative estimates of biomass. These chapters give a careful development of the acoustical theory at the level of first-year calculus and physics. Most recently, a comprehensive bibliography of hydroacoustics has been published through the Food and Agriculture Organization of the United Nations (Venema 1982).

12.1.2 Advantages and Limitations

Any discussion of advantages and limitations must start with a warning. The technology of acoustical oceanography is changing very rapidly. Tasks that were difficult or "impossible" a few years ago are simple with the new technology. The more important changes are (1) microprocessors have revolutionized our ability to acquire, process, and display data; (2) multibeam sonar systems and multiplexed transducer arrays have greatly increased sampling power and target resolution; (3) new sonars can have greatly increased capability for size discrimination; and (4) digital systems have much better displays. The combination of these and new signal-processing techniques may solve many of the most vexing problems in the application of acoustics to fisheries.

Hydroacoustics has a number of advantages over other assessment techniques, as well as several limitations. The advantages are (1) independence from fishery catch statistics, (2) favorable time scale, (3) relatively low operational costs, (4) low variance, and (5) potential for absolute population estimation. Independence from catch statistics allows application to unexploited or poorly exploited stocks. It also frees acoustics from the long lag times associated with catch statistics, leading to the second advantage. Unlike fishery catch statistics, which are obtained only after the fishery harvest, hydroacoustic techniques can be applied prior to harvest. This feature is especially important for short-lived fishes.

As a result of the high sampling power and efficiency of hydroacoustic techniques, operational costs are relatively low. The major operational costs are shiptime and manpower for data collection and analysis. In all three categories, the costs are usually much lower than those of traditional sampling with nets.

The low variance associated with hydroacoustic techniques is also the result of high sampling capability. For the same cost, in time and money, more hydroacoustic samples can be collected. The sampling power of hydroacoustics is at least an order of magnitude higher than that of direct capture techniques. This is especially important because of the patchy distribution of most fish.

The last advantage is the potential for absolute measurements. This advantage is not critical because most fishery management is based on relative indices such as catch-per-unit-effort (CPUE). However, the capability for absolute estimation allows reasonably precise management without a historical data base, and ultimately leads to a much better understanding of production processes.

The limitations of hydroacoustic assessment techniques are (1) poor species discrimination, (2) little or no sampling capability near bottom and surface, (3) relatively high complexity, (4) high initial investment, (5) lack of biological samples, and (6) potential bias associated with target strength and calibration.

The partition of acoustically derived abundance estimates into various species requires auxiliary information which is usually obtained by complementary sampling with nets. There are possibilities for hydroacoustic identification of species, but these depend mainly on establishing species-specific distributional patterns.

Echo sounding techniques are limited in their ability to survey near the surface and bottom of the water column and cannot detect targets on bottom. This limitation is a function of several parameters, especially the pulse length (pulse length is the amount of time the beam is transmitted, usually msec). Boat avoidance is an additional complication for near-surface fish. In some cases, this limitation can be greatly reduced by using a stationary transducer rather then a moving one; for example, Thorne (1980) was able to detect fish within 0.1 m of the bottom. Hydroacoustics is generally not applicable to surveys of demersal fish stocks because of this limitation. However, even fish classified as groundfish, such as walleye pollock, may be too far off bottom to be adequately sampled by bottom trawls, so that a combination of hydroacoustics and trawling are needed for assessment.

A third disadvantage of hydroacoustic techniques is their complexity. Because fisheries acoustics is a relatively recent discipline, training in these concepts is not usually included in the education of fisheries professionals. Therefore, fisheries biologists may be unfamiliar with the mathematical, electrical, and physical concepts of hydroacoustics.

Although operational costs are low, initial investment costs are high, even for basic equipment. The cost of acquiring a minimal system for scientific quality data collection and echo integration analysis is about $50,000, and more sophisticated equipment can be much more expensive. However, the cost of equipment is decreasing rapidly, while the quality is increasing as part of the current information revolution. The microcomputer "chips" and associated electronic technology permit the design and construction of systems that are more reliable and simpler to use than the best that were built a few years ago.

The fourth limitation is lack of biological samples that can be used for other purposes such as aging or stomach analysis. In general, this is a minor concern because complementary sampling with nets is needed for species identification in most cases.

The last limitation is bias due to uncertainty in the calibration or target strength value. Information on target strengths of fish is growing, but this area is one of the most important present research needs and lack of target strength information is currently the major limitation to more widespread use of hydroacoustic techniques. Calibration is a minor source of error if appropriate techniques are used. Unfortunately, hydroacoustic systems are often used without adequate calibration, resulting in substantial errors.

12.2 BASIC PRINCIPLES

12.2.1 Introduction

The literature of underwater acoustics is very extensive. Elementary principles are covered in many textbooks such as Urick (1975) and Clay and Medwin (1977) in far more detail than can be provided here, and specific principles of fisheries acoustics are detailed in Forbes and Nakken (1972) and Burczynski (1979). Major theoretical dissertations such as that of Ehrenberg (1973) on echo integration can be very intimidating to the non-mathematician, but basic concepts can be grasped without recourse to calculus. The important aspect for the fishery biologist is neither to underestimate the complexity or to be initimidated by it, but to strive for sufficient understanding to recognize personal limitations and, consequently, when and where to seek expert advice.

12.2.2 Equipment

Although a great variety of types and qualities of equipment are available, the basis for fisheries work is a sonar system. Usually the system is oriented vertically, in which case it is called an echosounder. A sonar or echosounding system basically consists of four parts: (1) transmitter, (2) transducer, (3) receiver-amplifier, and (4) control and display. The transmitter produces a burst or pulse of electrical energy which is converted by the transducer (or underwater loudspeaker) to an acoustical

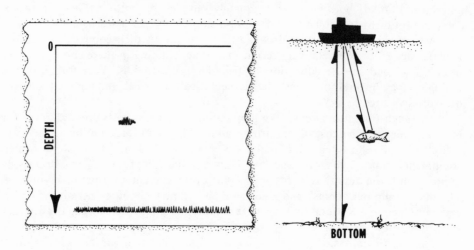

Figure 12.1 Acoustic echo sounding record showing midwater scatters. The sonar sends a "ping" or burst of sound energy down and then listens for echoes. Some of the energy is scattered back to the sonar by fish and the bottom. The length of time it takes for the echo to return corresponds to the depth.

signal in the form of a short "ping" or beep. This signal travels through the water, and when it hits a target, such as a fish or bottom, it is reflected towards the source as an "echo" (Figure 12.1). In most systems, the same transducer receives the echoes and converts them back into electrical signals. The receiver-amplifier increases and modifies the signal to a form suitable for display. The display is generally a paper recorder, but may be an oscilloscope or cathode ray tube (CRT). A recent development is a colorometric CRT, which has considerably increased dynamic range.

The pulse of electrical energy produced each time the transmitter is triggered has a certain width, frequency, and power. The maximum range of an echosounder is dependent on the power as well as some additional parameters. There is an upper limit to the amount of power which can be applied to a tranducer of a given size. An important part of the system is the range, that is, the approximate depth below the transducer at which the target fish are located. The range resolution of the echosounder is dependent on the length of pulse — the shorter the pulse length, the greater the resolution — but there are constraints related to the frequency in that a minimum number of cycles is required in the pulse to define the frequency. Thus, higher frequencies with shorter wave lengths can have smaller minimum pulse lengths. Pulse lengths of typical echosounders range from 0.1 to 5 msec, with frequencies typically 20 to 400 kHz.

The pulse of electrical energy from a transmitter is converted into acoustic energy by the transducer. This is accomplished by small vibrations which displace the particles of water back and forth, resulting in pressure changes. Similarly, pressure changes associated with the reflected sound wave are detected by the transducer and converted into electrical energy.

By analogy to the "beam of light" from a flashlight, the sound from a transducer is known as a sonar beam. This beam is often visualized as a cone, but in reality it is much more complex. Transducers concentrate or direct sound in a particular direction. The sound intensity is maximum at a direction perpendicular to the transducer surface, called the acoustic axis, and decreases with angle from the acoustic axis. The circular disk or piston transducer is commonly used. Its directional response is shown in Figure 12.2. The width of the beam is often defined by the angle where the intensity has decreased to one-half of the value on the acoustic axis. The width depends on the size of the transducer in relation to the frequency. Larger transducers and higher frequencies result in narrower beams (that is, the energy is more concentrated along the acoustic axis).

The amplitudes of the echoes are very small compared to the amplitudes of the transmitted signals. The primary function of the receiver-amplifier is to apply the appropriate amplification to the echoes received by the transducer. Almost all receivers have a manual gain control which increases or decreases the overall amplification of the receiver. Most also have some type of time- or depth- dependent gain function which is important for both display and quantitative processing because the amplitude of the signals decrease with range. This feature may be merely an initial suppression circuit to eliminate or decrease the amplitude of signals from near surface targets, especially those due to nonbiological factors such as turbulence from surface waves, or a time-varying-gain (TVG) which compensates for spreading and attenuation losses.

Control and displays for echosounders come in a variety of types and sizes. They have two basic functions: (1) to trigger the transmitter, and (2) to record the echoes as a function of depth. The number of transmissions per unit time, called the pulse repetition rate, is governed primarily by the range to be displayed and the depth of the bottom. Echoes generated by the previous pulse must return before a new pulse, or there will be confusion as to the true depth of the echoes; for example, a false bottom might appear on the display (Edgell 1935).

The primary display is usually a graphic recorder and it gives a paper record. The pen sweeps over the paper and blackens the paper when there is a signal voltage at the output of the receiver. In older and simple recorders, the recorder generates the trigger so that the transmission-reception cycle is synchronized with the recorder. On some, the display cycle can be triggered by an external trigger. The graphic records give a good visual history of what's been happening. However, the intensity range is small

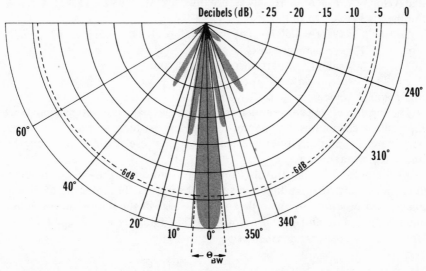

Figure 12.2 Directivity of a piston transducer. θ=0 is the acoustic axis. The shaded area is response in decibels (dB). The half power, -6 dB beam width is θ_{BW}.

and the graphic recorder gives a poor representation of the amplitude of fish echoes. It is sometimes helpful to operate standard cathode ray oscilloscopes in parallel with the recorder to observe echo amplitudes. Alternatively, one can record the echo sounding on magnetic tape.

The depth range displayed by a graphic recorder usually depends on the speed of rotation of the pen. For a given paper width, the faster the pen, the smaller the range displayed (or the greater the range resolution). Some more recent recorders have multiple pens and signal storage so that high range resolution can be obtained without the need for a high speed pen rotation.

Although a great variety of echosounders is available commercially, few are readily adaptable for scientific research for two reasons. First, their output device is usually an echogram, and the output is not easily accessed for more quantitative signal processing. Second, the range corrections are not usually based on any quantitative physical principles; this can lead to considerable bias when attempting to compare abundance at different depths. At present, there are only three commercial echosounders intended for quantitative fisheries research, the Simrad EK-400, Simrad EY-M, and the Biosonics Model 101 (Bayona 1982). The first two are manufactured in Norway, the latter in the United States. All three have output points for access to the signals for quantitative signal processing and have precision TVG for automatic range correction.

12.2.3   Sound Propagation and Reflection

A pulse of sound from the transducer travels through water at a speed which depends on the temperature, salinity, and depth, but is usually between 1400 and 1500 m/sec. The range (R) to target fish is determined by the time required for the sound to travel from the transducer to the target and back, that is

$$R = 0.5\ ct \tag{12.1}$$

where c = speed of sound in water, and
                    t = time.

The literature contains several empirical equations which may be used to calculate sound velocity (c); a brief review is provided by MacKenzie (1981). Urick (1975) recommends Leroy's formula:

$$c = 1449.34 + 4.56T - .046T^2 + (1.38 - .01T)\ (S\text{-}35) + D/61 \tag{12.2}$$

where T = temperature in C,
                    S = salinity in parts per thousand, and
                    D = depth in m,

giving a sound velocity in m/sec.

The display of an echosounder or sonar, as well as automatic signal processing equipment, incorporates an assumed value of c which will not be accurate if oceanographic conditions differ considerably from those assumed in the choice of the sound velocity parameter.

As the pulse of sound travels from the transducer to the target, its intensity is reduced by geometric spreading and sound absorption. The intensity of the reflected signal is also modified by the reflective properties of the target and by subsequent geometric spreading and sound absorption. The intensity of the signal (I_e) when it returns to the transducer can be described as:

$$I_e = \frac{kI_i 10^{-2\alpha R} b^2 (\Theta,\phi)}{R^4} \tag{12.3}$$

where k = a constant dependent on the reflective properties of the target

I_i = intensity of the signal at a unit range from the transducer

$\propto$ = attenuation coefficient (a function of signal frequency, water temperature, and salinity)

R = range of the target, and

$b(\theta, \phi)$ = directivity pattern function (a characteristic of the transducer).

Expressed in decibel units (the decibel is ten times the ratio of intensities expressed in logarithmic units), the above sonar equation is

$$EL = SL + TS - 40 \log R - 2 \propto R + 20 \log b(\theta, \phi) \qquad (12.4)$$

where EL (echo level) = $10 \log I_e$

SL (source level) = $10 \log I_i$, and

TS (target strength) = $10 \log k$.

The source level and directivity pattern are determined by calibration. The two-way propagation loss ($40 \log R - 2 \propto R$) can be calculated, and is usually incorporated in the TVG circuit of scientific echosounders.

The target strength *(TS)* depends on the size of the target and its reflective characteristics (Fig. 12.3). Considerable work has been done and remains to be done on the target strengths of fish. Often target strength is expressed in units of biomass, as that parameter is often the desired product of an assessment. An example of the variation of target strength per unit biomass for different fish lengths is shown in Figure 12.4. The swimbladder is a major source of signal scattering in fishes, probably accounting for about 90% of the echo. Fishes with swimbladders have target strengths some 10 dB higher than fish without swimbladders for corresponding lengths. The target strength also depends on the aspect angle, that is how the signal hits the fish. This variation is a major source of uncertainty in the practical application of target strength information in fish assessment because the variation in aspect of fishes *in situ* is poorly known. Variation in target strength with depth is another uncertainty which needs further definition. A special workshop on target strength (Anonymous 1978) concluded that fish with swimbladders appear to constitute a single class with respect to target properties.

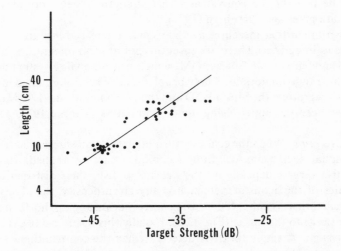

Figure 12.3 Observations on the maximum target strength of herring at 120 kHz as a function of fish length (after Nakken and Olson 1977).

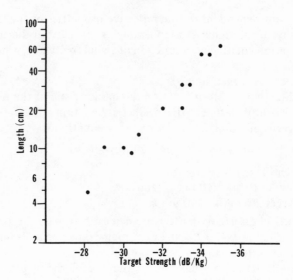

Figure 12.4 Relation between fish length and target strength per unit fish weight from various data sources.

12.2.4 Signal Processing

Echo Integration. Methods of extracting abundance information from echo signals depend either on measuring the strength of echoes or resolving and counting echoes from individual fish targets. The most widely used method for measuring signal strength is called echo integration. The technique is based on the principle that the acoustic intensity (amplitude squared) of a signal reflected from an underwater target is proportional to the number of target fish and the mean back scattering cross section of the fish, which is related to target strength. In order to obtain a biomass estimate from a concentration of fish, it is necessary to determine the total intensity of the acoustic signal reflected from the fish and to know the mean backscattering cross section of the fish (or the equivalent target strength value). Formulas for such a calculation are given by Ehrenberg (1973).

The instrument that measures average signal intensity for a group of targets is called an echo integrator. There are several types of echo integrators. The original commercial model was the Simrad QM, which was an analog system and suffered from dynamic range limitations. These problems have been eliminated in the present digital echo integrators, the Simrad QD and the Biosonics Model 120. More versatile systems have been developed using digital computers (Thorne 1973; Thorne et al. 1977).

Echo Counting. When the echoes returning to the transducer are predominantly from individual fish, echo counting techniques can be applied. Resolution of individual fish targets depends on the density of fish, their distribution, and the characteristics of the acoustic system. For any given density, the ability to resolve individual fish targets is enhanced by using narrow beam angles, shorter pulse lengths, and shorter range to the target (Fig. 12.5). Usually this means using a relatively high frequency system. A single target is detected when the combination of parameters described earlier in the sonar equation result in a target amplitude that can be detected above the noise level. For larger fish with greater target strengths, the effective

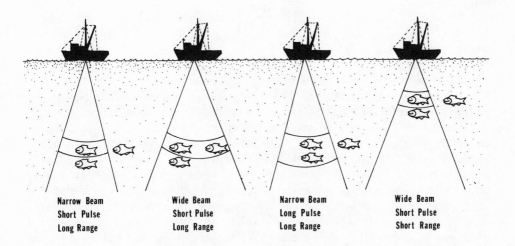

Narrow Beam
Short Pulse
Long Range

Wide Beam
Short Pulse
Long Range

Narrow Beam
Long Pulse
Long Range

Wide Beam
Short Pulse
Short Range

Figure 12.5 Geometry associated with resolution of individual fish targets. Note that at long range with a wide beam or a long pulse, more than one fish is included.

sampling angle of a transducer would be greater (Fig. 12.6). If the target strength of a fish or the mean target strength of a population is known, as well as the other system parameters, the sampling volume can be determined for any given threshold of background noise by calculating the angle of the directivity pattern where the echo level exceeds the threshold.

The sampling volume can also be determined by a technique called the duration-in-beam (Nunnallee 1980). The number of successive echoes from a given fish will depend on the dimensions of the transducer beam and the boat speed. For a circular transducer, the diameter of the sampling cross-section *(d)* at any given depth can be

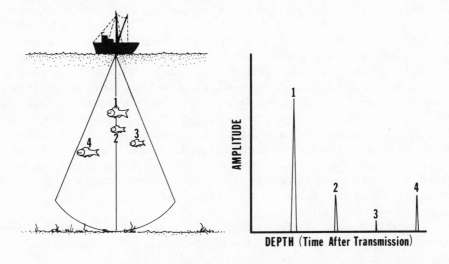

Figure 12.6 Echo amplitudes from fish targets of two sizes at various locations in the beam. The signal amplitudes have been corrected for range losses.

calculated as:

$$d = \frac{4s\bar{e}}{\pi r}$$

(12.5)

where s = boat speed,

$\bar{e}$ = average number of echoes per fish

and r = the pulse repetition rate of the echosounder.

The value $\bar{e}$ needs to be relatively large (greater than 3) to minimize a bias at low numbers. Values for the magnitude of the bias under various circumstances are tabled in Nunnallee (1980). The advantage of the duration-in-beam technique is that it eliminates the need for information on target strengths and even system parameters. However, the system parameters, including the choice of threshold, must be such as to incorporate virtually the entire range of target strengths (in other words, not eliminate targets with lesser strengths) while simultaneously excluding echoes from side lobes of the beam, which would invalidate the assumptions of the technique. The technique must be used cautiously when the sizes of fish targets differ greatly. A recommended procedure is to examine the fish echoes for the largest signals, then establish the threshold at a level just above that which would be expected from these large targets if they were located in the side lobes. However, since the target strength spread of even a single size of fish is considerable, the technique must be used cautiously, especially if the sizes of the fish targets differ considerably. Even for a single size of fish, the threshold should be at least 30 dB below the largest fish signals or there can be considerable bias.

The major advantage of echo counting techniques is that they are insensitive to error associated with target strength uncertainties. For example, a 6 dB error in the estimation of the mean target strength may result in about a 15% error in density measurements using echo counting. In contrast, the same error using echo integration would cause an error in the density estimation of 400%. However, large errors can result with echo counting techniques if smaller amplitude echoes are not properly accounted for, as described above.

Target strength. In addition to its importance for echo integration and echo counting techniques, target strength information can be used to estimate fish size or to distinguish among various size classes. For example, Thorne (1978) was able to discriminate between large adult sockeye salmon returning to Lake Washington and resident fish, and, consequently, use echo counting techniques to estimate the abundance of adults. Target strength information can be obtained either by controlled experiments, in which fish are confined, or by *in situ* measurements. In some controlled experiments, individual fish are suspended on a rigid structure, and detailed measurements are made as a function of fish orientation and acoustic frequency. The results illustrated in Fig. 12.3 were obtained from such an experiment. Another approach is to take measurements on individual or multiple fish that are confined in a pen. Burczynski (1979) gives considerable details on this type of cage experiment. The major disadvantage of controlled experiments is that there is some uncertainty about the amount of correlation between the acoustic properties of free swimming and confined fish.

The alternative to controlled experiments is to extract the target strength estimates from the acoustic returns *in situ*. The difficulty here is that the echo from a fish target depends not only on its target strength, but also on its position in the acoustic beam. Several procedures have been proposed for removing the beam pattern effect. These techniques fall into two categories: those that indirectly extract the effect of the pattern from a collection of echoes, and methods that directly remove the beam pattern from each individual echo. The appeal of the indirect techniques is that they can be implemented with the same single-transducer echo-sounding systems that are used for the surveys. The most commonly used method applies a series of linear equations to the echo strength data using information about the directivity pattern function in order to extract the target strength distribution (Craig and Forbes 1969). This technique is detailed in Forbes and Nakken (1972) and is the basis for a commercial target strength measuring system, the Simrad QD 200. Alternate and more powerful extraction techniques have been proposed by Ehrenberg (1972), Peterson et al. (1976), Robinson (1978), and Ehrenberg et al. (1981).

The alternative to the indirect method is to remove the beam pattern effect for each individual echo. This eliminates many of the problems involved in the indirect techniques, but requires some modification of basic assessment systems. The simplest configuration currently used for direct target strength estimation employs a dual beam transducer with axes aligned and two receiving channels. The acoustic signal is transmitted with a narrow beam transducer and received on both the narrow and wide beam receiver simultaneously. Because the beam pattern factor for the wide beam is approximately constant over the main lobe of the narrow beam, the beam pattern factor can be determined from the ratio of the intensities on the two channels. Further details are described in Ehrenberg (1974), Traynor and Ehrenberg (1979), Dickie et al. (1982), and Traynor and Williamson (1982).

12.3   APPLICATION

12.3.1   Requirements of Accuracy and Precision

As stressed earlier in this manual, the application of any sampling technique must consider the requirements for accuracy and precision, which in turn depend on the objectives of the study. This is especially true with hydroacoustics because the technique is highly quantitative and, unlike many direct capture techniques, the sources of error associated with hydroacoustics are predictable both in nature and magnitude. A different interpretation of both the magnitude of error and the typical requirements for accuracy in fisheries applications is part of the basis for the pessimism toward hydroacoustics expressed in Suomala and Yudanov (1981). In contrast, Suomala (1975) describes sampling with trawls as "well-established and effective." It can be argued, however, that the development of accurate hydroacoustic techniques is only recently revealing the considerable variability and bias associated with trawling techniques.

It is important to realize that the requirements for accuracy in the use of hydroacoustics depend on the type of data desired, ranging from information on presence or absence to estimation of absolute densities. Detection of presence or absence is the primary use of hydroacoustics in commercial fisheries and has considerable application in fisheries research, particularly for behavioral studies. An example of the type of information which might be obtained is illustrated in Figure 12.7, which shows an echogram from a transect upcurrent and downcurrent of a thermal discharge. The presence of fish downcurrent of the discharge demonstrates the attraction of fish to the thermal plume. Study of diel migration is another potential application at this level of accuracy, and, of course, information on presence or absence can be very valuable in the allocation of direct capture effort. This level requires neither an extensive knowledge of hydroacoustics nor sophisticated

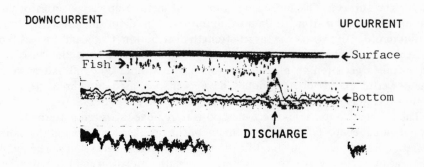

Figure 12.7 Echogram showing the distribution of fish upcurrent and downcurrent of the heated discharge from a power plant.

processing equipment, but merely an echogram and a system whose parameters are generally known. For example, an understanding of the performance of the system as a function of range would be sufficient for a study of diel migration.

Most fishery assessment studies require information on relative rather than absolute abundance. This requirement is more stringent than presence or absence and in most cases would require application of echo counting or echo integration processing and calibrated equipment. However, it would not require knowledge of target strength, but only the assumption that the target strength of the population was constant. Such an assumption is probably reasonable for studies of the same population through time. Figure 12.8, for example, shows the comparison of CPUE in

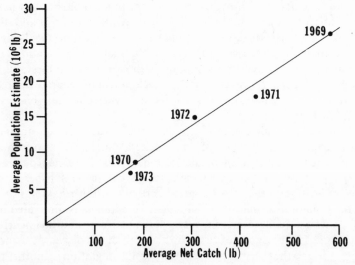

Figure 12.8 Relation between population estimates from acoustic surveys (in millions of pounds) and average net catches (in pounds) used as an index of abundance for hake, Port Susan, Washington.

trawls with acoustic population estimates of hake over a 5-year period. The assumption of a constant target strength is certainly as reasonable as that of a constant catchability coefficient.

The capability for absolute population estimation is not essential because fisheries management usually concerns the responses of a stock to exploitation on a relative basis. However, such information can be valuable. For example, in sockeye salmon management, it is possible to develop a relationship between relative abundance of presmolts and subsequent returns of adults. This relationship is an effective management tool, but its development requires several years of data. However, if an absolute estimate of the abundance of presmolts is available and the rate of marine survival is known, an immediate forecast of abundance of the adult run can be made. Conversely, if marine survival rate is unknown, then the capability to estimate the absolute abundance of presmolts and adults allows estimation of the marine survival rate.

Within these general requirements for accuracy, there is obviously considerable variation in requirements for precision. Pope (1981), for example, argues that for important fisheries, managers would like to have absolute population estimates with coefficients of variation of 10-20%. It is conceivable that for fishes which are very highly exploited and have very tight coupling between recruitment and population size, such precision might be needed. Most fisheries, however, do not warrant the cost of such precision even if achievable because of natural variation in the production process. Thus, Trumble et al. (1982) concluded that existing hydroacoustic data for management of herring stocks in the Gulf of Georgia resulted in a precision which exceeded knowledge of the production process, and no further refinement was needed.

12.3.2 Sources of Variability and Error

Sources of variability and error associated with hydroacoustic techniques can be categorized into three types: acoustical, distributional, and biological. The acoustical category includes two parts: (1) uncertainty associated with the acoustic system, and (2) uncertainty in theoretical relationships such as the target strength or the relation between scattering and biomass. Knowledge of the system parameters is important even for indication of presence or absence. Accurate determination of relative changes requires calibration. Techniques for calibration basically are of two types: those which use hydrophones and those which use standard targets. Hydrophones have the advantage of allowing measurement of the various transmitting and receiving characteristics, while standard targets lump these parameters. However, Blue (1982) warns against the general use of test hydrophones, arguing that hydrophones that have not been calibrated by reciprocity may have errors of several decibels. Unless the user has access to naval calibration facilities, the most practical procedure is the use of a standard target. Considerable investigation has been made recently of various types of standard targets for fisheries hydroacoustics. Ping pong balls provide a convenient target (Drew 1980), but there is some evidence of instability at certain frequencies (Dunn 1982). Foote et al. (1982) strongly recommends the use of copper spheres, claiming accuracy to 0.1 dB.

The physical procedure of making the calibration requires that the standard target or hydrophone be on the acoustic axis of the transducer. Also, the target should be more than 10 m from the sonar transducer to reduce "near-transducer effects." One needs positioning and mounting structures to hold the transducer and targets while making calibration measurements. It can be difficult to position the targets under a ship for hull-mounted systems.

Estimation of absolute abundance requires target strength information unless a technique like the duration-in-beam is used. The need and techniques for target strength measurement have already been discussed. Again, the considerable difference in sensitivity to error in this parameter between echo counting and echo integration should be noted.

Non-linearity in the relationship between acoustic scattering and biomass is another uncertainty. Non-linearities at very high fish densities have been demonstrated (Rottingen 1976), and theoretical models have been developed to compensate for this (Maxwell 1980). In general, however, this is an unusual occurrence and a minor concern. The possibility of coherent scattering from a lattice-like distribution in schooling fishes also has been suggested (Yudanov and Kalikmin 1981), and is frequently mentioned as a reason for distrust of acoustic results (Suomala and Yudanov 1981). A simple consideration of the required precision in positioning of the fish (on the order of millimeters) shows this concern to be without merit (Swingler and Hampton 1981). Other potential sources of error include noise from propellers, surface turbulence, or power supplies, and scattering from plankton layers. Echo integration is especially sensitive to this type of problem. Noise from non-biological sources can be eliminated with care, although sometimes surveys cannot be run because of turbulence from rough water conditions. Plankton layers are usually easily distinguished from fish targets.

The high sampling power of hydroacoustics has revealed the magnitude of uncertainty associated with the non-uniformity of fish distributions in a way that was previously impossible with other techniques (Thorne 1977). A recent illustration of the effect of fish distribution on precision was made by Thomas (1982) who examined the variability between replicate transects over a small area (a circle of 300 m radius) as a function of time between replicates. Thomas showed that replicates within 30 minutes produced coefficients of variation of about 0.2 and reflected the small-scale patchiness of the fish distribution, whereas replicates separated by longer time intervals had much larger variation caused by fish movement. The subjects of survey design and statistical treatment of hydroacoustic data are covered in Thorne et al. (1971), Bazigos (1975, 1976), Shotton and Dowd (1975), Mathisen et al. (1978), Kimora and Lemberg (1981), Gunderson et al. (1982), Shotton and Bazigos (1982), and Williamson (1982).

Despite the many arguments over the fine points of acoustic theory, the primary source of uncertainty in most hydroacoustic surveys is biological (Saville 1978). For hydroacoustics, this category includes uncertainty concerning availability of the fish to the acoustic gear (near surface and near bottom distribution and vessel avoidance), behavioral changes which might affect the mean back-scattering cross section, uncertainty as to species composition, and questions of stock identity and migratory behavior. Often surveys need to take advantage of fish behavior, such as diel migration or seasonal changes, in order to optimize conditions for hydroacoustic techniques (Thorne 1982; Thorne et al. 1982), and it is necessary to gain some general understanding of fish behavior before optimal surveys can be designed.

12.3.3 Complementary and Ground Truth Techniques

A danger in all manuals of sampling techniques, as a natural result of their organization, is treatment of techniques in isolation. It is important in considering hydroacoustics to realize that the technique is virtually always used in conjunction with some type of direct capture technique. As noted above, one of the limitations of hydroacoustic techniques is poor species discrimination. In most cases, complementary techniques, usually sampling with trawls, are required to identify species. There are many clues to species identification in hydroacoustic data, including target size and distributional characteristics. However, even these need to be established initially through comparisons with direct capture or other means. The importance of complementary techniques for species identification depends on the complexity of the species assemblages. In some cases a single species is predominant and the need for subsampling with nets is reduced or absent. In other cases the species assemblage may be very complex: for example, samples of fish entrapment at coastal power plant cooling water intakes off southern California typically include 30 species (Thorne 1979; Thomas 1980).

While the importance of using complementary techniques in combination with hydroacoustics for species identification and biological information is widely appreciated, the use of hydroacoustics to evaluate the performance of nets is just beginning to be recognized. In the *1973 Symposium on Hydroacoustics Research in Fisheries* (Margetts 1977), several papers used net catches to provide "ground truth" for hydroacoustical data, and such comparisons still provide a valuable check on performance. However, with the increased accuracy of hydroacoustic techniques, especially echo counting, the possibility exists to determine the catchability coefficients of various gears directly. Pearcy (1980), for example, compared the performance of five nets and was able to account for changes in fish density among trials by using echo integration. An example of a direct use of hydroacoustics to determine net efficiency is shown in Figure 12.9, where simultaneous hydroacoustic density measurements and tow net catches of juvenile sockeye salmon were compared to determine the efficiency of the net. Hydroacoustics also can be used to evaluate changes in the catchability coefficients of a gear under different environmental conditions. For example, Robinson and Barraclough (1978) used hydroacoustic techniques as a relative standard to measure the effect of changing ambient light on the performance of a trawl for juvenile sockeye in Great Central Lake, British Columbia.

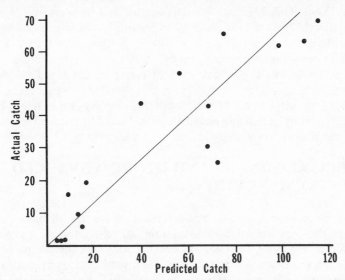

Figure 12.9 Comparison of actual catches of sockeye salmon in trawls with predicted catches based on hydroacoustic estimate of absolute abundance. The efficiency of the net in this case (that is, the slope of the line) was 69%.

They determined that catching efficiency for clear sky and dark of moon was only 47% of that for overcast sky and dark of moon, and catching efficiency for clear sky and full moon was only 6.4%. A similar comparison was made on the performance of gill nets over the summer in the Beaufort Sea (Thorne unpublished data). The hydroacoustic techniques showed a major decrease in the catchability coefficient during the summer as a result of reduced swimming speed, which in turn appeared to be correlated with temperature.

12.3.4 Advantages for Enviromental Impact Research

It is becoming increasingly apparent that hydroacoustic techniques offer special advantages for environmental impact research. There are three main reasons for this: (1) hydroacoustic techniques not only measure fish abundance, but monitor behavior, (2) hydroacoustic techniques are not affected by changes in environmental parameters, and (3) hydroacoustic techniques do not influence behavior or impact through removals. Historically, the interest in hydroacoustic techniques has been associated with fisheries resource assessment because most use and developmental work occur in commercial fisheries or outside the United States where environmental impact research has received less emphasis. Consequently, the application of quantitative hydroacoustic techniques to environmental impact studies has been minimal. The capability of hydroacoustics to monitor fish behavior, however, has been well recognized and is increasingly applied in environmental impact studies. Examples are Thorne (1979) and Thorne (1980), in which hydroacoustics was used to study the effect of a thermal discharge, the response of fish to underwater light associated with underwater television monitoring at a power plant water intake, and the abundance and behavior of fish under the ice in the Beaufort Sea.

As pointed out above, hydroacoustic techniques can be used to study the performance of direct capture techniques including changes in catchability coefficients with changes in environmental parameters. These changes can be very serious because the environmental alteration under study may affect catchability coefficients. Thus, CPUE results are confounded and inconclusive. Hydroacoustic data are unaffected by such changes and can provide conclusive results or can be used to evaluate the direction and effect of biases in direct capture techniques.

The third advantage, no removal of fish, is two-fold. First, there is no mortality. Ironically, it is possible in some cases for the major impact of a development under evaluation to be the mortality caused by the environmental impact research. Second, fish behavior is not affected by the presence of the sampling gear or removal of the fish. For example, a study of fish movement could be biased by removal of a considerable portion of the population whose movement is under study.

12.4 SPECIAL SONARS AND OTHER ADVANCED INSTRUMENTATION

The techniques and applications described above are all basic echo-sounder systems with the commonly used echo counting or echo integration processing. The principles described apply to sonars in general, although the complexity is greatly increased when the beam is directed other than vertical (Hewitt et al. 1976; Smith 1977). More complex instrumentation is applicable to fisheries problems, but because of their cost and specialized nature, these techniques are seldom used at this time.

12.4.1 Sector Scanning Sonar

A conventional echosounder or sonar transmits sound on a single transducer and usually receives the echoes on the same transducer. The angle of the sound beam depends on the ratio of the wave length to the transducer size, and thus, on the frequency of the sound and the transducer dimensions. For any given frequency, the larger the transducer the smaller the beam angle. Because there is only a single transducer with a fixed beam angle, there is always a trade-off between resolution and sampling volume.

Sector scanning or multi-beam sonar eliminates this limitation. The sonar transducer comprises several separate elements. Sound is transmitted simultaneously through all elements to form a large angle transmitted beam, but the individual elements are electronically scanned at a high rate (within one pulse length) so that, in effect, the receiver functions like many narrow beam transducers, thus achieving both large sampling power and high angular resolution.

Considerable research on the sector scanning sonar has been conducted in Great Britain, and several commercial models have become available (C-Tech Omni, Furuno 75A and B, Krupp Atlas Fishing Sonar 950, and Simrad SM600). One U.S. study (Gunderson et al. 1982) investigated the advantages of the sector scanning sonar as a large sample search tool. The potential use of a multiple channel echo integrator in conjunction with the sector scanning sonar has been explored by Ehrenberg (1980), who demonstrated that the sector scanning sonar can provide estimates with significantly smaller error than a conventional single transducer system. The use of the sector scanning sonar for target strength has also been explored by Ehrenberg (1979).

12.4.2 Split Beam Sonar

The split beam sonar differs from the ordinary sonar in that the transducer is divided into four quadrants. Echoes from each quadrant are independently processed. It differs from the sector scanning sonar in that information on the location of targets can be obtained by analysis of the phase relationships among the various quadrants. The advantages of this capability are twofold. First, knowledge of the location of the target in the transducer beam allows direct measurement of target strength. Evaluation of the split beam sonar as a target strength measuring device has been conducted by Carlson and Jackson (1980). Second, because the split beam sonar allows location of a target in three dimensions, it is possible to determine position, size, and target strength of fish targets. The major disadvantages of the split beam sonar are cost and complexity. Costs of data processing must decrease considerably before its use is practical for routine fisheries assessment work, although it has present potential for special studies.

12.4.3 Doppler Sonar

A Doppler shift is an apparent change in the frequency of sound as a result of motion of the target. The classic example is the change in pitch of a train whistle as it passes by. Doppler shift has many uses for speed measurement, including measuring ship velocity relative to the bottom or to water, and measuring auto speed in traffic control. There are two potential applications of Doppler shift in fisheries research. One is the measurement of swimming speeds, either of individual fish or schools. The possibility of using swimming speeds of schools as a key to identification was explored by Holliday (1977). A second potential is the detection and enumeration of upstream migrating fishes. Doppler sonars are much less affected by boundaries and can distinguish upstream migrating fish from debris or turbulence. The major drawback with Doppler sonars is that long pulse lengths are necessary to achieve the frequency resolution required for detection. Thus, the system has poor range resolution and requires low densities for counting.

12.4.4 Resonance Frequency Analysis

There are two potential applications of resonance frequency analysis in fisheries. The first is as a key to species identification. Resonance can be used to infer size and the presence or absence of a swim bladder. The other potential application is an indication of year-class strength. This potential is especially applicable for short-lived fishes. In this case the use of resonance would be similar to length-frequency analysis, but may be less biased by selectivity.

12.4.5 Stationary System

Hydroacoustic systems are usually deployed from a moving vessel. Application in a stationary mode results in greatly reduced sampling power. However, when interest is focused on a particular area, as is often the case with environmental impact research, there are many advantages to stationary deployment. A stationary deployment virtually eliminates the problem of resolution of fish near boundaries, greatly improves signal to noise characteristics, and obtains much more behavioral information including fish swimming speeds. The use of several multiplexed stationary transducers overcomes much of the sampling power limitation, and this type of system is particularly suited to remote (unattended) application. Its use is presently limited by lack of development of automatic data collection and processing capability. However, such development is well underway, and deployment of completely automatic multiple transducer systems could completely revolutionize our approach to environmental impact studies.

12.5 REFERENCES

Bayona, J. D. 1982. General standards of acoustic assessment systems for use in developing countries. Paper number 116, *Symposium on Fisheries Acoustics,* 21-24 June, 1982, Bergen, Norway.

Bazigos, G. P. 1975. *The statistical efficiency of echo surveys with special reference to Lake Tanganyika.* Food and Agriculture Organization of the United Nations, Technical Paper 139, Rome, Italy.

Blue, J. E. 1982. Physical calibration. Paper number 2, *Syposium on Fisheries Acoustics,* 21-24 June, 1982, Bergen, Norway.

Burczynski, J. 1979. *Introduction to the use of sonar systems for estimating fish biomass.* Food and Agriculture Organization of the United Nations, Fisheries Technical Paper 19, Rome, Italy.

Carlson, T. J., and D. R. Jackson. 1980. *Empirical evaluation of the feasibility of split beam methods for direct in situ target strength measurement of single fish.* Applied Physics Laboratory, University of Washington (APL-UW 8006), Seattle, Washington, USA.

Clay, C. S., and H. Medwin. 1977. *Acoustical oceanography: principles and applications.* John Wiley and Sons, New York, New York, USA.

Craig, R. E., and S. T. Forbes. 1969. Design of a sonar for fish counting. *Fiskeridirektorateta Skrifter (Havundersokelser)* 15(3):210-219.

Cushing, D. H. 1973. *The detection of fish.* Pergamon Press, Oxford, England.

Dickie, L. M., R. G. Dowd, and P. R. Boudreau. 1982. The ECOLOG acoustic detection system giving numerical field information on demersal fish size and density. Paper number 5, *Symposium on Fisheries Acousitcs,* 21-24 June, 1982, Bergen, Norway.

Drew, A. W. 1980. Initial results from a portable dual-beam sounder for *in situ* measurements of target strength of fish. Pages 376-380 *in Oceans '80,* Institute of Electrical and Electronic Engineers, Piscataway, New Jersey, USA.

Dunn, J. R. 1982. The absolute acoustic calibration of table-tennis balls as standard targets. Paper number 89, *Symposium on Fisheries Acoustics,* 21-24 June, 1982, Bergen, Norway.

Edgell, J. A. 1935. Fake echoes in deep water. *International Hydrographic Review* 12(1):19-20.

Ehrenberg, J. E. 1972. A method for extracting the fish target strength distribution from acoustic echoes. *Proceedings, 1972 Institute of Electrical and Electronic Engineers Conference on Engineering in the Ocean Environment* 1:61-64.

Ehrenberg, J. E. 1973. *Estimation of the intensity of a filtered Poisson process and its application to acoustic assessment of marine organisms.* University of Washington Sea Grant Publication WSG 73-2, Seattle, Washington, USA.

Ehrenberg, J. E. 1974. Two applications for a dual-beam transducer in hydroacoustic fish assessment systems. *Proceedings, Oceans 1974, Institute of Electrical and Electronic Engineers Conference on Engineering in the Ocean Environment* 1:152-155.

Ehrenberg, J. E. 1979. The potential of the sector-scanning sonar for *in situ* measurement of fish target strengths. *In Progress in Sector Scanning Sonar,* Lowestoft (UK), 18 December, 1979, Bath, England.

Ehrenberg, J. E. 1980. Echo counting and echo integration with a sector scanning sonar. *Journal Sound Vibration* 73(3):321-332.

Ehrenberg, J. E. et al. 1981. Indirect measurement of the mean acoustical backscattering cross section of fish. *Journal of the Acoustical Society of America* 69(4):955-962.

Food and Agriculture Organization/Advisory Committee on Marine Resources Research. 1978. *Report on the meeting of the working party on fish target strength.* Aberdeen, Scotland, 13-16 December, 1977. Food and Agriculture Organization of the United Nations, Advisory Committee on Marine Resources Research, 9/78/Inf. 14, Rome, Italy.

Foote, K. G., H. P. Knudsen, and G. Vestnes. 1982. Standard calibration of echo sounders and integrators with optimal copper spheres. Paper number 40, *Symposium on Fisheries Acoustics,* 21-24 June, 1982, Bergen, Norway.

Forbes, S. T., and O. Nakken, editors. 1972. *Manual of methods for fisheries resource survey and appraisal.* Part 2, the use of acoustic instruments for fish detection and abundance estimation. Food and Agriculture Organization of the United Nations, Manual in Fisheries Science 5, Rome, Italy.

Gunderson, D. R., G. L. Thomas, and P. J. Cullenberg. 1982. Combining sector scanning sonar and echosounder data from acoustic surveys. Paper number 101, *Symposium on Fisheries Acoustics,* 21-24 June, 1982, Bergen, Norway.

Hewitt, R. P., P. E. Smith, and J. C. Brown. 1976. Development and use of sonar mapping for pelagic stock assessment in the California United States current area. *Fisheries Bulletin* 74:281-300.

Holliday, D. V. 1977. Two applications of the doppler effect in the study of fish schools. *Rapports et proces-verbaux des reunions conseil international pour l'exploration de la mer* 170:21-30.

Kimura, D. K., and N. A. Lemberg. 1981. Variability of line intercept density estimates (a simulation study of the variance of hydroacoustic biomass estimates). *Canadian Journal of Fisheries and Aquatic Sciences* 38:1141-1152.

MacKenzie, K. V. 1981. Discussion of seawater sound speed determinations. *Journal of the Acoustical Society of America* 70:801-806.

MacNeill, I. B. 1971. Quick statistical methods for analyzing the sequences of fish counts provided by digital echo counters. *Journal of the Fisheries Research Board of Canada* 28:1035-1042.

Margetts, A. R., editor. 1977. *Symposium on hydroacoustics in fisheries research, Bergen, Norway, June 19-22, 1973.* Rapports et proces-verbaux des reunions conseil international pour l'eploration de la mer, 170.

Mathisen, O. A., R. E. Thorne, R. J. Trumble, and M. Blackburn. 1978. Food consumption of pelagic fish in an upwelling area. Pages 112-123 *in* R. Boje and M. Tomczak, editors. *Upwelling Ecosystems.* Springer-Verlag, New York, New York, USA.

Maxwell, D. R. 1980. *Hydroacoustic estimation of dense or extensive fish populations.* Doctoral dissertation. University of Washington, Seattle, Washington, USA.

Nakken, O., and K. Olsen. 1977. Target strength measurements of fish. *Rapports et proces-verbaux des reunions conseil international l'exploration de la mer* 170:52-69.

Nunnallee, E. P., Jr. 1980. *Application of an empirically scaled digital echo integrator for assessment of juvenile sockeye salmon, (Oncorhynchus nerka Walbaum) populations.* Doctoral dissertation. University of Washington, Seattle, Washington, USA.

Pearcy, W. G. 1980. *Comparisons of the catches of euphausiids in five nets.* Proceedings SCOR working group 52 Symposium on Assessment of Micronekton, April 27-30, 1980.

Peterson, M. L., C. W. Clay, and S. B. Brandt. 1976. Acoustic estimates of fish density and scattering function. *Journal of the Acoustical Society of America* 60(3):618-622.

Pope, J. G. 1981. *Practical guidelines for the precision of assessment data.* International Council for the Exploration of the Sea CM 1981/G:13.

Robinson, B. H. 1978. *In situ* measurement of fish target strength. *In Proceedings of 1978 Institute of Acoustics Specialists Meeting on Acoustics in Fisheries, Hull, England, 1978.* University of Bath, Bath, England.

Robinson, D. G., and W. E. Barraclough. 1978. Population estimates of sockeye salmon *(Oncorhynchus nerka)* in a fertilized oligotrophic lake. *Journal of the Fisheries Research Board of Canada* 35:851-860.

Rottingen, I. 1980. On the relation between echo intensity and fish density. *Fiskeridirektorateta Skrifter (Havundersokelser)* 16(9):301-314.

Saville, A., editor. 1977. *Survey methods of appraising fishery resources.* Food and Agriculture Organization of the United Nations, Fisheries Technical Paper 171, Rome, Italy.

Shotton, R., and R. G. Dowd. 1975. *Current research in acoustic fish stock assessment at the Marine Ecology Laboratory.* International Commission for North Atlantic Fisheries Research Document 75/16, Serial Number 3468.

Shotton, R., and G. P. Bazigos. 1982. Techniques and considerations in the design of acoustic surveys. Paper number 4, *Symposium on Fisheries Acoustics,* 21-24 June, 1982, Bergen, Norway.

Smith, P. E. 1977. The effects of internal waves on fish school mapping with sonar in the California current area. *Rapports et proces-verbaux des reunions conseil international pour l'exploration de la mer,* 170:223-231.

Suomala, J. B. 1975. *A short note on the applicability of hydroacoustic methods for demersal fish counting and abundance estimates.* International Commission for North Atlantic Fisheries Research Document 75/95, Serial 3575.

Suomala, J. B., and K. I. Yudanov, editors. 1980. *Meeting on hydroacoustical methods for the estimation of marine fish populations, 25-29 June 1979,* volume 1. Findings of the scientific and technical specialists: a critical review. The Charles Stark Draper Laboratory, Cambridge, Massachusetts, USA.

Swingler, D. N., and I. Hampton. 1981. Investigations and comparison of current theories for the echointegration technique of estimating fish abundance and their verification by experiment. *In* J. B. Suomala, editor. *Meeting on hydroacoustical methods for the estimation of marine fish populations, 25-29 June 1979,* volume 2, part A. Charles Stark Draper Laboratory, Cambridge, Massachusetts, USA.

Thomas, G. L. 1979. The application of hydroacoustic techniques to determine the spatial distribution and abundance of fishes in the nearshore area in the vicinity of thermal generating stations. Pages 61-63 *in Oceans '79,* Institute of Electrical and Electronic Engineers, Piscataway, New Jersey, USA.

Thomas, G. L. 1982. Complementary hydroacoustic and net sampling techniques for determining the vulnerability of fish to power plant entrainment. Paper number 102, *Symposium on Fisheries Acoustics,* 21-24 June, 1982, Bergen, Norway.

Thorne, R. E. 1973. Digital hydroacoustic data-processing system and application to Pacific hake stock assessment in Port Susan, Wahington. *Fisheries Bulletin* 71:837-843.

Thorne, R. E. 1977. Acoustic assessment of Pacific hake and herring stocks in Puget Sound, Washington and southeastern Alaska. *Rapports et proces-verbaux des reunions conseil international pour l'exploration de la mer* 170:265-278.

Thorne, R. E. 1979. Hydroacoustic estimates of adult sockeye salmon *(Oncorhynchus nerka)* in Lake Washington, 1972-1975. *Journal of the Fisheries Research Board of Canada* 36:1145-1149.

Thorne, R. E. 1980. The application of hydroacoustics to tropical small-scale fishery stock assessment. *In P. Roedel and S. Saila, editors. Proceedings, Workshop on Tropical Small-scale Fishery Stock Assessment.* University of Rhode Island, International Center for Marine Resource Development, Kingston, Rhode Island, USA.

Thorne, R. E. 1982. Application of stationary hydroacoustic systems to studies of fish abundance and behavior. Pages 381-385 *in Oceans '80.* Institute of Electrical and Electronic Engineers, Piscataway, New Jersey, USA.

Thorne, R. E. 1982. Application of hydroacoustic assessment techniques to three lakes with contrasting fish distributions. Paper number 97, *Symposium on Fishery Acoustics,* 21-24 June, 1982, Bergen, Norway.

Thorne, R. E. (In press.) Assessment of population abundance by echo intergration. *Journal of Biological Oceanography* 2:253-262.

Thorne, R. E., J. E. Reeves, and A. E. Millikan. 1971. Estimation of the Pacific hake *(Merluccius productus)* population in Port Susan, Washington, using an echo integrator. *Journal of the Fisheries Research Board of Canada* 28:1275-1284.

Thorne, R. E., O. A. Mathisen, R. Trumble, and M. Blackburn. 1977. Distribution and abundance of pelagic fish off Spanish Sahara during coastal upwelling ecosystem analysis expedition (Coastal Upwelling Ecosystem Analysis) Joint-I. *Deep-Sea Research* 24(1):75-82.

Thorne, R. E., G. L. Thomas, W. C. Acker, and L. Johnson. 1979. *Two applications of hydroacoustic techniques to the study of fish behavior around coastal power generating stations.* University of Washington Sea Grant Publication WSG 79-2, Seattle, Washington, USA.

Thorne, R. E., R. J. Trumble, N. A. Lemberg, and D. Blankenbeckler. 1982. Hydroacoustic assessment and management of herring fisheries in Washington and southeastern Alaska. Paper number 98, *Symposium on Fisheries Acoustics,* 21-24 June, 1982, Bergen, Norway.

Traynor, J. J., and J. E. Ehrenberg. 1979. Evaluation of the dual beam acoustic fish target strength measurement system. *Journal of the Fisheries Research Board of Canada* 36:1064-1071.

Traynor, J. J., and N. J. Williamson. 1982. Target strength measurements of walleye pollock, *Theragra chalcogramma,* and a simulation study of the dual beam method. Paper number 94, *Symposium on Fisheries Acoustics,* 21-24 June, 1982, Bergen, Norway.

Trumble, R. J., R. E. Thorne, and N. Lemberg. 1982. The Strait of Georgia herring fishery, a case history of timely management by means of hydroacoustic surveys. *Fishery Bulletin* 80:381-398.

Urick, R. J. 1975. *The principles of underwater sound for engineers,* 2nd edition. McGraw-Hill, New York, New York, USA.

Venema, S. C. 1982. A selected bibliography of acoustics in fisheries research and related fields. Food and Agriculture Organization of the United Nations Fisheries Circular 748, Rome, Italy.

Williamson, N. J. 1982. Cluster sampling estimation of the variance of abundance estimates derived from quantitative echo sounder surveys. *Canadian Journal of Fisheries and Aquatic Sciences* 39:229-231.

Yudanov, K. I., and I. L. Kalikhman. 1981. Sound scattering by marine animals. Pages 53-96 *in* J. B. Suomala, editor. *Meeting on hydroacoustical methods for the estimation of marine fish populations,* 25-29 June 1979, volume 2, part A. Charles Stark Draper Laboratory, Cambridge, Massachusetts, USA.

Chapter 13
Fish Kill Investigation Procedures

DONLEY M. HILL

13.1 INTRODUCTION

In its 1977 summarization of fish kill statistics, the Environmental Protection Agency (EPA) reported a total of 503 fish kills in which 16.5 million fish died (EPA 1980). The largest reported fish kills took place in 1969 in Lake Thonotosassa at Plant City, Florida, when more than 26 million fish died over a period of nine days because of food products dumped into the lake, and in 1974 in the Back River near Essex, Maryland, when 47 million fish were killed by a discharge from a sewerage system. Less spectacular kills involving hundreds or thousands of fish occur almost daily and contribute significantly to the "nonangler" mortality component of our nation's fisheries.

In reporting statistics on fish kills, EPA categorizes the fish kill according to the source of the toxicant, i.e., municipal operations (sewerage systems, power plants, refuse disposal, water supply systems, and swimming pools), agricultural operations (pesticide and fertilizer application and manure-silage drainage), industrial operations (chemicals, petroleum, mining, food products, paper products, and metals), transportation, "other" operations, and unknown. Municipal operations were the largest identifiable cause of fish kills in the United States in 1977, with industrial and agricultural operations close behind (Fig. 13.1). The large percentage of unknown kills is testimony to the difficult task facing a fish kill investigator. This is easily appreciated when one considers that although fish kills often have a duration of several days, the event may be triggered by something as transient as a toxic slug in a fast moving stream or a predawn oxygen deficit or as hard to identify as a naturally occurring viral or bacterial infection.

The fact that the number and severity of reported fish kills has steadily decreased each year since 1971, with the exception of 1976 (EPA 1980), can probably be attributed to increased public awareness of water pollution consequences, public demand for clean water voiced through legislative acts such as the Clean Water Act, and the adoption and enforcement of water quality criteria and standards by state and federal agencies. Carefully conducted investigations of water pollution incidents and of fish kills by well-trained, knowledgeable individuals have played a significant role in each of the above developments.

Beginning in the late 1960s committees of the Southern and North Central Divisions of the American Fisheries Society recognized the need to apply monetary values to fish lost during water pollution incidents and to standardize fish kill investigation techniques. As a result, these groups prepared publications on monetary values of fish. Together with fish kill counting guidelines prepared by the Southern Division's Pollution Committee, these documents have stood the test of legal courts as

261

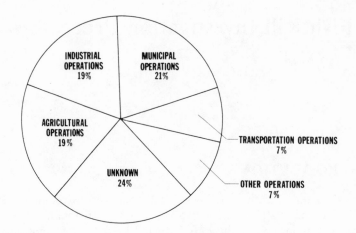

Figure 13.1   Percentage of fish kills caused by various types of pollutants in the United States during 1977 (EPA 1980).

a basis for assessing damages against polluters.

The purpose of this chapter is to describe basic procedures to be followed during fish kill investigations. The chapter is an overview and should be complemented by more specifically designed information, such as the American Fisheries Society publications and appropriate individual agency documents.

13.2   COORDINATING THE INVESTIGATION

One of the most important aspects of a fish kill investigation is coordinating the involvement of appropriate agencies or organizations. Although responsibility for fish kill investigations varies greatly from state to state and among EPA regions, some generalizations can be stated.

13.2.1   State Agency Responsibilities

Because fish are generally recognized as the property of the state in which they reside, most states have assumed principal responsibility for investigating fish kills and for coordinating the investigation and its results with other appropriate agencies and organizations (Fig. 13.2). Exceptions are made for spills of oil or hazardous materials, in which case the EPA or the United States Coast Guard may assume the principal role.

In the case of routine pollutants, the state water quality control agency usually makes the pollution tests and detection surveys, and the state fish and wildlife agency usually evaluates the damage to the fishery resource. The state water quality control agency coordinates other agency involvement and enforces necessary regulatory or damage assessment actions. As might be expected, the most successful investigative programs are those in which the various state agencies have evolved a cooperative interdisciplinary approach such as is depicted in the hypothetical investigation outlined in Figure 13.3.

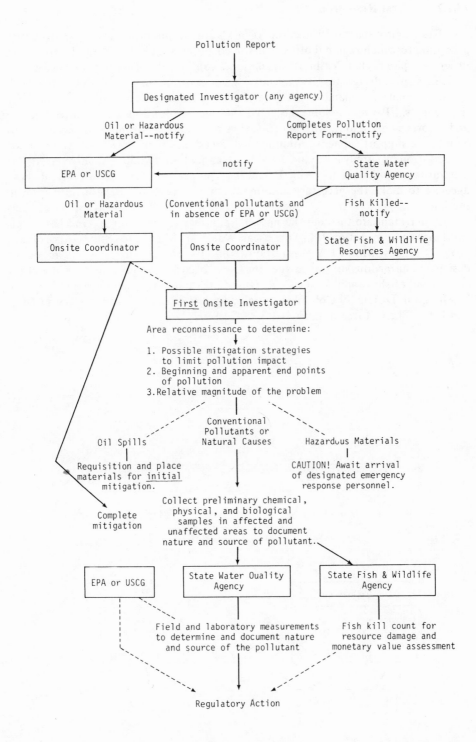

Figure 13.2 Flow chart for coordination and performance of a fish kill investigation.

13.2.2 Federal Responsibilities

The Environmental Protection Agency's role in pollution investigations varies according to which regional office is involved and in which state the pollution event occurs. In spite of that flexibility, the national contingency plan for oil and hazardous substance spills designates EPA as the on-scene coordinator for investigation of such spills in the nation's inland or estuarine waters. Because of its mandated authority for these events, EPA may assume the lead in the investigation and subsequent mitigation and enforcement actions, with appropriate state and other agencies or organizations acting in a support capacity. An individual investigator can initiate the coordinated investigation by simply notifying the state water quality or wildlife resources agency of a hazardous spill or a fish kill. Established procedures then allow staff of those agencies to make the necessary additional contacts and to set the investigation in motion.

Under mandate through the national contingency plan for oil and hazardous substance spills in section 311 of the Clean Water Act, the United States Coast Guard (USCG) shares with EPA the responsibility for responding to spills of oil or designated hazardous substances (i.e., the Coast Guard is the on-scene coordinator for coastal and offshore spills). The Coast Guard operates a National Response Center in Washington, District of Columbia, to relay reports of such incidents to both USCG and EPA. The toll free number is 1-800-424-8802.

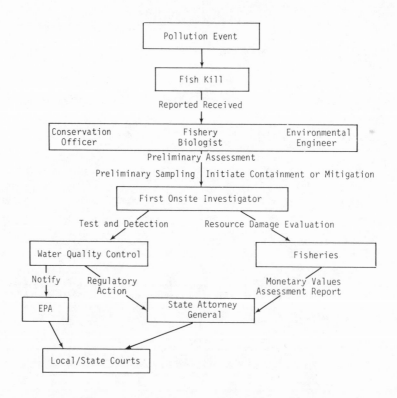

Figure 13.3 Hypothetical lines of responsibility among agencies concerned with a pollution event.

13.2.3 Importance of Timely Action

Although the "typical" fish kill may take place over a period of several days, prompt response to the event is critical for several reasons. First, the cause of the event may be a transient phenomenon detectable for a limited period of time. Second, it is important to examine dying or freshly dead fish to separate natural die-offs from man-caused pollution. Finally, a more realistic assessment of the quantity of fish killed can be made when counts are conducted after the kill is completed but before scavengers and decomposition remove the fish from the countable population.

It is important that individuals or groups assigned responsibility for responding to fish kills have certain basic equipment available at all times. This equipment should include a thermometer, dissolved oxygen test kit, conductivity meter, pH meter (or a general water chemistry kit), biological sampling equipment, sampling bottles, and other specimen bottles (APHA 1980). In addition, the equipment should include maps suitable for locating access to various areas on the body of water and field forms for recording observations. Given these basic materials, a trained investigator can make the initial response, begin the investigation, and at the same time lay the groundwork for a more intensive or broader scale survey.

13.3 INVESTIGATIVE STRATEGIES

13.3.1 Protocol of Action

With information on hand which defines lines of communication and coordination within the designated investigator's agency or organization and with appropriate state and federal agencies, the first (and often most important) action the investigator must take is to record accurately a reported fish kill. This is best accomplished when a standard form is available. Information to be recorded should include at least: (1) name, address, and telephone number of the person(s) reporting the fish kill, (2) time and date of the report, (3) location of the fish kill (name of body of water and location relative to specific roads, landmarks, or other identifiable areas), (4) magnitude of the kill, (5) suspected cause of the kill and any additional information the informant may have, and (6) time, date, and person in other agencies or organizations to whom the designated investigator relays the reported information. The importance of this step in the fish kill response process cannot be overstated because it constitutes both the beginning of the legal documentation of the pollution event and a record of information necessary for a properly coordinated field response.

With this initial information on hand, the designated investigator who receives the report can decide (1) who should receive first notification (oil spills or hazardous materials—EPA or USCG; conventional pollutants—state water quality and/or state fish and wildlife resources agency), and (2) who should make the first field response. If hazardous materials are involved, EPA or the USCG should be notified, and they, in turn, will alert the appropriate emergency response group. It cannot be overemphasized that in the case of spills of hazardous materials, public and individual safety must be the first concern and can best be guarded by following the lead of the group in charge of those emergencies.

After documenting the fish kill report and communicating the information to appropriate groups, the designated investigator assembles his materials and equipment and travels to the kill area where his first priority should be to minimize the extent of the kill by (1) notifying the polluter of what is happening if he is not aware, and (2) beginning mitigative strategies where appropriate and where authorized (for example, placing booms of straw or other absorbent material across a stream below an oil spill or using dikes to contain a pollutant). Additional actions should be directed at determining the cause and magnitude of the kill. These procedures are described in the following section.

13.3.2 Defining the Nature and Extent of a Fish Kill

Probably the single most important field observation in determining the cause of a fish kill is to determine the area involved as evidenced by dead or distressed fish or other aquatic organisms. This, in turn, limits the boundaries of potential pollution sources and prescribes "above, within, and below" areas to be sampled for appropriate field parameters. Before intensive documentary sampling begins, however, it may be possible to complement visual observations with qualitative or reconnaissance-type chemical and physical sampling which may more accurately delimit the area of concern. In the absence of evidence of the probable nature of the pollutant(s), tests of routine limnological parameters such as dissolved oxygen, pH, conductivity, and temperature will often reveal the presence of a pollutant by producing values outside the "normal" range. For example, municipal or industrial discharges may have a different temperature than ambient waters, fertilizers and other salts may increase the conductivity of receiving waters, acid spills or drainages may influence the pH of water receiving it, and large loads of oxygen demanding materials may depress dissolved oxygen concentrations. Although specific types of sampling and analyses must be determined on a case by case basis by the onsite investigator, the essential task is to collect documentary samples from within and outside the affected area. At the minimum, samples should be collected above, within, and below the kill area. Otherwise, the data will probably not withstand either legal or scientific scrutiny.

13.3.3 Water Quality Sampling

In addition to routine or screening limnological tests, detecting causes requires sampling for evidence of excessive concentration of specific pollutants. Some of the categories of pollutants which share common handling procedures are oxygen demanding materials (nontoxic organics such as sewage, etc.) heavy metals (copper, zinc, cadmium, mercury, etc.), nutrients (ammonia, nitrates, nitrites, phosphorus, etc.), pesticides or other toxic organics, and cyanides. Because most agencies have laboratories well-equipped for these analyses, only means of collecting and handling will be discussed here. If an individual does not have access to a laboratory, procedures explained in *Standard Methods* (APHA 1980) or *Methods for Chemical Analysis of Water and Wastes* (EPA 1979) describe the appropriate analytical techniques. A trained investigator needs only a general water sampling kit, properly cleaned glass and plastic bottles, concentrated nitric and sulfuric acids, concentrated sodium hydroxide, and an ice chest to collect a full spectrum of water quality samples.

Samples of oxygen demanding materials may be subjected to either biochemical (BOD) or chemical (COD) oxygen demand tests. Use either glass or plastic bottles for these samples. If COD is to be measured, add 1 ml of 1+4 sulfuric acid (1 volume of concentrated acid + 4 volumes of water) to a 125-ml bottle of sample water. If BOD is to be measured, collect one liter of sample water, store on ice, and transport to the laboratory within 24 hours.

Collect water samples to be analyzed for heavy metals in linear polyethylene

(LPE) bottles to avoid absorption losses onto glass. Acidify a sample of at least 100 ml of water with redistilled nitric acid to a pH lower than 2. Toxic levels for metals frequently exist below or near the 1 ppb level, which requires the use of redistilled nitric acid to avoid adding metals to the sample. If "wet chemistry" analytical techniques are to be used, collect at least a one liter sample.

To analyze for nutrients such as ammonia, collect a 1000 ml water sample in a plastic or glass bottle, acidify with sulfuric acid to a pH lower than 2, store on ice, and transport to the laboratory for analysis within 24 hours.

Pesticides and other toxic organics require meticulous, specialized handling. Collect one liter of sample water in a one-liter glass bottle which has been washed in hexane. The bottle must have a glass or teflon lid, and the sample must be stored on ice. The sample must not come into contact with plastic at any time. If you suspect that volatile organics are present, the sample bottle must be completely filled, leaving no air spaces. Transport samples to the laboratory for analysis within 24 hours.

Collect samples suspected of containing cyanide in a one-liter plastic or glass bottle to which 2 ml of 10-normal sodium hydroxide is added, giving a pH of 12. Store the sample on ice and transport to the laboratory for analysis within 24 hours.

Although dissolved oxygen depletion and ammonia are the predominant causes of municipal fish kills, anytime a fish kill occurs in the vicinity of a municipal waste treatment plant or an electrical power generating plant, the investigator should analyze for chlorine as a pollutant. Because the sample cannot be stored, this analysis is best accomplished in the field according to procedures in *Standard Methods* (APHA 1980).

In addition to water samples from within and outside of the kill area, additional evidence of the cause of the kill and documentation of the area affected can be gained by sampling biota from the same areas. Collect fish live or freshly dead and store on ice for parasitological and other analyses. Sample bottom fauna and plankton before recolonization takes place to allow comparison of community characteristics in affected and unaffected areas.

At this point, the investigator has coordinated the fish kill report, delimited the area of concern, and sampled the water for documentary evidence of the cause of the fish kill. It is important to note that unless one is able to document that the samples were not and could not have been tampered with from the time of collection to analysis, the data resulting from the samples may be useless in court. A chain of custody form (Fig. 13.4) which documents the number of samples collected, number received in the laboratory, personnel who had custody at various times, whether the sample container was locked if shipped by common carrier, etc., will usually suffice to substantiate that the data are valid.

Photographs are useful pieces of evidence in a fish kill investigation. Take photographs of people collecting biological and chemical samples and of witnesses observing the kill. Pictures of dead or dying fish are very effective. Record in field notes the photographer's name, type of camera and film, date, and time of day for all photographs. Also list camera settings, the facing direction of the photographic exposure, and the subject photographed.

By following these procedures, the investigator will have a reasonable probability of determining the cause of the fish kill and who is responsible for it and of enhancing the outcome and restitution from possible litigation. The remaining task is to estimate the number of fish killed and to compute their monetary value in order that damages can be assessed against the polluter.

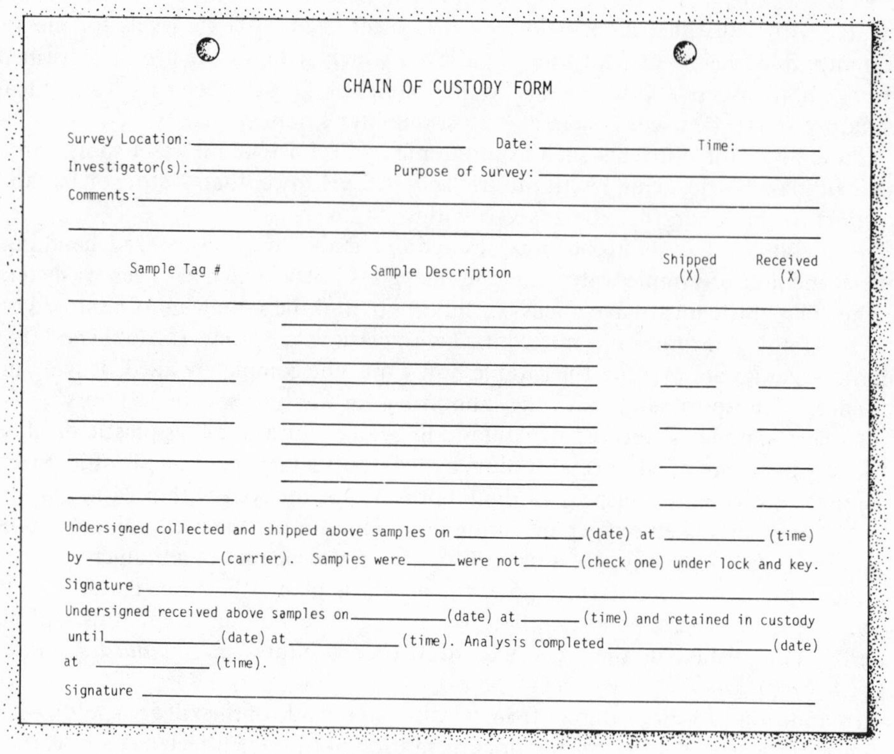

CHAIN OF CUSTODY FORM

Survey Location:_____   Date: _____   Time:_____

Investigator(s):_____   Purpose of Survey: _____

Comments:_____

Sample Tag #	Sample Description	Shipped (X)	Received (X)

Undersigned collected and shipped above samples on _____(date) at _____(time)

by _____(carrier). Samples were_____were not_____(check one) under lock and key.

Signature _____

Undersigned received above samples on_____(date) at _____(time) and retained in custody

until_____(date) at_____ (time). Analysis completed_____(date)

at _____ (time).

Signature _____

Figure 13.4   A chain of custody form for water quality samples.

13.4   FISH KILL COUNTING GUIDELINES

The Pollution Committee of the Southern Division, American Fisheries Society, published procedures for estimating numbers of fish killed in different habitats in 1970 and 1979. More recently, the American Fisheries Society incorporated monetary values of freshwater fish and fish kill counting guidelines into a single source document (AFS 1982). The following procedures are abstracted highlights of these detailed procedures.

Fish kill counting procedures are needed because it is often impossible to collect all the fish within a kill area. The procedures below provide for partial sampling assessments. As is stated in the most recent AFS Guidelines, " . . . certain principles of area sampling must be followed to produce unbiased results: 1) *Sampling units are areas.* Where dead fish are scattered over an extended area (stream, lake shore, or open water) their numbers may be estimated by expanding collections and counts from sample areas to the entire affected area." Thus, areas rather than fish are being sampled. "2) *Sampling units must be chosen at random...* The more likely the results are to be exposed to hostile criticism, the more mandatory that the sampling be at random and therefore as defensible as possible. 3) *Precision depends on sample size and number of fish collected.* The larger the sample size (number of sample areas counted) and the greater the number of dead fish counted, the more reliable will be the estimate of the total number killed." (AFS 1982).

The following sampling procedures are designed for a one-time pickup and, consequently, will invariably underestimate the total number of fish killed because: (1) most fish kills do not occur instantaneously, (2) fish die at differing rates, and once dead they may float or sink on different schedules, and (3) many fish lie too deep in the water to be seen, are hidden by debris, have been taken by scavengers, have decomposed, or are visible but overlooked (AFS 1982). In a test of these guidelines, Hayne et al. (1980) determined that estimates made early the second morning equalled only about half of the dead fish found that day and about 40% of those found during three days.

13.4.1 Narrow Streams

A narrow stream "is one where the investigator can traverse each sample section either by boat or by walking one or both banks and can count every visible dead fish . . . both along the banks and out in the stream." (AFS 1982). The sampling procedure is to determine the total length of stream involved in the kill and then to count all dead fish in randomly selected 100-yard segments in each half-mile section of the affected area. Specifically, the guidelines state that "the first 100-yard appraisal is made within the first half-mile section of the kill area, and starts at a point chosen randomly. Additional 100-yard appraisals are made at each successive half-mile interval, measured from the start of one segment to the start of the next, throughout the region of the kill." (AFS 1982). Because there are approximately eight 100-yard sections in a half-mile stream reach, one simply selects a number from one to eight from a single digit random numbers table, uses that number to identify the first sample segment, moves downstream (or upstream if preferred) one-half mile, counts fish in the second 100-yard segment, etc.,until the entire reach has been sampled. Thus, in a two-mile length of stream affected by a fish kill, there will be four sample sections (one each half-mile), the first of which is chosen randomly, and the remainder of which are evenly spaced at half-mile intervals (Fig. 13.5). The sum of the segment counts is expanded to the whole length of affected stream by using an expansion factor *(E)* calculated according to the following formula:

$$E = (1760 \text{ yard/mile}) \frac{T}{L} \qquad (13.1)$$

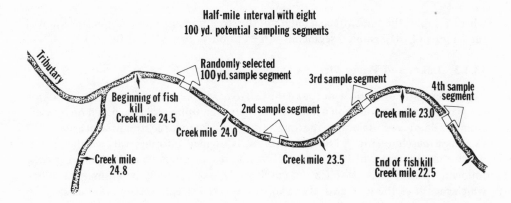

Figure 13.5 Fish kill counting segments for a narrow stream (measurement not to scale).

Box 13.1   A Fish Kill in a Narrow Stream

To estimate the number of dead fish from a kill that extended along two miles of stream, four 100-yard segments were examined. The following numbers of bluegill were counted:

Inch	Segment				
Group	1	2	3	4	Sum
1	140	100	40	0	280
2	120	80	30	10	240
3	60	40	20	5	125
4	30	20	15	10	75
5	25	15	25	15	80
6	30	15	20	5	70
7	10	5	10	0	25
8	5	5	5	0	15
Sum	420	280	165	45	910

For calculating the expansion factor, equation 13.1 is used, in which $T = 2$ and $L = 400$:

$$E = 1760 \text{ yds/mi} \times \frac{2 \text{ mi}}{400 \text{ yds}} = 8.8.$$

For estimating the total numbers of bluegill killed, the number counted is multiplied by E:

Total kill: 8.8 x 910 = 8008; rounded 8,000
1-inch kill: 8.8 x 280 = 2464; rounded 2500
2-inch kill: 8.8 x 240 = 2112; rounded 2100.

This example is adapted from AFS (1982) with permission.

---

where T equals the total affected length of the stream in miles and L equals the sum of the lengths of the sampled segments. An example is provided in Box 13.1.

13.4.2   Lakes and Wider Streams

When a stream is too wide for all the fish to be visible from a boat or while walking one bank or when a kill is on a lake or reservoir, a sampling scheme based on transect counts is employed. Basically, the long axis of the lake or large stream is measured and used as a baseline (Fig. 13.6). The baseline is divided into intervals within each of which one transect will be sampled. The number of intervals is determined by the following process: (1) establish the baseline—this will usually be an imaginary line which measures the long axis of the lake or affected stream section, (2) measure the baseline length, (3) decide how many transects can be counted with available manpower, (4) divide baseline length by the number of transects selected to determine

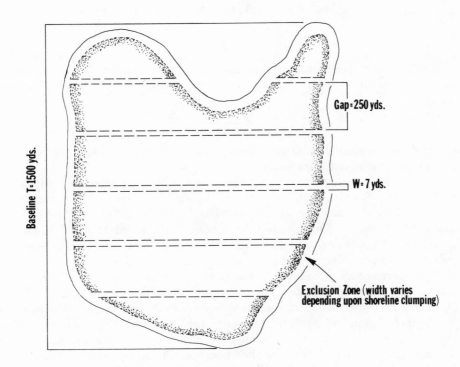

Figure 13.6 Fish kill counting segments for a lake, including baseline, open water transects, and shoreline exclusion zone. Adapted from AFS (1982).

the number of intervals and their length. The next step is to choose where the transect will occur within each interval. Divide the interval distance by transect width (usually seven yards because one can dip fish ten feet from the center of a boat on either side) to determine the number of transects possible in each interval. Number the possible transects consecutively, and select the number of the initial transect from a random numbers table. The other transects are located at the same position within each of the intervals. In other words, space additional transects uniformly along the baseline at distances equal to the length of the intervals, starting from the location of the first transect. Sampling occurs from the center of each transect.

For the example in Figure 13.6, the baseline distance is 1500 yards, six transects can be counted, and the transect width is 7 yards. Thus, the interval distance is 250 yards (1500 ÷ 6), and there are approximately 36 possible transects in an interval (250 ÷ 7 = 35.7). The two-digit number drawn from a random number table (33 in this example) designates the 33rd transect of the 36 possible as the initial transect. Sampling will occur at about 228 yards from the end of the baseline (32.5 x 7 = 227.5). Each of the other transects is located 250 yards farther along the baseline, at 478, 728, 978, 1228, and 1478 yards.

Transects are run by boat, picking up all fish within the transect boundary but none outside it. The total number of fish killed is determined by multiplying the total number of dead fish collected by an expansion factor. Expansion factors may be calculated from baseline length or area of the body of water. The latter method requires a good map and planimeter or other means of area measurement. When a good map is not available, the expansion factor *(E)* is calculated as

$$E = \frac{T}{WN} \tag{13.2}$$

where *T* equals baseline length, *W* equals transect width, and *N* equals the number of transects sampled. If a good map and measuring tools are available, the expansion formula is

$$E = \frac{A}{WL} \tag{13.3}$$

where *A* equals the area of the water body, *W* equals transect width, and *L* equals the sum of the lengths of the transects.

The transect method described above is for either open water areas of lakes or reservoirs, in which some of the dead fish are floating in the open waters and some are clumped along the shoreline, or for those kills in which the fish have not collected along the shoreline. Where shoreline clumping occurs, the shoreline should be sampled much the same as a narrow stream is sampled. The results of these counts are then added to the transect counts for an estimate of a total kill.

To make shoreline counts, (1) measure the shoreline included in the kill, (2) choose the number and length of sampling segments to be examined (at least three), (3) calculate the interval distance between segments by dividing the shoreline length by the number of sampling segments, (4) choose the first segment at random in the first interval, and (5) locate successive sampling segments uniformly at distances equal to the interval calculated above. The shoreline zone should be as wide as the largest clumping of fish on the shoreline. The transect counts used for open water should begin and end at the boundary of a shoreline "exclusion" zone which corresponds to the width of the shoreline zone.

Total number of dead fish in the shoreline zone is determined by multiplying the number of fish counted in the sampling segments by an expansion factor. The expansion factor *(E)* is calculated from the formula

$$E = \frac{T}{L} \tag{13.4}$$

where *T* equals the total length of the affected shoreline and *L* equals the sum of the lengths of the segments counted. Shoreline counts are kept separate from transect counts until the two estimates are combined for a total kill estimate. An example is provided in Box 13.2.

The above procedures provide a basis for statistically estimating the number of dead fish in typical habitats at one point in time. Refer to the AFS guidelines for more detail and for special or unique circumstances. Because of field conditions and constraints on the procedures, the estimates are usually conservative, a fact the field investigator should stress when reporting findings during both judicial proceedings and evaluation of resource damage.

13.5 ASSESSMENT OF MONETARY VALUE

In 1970 and again in 1975, the Pollution Committee of the Southern Division, American Fisheries Society, published a listing of monetary values of fish. This innovative publication was based on the conviction of fishery professionals that every fish has a value (both intrinsic and monetary) and that polluters should compensate the public for the loss of fish killed as a consequence of their actions. The monetary

Box 13.2   A Fish Kill in a Lake

In this example, a fish kill affected an entire lake. For estimating total kill, the lake was divided into an open water zone and a shoreline zone.

Open water counts. The open water area is the same as that described in the text and shown in Figure 13.6. Counts were made of bluegills in each of six transects with the following results.

				Transect			
Inch Group	1	2	3	4	5	6	Sum
1	200	260	400	220	300	100	1480
2	100	30	170	150	20	40	510
3	120	30	0	200	170	30	550
4	30	40	20	80	50	10	230
5	50	40	0	80	60	0	230
Sum	500	400	590	730	600	180	3000

For calculating the expansion factor, equation 13.2 is used, in which $T = 1500$, $W = 7$, and $N = 6$.:

$$E = \frac{1500}{7 \times 6} = 35.7.$$

For estimating the total numbers of bluegill killed in open water, the number counted is multiplied by E:

$$\text{Total kill} = 35.7 \times 3000 = 107,100.$$

Shoreline counts. For the shoreline count, three 25-yard segments are sampled out of the entire 5000-yard shoreline. Counts of bluegills give the following:

		Segment		
Inch Group	1	2	3	Sum
1	330	880	410	1620
2	120	430	180	730
3	170	50	220	440
4	20	60	300	380
5	60	20	0	80
Sum	700	1440	1110	3250

For calculating the expansion factor, equation 13.4 is used, in which $T = 5000$ and $L = 75$:

Continued

$$E = \frac{5000}{75} = 66.7.$$

For estimating the total numbers of bluegill killed in the shoreline zone, the number counted is multiplied by E:

Total kill = 3250 x 66.7 = 216,775.

Total kill. The total estimated kill is the sum of the open water and shoreline estimates:

Total = 107,100 + 216,775 = 323,875.

values published in 1970 represented a comprehensive survey of major hatcheries in the Southeastern United States and, for the first time, constituted an authoritative, defensible source document from which monetary value could be placed on virtually any fish found in the inland waters of the Southeast. Most recently, the American Fisheries Society has published monetary values of freshwater fish throughout the United States (AFS 1982).

Monetary values are listed individually by inch-group and by the pound for each species of fish. Thus, once a designated investigator has made counts of dead fish in an affected area, the individual fish should be sorted by species and grouped according to size or weight so that values can be assessed.

13.6 REFERENCES

AFS (American Fisheries Society). 1982. *Monetary values of freshwater fish and fish-kill counting guidelines.* American Fisheries Society Special Publication Number 13.

APHA (American Public Health Association). 1980. *Standard methods for the examination of water and wastewater,* 15th edition. American Public Health Association, Washington, District of Columbia, USA.

EPA (Environmental Protection Agency). 1979. *Methods for chemical analysis of water and wastes.* United States Environmental Protection Agency EPA 600/4-79-020, Washington, District of Columbia, USA.

EPA (Environmental Protection Agency). 1980. *Fish kills caused by pollution in 1977.* United States Environmental Protection Agency, Office of Water Planning and Standards, Publication Number EPA 400/4-80-004, Washington, District of Columbia, USA.

Hayne, D. W., R. D. Ober, L. E. Schaaf, and D. G. Scott. 1980. Comparisons of second-day pickup with numbers estimated by pollution committee counting guidelines—Barkley Lake study. *Proceedings of the Annual Conference Southeastern Association of Fish and Wildlife Agencies* 33:738-752.

Chapter 14
Reference Collections and Faunal Surveys

RICHARD L. HAEDRICH

14.1 INTRODUCTION

This chapter, a general discussion of collections of fishes, will deal with the uses of collections, the types of collections that may be found, what material should go into a collection, the organization and arrangement of collections, technical matters associated with the preservation of collections, and a brief mention of particularly important collections and regional keys. These subjects and related matters are also treated in a recent excellent review by Crossman (1980).

14.2 USES OF COLLECTIONS

Collections provide the fundamental materials for research concerning fishes. Classically, ichthyologists have based their studies on major fish collections. Their work has been primarily concerned with the naming of fishes and detailed studies of morphology, evolution, and zoogeography.

For fisheries biologists, nomenclature is a matter of concern. Names become particularly important when widespread commercial species are involved. Reference collections that have specimens from the entire range where the species occurs can be used to assess the relationships of the various named forms and to determine their proper names. Scientific names may also be associated with certain stocks, as with herring.

Zoogeographic studies are based on research collections. An extensive collection is a rich source of material for determining ranges. Detailed study of distributions can suggest habitat requirements and the environmental factors which limit species. Collections which have been in existence for a long time may contain specimens collected from localities where the species no longer occurs. Such material is important in documenting change, and it may be possible to ascribe such changes to either climatic variation or to the activities of man.

Another topic in zoogeography is determining the distribution of faunal assemblages or communities. Because of the arrangement of collections (section 14.4), it is usually difficult to determine exactly which species were collected together from any one locality. Nonetheless, collections and particularly original data records can be used to reconstruct faunal groupings from the past or present. An investigator may also be interested in compiling checklists of species found in certain geographic areas. The fundamental material for preparing such checklists comes from collections.

Collections with large numbers of an individual species are useful for autecological studies. Food habits studies are one obvious area. Well-preserved specimens will have their stomach contents well-preserved as well. Other aspects of a species' biology, especially reproduction, are readily studied from preserved material.

The data that fishery managers seek are usually available from collections. Where full series have been preserved, length frequencies are available. Information on aging, derived from scales, otoliths or bones, can be obtained from preserved material and related to the length. Length-weight relationships can also be derived, but these can be applied to living material only if the specimens have been measured fresh and prior to fixation.

In addition to direct research purposes, collections are also maintained for the particular services they can provide. A well-curated collection, with up-to-date names and correct identifications, can be used as reference material with which new specimens can be compared to see if they have been identified properly. Sometimes the easiest way to identify a fish is to compare it with specimens already identified. With comparative material, special keys and identification sheets can be prepared for field workers.

Collections can help satisfy certain legal requirements. Environmental impact legislation can require that the material from a baseline survey be preserved as documentation. The preparer of the impact assessment is often in a poor position to take care of the material which formed a part of the baseline, but if the material is deposited in a well-curated collection, it can be preserved indefinitely.

Finally, collections are of great importance in teaching because they can be used year after year to show the diversity of fishes and aspects of their evolution. Excess material held in a collection may be used in laboratory dissections and discarded after use. Museum displays, intended to educate the public-at-large, may use actual specimens, but the displays in most museums are based on models constructed through careful reference to the material in collections.

14.3   TYPES OF COLLECTIONS

Collections range in size from very small to extremely large collections which may contain several million specimens. Regardless of their size, they can be classified as either permanent or semipermanent/temporary collections. The American Society of Ichthyologists and Herpetologists has identified a number of important fish collections throughout North America (Collette and Lachner 1976; Lachner et al. 1976), and some of these are listed in Table 14.1. While universities are the main locations of collections, many exist also in state or federal laboratories and private research organizations.

Permanent collections are those that in general have been around for a long while and appear to have a long future. These include many university-based reference and teaching collections and the collections of natural history museums and systematics centers.

The most important collections are those in the major research museums and systematics centers. These are the preferred final depositories for all important material and for any that should be guaranteed preservation, including type specimens (the material upon which the first description of a species is based) and material of special scientific or historic interest.

It is important to distinguish between the research and the teaching portions of a university collection and to keep the two portions separate. Teaching material should be available for any instructional use, including destructive dissection, and damaged specimens should be replaced as needed. For this reason, material in teaching collections is often not entered in a catalogue (section 14.4). Research material,

Table 14.1. Some important fish collections in the United States and Canada (Adapted from Collette and Lachner 1976).

Name, location, and acronym	Number of specimens	Number of types
University of Michigan Museum of Zoology Ann Arbor, MI (UMMZ)	3,685,000	650
National Museum of Natural History Washington, DC (USNM)	3,000,000	5,500
Tulane University New Orleans, LA (TU)	2,867,000	61
Academy of Natural Sciences of Philadelphia Philadelphia, PA (ANSP)	2,000,000	3,500
Los Angeles County Museum of Natural History Los Angeles, CA (LACM)	2,000,000	143
Cornell University Ithaca, NY (CU)	1,500,000	112
Scripps Institution of Oceanography La Jolla, CA (SIO)	1,200,000	131
California Academy of Sciences San Francisco, CA (CAS)	1,000,000	2,800
Field Museum of Natural History Chicago, IL (FMNH)	1,000,000	1,400
Royal Ontario Museum Toronto, ON (ROM)	800,000	40
University of British Columbia Vancouver, BC (UBC)	800,000	29
Museum of Comparative Zoology Cambridge, MA (MCZ)	650,000	1,800
American Museum of Natural History New York, NY (AMNH)	500,000	388
Florida State Museum Gainesville, FL (UF)	460,000	85
National Museum of Canada Ottawa, ON (NMC)	235,000	21

however, should have a permanence guaranteed by strict regulations concerning use. Ecological studies, for example, may harm specimens because such studies usually involve dissection. If dissection is done carefully, however, damage to the specimens can be minimized. In all cases, the specimens must be retained in the research collection even if they are considered damaged. Future investigators may need to confirm either the observations themselves or the identity of the specimens upon which they were based.

Semipermanent/temporary collections are usually those which have been made by individuals for their own purposes and are maintained at that individual's expense. Little provision other than the individual's effort is given to maintain the collection. Such collections are generally poor places for depositing material of any importance.

In addition to the collections mentioned above which comprise large series of fishes representing a broad diversity of species, special-focus collections also exist. Such collections can either be permanent or semipermanent/temporary. Examples of special-focus collections are (1) skeltons, either dried or cleared-and-stained (in glycerin), (2) otoliths, which may be used for identification purpose, (3) scales, which may be used for identification or aging, and (4) various tissues such as gonads, muscle, etc. There are also collections of parasites from fishes, diseased tissues, cell types or chromosomal preparations on microscope slides, electrophoretic plates or gels, and photographs and radiographs.

14.4   ORGANIZATION OF COLLECTIONS

Most permanent collections are arranged according to some systematic order, with the fishes placed on the shelves following a phylogenetic scheme. The classification of Berg (1940) has been one of the most frequently followed, although with certain variations. The 1929 classification of Regan in *Encyclopedia Britannica* is the arrangement used by the British Museum. Another important classification is that of Jordan (1923). Most recently, the classification of Greenwood et al. (1966) has found wide usage. Because of the problem of moving large amounts of material, the physical arrangements of most collections have remained as originally established even though the classification system may have been changed.

Collections can also be kept according to an alphabetic or arbitrary numerical code order. Such collections are arranged strictly by an artificial system and have no roots in a biological arrangement. The advantage of such a system is rapid access by non-ichthyologists, but a disadvantage is that closely related forms are not likely to be found together. Collections kept in this style are usually semipermanent/temporary.

A vitally important part of keeping a collection orderly is proper data keeping. This begins with the original field data sheets made at the time of collecting. When material is brought to a collection, a corresponding entry is made in a list of accessions. This later becomes the basis of a catalogue entry, which assigns a unique catalogue number to the material. Catalogue entries contain the data from the original logs and indicate when the material was brought to the collection, who identified the material, and where the material is located in the collection. The best collections will thus contain (1) a file of original field log sheets or notes (numbered and systematically kept), (2) a file of accessions, (3) a permanently bound catalogue, and (4) the specimens themselves with appropriate labels. Lists of holdings by species or by geographic region are derived from the catalogue.

Labels associated with the material are extremely important. When fishes are kept in a wet-preserved condition, as most are, such labels are usually contained in the jar rather than being affixed directly to the specimen. The label paper should be 100% rag and must be resistant to shredding or deterioration in the preservative. Most museum curators recommend Byron-Weston Resistall Linen Ledger, substance 36, short grain as label paper. Individual specimens of a series may be identified with separate numbered tags sewn to the caudal peduncle or gill cover; Dennison Company linen tag stock or the Resistall paper is preferred.

Labels should be written with a permanent black carbon ink such as Higgins Engrossing no. 893. The best practice is to have separate name and data labels. The name label should contain (1) the full scientific name, including author and date, (2) the number of specimens and range of standard lengths, and (3) the name of the identifier and the date identified (customarily following the abbreviation "det" or if material is subsequently reidentified, the abbreviation "redet"). Name labels must always remain with the specimen even if the name is changed; the dated labels are a part of the specimen's history. Information recorded on the data label should include, at a minimum: (1) the precise locality of capture, (2) the date of capture, (3) the name of the collector, and (4) the catalogue number.

When very large series are being identified or when there are numerous and diverse specimens from only a few station localities, preprinted labels may be used. It is important that the ink used in the printing be resistant to preservatives and particularly to the dissolving effects of alcohol. Experimentation may be necessary to establish the suitability of printed labels.

Computers are coming into increasing use in museum work and can be used for record keeping at all stages in the data entry and cataloguing process. They may also be used for printing labels if, as noted above, the printer ink will not dissolve in alcohol. Once information is entered and verified, computers can be used to rearrange the data in any manner needed by the researcher. This may include printing a catalogue, sorting species' records and associated data, and listing of material in various combinations. Such an operation may be exceedingly important for resynthesizing the species assemblages that were taken at a particular locality. Likewise, a computer can be used to search catalogue entries and prepare checklists for any specific region.

When collections are small, they are rather readily put onto computers, but as they grow larger, the task becomes more burdensome. People just setting up a collection, even if it is a very small one, should consider the use of a computer system, for ultimately it will make the collection far more useful. There are a number of schemes in use today, such as that described by McAllister (1978).

14.5 WHAT TO COLLECT

A fish collection is not like a stamp collection. A "one-of-a-kind" approach is suitable only for the most elementary reference collection. A proper collection should contain series which include a large range of sizes, both sexes, and material from as many localities, as possible. For some fishes, especially those which undergo transformations at spawning time, collections from several seasons may be desirable. Larval fishes have their place, too, but their identification is difficult, and, thus, they are not commonly found in most collections. Surprisingly, some of the most common species may be poorly represented in collections. It is an error to assume that such species can be collected when needed.

It makes sense both for scientific and conservation reasons to preserve and keep all material collected rather than just a representative sample. A complete sample means a more accurate reconstruction of a fauna in terms of relative abundance and also is more likely to contain satisfactory series. Seemingly excess material provides for future needs and destructive analysis and will not require killing more fish. Such excess also can be used in teaching collections (section 14.3) and can be exchanged with other universities to obtain teaching material. Situations do arise, however, where it is impossible to keep everything collected. In such cases, what is kept will depend on the immediate concern of the collector with the general rule-of-thumb to include (1) representatives of all species, and (2) size series from the smallest to the largest specimens for each species.

14.6 PRESERVING COLLECTIONS

After specimens are first collected, they must be fixed. For general purposes, 10% formalin is always used. Formaldehyde gas (HCHO) in stock 40% aqueous solution is diluted 9:1 with water to make 10% formalin. Since the volume of the fishes themselves must be taken into account upon fixation, formalin for field use should be stronger than 10%, and even 20% will not hurt. Kill fish by placing them directly in the formalin. Fish should be completely covered by formalin and occupy no more than half of the volume of the container. Be certain that the abdominal cavity of specimens longer than 150 mm are flooded with strong formalin, either by injecting formalin or by cutting a small slit in the body wall on the right hand side of the specimen and puncturing the swimbladder. Also inject formalin into the musculature of very large specimens.

To prepare the very best specimens, use freshly killed fish and clean the body surface carefully before placing them flat in fixative. Avoid curling by putting the specimen jar on its side or by fixing the fish in a flat tray. Pin the fins open while the specimen is being fixed, and they will remain upright when the pins are removed.

Formalin is a very hazardous substance—poisonous, destructive of tissue, and perhaps carcinogenic. Be cautious when using formalin. Avoid breathing the fumes, wear safety glasses, and wash immediately if formalin comes in contact with the skin or especially the eyes. To avoid splashing, always tightly cover the containers in which specimens are placed.

Formalin oxidizes to make formic acid, which will decalcify bones and especially otoliths (McMahon and Tash 1979). Buffer the formalin to avoid acidification. Calcium carbonate (marble chips or ground limestone) is the best buffer and in a saturated solution will maintain formalin at a pH around 6.0 (Taylor 1977).

After about three days in fixative or even less for very small specimens, transfer the material to a preservative. Alcohol preservation keeps specimens more pliable than formalin and makes working with them easier and more pleasant. Ethanol (CH_3CH_2OH) in 70-75% strength and isopropanol (($CH_3)_2CHOH$) in 45-50% are most commonly used and are considered equally good. It is customary to soak the specimens in water for a day or more to remove the fixative and then to transfer specimens consecutively through a series of dilutions (i.e., 2.5:1, 2:1, 1.5:1, 1:1) into preservative. Not all ichthyologists consider this protocol necessary. Well-buffered 10% formalin itself can be used as a preservative and is preferred over alcohols for some soft-bodied fishes. Larval fishes should be kept in 5% formalin. Transferring specimens from one preservative to another can be damaging to specimens and should be avoided.

Color retention is a problem no matter what preservative is used. To retain a record of color, photograph the specimens when they are collected.

Screw-top jars with polypropylene lids and polypropylene liners are good containers and can be purchased at reasonable price in large lots directly from the manufacturer. Bakelite or metal lids and cardboard or foil liners are not recommended. Bail-top jars are the best because of their tight closure, but they are expensive. These should have gaskets made of Buna-N synthetic rubber.

Dried skeletons are best kept in closed boxes, with a few mothballs added if pests are a problem. Articulated skeletons, held together by dried connective tissue, are quite fragile and handling should be limited. Otoliths can either be kept dry in compartmented microscope slides or wet in 95% ethanol in vials.

Frozen material, increasingly required in ichthyological studies, must be kept in freezers and needs extremely expensive backup support and day-to-day care. Frozen collections should be considered for temporary use only, except under special circumstances.

14.7 MAINTENANCE OF COLLECTIONS

The person who is responsible for a collection is its curator. The curator makes sure that material in the collection is properly documented, identified, and maintained. Most of the curator's time may be spent at the microscope identifying and cataloguing new acquisitions. The curator must also make the collection and its data readily available to other investigators. Access may be provided at the place where the collection is kept, but it is common for specimens to be loaned through the mail.

Routine maintenance involves periodic inspection of the collection as frequently as possible but at least once a year to guard against evaporation of alcohol, deterioration of labels and jar closures, and the like. "Topping-off" to replace lost alcohol is a questionable practice. It is impossible to ascertain the strength of the "topped-off" solution, and the process can introduce oxygen which may contribute to specimen decomposition. The curator must also consider the conditions of light and temperature and how they affect preservation.

A collection should have separate areas for specimen storage and for work. The storage environment should be as cool and dark as possible to reduce alcohol evaporation and specimen fading and deterioration. Fluorescent lighting is best, used only when needed and with individual switches for each aisle; sunlight should be excluded entirely by covering windows with black paint such as Insulux. Temperature of the collection room should be uniform throughout the year and is probably best held around 15-18 C (60-65 F). Freezing must be avoided; this is especially critical for formalin-preserved material because the strength of formalin is reduced at low temperatures through formation of a milky paraformaldehyde precipitate.

Shelves must be strong enough to bear the heavy weight of wet-preserved material yet open and clear enough for easy access. Most modern collections are kept on steel shelving. Dust problems will be reduced or eliminated if closed cabinets are used, but these can be troublesome because access and lighting are difficult.

An adequate work area in a collection includes sinks with running water and microscope benches set at a comfortable level. Good lighting is necessary. Other facilities, such as a darkroom for the development of radiographs and film, will add to the usefulness of the work area.

In both the storage and the work areas of a collection, safety is a primary consideration. Government regulations may require certain safety equipment to be installed. Because alcohol is an inflammable liquid, overhead sprinkler systems may be required throughout a collection. Good ventilation and a fume hood are necessary wherever formalin is used.

14.8 KEYS

Keys are a major tool of the person maintaining a collection. These can range from simple checklists to full and detailed systematic treatments. Original publication of systematic work and new keys is found in professional journals, such as *Copeia*, but this information is eventually incorporated into ichthyological books which treat whole faunas. Only a few examples are mentioned below, but others may be found in the American Fisheries Society List of Regional Guides.

The Sea Fishes of South Africa (Smith 1953) is an often reissued book that, because of the great diversity of the South African fish fauna, is of general worldwide application in temperate and tropical marine waters. The Food and Agriculture

Organization identification sheets, although they tend to emphasize the commercial species, are excellent aids, useful over broad geographic areas (e.g., Eastern and Western Central Atlantic, Western Indian Ocean). Two classic works of more local regions are *Fishes of the Great Lakes Region* (Hubbs and Lagler 1949) and *Fishes of the Gulf of Maine* (Bigelow and Schroeder 1953). These two books established a basic, clear, and readable format that continues in such books as *Freshwater Fishes of Canada* (Scott and Crossman 1973) and *Fishes of Kentucky* (Clay 1975).

14.9 REFERENCES

Anderson, R. M. 1965. Methods of collecting and preserving vertebrate animals, Revised 4th edition. *National Museum of Canada Bulletin* 69, Biological Series 18.

Barr, S. W. 1974. *Application of systematics collections: The environment.* Symposium proceedings, Association of Systematics Collections, Second Annual Meeting, Texas Technological University, Lubbock, Texas, USA.

Berg, L. S. 1940. Classification of fishes, both recent and fossil. *Travaux de l'Institut Zoologique de l'Academie des Science de l'URSS* 5(2):87-348. (Reprinted with English translation by J. W. Edwards, Ann Arbor, Michigan, 1947).

Bigelow, H. B., and W. C. Schroeder. 1953. Fishes of the Gulf of Maine. *Fishery Bulletin* 53:1-577.

Clay, W. M. 1975. *The fishes of Kentucky.* Kentucky Department of Fish and Wildlife Resources, Frankfort, Kentucky, USA.

Collette, B. B., and E. A. Lachner. 1976. Fish collections in the United States and Canada. *Copeia* 1976:625-642.

Cross, F. B. 1962. *Collecting and preserving fishes.* University of Kansas, Museum of Natural History, Miscellaneous Publication No. 30:41-44.

Crossman, E. J. 1980. The role of reference collections in the biomonitoring of fishes. Pages 357-378 *in* C. H. Hocutt and J. R. Stauffer, Jr., editors. *Biological monitoring of fish.* Lexington Books, Lexington, Massachusetts, USA.

Fink, W. L., K. E. Hartel, W. G. Saul, E. M. Koon, and E. O. Wiley. 1979. *A report on current supplies and practices used in curation of ichthyological collections.* American Society of Ichthyology and Herpetology, Washington, District of Columbia, USA.

Greenwood, P. H., D. E. Rosen, S. H. Weitzman, and G. S. Myers. 1966. Phyletic studies of teleostean fishes with a provisional classification of living forms. *American Museum of Natural History Bulletin* 131:339-456.

Hubbs, C. L., and K. F. Lagler. 1949. *Fishes of the Great Lakes region.* University of Michigan Press, Ann Arbor, Michigan, USA.

Irwin, H. S. 1974. Environmental protection: a legitimate focus for systematic institutions. Pages 18-24 *in* S. W. Barr, editor. *Applications of systematics collections: The environment.* Symposium proceedings, Association of Systematics Collections, Second Annual Meeting, Texas Technological University, Lubbock, Texas, USA.

Jordan, D. S. 1923. A classification of fishes including families and genera as far as known. *Stanford University Publications, University Series, Biological Sciences* 3(2):77-243.

Lachner, E. A., J. W. Atz, G. W. Barlow, B. B. Collette, R. S. Lavenberg, C. R. Robins, and R. S. Schultz. 1976. A national plan for ichthyology. *Copeia* 1976:617-625.

McAllister, D. E. 1978. The complete minicomputer cataloguing and research system for a museum. *Curator* 21(1):63-91.

McMahon, T. E., and J. C. Tach. 1979. Effects of formalin (buffered and unbuffered) and hydrochloric acid on fish otoliths. *Copeia* 1979:155-156.

Peters, J. A., and B. B. Collette. 1968. The role of time-share computing in museum research. *Curator* 11(1):65-75.

Scott, W. B., and E. J. Crossman. 1973. Freshwater fishes of Canada. *Fisheries Research Board of Canada Bulletin* 184.

Smith, J. L. B. 1965. *The sea fishes of South Africa.* Central News Agency Ltd., Johannesburg, South Africa.

Taylor, W. R. 1977. Observations on specimen fixation. *Proceedings of the Biological Society of Washington* 90:753-763.

Chapter 15
Length, Weight, and Associated Structural Indices

RICHARD O. ANDERSON AND STEVEN J. GUTREUTER

15.1 INTRODUCTION

Length and weight data provide statistics that are cornerstones in the foundation of fishery research and management. The numbers and sizes of available fish in a population determine its potential to provide benefits for commercial or recreational fisheries. These measurements also provide the basis for estimating growth, standing crop, and production in natural waters as well as in fish hatcheries and laboratories. The objective of this chapter is to present information on methods of measurement and the calculation of structural indices from such measurements.

15.1.1 Why Length Measurements?

Rate of change in length of individuals and length-frequency distributions are key attributes of fish populations. Fishes, in general, have indeterminate growth; all ages and sizes have some inherent growth capacity. Evaluations of length at age and annual length increments are approaches to measuring the growth process. The growth history of individuals and apparent averages for age groups are determined from measured lengths at capture and back-calculated lengths, which are based on measurements of annular marks on scales, spines, or bones (Ricker 1975; *see* chapter 16). Samples from a population plotted as a length-frequency distribution often provide important information for the interpretation of age and growth, especially for young fish; length-frequency distributions also provide an important description of population structure. In addition, length is often better than age as an indicator of maturity.

Length is also important to the recreational angler. Weithman and Anderson (1978) and Weithman and Katti (1979) developed an index, Fish Quality, which describes the value of a captured fish in terms of the world record size for that species. In many fisheries, length is used to define legal size for harvest.

15.1.2 Why Weight Measurements?

Weights of individuals and populations are also key attributes of populations. The production process results in the creation of tissue by individuals and populations. Although production can be expressed as calories or weights of carbon, protein, or dried tissue, all these measures are normally based on a measurement of wet or live weight. From a fishery perspective, total weight or weight per unit of area is the statistic normally reported for harvest or standing crop. Weight is the common basis of reporting catches, whether made by anglers or by commercial fishermen. Weight at age and annual weight increments are other measurements that describe the growth process. Annual weight increments best reflect how fish of various sizes are gaining in value to the fishery; annual weight gains and appropriate growth efficiency values can be used to estimate consumption of prey. Length and weight data can also be used to calculate condition factors or indices of well-being.

15.2 SAMPLING AND MEASURING CONSIDERATIONS

All methods of sampling fish populations have some inherent bias. The bias for or against certain sizes or a sex is related to the sampling gear, factors that can influence fish distribution in space, and factors that influence vulnerability to the gear as affected by time within a day or among seasons (*see* chapters 6, 7, and 8). When sampling methods are size or sex selective, it is important to acknowledge the bias and, if possible, to correct for it.

The number of fish to be measured and weighed is influenced by sampling objectives, the variability among individuals and populations studied, and the number and sizes collected. Often only subsamples can or need to be measured or weighed. Research often requires larger sample sizes than are necessary for management surveys. In general, it is desirable to measure more fish for a length-frequency distribution than need to be weighed for determining weight-length relations. For many species, sexes should be distinguished because males and females may differ in weight-length relations. The best measures of variability and confidence can usually be determined from the length and weight of individuals rather than from the total and average weights for a group within a size class.

In fishery management and stock assessment studies, it is desirable to follow standard practices. The justification is not that one measure is inherently superior to another, but that convention facilitates communication and reduces confusion. There are two conventions for designating a length class in length-frequency histograms: for example, a 10-cm size may include fish from 10.0 to 10.99 cm or from 9.5 to 10.49 cm. This difference can lead to problems in communication and data comparison. The first convention is recommended because size limits, when they are established, are based on whole numbers, as are the defined stock and quality sizes (Anderson 1976) used for determination of length-frequency indices. Many scientific and working groups have adopted this recommended convention (Holden and Raitt 1974).

15.2.1 Length Measurements

Two approaches are used to measure fish length: measurement of the whole body and measurement of body parts. Whole-body measurements are most commonly used in fishery investigations although partial measurements are useful in food studies or for fish that have been beheaded or have damaged tails or heads.

Three common whole-body measures of fish are total, fork, and standard lengths (Fig. 15.1). Total length is defined as the length from the anterior-most part of the fish to the tip of the longest caudal fin rays. There are two common conventions for measuring total length. Maximum total length is measured when the lobes of the caudal fin are compressed dorso-ventrally. Natural total length, the conventional measurement in Europe, means measurement to the tip of the tail in a "natural" position. Investigators should always describe what length is used and be wary of published measurements in which the unqualified phrase "total length" is used. In

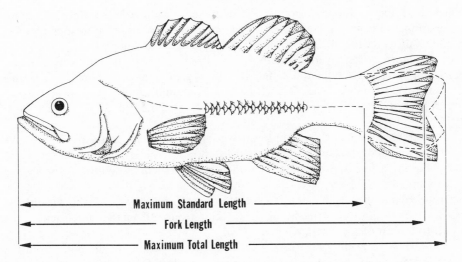

Figure 15.1. Common measurements of fish length—maximum standard, fork, and maximum total.

Canada during the 1920s, "total length" meant fork length (W. E. Ricker, personal communication). Maximum total length is the convention of choice in fishery investigations in the United States.

Fork length is the length from the most anterior part of the fish to the tip of the median caudal fin rays. Fork length is commonly used in fishery studies in Canada and for certain species such as tunas.

Standard length is the length from the tip of the upper jaw to the posterior end of the hypural bone. In practice it may be measured to some external feature such as the position of the last lateral line scale, the end of the fleshy caudal peduncle, or the midline of a crease that forms when the tail is bent sharply. Standard length is most commonly used in taxonomic studies because it is unaffected by caudal fin anomalies. Maximum standard length, used by some investigators, is measured from the most anterior part of the fish; this is longer than standard length for fish that have protruding lower jaws. Standard length is a less convenient measure than either fork or total length.

Less common measuring techniques are used for special situations. Measures of specific body parts are used when intact fish are not obtainable. Pectoral length, the distance from the posterior insertion of the pectoral fin to the posterior margin of the longest caudal fin rays, is used when fish have been beheaded (Laevastu 1965). Head length is also used when body damage impedes other measurements. For spawning or recently dead Pacific salmon, which often have frayed tails and enlarged or damaged jaws, measurements are made from the eye to the hypural bone. Unfortunately, two measurements are made—from the middle of the orbit (midorbital-hypural) in the United States and from the hind margin of the orbit (postorbital-hypural) in Canada.

Because different measurements of length are used, it is sometimes necessary to convert one to another to make comparisons. Such conversion functions are usually linear and often reduce to a constant of proportionality. These functions may be estimated for any sample by measuring any combination of total, fork, standard, or other lengths and applying regression techniques. Geometric mean (GM) regressions give much better results than ordinary regressions when projected over a wide range or when used for a different population. Any new sample, even from the same population, is likely to include a different range or distribution of lengths and will require a different line. If the complete range of lengths in a population is used, and they are more or less evenly distributed by length classes, the GM and ordinary regressions are practically the same. But in a great majority of cases, a simple proportion is adequate (W. E. Ricker, personal communication). An alternative is to use published conversion factors for the species concerned (e. g., Carlander 1969, 1977).

Other measurements are employed for Mollusca and Crustacea. For bivalve molluscs, length of the shell is the greatest distance in an anterior-posterior direction, usually along a line approximately parallel to the hinge axis. Width of the shell is usually measured as the greatest distance in a dorso-ventral direction, usually perpendicular to the length measurements (Loosanoff and Nomejko 1949). Gastropod molluscs are usually measured along the greatest distance from the tip or focus of the whorl to the tip of the shell (Warren 1958).

Lobsters are usually measured in a straight line parallel to the body from the posterior margin of the eye socket to the posterior edge of the carapace (Holden and Raitt 1974). Carapace length, used for shrimp and freshwater crayfish, is the distance from the anterior tip of the rostrum to the posterior margin of the carapace along the dorsal median (Hobbs 1976). Total length, which is also used, is the distance from the tip of the rostrum to the posterior edge of the median uropod. Crabs are usually measured across the width of the carapace (Holden and Raitt 1974).

Measuring devices. Various devices used to measure fish length include measuring boards, tapes, calipers, and plastic or foil sheets. The choice depends in part on the sizes and species of fish measured. Often one person makes the measurements and a second records them. Tape recorders are sometimes used to record data.

Measuring boards consist of a linear scale on a board with a rigid head piece (Fig. 15.2). Finfish are usually measured with the mouth closed with slight pressure against the head piece and the body positioned on its right side, with the head facing the observer's left. Herke (1977) described a plexiglass measuring board that is easily fabricated. Low-cost boards may also be made from pine stock, door moulding, a ruler, screws, waterproof wood glue, and varnish which has a urethane or epoxy base. Aluminum rulers have an advantage of durability. Boards are also available from commercial sources. When pectoral lengths are measured, the head piece is undesirable. Measuring boards should be maintained in good condition. Warped or loose head pieces and worn or faded markings can be a cause of error.

Tapes are desirable for measuring large marine fish. Calipers may be used for small fishes; although precise, they are more time-consuming and cumbersome than other devices.

Plastic or foil sheets are useful when lengths are the only data recorded for individual fish (Holden and Raitt 1974). Removable sheets are fixed across the measuring board. Instead of recording fish length, a pin hole is made at the terminus of the measure. Many fish can be measured on the same sheet and length-frequency distributions easily determined.

Accuracy and precision. Accuracy refers to the difference between any measurement and the actual value. Precision refers to the reproducibility of a measurement; it is expressed in terms of the standard deviation. Accuracy of length measurements is determined by the condition of fish, ability of the investigator, and technique used. Fish should be measured fresh if possible. Rigor mortis, drying, and

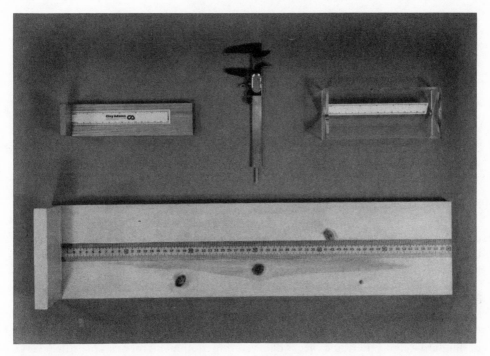

Figure 15.2. Devices for measuring fish.

preservation result in shrinkage. If rigor mortis has set in, fish should be flexed several times to relax the musculature. Accuracy can be influenced by any numerical bias for even, odd, or other specific divisions on a scale by an investigator. The consistency with which fish are pressed against the head piece and with which caudal lobes are compressed affects precision. Fish size and morphology also influence precision. The quality of length and weight data and the speed of measurement are influenced by working conditions. Whatever is practical should be done to maintain good physical conditions; adequate lighting and comfort of workers are of primary concern.

15.2.2 Weight Measurements

It is more difficult and time-consuming to weigh fish than to measure their length. Weighing fish under field conditions presents special challenges. Spring balances are most commonly used for weighing individual fish (Fig. 15.3). Scale sensitivity should be about 1% of fish weight. Two or more scales with a range of sensitivities are desirable to weigh fish of a wide range of sizes. Scales with dashpot temperature compensators are available. Electronic scales with digital readout are available for field use. Large individuals or bulk lots of fish are usually weighed on hanging or platform scales. Live fish are often weighed in a tared volume of water.

Weighing fish in the laboratory circumvents some problems imposed by field conditions. A wide array of equipment offering greater accuracy can be used. Electronic balances, which would be unsatisfactory under field conditions, can be used with ease in the laboratory.

Because most fish maintain near-neutral buoyancy in water throughout their lives, their specific gravity is close to 1.0. Body volume is proportional to weight and

Figure 15.3.   Scales for weighing fish.

can provide a substitute measure. Volumetric measure of water displaced can prove useful for weighing large numbers of live fish as a lot—a need that often arises in fish hatcheries.

Accuracy and precision. The accuracy associated with weight measurements is determined by the accuracy of the weighing device, the amount of moisture on fish, and changes caused by death or preservation. Of these variables, the accuracy of the weighing device is easiest to control. Accuracy of a scale can be determined with a set of standard weights. Periodic checks of measuring equipment, for example, after every tenth fish, are recommended.

All fish carry with them a quantity of water on the body surface and in the buccal cavity. The error caused by surface water is inversely related to fish size. Small fish have a greater surface area for their volume or weight than do large fish. Because this quantity of water is variable among fish it results in a loss of precision. To increase precision, allow water to drip from the fish or blot the fish before they are weighed. Parker (1963) reported that blotting of fish before they were weighed improved the accuracy but increased the standard deviation of weights slightly as compared with drip drying. Damp chamois or cloth can be used to blot much of the surface moisture from fish without causing damage. When wet fish are being weighed in the field, water inevitably accumulates on the scale. Ensure proper drainage or compensate for water accumulation by adjusting the tare as often as necessary. When platform scales are used, take care to avoid drops of water on the balance beam.

Some measurements of body weight have been made after removal of the stomach contents. In laboratory studies, fish are often fasted before they are weighed. Variations in the degree of hydration of sex products also can influence body weight. Usually body weight should include the weight of sex organs because the energy and materials of these organs are part of fish production.

Accuracy and precision of weight measurements are determined by motion and by water on the fish. Weighing devices are sensitive to motion from the wind, the boat, and the fish. Wind problems can be avoided or reduced by using a suitable windbreak. The problem of boat movement can be circumvented by weighing fish on shore. The problem of fish movements can be alleviated by either sacrificing or anesthetizing fish. Anesthesia also helps reduce handling stress (*see* chapter 5).

Preservation effects. The effects of preservation on length, weight, and indices of well-being deserve special attention. The method and duration of preservation affects both length and weight. The effects of preservation in formalin have been studied most

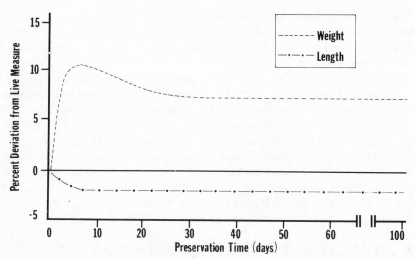

Figure 15.4. Typical pattern of effects on length and weight that might be observed for fishes preserved in 10% freshwater formalin.

frequently; the pattern of effects on length and weight that has emerged from these studies is depicted in Figure 15.4. Lengths of fish tend to decrease rapidly and then stabilize; weights tend to increase to a maximum and then decline to a value somewhat greater than live weight. Parker (1963) stated that initial weight gains probably result from cessation of osmoregulation; he detected this pattern in age-0 sockeye, pink, and chum salmon. The direction and degree of weight changes were different in freshwater and seawater formalin solutions. The combined effects of preservation on length and weight in freshwater formalin resulted in relative condition factors universally greater than 116% of live values; in seawater formalin, values ranged between 96 and 112%. Jones and Geen (1976), working with spiny dogfish, found that lengths decreased as salinity of the preservative increased; weights of embryos preserved in isotonic formalin approximated live values. Theilaker (1980) found that length of northern anchovies decreased in 10% formalin but not in 80% ethanol.

The results of these and several other studies demonstrate the need to establish appropriate conversion factors to convert length and weight of dead or preserved fish to equivalent live values. A subsample of live fish should be weighed and measured and individually marked before preservation.

15.2.3 Body Composition

Dry weight. For some kinds of research, particularly in bioenergetics, investigators use dry weight—the tissue weight minus the weight of free water and volatile organics. Use of dry weight improves precision because the moisture variable is removed. However, dry weights are not intuitively comparable to live or wet weights for most workers. Dry weights may be obtained by drying an organism, organ, or quantity of tissue in an oven to a constant weight. Temperatures most commonly used range between 60 and 100 C (Dowgiallo 1975). Lower temperatures conserve more of the volatile organics but prolong drying time. Nimii (1972) obtained dry weights by freeze-drying whole-fish homogenates. Freeze-drying reduces the problems associated with the conversion of fats to fatty acids and volatilization at high temperatures.

Caloric content. Sometimes it is desirable to convert weights to energy equivalents, usually expressed as Kcal/g dry weight. Energy content is measured with a bomb or microbomb calorimeter. A small quantity of dried tissue is combusted in the calorimeter, and the heat of combustion is measured (Gorecki 1975; Prus 1975). Caloric content also can be estimated from protein and lipid composition.

Proximate composition. Analysis of proximate composition or major tissue constituents is an important aspect of bioenergetic research. The usual quantities of interest are the weights of the ash, fat, and protein fractions (McComish et al. 1974). Carbohydrate is a minor composite of fish and is usually not considered except in liver studies. Lipid content is determined by extraction from dry homogenate, usually with ether or chloroform and methanol as the solvents. Ash content is determined by incinerating weighed, desiccator-dried samples for four hours or more, at 600 C. Protein content is determined by the Kjeldahl method (Association of Official Analytical Chemists 1980). Particularly useful information on the analyses of proximate composition was provided by Dowgiallo (1975).

15.3 STRUCTURAL CHARACTERISTICS AND INDICES

Length-frequency distributions and population biomass are measurable structural characteristics of fish populations. Weight-length relations describe structural characteristics of individuals within populations. Several fish population and fish community models are based on length-frequency and biomass data. Weight-length data provide the basis for calculating power functions that are good models of weight as a function of length and for determination of condition factors.

15.3.1 Length-Frequency Histograms

Length-frequency distributions reflect an interaction of rates of reproduction, growth, and mortality of the age groups present. These distributions and changes in distributions with time can help in understanding the dynamics of populations and in identifying problems such as year-class failures or low recruitment, slow growth, or excessive annual mortality.

For histograms and general stock assessment purposes, sample at least 100 adult or stock-size fish. The width of size groups for the histogram depends on maximum fish length—use a 1.0-cm interval for species that reach 30-cm, a 2.0-cm interval for 60-cm species, and 5.0-cm interval for 150-cm species. (Remember the recommended convention that fish from 10.0 to 10.99 cm should be noted as 10 cm.)

Rarely should the number of fish collected in each size group be plotted. For comparisons over time and with variable sampling effort, plot the catch per unit of sampling effort (CPUE) as an important scale that can illustrate changes in relative density. When CPUE may not reflect density or when samples from more than one collection method are combined, plot percentage distribution for each sample and method.

15.3.2 Length-Frequency Indices

The concept of length-frequency indices evolved from analyses of populations of largemouth bass and bluegills (Anderson 1976; Wege and Anderson 1978). In subsequent work, Anderson and Weithman (1978), Anderson (1980), and Gabelhouse (1981) defined minimum lengths for stock, quality, preferred, memorable, and trophy

Table 15.1   Proposed maximum total length (cm) for minimum stock, quality, preferred, memorable, and trophy sizes based on percentages of world record lengths (from Gabelhouse 1983).

Species	Size designation				
	Stock	Quality	Preferred	Memorable	Trophy
Largemouth bass	20	30	38	51	63
Smallmouth bass	18	28	35	43	51
Spotted bass	18	28	35	43	51
Walleye	25	38	51	63	76
Sauger	20	30	38	51	63
Muskellunge	51	76	97	107	127
Northern pike	35	53	71	86	112
Chain pickerel	25	38	51	63	76
Blue catfish	30	51	76	89	114
Channel catfish	28	41	61	71	91
Flathead catfish	28	41	61	71	91
Bluegill	8	15	20	25	30
Green sunfish	8	15	20	25	30
Pumpkinseed	8	15	20	25	30
Redear sunfish	10	18	23	28	33
Rock bass	10	18	23	28	33
Black crappie	13	20	25	30	38
White crappie	13	20	25	30	38
White bass	15	23	30	38	46
Yellow bass	10	18	23	28	33
White perch	13	20	25	30	38
Yellow perch	13	20	25	30	38
Black bullhead	15	23	30	38	46
Common carp	28	41	53	66	84
Freshwater drum	20	30	38	51	63
Bigmouth buffalo	28	41	53	66	84
Smallmouth buffalo	28	41	53	66	84
Gizzard shad	18	28	—	—	—

sizes of fish (Table 15.1). A variety of length-frequency indices are in common use today.

Proportional Stock Density (PSD). The index *PSD* is the proportion of fish of quality size in a stock. The value, expressed as a percentage, is calculated as

$$PSD = \frac{\text{number} \geq \text{minimum quality length}}{\text{number} \geq \text{minimum stock length}} \times 100 \qquad (15.1)$$

Index values range from 0 to 100. For example, if 50 largemouth bass are ≥ 30.0 cm long (quality size) in a sample of 100 fish ≥ 20.0 cm long (stock size), the *PSD* is 50. All expressions of *PSD* should be rounded to the nearest whole number; decimals represent unfounded accuracy.

When the sampling objective is to determine if population *PSD* is within an objective range, a sequential sampling approach can reduce sampling effort (Weithman et al. 1979). Fish are measured as they are collected or in batches. Sampling can be terminated as soon as collections permit a stock to be classified as below, above, or within a target range of *PSD*. Based on an electrofishing sample of largemouth bass, minimum expected sample sizes required to evaluate stock structure are 20 fish when true *PSD* is below 20 or above 80, 30 fish when true *PSD* is 20-30 or 70-80, and 70 fish when *PSD* is 40 to 60.

Relative Stock Density (RSD). The index *RSD* is the proportion of fish of any designated size group in a stock (*PSD* is a special case of *RSD*). Dividing length-frequency distributions into three or more blocks is more sensitive in recognizing quality of the stocks than is use of *PSD* alone. For example, two populations of largemouth bass may have a *PSD* of 50. The percentage of fish $\geq$ 38 cm long might be 0 (designated as *RSD-38* or *RSD-P* = 0) or 15 (*RSD-38* or *RSD-P* = 15). The latter population would probably provide better quality fishing than the former.

Young-Adult Ratio. The index Young-Adult Ratio *(YAR)* proposed by Reynolds and Babb (1978) may provide a relative measure of reproductive success and population structure. The ratio of numbers in length groups can be used to calculate the index. For example, in samples of largemouth bass collected in late summer or fall, *YAR* is defined as the number $\leq$ 15.0 cm divided by the number $\geq$ 30.0 cm.

15.3.3 Population and Community Models

Biomass models. One key to satisfactory fishing success is a good fish population. A good or balanced fish population was defined by Swingle (1950) as a population that can sustain a harvest of good-sized fish in proportion to the productivity of the water. Structural characteristics based on biomass values were developed empirically for fish populations in small impoundments.

Swingle characterized fishes that often feed on fish as carnivores or C species and those that often feed on invertebrates and serve as food for carnivores as forage or F species. Fish were designated as small, intermediate, or large on the basis of relative sizes present rather than on the basis of specifically defined lengths or weights. The F/C ratio is the weight of F species divided by weight of C species. The most desirable range of F/C is 3.0 to 6.0.

The Y value is the total weight of fish in the F group small enough to be eaten by the average-sized adult in the C group. The Y/C ratio is the total weight of Y divided by the total weight of C. The desirable range is 1.0 to 3.0.

Jenkins and Morais (1978) developed the ratio "available prey/predator" (AP/P), which is similar in concept to Swingle's Y/C ratio. For the determination of AP/P, all sizes of predators are equated to the appropriate-sized largemouth bass on the basis of maximum size of prey that each can consume. Biomass of prey small enough to be eaten by a particular size of predator is plotted as a function of the cumulative biomass of predators on log-log scales (Fig. 15.5). The AP/P ratio has been successful for documenting shortages and surpluses of available prey and for documenting changes in the seasonal availability of prey (Jenkins 1979; Timmons et al. 1980).

Another of Swingle's models is A, the percentage of total weight of a fish population composed of fish of harvestable size. The most desirable range of A for largemouth bass and bluegill populations is 60 to 85%. Swingle's E value for a species or group is the percentage of weight of the entire fish community composed of that species or group. The desirable range of E for largemouth bass in small impoundments is 14 to 25%.

Length-frequency models. Fish community structure can be visualized by plotting *PSD* of bluegills or panfish as a function of *PSD* of largemouth bass or game

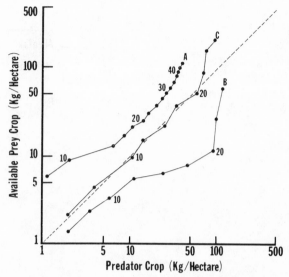

Figure 15.5. Logarithmic plot of cumulative available prey/predator (AP/P) for three general conditions: *(A)* prey excesses for all sizes of predators; *(B)* prey deficiencies for all sizes of predator; and *(C)* prey adequacy for small predators but excess for large (>20 cm) predators. Diagonal line indicates minimum desirable AP/P ratio (from Noble 1981).

fish (Fig 15.6). When desired ranges of *PSD* are defined for both groups, balanced fish communities are located within the rectangle bounded by the two desired ranges. Community structures in the lower left reflect low quality of both game fish and panfish. Low *PSD* of panfish populations reflects low density of game fish, ineffective predation associated with habitat problems, or both. A community in the lower right reflects depleted game fish stocks and low recruitment to stock size. A community in the upper left reflects the opposite condition of high game fish density and relatively low recruitment of panfish to quality size. Anderson and Gutreuter (1981) suggested that community structures in the upper right reflect a "balance of nature," with game fish at or near carrying capacity and panfish or food fish at or near 50% of carrying capacity for these species.

15.3.4 Weight-Length Relations

A variety of useful concepts, centering on the body shape of individual fish, arises from the consideration of combined weight-length data.

The power function

$$W = aL^b \tag{15.2}$$

where W = weight, L = length, and a and b are parameters, has proved to be a useful model for weight as a function of length (Fig. 15.7). In general, b less than 3.0 represents fish that become less rotund as length increases, and b greater than 3.0 represents fish that become more rotund as length increases. For many species, b is greater than 3.0. If b equals 3.0, growth may be isometric, meaning that the shape does not change as fish grow. However, it is possible for shape to change when b equals 3.0.

The parameters a and b may be estimated by taking logarithms (base 10) of both sides of equation (15.2) giving

$$\log W = \log a + b \log L. \tag{15.3}$$

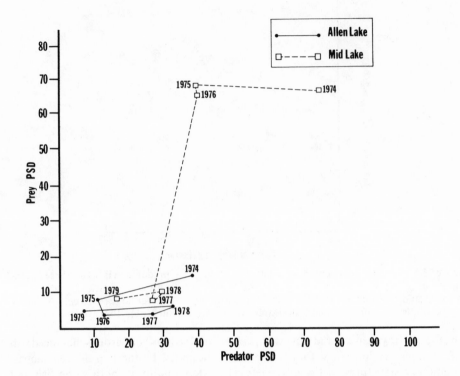

Figure 15.6. Phase-plane graphs showing the time trajectory of weighted *PSD* for game fish (largemouth bass, northern pike) and panfish (bluegill, yellow perch, pumpkinseed) for two Wisconsin lakes. Mid Lake was opened to angling in 1976 after having been closed for 20 years. Allen Lake had been continuously exploited (from Goedde and Coble 1981).

Log *a* and *b* may be estimated for a series of pairs of lengths and weights by least squares or GM regression techniques (Ricker 1973, 1975). The GM method is recommended. Always report the regression method used.

When collecting data for weight-length relations, measure individual fish to the nearest millimeter and weigh them on scales with adequate sensitivity. Always report the length measurement used because values will vary among total, fork, and standard lengths. Fish less than 30 cm long should be blotted on a damp cloth or chamois to remove surface water; allow water on larger fish to drip off. Five fish per size interval (e. g., 1-cm) usually makes an adequate sample. Measure more fish if males and females have different weight-length relations. Do not include small fish in weight-length data if either accuracy or precision of weight measurements are low.

Indices of well-being. Indices of well-being or condition are more easily interpreted and compared than *a* and *b* in weight-length relations. There are three basic variations of indices of well-being for whole fish—Fulton-type condition factor, relative condition factor, and Relative Weight.

Fulton-type condition factors are of the form

$$K \text{ or } C = \frac{W}{L^3} X \tag{15.4}$$

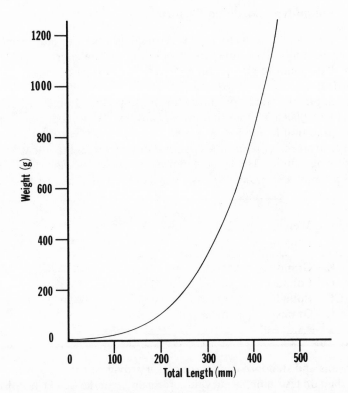

Figure 15.7 Typical weight-length relation for largemouth bass. Note that the slope of the curve increases continuously with increasing length.

where K and C are the conventional labels for metric and English units, respectively, W equals weight, L equals length, and X is an arbitrary scaling constant which varies with units of measure (Box 15.1).

A practical problem exists because Fulton-type condition factors vary for the same fish, depending on whether metric or English units are used. Also, because K and C increase with length for fish with $b > 3$, comparisons should be limited to fish of similar lengths. Comparison of K between species is usually impossible because different fishes have different shapes. For example, Bennett (1970) suggested that largemouth bass and bluegills are in "normal" or "average" condition when C equals 4.6 to 5.5 for largemouth bass and 7.1 to 8.0 for bluegills.

Relative condition factor *(Kn)* compensates for allometric growth, i.e., when shape changes as fish grow (LeCren 1951). It is calculated for each individual fish as

$$Kn = \frac{W}{aL^{b}} \qquad (15.5)$$

where W is weight of the individual, L is length of the individual, and a and b are the constants from the weight-length relation.

Relative condition factor was used to advantage by LeCren (1951) to compare males and females collected in different seasons within one population. The concept of *Kn* was expanded by Swingle (1965) and Swingle and Shell (1971) by establishing a state-average weight-length relation for several fishes in Alabama. An advantage of

Box 15.1.   Fulton-type Condition Factors

Fulton-type condition factors are very small numbers for most combinations of measurement units. Therefore, an appropriate scaling constant (X) is used to convert small decimals to mixed numbers so that the numbers can be more easily comprehended.

For example, the condition factor for the 15-pound, 13-inch fish used in the table would be 0.00068 without the scaling constant. With the scaling constant of 10,000, the condition factor becomes 6.8.

Another problem with Fulton-type condition factors is that the value changes with measurement units. The last column of the table lists values for a fish that weighs 1.5 pounds (681 grams) and is 13 inches (330 millimeters) long.

Index	Weight units	Length units	X	Example
K	Grams	Millimeters	100,000	1.89
C	Pounds	Inches	10,000	6.8
CF	Pounds	Inches	100,000	68.0
R	Grams	Inches	10	13.4

Kn is that means and standard deviations of Kn provide a better basis for statistical comparison than do tests comparing values for a and b in the weight-length equation. A practical advantage of Kn is that average fish of all lengths and species have a value of 1.0, regardless of the species or units of measurement. A disadvantage of Kn is that averages may not describe fish in good condition.

Relative Weight *(Wr)* represents refinement of the concept Kn (Wege and Anderson 1978); Wr is given by the equation

$$Wr = \frac{W}{Ws} \times 100 \qquad (15.6)$$

where W is the weight of an individual and Ws is a length-specific standard weight. The standard weight functions are of the form

$$Ws = a' \, L^{b'} \qquad (15.7)$$

where a' and b' ideally account for genetically determined shape characteristics of a species and yield Wr values of 100 at particular times of the year for fish that have been well fed (Box 15.2). Several approaches have been used to estimate a' and b' (Wege and Anderson 1978; Anderson and Gutreuter 1981: Hillman 1982).

In concept, a mean Wr of 100 for a broad range of size groups may reflect ecological and physiological optimality for populations. Wege and Anderson (1978) demonstrated that largemouth bass in both of two size groups in small impoundments in late summer and fall were likely to have mean Wr values near 100 in balanced populations with a *PSD* of about 60. When mean Wr values are well below 100 for a size group, problems exist in food and feeding relationships. When mean Wr values are well above 100 for a size group, fish may not be making the best use of available prey.

Box 15.2 Computing *Ws*

Working *Ws* equations have been proposed for only a few species (Anderson 1980). Those that have been proposed and comments are as follows:

Northern largemouth bass: $\log Ws = -5.316 + 3.191 \log L$.

This equation appears to have applicability over a wide geographic area.

Bluegill: $\log Ws = -5.374 + 3.316 \log L$.

This relation developed by Hillman (1982) from data on a few ponds in Missouri needs broader evaluation.

Gizzard shad: $\log Ws = -5.376 + 3.170 \log L$.

The equation as proposed by B. Schonhoff and published in 1980 has been revised as above.

Equations for black crappie, white crappie, rainbow trout, and other salmonids are now under review and subject to change. Obviously, much more research is needed to develop the concept of *Wr*. Coordination and cooperation will be needed to establish standard weight-length parameters.

For rapid estimation of *Wr* in the field, charts can be constructed, and interpolated points can be plotted (Fig. 15.8). Such charts are helpful in recognizing measurement errors or in providing immediate estimates when needed.

Organ weight indices. The most common organs used as indicators of well-being are gonads and liver. Gonadosomatic indices (*GSI, see* chapter 9) and hepatosomatic (liver) indices are usually expressed as organ weight as a percentage of body weight. A

LARGEMOUTH BASS RELATIVE WEIGHT (grams)

Lake ______________________ Date ______________

Remarks: ______________________________________

Total Length (cm)

Relative weight	20	22	24	26	28	30	32	34	36	38	40	42	44	46	48	50	52	54	56	58
130	138	187	247	320	404	504	619	751	902	1071	1262	1474	1711	1972	2258	2573	2916	3289	3693	4131
120	127	173	228	295	373	466	571	694	833	989	1165	1361	1579	1820	2084	2375	2692	3036	3409	3814
110	117	158	209	271	342	427	524	636	763	906	1068	1247	1448	1669	1911	2177	2467	2783	3125	3496
100	106	144	190	246	311	388	476	578	694	824	971	1134	1316	1517	1737	1979	2243	2530	2841	3178
90	95	130	171	221	280	349	428	520	625	742	874	1021	1184	1365	1563	1781	2019	2277	2557	2860
80	85	115	152	197	249	310	381	462	555	659	777	907	1053	1214	1390	1583	1794	2024	2273	2542
70	74	101	133	172	218	272	333	405	486	577	680	794	921	1062	1216	1385	1570	1771	1989	2225

Figure 15.8. Relative Weight *(Wr)* chart for largemouth bass designed for field use.

problem with *GSI* as an index of fecundity is that two fish having identical ovary weights and fecundities but different body weights have different *GSI* values. This problem led Legler (1977) to develop Relative Gonad Weight *(Gr)* as

$$Gr = \frac{\text{Ovary weight x 100}}{Ws} \qquad (15.8)$$

where *Ws* is the length-specific standard weight. In the absence of *Ws* values, an alternative index *(Gn)* calculated on the basis of the denominator in equation (15.5) may be a better statistic than *GSI*. Both *Gr* and *Gn* fix the denominator for fish of the same length.

The liver, which functions as highly dynamic storehouse of glycogen, is sensitive to the rate of feeding over short periods of time and can provide an index of nutritional state (Novinger 1973; Tyler and Dunn 1976). The hepatosomatic index *(HSI or LSI)* suffers the same limitations as *GSI*. Hence, Legler (1977) proposed the index Relative Liver Weight *(Lr)* as

$$Lr = \frac{\text{Liver weight x 100}}{Ws} \qquad (15.9)$$

which offers the same advantages as *Gr*. *Ln* is the liver-weight index that corresponds to *Gn*.

15.4 REFERENCES

Anderson, R. O. 1976. Management of small warm water impoundments. *Fisheries* 1 (6): 5-7, 26-28.

Anderson, R. O. 1980. Proportional Stock Density *(PSD)* and Relative Weight *(Wr)*: interpretive indices for fish populations and communities. Pages 27-33 *in* S. Gloss and B. Shupp, editors. *Practical fisheries management: more with less in the 1980's.* Proceedings of the 1st Annual Workshop of the New York Chapter American Fisheries Society. (Available from New York Cooperative Fishery Research Unit, Ithaca, New York, USA).

Anderson, R. O., and S. J. Gutreuter. 1981. *Evolution and evaluation of structural indices of fish populations* (Abstract). Midwest Fish and Wildlife Conference. (Available from the Missouri Cooperative Fishery Research Unit, Columbia, Missouri, USA).

Anderson, R. O., and A. S. Weithman. 1978. The concept of balance for coolwater fish populations. *American Fisheries Society Special Publication* 11:371-381.

Association of Official Analytical Chemists. 1980. *Methods of analysis.* 13th Edition. Association of Official Analytical Chemists. Washington, District of Columbia, USA.

Bennett, G. W. 1970. *Management of lakes and ponds.* Van Nostrand Reinhold, New York, New York, USA.

Carlander, K. D. 1969. *Handbook of freshwater fishery biology,* volume 1. Iowa State University Press, Ames, Iowa, USA.

Carlander, K. D. 1977. *Handbook of freshwater fishery biology,* volume 2. Iowa State University Press, Ames, Iowa, USA.

Dowgiallo, A. 1975. Chemical composition of an animal's body and of its food. Pages 160-184 *in* W. Grodzinski, R. Z. Klekowski, and A. Duncan, editors. *Methods for ecological bioenergetics.* International Biological Programme Handbook 24, Blackwell Scientific Publications, Oxford, England.

Gabelhouse, D. W., Jr. 1983 (In press). A length categorization system to assess fish stocks. *North American Journal of Fisheries Management.*

Goedde, L. E., and D. W. Coble. 1981. Effects of angling on a previously fished and an unfished warmwater fish community in two Wisconsin lakes. *Transactions of the American Fisheries Society* 110:594-603.

Gorecki, A. 1975. The adiabatic bomb calorimeter. Pages 281-288 *in* W. Grodzinski, R. F. Klekowski, and A. Duncan, editors. *Methods for ecological bioenergetics.* International Biological Programme Handbook 24, Blackwell Scientific Publications, Oxford, England.

Herke, W. H. 1977. An easily made, durable fish-measuring board. *Progressive Fish-Culturist* 39:47-48.

Hillman, W. P. 1982. *Structure and dynamics of unique bluegill populations*. Master's thesis. University of Missouri, Columbia, Missouri, USA.

Hobbs, H. R., Jr. 1976. *Crayfishes (Astacidae) of North and Middle America*. United States Environmental Protection Agency, Water Pollution Control Research Series 18050, Cincinnati, Ohio, USA.

Holden, M. J., and D. F. S. Raitt, editors. 1974. *Manual of fishery science, part 2: Methods of resource investigation and their application*. Food and Agriculture Organization of the United Nations, Fisheries Technical Paper 115, Rome, Italy.

Jenkins, R. M. 1979. Predator-prey relations in reservoirs. Pages 123-134 *in* H. Clepper, editor. *Predator-prey systems in fisheries management*. Sport Fishing Institute, Washington, District of Columbia, USA.

Jenkins, R. M., and D. I. Morais. 1978. Prey-predator relations in the predator-stocking-evaluation reservoirs. *Proceedings of the Annual Conference Southeastern Association of Fish and Wildlife Agencies* 30:141-157.

Jones, B. C., and G. H. Geen. 1977. Morphometric changes in an elasmobranch *(Squalus acanthias)* after preservation. *Canadian Journal of Zoology* 55:1060-1062.

Laevastu, T. 1965. *Manual of methods in fisheries biology*. Food and Agriculture Organization of the United Nations. Manual in Fishery Science Number 1, Rome, Italy.

LeCren, E. D. 1951. The length-weight relationship and seasonal cycle in gonad weight and condition in the perch *Perca fluviatilis*. *Journal of Animal Ecology* 20:201-219.

Legler, R. E. 1977. *New indices of well-being for bluegills*. Master's thesis. University of Missouri, Columbia, Missouri, USA.

Loosanoff, V. L., and C. A. Nomejko. 1949. Growth of oysters, *Ostrea virginica* during different months. *Biological Bulletin of the Marine Biology Laboratory, Woods Hole* 97:82-94.

McComish, T. S., R. O. Anderson, and F. G. Goff. 1974. Estimation of bluegill *(Lepomis macrochirus)* proximate composition with regression models. *Journal of the Fisheries Research Board of Canada* 31:1250-1254.

Nimii, A. J. 1972. Changes in the proximate body composition of largemouth bass *(Micropterus salmoides)* with starvation. *Canadian Journal of Zoology* 50:815-819.

Noble, R. L. 1981. Management of forage fishes in impoundments of the southern United States. *Transactions of the American Fisheries Society* 110:738-750.

Novinger, G. D. 1973. *The effect of food quantity on ovary development and condition of female bluegill*. Master's thesis. University of Missouri, Columbia, Missouri, USA.

Novinger, G. D., and R. E. Legler. 1978. Bluegill population structure and dynamics. Pages 37-49 *in* G. D. Novinger and J. G. Dillard, editors. *New approaches to the management of small impoundments*. North Central Division, American Fisheries Society, Special Publication 5.

Parker, R. R. 1963. Effects of formalin on length and weight of fishes. *Journal of the Fisheries Research Board of Canada* 20:1441-1455.

Prus, T. 1975. Calorimetry and body composition. Pages 149-160 *in* W. Grodzinski, R. Z. Klekowski, and A. Duncan, editors. *Methods for ecological bioenergetics*. International Biological Programme Handbook 24, Blackwell Scientific Publications, Oxford, England.

Reynolds, J. B., and L. R. Babb. 1978. Structure and dynamics of largemouth bass populations. Pages 50-61 *in* G. D. Novinger and J. G. Dillard, editors. *New approaches to the management of small impoundments*. North Central Division, American Fisheries Society, Special Publication 5.

Ricker, W. E. 1973. Linear regressions in fishery research. *Journal of the Fisheries Research Board of Canada* 30:409-434.

Ricker, W. E. 1975. *Computation and interpretation of biological statistics of fish populations*. Fisheries Research Board of Canada, Bulletin 191, Ottawa, Ontario, Canada.

Swingle, H. S. 1950. *Relationships and dynamics of balanced and unbalanced fish populations*. Auburn University Agricultural Experiment Station Bulletin 274, Auburn, Alabama, USA.

Swingle, H. S. 1956. Appraisal of methods of fish population study—part IV. Determination of balance in farm fish ponds. *Transactions of the North American Wildlife Conference* 21:299-318.

Swingle, H. S., and W. E. Swingle. 1967. Problems in dynamics of fish populations in reservoirs. Pages 229-243 *in Reservoir fishery resources symposium.* Southern Division, American Fisheries Society, Bethesda, Maryland, USA.

Swingle, W. E. 1965. *Length-weight relationships of Alabama fishes.* Fisheries and Allied Aquacultures Departmental Series Number 1, Auburn University, Auburn, Alabama, U.S.A.

Swingle, W. E., and E. W. Shell. 1971. *Tables for computing relative conditions of some common freshwater fishes.* Agricultural Experiment Station, Circular 183, Auburn, Alabama.

Theilacker, G. H. 1980. Changes in body measurements of larval northern anchovy, *Engraulis mordax,* and other fishes due to handling and preservation. *United States National Marine Fisheries Service Fishery Bulletin* 78:685-692.

Timmons, T. J., W. L. Shelton, and W. D. Davies. 1980. Differential growth of largemouth bass in West Point Reservoir, Alabama-Georgia. *Transactions of the American Fisheries Society* 109:176-186.

Tyler, A. V., and R. S. Dunn. 1976. Ration, growth, and measures of somatic and organ condition in relation to meal frequency in winter flounder, *Pseudopleuronectes americanus,* with hypotheses regarding population homeostasis. *Journal of the Fisheries Research Board of Canada* 33:63-75.

Warren, P. J. 1958. Advice for rapid single-handed measurement of shellfish. *Journal du Conseil Permanent International pour l'Exploration de la Mer* 23:440-442.

Wege, G. J., and R. O. Anderson. 1978. Relative Weight *(Wr):* a new index of condition for largemouth bass. Pages 79-91 *in* G. D. Novinger and J. G. Dillard, editors. *New approaches to the management of small impoundments.* North Central Division, American Fisheries Society, Special Publication 5.

Weithman, A. S., and R. O. Anderson. 1978. A method of evaluating fishing quality. *Fisheries* 3 (3):6-10.

Weithman, A. S., and S. K. Katti. 1979. Testing of fishing quality indices. *Transactions of the American Fisheries Society* 108:320-325.

Weithman, A. S., J. B. Reynolds, and D. E. Simpson. 1979. Assessment of structure of largemouth bass stocks by sequential sampling. *Proceedings of the Annual Conference of Southeastern Association of Fish and Wildlife Agencies* 33:415-424.

Chapter 16
Age Determination

AMBROSE JEARLD, JR.

16.1 INTRODUCTION

"Why age fish?" is a question often asked by fisheries students. "How can fish be aged?" was the question asked by the scientists who first began utilizing fishery resources. Both questions are important and serve as the foundation of this chapter. A basic knowledge of how quickly fish grow and the relative numbers of juvenile and mature fish in a population is required to help answer questions about how fishing affects the population. It is helpful to know at what size and age a particular species reaches sexual maturity. Fishing can then be restricted so that sufficient numbers of fish can reproduce before being exposed to sustained fishing pressure. Also, it is often necessary to hold fish in a hatchery until they reach an age capable of reproducing. Knowing the average size and the size variation at different ages over several years is also important for basic comparison studies. Changes in these may be normal or may reflect a change in the suitability of the environment. This is either for the worse, as is the case with the addition of some contaminants, or for the better, as after clean-up efforts have taken effect.

16.1.1 Approaches to Aging

Three basic approaches to age determination have evolved. They can be categorized as follows: (1) an empirical approach based on direct observation of individual fish held in confinement or of fish marked and recaptured, (2) a statistical approach based on length-frequency distributions, and (3) an anatomical approach based on aging individual fish from scales, bones, and other structures.

The empirical approach to aging was initially used by fish culturists and is by far the oldest method of age determination. It relies on measuring the size of confined fish at different ages. A logical extension of this method was the discovery that the ages of wild fish could be determined if the fish had been marked, released, and recaptured (*see* chapter 11 for marking and tagging). Both of these methods depend on observing individuals and extrapolating the results to populations.

The empirical approach to aging is the least used of the three basic approaches because the cost-benefit ratio is very high. In confinement studies the amount of space needed for holding several individuals is a limiting factor. Also, confined fish seldom experience the same conditions of temperature, day length, food, etc. that fish in natural conditions experience. As a result, captive fish often have different sizes at a given age than wild fish. Furthermore, capturing, handling, and recapturing marked fish are time consuming and risky. Many fish do not survive being marked and those that do survive may not behave normally or grow as fast as unmarked fish. However, the benefits of the empirical method may outweigh the costs in studies concerned with growth, migration, and stock identification. Because this method is not widely used today, it is not discussed further in this chapter.

The second basic approach, the statistical analysis of length-frequency distributions, has been used to estimate the age of fish since the late 19th century. In 1892 the Danish biologist C. G. John Petersen showed that when the fish in a large sample are separated by size and the number of fish of each size is plotted, distinct peaks emerge. The number of different age groups were determined by counting the number of peaks. Since Petersen's work, more sophisticated methods of modal (peak) analysis have evolved, as described in section 16.7.

The third basic approach is anatomical, using body parts to age fish. The first serious account of the theoretical and empirical basis of this method was published in 1759 by the Reverend Hans Hederstrom, who demonstrated that the age of a fish may be discerned from the marks on its vertebrae. Because this important article was overlooked for almost two centuries, Dahl (1910) gave credit for the discovery of age marks on hard fish parts to C. Hoffbauer and J. Rebisch. Hoffbauer utilized scales to age common carp in 1898, and Rebisch utilized otoliths (ear bones) to age plaice in 1899. Since Dahl's review many independent studies and reviews have been published (e.g., Bagenal and Tesch 1978; Ricker 1979). A quick tabulation of various studies clearly shows that the anatomical approach is the most widely used and preferred method of age determination.

Basic to the anatomical method of aging fish is the existence of regular periodic growth markings in hard body parts to which a regular time scale can be assigned. This is a concept analogous to determining the age of a tree by counting annual rings in a cross section through the base of the trunk. As in trees, seasonal changes in the growth of fishes in temperate waters are generally recorded as contrasting bands in body parts such as scales, otoliths, fin rays, spines, and bones. In bivalve molluscs the contrasting bands can be found in the shells.

Many recent developments in aging are concentrated on the anatomical approach. Considerable effort is being placed on developing a technique for automatic age determination of fish. This technique uses sophisticated image-analysis instruments coupled with a computer to read the marks on a fish scale or otolith. This technique assumes that light-sensitive marks on anatomical parts can be recognized and will appear in a predictable pattern. Advantages of the technique include increased objectivity, quicker assessment of many time-consuming measurements used in fishery analysis, and rapid aging of large volumes of age samples. *See* Fawell (1974) for a discussion of some aspects of image analysis.

Since Panella (1971) discovered that the otoliths of some tropical and temperate fishes contained daily growth rings (or DGIs—daily growth increments), there has been increasing interest in the use of these rings for age determination. In temperate regions, most work by this method has been done with larvae. Information gained from accurately aging wild larvae can be applied to studies of the growth, mortality, and hatch dates of wild populations. The study of daily growth is also useful for adult fish in tropical regions. DGIs and rings formed in response to lunar activity have been noted on their otoliths. Although the structures of tropical species do not exhibit distinct seasonal zones, annual or subannual slow growth zones may form as the result of spawning activity (as is also true for temperate species) or, possibly, in response to changing conditions during rainy and dry seasons (Panella 1974).

Examination of otoliths from larval fish requires the use of high magnification or scanning electron microscopy (SEM) to reveal the fine rings. Preparation for SEM studies may include grinding, polishing, and etching with decalcifiers (*see* Radtke and Dean 1981). Otherwise the whole otolith is viewed in resin at 600x-1000x. Because DGIs may become evident at different times (e.g., at hatching or yolk sac absorption) in larval development depending on species, a full time series from hatching should be studied. When the time of first DGI formation is established, the daily nature of the rings formed thereafter can be verified (e.g., Barkman 1978; Brothers et al. 1976; Lough et al. 1982).

16.1.2 Age Terminology

Writings about age determination of fish and shellfish are filled with disagreements and conflicting information. Some of the confusion is caused by inconsistent terminology used to report results. Careful consideration must be given by age readers to standard terminology and notation (Box 16.1). *See* Bagenal and Tesch (1978, pages 105-106) and Jensen (1965) for a discussion of terminology used among fish age investigators.

The basis of most aging is the annulus or year mark. The annulus is the result of a slowing of the growth rate in response to such factors as colder winter temperatures. Environmental and physiological factors cause variations in time of annulus formation. The response to these factors (change in growth rate) may vary among individuals, populations, and species.

There are intra- and interspecific differences as well as geographical differences in the time of year when an annulus appears. Since biologists must be able to compare their findings, it has become necessary to standardize the birth date of fish. Internationally accepted convention is that January 1 is designated as the birth date for fish in the Northern Hemisphere. Therefore, a winter growth zone forming on the edge of scales, otoliths, fin rays, spines, etc. is designated as an annulus on January 1, even if the zone is not complete. July 1 is the corresponding date for year-mark formation on hard parts of fish in the Southern Hemisphere. The assignment of an arbitrary birth date, other than the biological birth date, may not be the best system in some instances. In some fish, annuli may appear months after the assigned birth date and may be misinterpreted by different workers.

Reporting the age of fish can also be confusing. By convention, some fishery biologists designate the age of a fish in Roman numerals, e.g., age I for a one-year-old, age II for a two-year-old, etc. However, the less cumbersome procedure of designating the age of a fish by using Arabic numerals is becoming more prevalent (Ricker 1975). In either case it is consistency that is important.

Ages of fish are expressed by numerals corresponding to the number of annuli or completed years of life. Fish in their first year of life, before their first January 1st birthdate (those whose calcareous structures are without an annulus), are designated as members of age-0 group. A fish in age group I has completed one year or less of growth from time of hatching to the January 1st birth date and has entered its second growth season. During the growth season after annulus formation, any growth of the age structure between the last annulus and the edge is termed "+" growth. If the age group I fish is collected in the summer or fall, its age will, therefore, be I+. If the same fish is caught during the next winter, after January 1, but before the second annulus is complete, it will be assigned to age group II even if the edge still only indicates "+" growth. Using a conventional birth date regardless of minor variation in time of annulus formation ensures that fish hatched in the same calendar year are members of the same age group (also called year class, brood, or cohort). For example, the 1980 V

Box 16.1   Terminology of Aging

TERMS	SYNONYMS	DEFINITION

Age group	- age class
	  cohort
	  year class
		Fish of the same calendar age, hatched in the same year.

Annulus	- band
	  ring
	  year or age mark
	  zone (winter zone)
		Slow-growth zone on age structure, considered to form annually and counted for age determinations.

Center	- central kernel
	  nucleus (otolith)
	  origin
	  focus (scale)
	  central lumen (spine)
		Point of origin of age structure.

Check	- accessory ring or mark
	  false ring or annulus
	  secondary ring or zone
		Zone or ring on age structure, considered to form subannually; not counted for age determinations.

Circulus	- ridge	Raised, mineralized plate-like structure on the surface of a scale (appears as ring around focus).
Edge	- margin	Outer periphery of an age structure; represents most recent growth.

Opaque	- summer zone or ring
	  fast-growth zone
	  optically dense zone
		Optically dense zone on age structures formed during periods of active growth.

Radius		Groove-like depression radiating from the focus to the edge on some scales.
Regenerated scale		Scale formed to replace one previously lost.
Split (annulus)	- double ring	Annulus composed of two or more closely-spaced zones formed within one winter season.

Translucent	- winter zone or ring
	  hyaline zone
	  slow-growth zone
		Zone of low optical density formed during periods of slow growth.

cohort, the 1981 VI cohort, and the 1983 VIII cohort are all members of the 1975 year class.

Much diversity is found in the designation of ages and the methods of designating age groups for fish that migrate between fresh and salt waters. Special notation is used to designate the time in each environment. In the United States, salmon age designation usually follows either the European method or the Gilbert and Rich method. Under the European method, age 2.3 indicates that the fish had lived two years of freshwater life and three years of saltwater life or five years total. With the Gilbert and Rich method, the same fish would be age 5_2. The 5 indicates the fish had lived 5 winters from the time its parents spawned to the time it was captured. The subscript 2 indicates the number of winters between the time that the parents spawned and the time that fish migrated seaward. Many other terminologies have been used in the past, and much confusion still exists. The special problem associated with aging salmon and using appropriate terminologies is thoroughly discussed in Koo (1962) and Mosher (1968).

16.2 COLLECTING ANATOMICAL STRUCTURES FOR AGING

The value of a thorough literature review on the life history of the species selected for investigation cannot be overemphasized. The literature will indicate (1) whether or not a method of age determination has been established for the species in question, (2) which age determination approaches have been utilized, and (3) which hard structures have been used.

If the literature search indicates previous aging studies have already been done on the species of interest, then a thorough review of these papers should reveal the best method for the purposes required. For many fish, certain body parts have become accepted as the most useful for age determination (Table 16.1).

If the literature search shows that no age determination process yet exists for the species in question, then you must make an independent decision. Traditionally, most people start with the anatomical approach. Next you must decide which of the hard body structures (e.g., scales, otoliths, vertebrae, spines, fin rays, cleithra) best shows the periodic changes in growth and most clearly reflects the age of the fish. You must examine each structure to determine if it shows a recognizable pattern and if a regular time scale can be assigned to that pattern, beginning as early in the life of the individual as possible. After a random sample of the fish has been obtained using methods discussed in chapter 1 of this book, choose various hard structures from several (5-10) fish of each age group (larvae, age-0, yearlings and older juveniles, young adults, and old adults). When deciding which hard structure is most suitable, you must consider the time and effort involved in collecting and preparing the structure.

16.2.1 Scales

Collection. Although scale samples are the easiest hard structure to collect from fish and, therefore, the most popular, removal must be done carefully and by a standard procedure. Some useful instruments are a pair of forceps, a blunt knife (table variety), and a sharp knife (preferred by some studying freshwater and tropical species).

Only scales from particular areas on a fish are suitable for aging (Fig. 16.1). The area generally used for scale sampling is the middle region of the side of the body

Table 16.1 Anatomical structures commonly used for age determination

Family	Common Name	Structure
Squalidae	Dogfish sharks	Dorsal spine
Rajidae	Skates	Vertebral centrum
Acipenseridae	Sturgeons	Fin ray
Elopidae	Tarpons	Scale
Anguillidae	Freshwater eels	Otolith
Clupeidae	Herrings	Otolith
Engraulidae	Anchovies	Otolith
Salmonidae	Trouts	Scale
Esocidae	Pikes	Scale, cleithra
Cyprinidae	Minnows and carps	Scale
Catostomidae	Suckers	Scale, fin ray
Ictaluridae	Bullhead catfishes	Pectoral spine
Batrachoididae	Toadfishes	Otolith
Gadidae	Codfishes	Otolith
Cyprinodontidae	Killifishes	Scale
Atherinidae	Silversides	Otolith
Percichthyidae	Temperate basses	Scale
Serranidae	Sea basses	Scale
Centrarchidae	Sunfishes	Scale
Malacanthidae	Tilefishes	Otolith
Haemulidae	Grunts	Otolith
Sparidae	Porgies	Scale
Sciaenidae	Drums	Scale
Cichlidae	Cichlids	Scale
Scombridae	Mackerels and tunas	Otolith, fin ray
Bothidae	Lefteye founders	Otolith, scale
Pleuronectidae	Righteye flounders	Otolith, scale
Mugilidae	Mullets	Scale
Lophiidae	Goosefishes	Vertebrae, ray
Carangidae	Jacks	Otolith

(Bagenal and Tesch 1978). For some species, however, the preferred area is below the lateral line, near the point of the pectoral fin when the fin is pressed to the body (Carlander 1982). You should experiment by selecting and examining scales from several areas to determine where consistently large and symmetrical specimens occur. An area likely to shed scales or that has irregularly shaped scales is a poor site choice for collecting scale samples. *See* Lagler (1956) and Bagenal and Tesch (1978) for additional discussions of scale collection methods.

Before collecting scales, remove mucus, dirt, and epidermis from the area by gently but thoroughly wiping in the direction of the tail with a blunt-edged knife. Loosen the scales by a quick, firm scraping motion in the direction of the head, and remove the scales on the blade of the blunt knife. Then insert the knife blade between the liners of a scale (coin) envelope to remove the scales. Envelopes without glue on the flap are preferred. An alternate method is to place the point of a sharp knife firmly on a scale and push toward the tail to remove the scale. With this method, as many scales as desired may be collected one at a time without scales being lost. With either method, clean instruments of all scales from one fish before you start on another.

Preparation. Aging is done with either raw scales or scale impressions. Raw scales are generally less desirable because they: (1) are covered with dried and pigmented

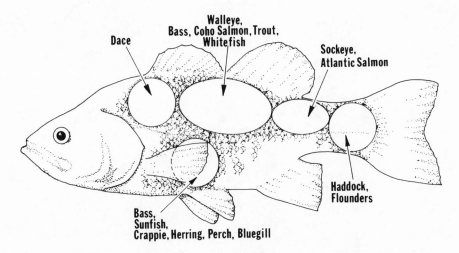

Figure 16.1. General areas on a fish where scales of various species may be removed.

residue from the fish and miscellaneous dirt picked up through handling, (2) are generally translucent rather than transparent, thus interfering with viewing under transmitted light, and (3) are not flat, causing problems of illumination during microscopic examinations. In situations where raw scales must be used, mount 6 to 10, sculptured side up, between glass slides labeled and held together with masking tape.

A more satisfactory method is to make an impression of the outer surface of the scale on a plastic slide. Scales are placed between plastic slides and passed through manual or power-driven roller presses. Clear cellulose slides approximately 1 mm thick are generally used (Smith 1954), but the National Marine Fisheries Service Woods Hole Laboratory currently uses a double plastic film (laminated plastic) with a thin (2 mils) soft polyethylene layer over a thick (6-8 mils), harder vinyl substrate. This material results in accurate scale impressions because it is soft enough to make impressions without the use of heat, heavy pressure, or softening chemicals.

Scale impressions may be magnified and viewed with transmitted light and have several advantages over direct use of scales. Several scales may be impressed at the same time on one slide, and the impressions showing the clearest scale features can be aged. The impressions are clean, even if the original scale was not, and they are easy to store and handle. (One option is to slip the plastic impression into the already properly labeled coin envelope.) The image of the scale is generally flatter than the original scale, although it does retain enough depth features to cause depth-of-focus problems at high magnifications. Other disadvantages of using cellulose acetate are that: (1) the technique can be time-consuming, and (2) scales having delicate and shallow sculpturing (e.g., yellowtail flounder) can be problematic. If laminated plastic is used, thick scales can cause distorted images.

16.2.2 Otoliths

Otoliths, or earstones, of which there are three pairs of varying size in fish, are flat-oval to spindle-shaped structures found in the heads of bony fishes. The largest pair, called the sagittae or saccular otoliths, is preferred for determination of annual year marks. More recently all three pairs have been used for counting daily age marks.

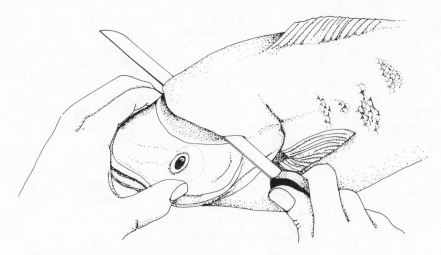

Figure 16.2. Typical method of cutting head for otolith removal.

Collection. Collecting otoliths necessitates removing the otoliths from behind the brain. For most fish this may be done in the following manner. Grip the head firmly by the eye sockets with one hand and cut the top of the skull slightly behind the eyes, down and back to the upper edge of the gill cover (Figure 16.2). A strong, sharp knife is sufficient for small to large fish, but a saw is sometimes necessary for larger or hard-headed fish, such as pollock and flathead catfish. For fish with poisonous spines, e.g., redfish, catfish, and some tropical fish, wear a heavy glove to hold the fish firmly by the head while cutting the skull. Open the head by pressing down quickly on the nose. If the angle of the cut is correct, the large sacculus otoliths should be visible behind the brain. Searching is necessary if the cut does not reveal the otolith.

Making an accurate cut for removing otoliths from flatfish is more difficult. First, locate the bony ridge between the eye and the edge of the gill cover, and imagine a line

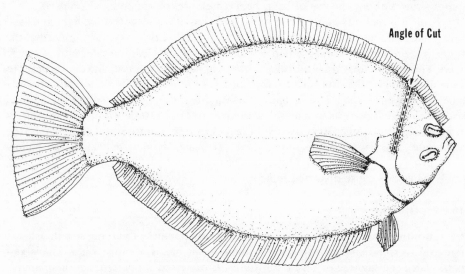

Figure 16.3. Typical flounder showing where head may be opened to remove otoliths.

extending from the end of this ridge to the dorsal fin (Figure 16.3). Open the skull by pressing a sharp blade along this line until the bone is penetrated. Do not press down too far or you will shatter the otoliths. Open the head by bending down the nose of the fish. One otolith should be visible; the other can be found underneath by probing with tweezers.

After removing the otoliths, store them dry in a coin envelope or in alcohol or glycerin. Storing otoliths in a medium of 2:3 glycerine:alcohol will clear thicker otoliths, enabling the viewing of rings at a later date. Do not store otoliths in formalin or acid; these chemicals dissolve otoliths.

Advantages of using otoliths for aging are that: (1) they form during the embryonic period and, therefore, reflect all life history events; (2) in some cases, they show age more clearly than scales and are often better than scales for aging older fish; (3) a rather small sample size can be used; and (4) all fish of a species have otoliths that are similar in shape.

A major disadvantage of using otoliths is that their removal requires killing the fish. This becomes a problem when fish are valuable as commercial fish, sport fish, trophies, or endangered species. There are only two large otoliths per fish, and they crack easily. Collecting otoliths requires more skill and time than collecting scales.

Preparation. The technique for preparing otoliths is variable, and in most instances it is modified for the convenience of the investigator. Some otoliths may be viewed in glycerin or alcohol while others must be sectioned. Tiny otoliths, such as those of herring, mackerel, and many freshwater species, can be imbedded in resin using molded black plastic trays with rows of circular depressions. Assuming that the ring structure on the whole otolith can be seen, a resin will enhance the contrast between summer and winter zones to an extent not possible with the simple use of alcohol.

The earlier-formed annuli on otoliths may be obscured by subsequent calcium deposition. Therefore, it may be necessary to examine a cross section of the structure. For some species with large otoliths (e.g., cod and haddock), the process is to break the otolith in two. Break the otolith at the sulcus (nucleus center) by applying pressure with the thumb or with a pair of pliers. A variation of this technique is to bake the otolith before breaking. Baking enhances the annual marks.

More involved sectioning techniques require the use of sandpaper or a saw (e.g., a jeweler's saw). Otoliths may be mounted on a glass slide with a thermoplastic cement. Otoliths that do not clearly reveal all annuli under low power magnification using reflected light can be ground down with fine sandpaper or a dentist's drill until all annuli are visible.

Many otoliths with complex growth patterns must be carefully cut into very thin sections for accurate age determination (Figure 16.4). Nichy (1977) developed a method that takes less than two minutes to cut a 0.175-mm section from an otolith using a low speed diamond-blade saw. Other reliable techniques for sectioning otoliths are available, as reported in Bagenal and Tesch (1978).

16.2.3 Spines and Fin Rays

In bony fish without scales or satisfactory otoliths for aging, some other aging structure must be selected. In catfish, for example, the choice is usually made between the pectoral spines (dorsal in some studies) and the vertebrae (Marzolf 1955). Pectoral spines are preferred over vertebrae, however, because collection does not require killing the fish, collection in the field and laboratory is easier, preparation is quicker,

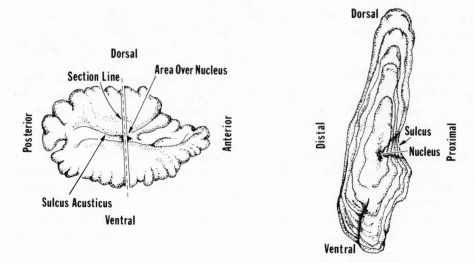

Figure 16.4. (Left) Redfish otolith. (Right) Transverse section from a redfish otolith.

and fewer false annuli are present. The only disadvantage of using the pectoral spine is that the first annulus may be obscured in older fish. For some fishes, such as suckers and sturgeons, the best structure for age determination is the fin ray.

Collection. Spines can usually be removed free of tissue, except for a thin layer of skin and blood, and require little special treatment or preparation. Two methods are used to detach the spine (left or right consistently). The first method is to simultaneously twist and depress the spine toward the body of the fish at the articulating process. The second method is to grasp the spine with a pair of pliers for large fish (or forceps for small fish), pull outward to loosen the joint, and rotate clockwise for left spines and counter-clockwise for right spines. For larger fish it may be necessary to cut the muscles surrounding the spine. Dry the spines in open air before storing in coin envelopes.

To collect a fin ray, remove it just below the point of articulation with a scissor, knife, or pliers. Remove excess membranous tissue by scraping, and then soak the fin ray in household bleach to remove any remaining tissue traces. Dry in air and store as for spines.

Sectioning. For sectioning spines, the position of the cut is critical (Figure 16.5) in order to retain annuli. Channel catfish pectoral spines should be sectioned at the distal end of the basal groove (Sneed 1951). Sections cut through the articulating process will help distinguish annuli in the peripheral region of basal groove sections. For flathead catfish, the articulating process of the pectoral spine is the preferred location for sectioning (Turner 1980). Instruments used for sectioning may be similar to those described by Witt (1961). Section thickness depends on many factors, but, in general, the thinnest sections are most useful if they will be viewed with transmitted light. Some cross sections may be examined under water in a shallow dish. Other cross sections will require grinding and polishing before microscopic examination and/or measurement. Grinding and polishing may be performed on an individual cross section held with the fingers or several cross sections mounted on glass or plastic slides. *See* Sneed (1951) for an excellent discussion of spine methodology.

Fin rays should be sectioned near the base with a jeweler's saw. Those to be read

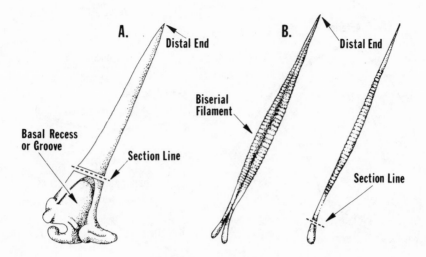

Figure 16.5. *A.* Channel catfish spine showing location of cut for section. *B.* Diagram of a fin ray (biserial) showing the location of cut for section, after filaments are separated.

without grinding should be 0.4-0.6 mm thick. Sections can be mounted on glass or plastic slides with a cement. Thicker sections, e.g., 1.0 mm or less, can be polished with a fine grit sandpaper. *See* Skidmore and Glass (1953) and Pycha (1955) for more information on preparing fin rays for age determination.

Both spines and fin rays can be oven-dried before sectioning. Dried structures are then soaked in Axion (household detergent with enzyme) and rinsed in ammonia (to stop the enzyme action). This procedure completely cleans the structures, removes grease and oil, and speeds sectioning time.

16.3 COLLECTION AND PREPARATION OF BIVALVE SHELLS

Age determination of molluscs has evolved from a relatively simple visual examination of external shells of bivalves to rather complex microstructural examinations (Lutz and Rhoads 1980). Seasonal changes in the growth rate are often reflected in zones or bands in shell structures of clams, oysters, scallops, and mussels. Zonation is similar to that found in hard structures of finfish, with a light band forming in the early part of the growing season followed by a narrow, dark band in the winter. As in finfish, the first annulus of shellfish is difficult to discern or lacking, and accuracy of age determination is lower in very old molluscs.

Because bivalves spend most of their lives as benthic organisms, they are usually collected with dredging apparatus, and the shells must be cleaned before they can be aged. Shells should be shucked of their meats and placed in a household bleach solution before aging. Soaking can range from a brief dip for relatively clean shells to a lengthy storage period. Check shells frequently during long soaking periods to avoid over-bleaching annual marks. After bleaching, rinse shells in tap water and, for some shells, scrub with a brush. Let shells dry before storing for later sectioning. Both halves of the shells are usually cupped together for storage. This permits aging of both halves if it is necessary and reduces the volume and breakage of the shells.

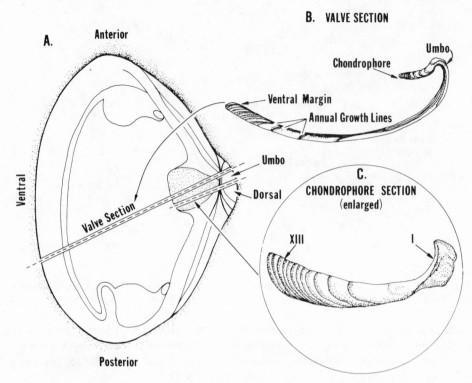

Figure 16.6. *A*. Inner surface of the valve of the surf clam indicating location of cut for chondrophore and valve sections. *B*. Valve section. *C*. Chondrophore section.

Examination of the internal shell structure requires that it be sectioned. Ropes and O'Brien (1979) found a correspondence between the number of annuli in the chondrophore and in the shell of surf clams (Figure 16.6). Their findings formed the basis for a unique and expedient means of accumulating age data for this species. Briefly, the methodology for preparing thin sections of a chondrophore is as follows:

1. Using a pair of diamond blades, excise a thick piece of the chondrophore from the right valve hinge.
2. Using wettable silicon carbide paper, grind and polish the anterior cut surface of the piece down to the umbo.
3. Air dry the piece and mount on a glass slide by applying a two-part epoxy glue to the polished surface.
4. Using a low speed (0-300 rpm) saw, cut a section about 0.02 mm thick.

16.4 AGING PROCEDURES

After anatomical samples have been collected and prepared, the next step is to determine their age. One important consideration is time of annulus formation. In order to establish this time, it is helpful to supplement interpretation of the structure with information on spawning, migration, and feeding habits of the sampled fish population, as well as environmental data such as latitude and water temperature range. All of these factors influence growth rate and, therefore, zone formation on hard structures. Depending on latitude and environmental conditions, seasonal growth may shift from the general pattern and result in annuli forming in the spring or fall, rather than in the winter months. For species of the Temperate Zone where there are distinct seasonal changes, clear annular zones form during the colder months of the winter. This is especially true for species in freshwater environments where changes in water temperature and chemistry are more radical than in marine environments. Time of annulus formation may also vary with age. Typically, the younger fish of a population resume growth earlier than older individuals and may begin annulus formation earlier as well.

Another consideration in aging hard structures is the formation of anomalous rings such as checks or split annuli, which occur in response to physiological changes or stresses that slow growth. Checks are generally formed during rapid growth periods, whereas split annuli are formed during slow growth periods. Often these accessory marks are difficult to distinguish from true annuli and can lead to overestimating age. Validation is important, therefore, to verify which age marks are true annuli.

Biases common to all structures used to determine the age of a fish are errors associated with (1) missing the first annulus, (2) crowding of annuli with increasing age, (3) overestimation of age due to the presence of anomalous rings, and (4) loss of peripheral annuli due to resorption or erosion.

16.4.1   Aging Scales

To determine age you will need to magnify the scale. Scales or scale impressions may be examined with a low-power microscope, a microfiche reader, or with a microprojector, the most popular technique. To project a scale, select a magnification that will accommodate all scales of a sample. Scales or impressions should be oriented with the sculpted surface toward the light source. Be sure the scale or impression is held flat during projection and measurement. Usually the preferred orientation is anterior field up, posterior field down.

Pattern recognition. Scale features used for aging are fine ridges called circuli (dark lines). Circuli are laid down in a circular pattern around the scale center or focus. Several circuli are added to the scale each year. During the cold months growth is slow and ridges are laid down close together. Fish continue to grow throughout their life; therefore, this pattern is repeated each year. The outer edge of a group of closely spaced circuli indicates the termination of that year's growth, and this point is called the year mark or annulus. The age of a fish is determined by counting the number of annuli. *See* Box 16.2 for a fuller discussion of scale pattern recognition.

Scale types of most fish are either cycloid or ctenoid (Fig. 16.7 and 16.8). Cycloid scales have circuli which extend completely around the scale edge as growth continues. The anterior field of the scale, which is embedded in the skin and covers most of the surface area, is usually used for aging. Fish with cycloid scales include cod, haddock, salmon, trout, whitefish, pike, minnows, and most other soft-finned fish. On ctenoid scales, the field posterior to the focus has no clearly defined circuli and may be obscured by prominent spines, or ctenii. The anterior field is usually used for age determination. Fish with ctenoid scales include bass, sunfish, perch, some flounder, and most other spiny-finned fishes.

Regenerated scales, which develop as the result of prior scale loss, should not be aged. They do not contain the circuli present in the original scales. The central part of regenerated scales contains a clear, window-like area surrounded by irregular circuli.

Box 16.2   Scale Growth Pattern Recognition

The following criteria are used to identify a true scale annulus and are readily recognized after experience:

1.   Relative spacing of the circuli. *See* text.
2.   "Cutting over" or "crossing over" of the annulus across previously deposited circuli, particularly on the lateral edges. On some scales, the outer circuli tend to flare outward or end abruptly on the side of the scale. When rapid growth begins again, the new circulus is complete and appears to cut across earlier circuli. Often associated with crossing over is a thin clear zone with no circuli ridges which extends across the anterior field of the scale. The erosion or absorption of the scale edge during slow growth periods also may result in cutting over. (Fig. 16.9)
3.   Bending or waviness of unsegmented circuli (Clupeidae). A similar description for sunfish is "bell marks," which often form at the radii. These appear as bell-shaped blank spots located at the junction of the radii and the annulus.
4.   Circuli counts. An average number of rows of circuli may be associated with a given annulus. This number decreases with age.
5.   Changes in circulus shape. Circulus segments may be thicker, more wavy, or fragmented during active growth. During periods of slow growth these segments are thin, straight, and less fragmented (Fig. 16.8). Changing of focus on the microprojector will show the change in circulus thickness that is present in the annulus.
6.   Radii. Radii are scale flexion lines extending in an anterior/posterior direction. New radii may form at the outer edge of an annulus or existing radii may bend or branch (Fig. 16.9). This criterion is not generally reliable.

The following criteria are used to identify accessory rings or checks:

1.   A ring with closely spaced rows of circuli or "cutting over" that is discontinuous around the edge of the scale.
2.   A ring with fewer rows of circuli than are present in obvious annuli.
3.   Circuli of the wrong type (broad rather than narrow as found in a true annulus) make up the ring.

A split annulus may be identified by:

1.   Unusual spacing of rings, especially in a paired pattern.
2.   Observation of fast growth on the edge during the winter months.

16.4.2   Aging Otoliths

Otoliths of many species can be viewed with low magnification (15-20x). Improvise with lighting, magnification, and immersion in various clearing fluids (e.g., Fotoflo, clove oil, alcohol, or glycerin) to enhance the clarity of age marks. Clove oil will enhance contrast between opaque and translucent zones if the opaque zones are weakly defined.

Otoliths from older fish and sections of otoliths are more difficult to examine. A recommended technique for microscopic examination of thin (e.g., 0.2 mm thick) transverse sections of otoliths is as follows: (1) mount the section on a dark background (e.g., black paper), (2) moisten with clove oil to enhance the contrast between opaque and translucent zones, and (3) view at an appropriate magnification

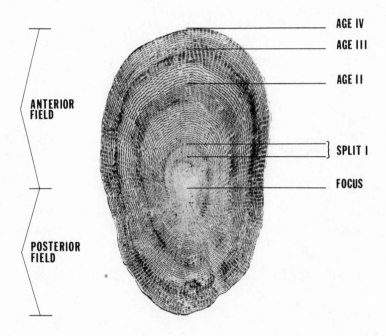

Figure 16.7. Cycloid scale of a haddock.

using reflected light. Use the part of the otolith where rings seem most distinct and condensed. If sections require transmitted light for best resolution, place them on a transparent or translucent surface.

Pattern recognition. Otolith form is specific and varies from a flat oval to a spindle shape (Blacker 1974). The most prominent external feature of the otolith is a central groove, or sulcus acusticus. It extends from the anterior to the posterior end of the inner (concave) surface of the otolith. This groove is very useful in locating the center or nucleus of the otolith (Fig. 16.4).

The formation of growth zones in otoliths and other bones follows the general pattern outlined earlier for scales. As in scales, growth is concentric around a central nucleus, and distinct bands are present. When whole otoliths or transverse sections are viewed under a microscope, the layers making up spring and summer months of active growth appear as white opaque bands (zones) under reflected light. Layers laid down during slow growth periods (usually the fall and winter months) appear as dark or translucent hyaline bands (zones). A light and dark band together represents one year of growth. Age in years is usually determined by counting the number of dark bands or annuli (Fig. 16.10). The following criteria should be considered when aging otoliths: (1) count the widest, strongest, and most distinct hyaline zones on the otolith—the width decreases with age, (2) hyaline zones should be consistent in formation around the periphery of the whole otolith, (3) the opaque spaces between hyaline zones become narrower from the nucleus to the edge, and (4) on transverse sections true annuli are usually continuous through the central groove.

As with scales, accessory markings such as checks, splits, and false annuli are common. Checks may appear as thin or discontinuous hyaline zones around the

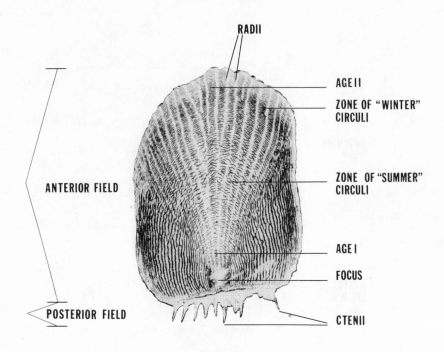

RADII

ANTERIOR FIELD

POSTERIOR FIELD

AGE II

ZONE OF "WINTER" CIRCULI

ZONE OF "SUMMER" CIRCULI

AGE I

FOCUS

CTENII

Figure 16.8.   Ctenoid scale of a yellowtail flounder.

periphery of the otolith. Abnormally shaped crystalline otoliths should not be used for age determination. On these otoliths either calcium has been resorbed or disruption of the otolith membrane has occurred (Bilton 1974).

16.4.3   Aging Other Hard Structures

Various bones other than the otolith have been used for aging different fishes (Menon 1950). Harrison and Hadley (1979) favored the cleithra of muskellunge over scales, Quinn and Ross (1982) found fin rays of white suckers to be more accurate than scales, and Marzolf (1955) found both pectoral spines and vertebrae useful for aging channel catfish. However, the time and materials required for a satisfactory preparation of permanent study sections has been a serious objection to the widespread use of these structures.

These structures may be examined with transmitted or reflected light. As with otoliths, a fluid may be required to enhance the resolution of the opaque and translucent bands. Annual bands in channel catfish spines appear as concentric rings around the hollow central cavity. Zonation in vertebrae is concentric around the core of the centrum. Both opercular and dentary bones show banding along their growing edge.

As for spines, the first ring on the fin ray in older fish may be faint. In cross sections made too far above the base, the first annulus may be difficult to discern or may be missing. Finding a fixed central point for making measurements is also a problem. Another drawback is that different rays of the pectoral fin may not show a growth pattern as sharply defined as others. Therefore, considerable practice is

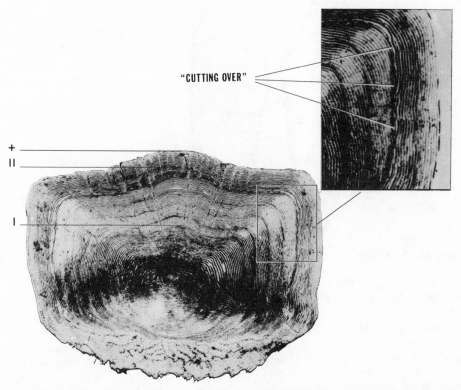

"CUTTING OVER"

+

II

I

Figure 16.9. Scale of a bluefish showing "cutting over" of circuli; insert shows "cutting over" on outer edge of second annulus.

necessary when selecting the best fin ray for aging.

Accessory checks may occur on these structures. These, as well as abnormalities in bone formation, should be noted and avoided. Bones are particularly prone to extra protrusions and growths resulting from diseases or injuries.

Elasmobranches. Most anatomical structures used for aging bony fishes are not useful for aging elasmobranches. In this group spines are uncommon, teeth are constantly renewed, scales are unsatisfactory, and most of the skeleton is cartilage. Mineralized vertebral centra are the only hard structures consistently present among elasmobranches (Gilbert et al. 1967), and these are not well-calcified. Zonation in the centrum of elasmobranches is similar to that found in the vertebrae of bony fishes. Annual marks are laid down as concentric rings reflecting fast or slow growth. A year zone consists of a lightly calcified opaque zone and a dark translucent zone when the centrum is viewed under reflected light. Accessory checks are frequently observed in cross section. Studies of elasmobranch aging include Ishiyama (1951), Richards et al. (1963), Holden and Vince (1973), and Waring (1980).

16.5 VALIDATION

Validation involves using several independent techniques to age the same fish. This provides a verification so that the final age can be assigned with some confidence. Although using hard structures in age determination is the most commonly used method, other methods can verify or challenge the original ages assigned. The

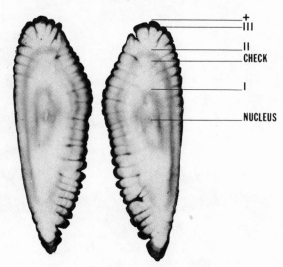

Figure 16.10. Whole otoliths, cleared in glycerin, of silver hake.

usefulness of a thorough literature search for validation of age marks cannot be overstated. For an excellent description of validation methods, *see* Brothers (1979, 1982).

Some standard methods for age validation are (1) length-frequency analysis, (2) modal-progression analysis (e.g., following the relative abundance of a dominant year class from year to year), (3) examination of known-age fish [ones marked and recaptured or grown in confinement (Taubert and Tranquilli 1982], (4) determination of periodicity of annual zone formation (i.e., examining the edges of samples taken at different times of the year), (5) comparison of ages derived from different hard structures (e.g., Marzolf 1955; Harrison and Hadley 1979), (6) comparison of backcalculated lengths-at-age determined from hard structures with lengths calculated from mark-recapture or length frequencies, and (7) comparison of age-0 and age-I fish from different sources to validate the interpretation of the first annulus. A more recent validation technique is to count the number of daily rings between successive annuli. For a detailed study of validation of otolith aging, *see* Mayo et al. (1981).

The results of validation studies are useful in establishing aging criteria for interpretation of annuli, as well as for pinpointing significant life history events. Occurrence of accessory marks between validated annual marks may be excellent indicators of spawning, extreme temperature changes (e.g., thermal discharge into a reservoir), lack of food, change in habitat, or the presence of pollution.

16.6 BACKCALCULATION

Thus far the discussion on the use of fish body structures has concentrated on aging. Another use of these structures involves measuring the dimensions of annual rings on the body parts to estimate the size of the individual fish at earlier ages. This process, called backcalculation, assumes that there is a proportionate relationship between how much the fish increases in length and how much the hard structure increases in size. All methods depend upon knowledge of the correlation between body length and age structure size. Two primary methods are the proportional and regression techniques. Lagler (1956) reports a summary of earlier methods and more

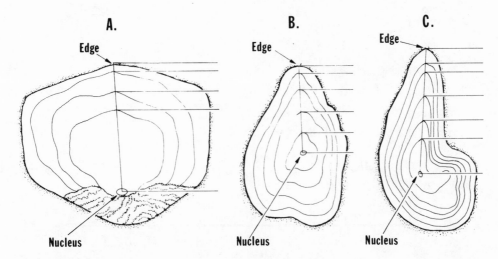

Figure 16.11. Typical scale *(A)*, whole otolith *(B)*, and fin ray section *(C)* showing radius of measurements for backcalculation.

recent discussions are found in Hile (1970), Bagenal and Tesch (1978), and Carlander (1981).

The whole otolith, rather than a section, is preferred for backcalculation because of the difficulty of sectioning otoliths precisely at the nucleus. Sections of fin rays and spines must expose the center of the structure. Measurements for backcalculations from otolith, spines, fin rays, and bones are made with an ocular micrometer. When scales are used, measurements are easiest using a ruler attached to a microprojector or a nomograph (Carlander and Smith 1944). In general, measure the radius of each ring from the center of the structure to the outer edge of each annulus or the total diameter of each ring through the center. Measure the radius or diameter along the longest axis (e.g., largest lateral lobe in spines), assuming that there are no irregularities on that part of the structure (Fig. 16.11). Indentations, erosion (crystalline areas), or unusual proturberances that are likely to interrupt growth or measurements should be avoided. Measure scales from the focus to the anterior edge along a consistent line. Standardizing the direction of measurement and plane of sectioning is essential. Variation in the point at which the section is taken causes variations in measurements that can invalidate any calculations. Scales taken from several points on the fish can also result in variations in backcalculated lengths.

There are many methods of calculating fish length at successive annuli. The earlier cited references, as well as Carlander (1950), and Whitney and Carlander (1956), provide formulas and information on the methods to use. It is important to note that, unless the age determinations are correct, the backcalculations will be misleading.

16.7   AGE DETERMINATION FROM SIZE FREQUENCY DISTRIBUTION

In the event that individual fish cannot be aged reliably, population size structure may be analyzed for indications of age groupings. This alternative method of age determination began with speculation that peaks in size-frequency distribution plots

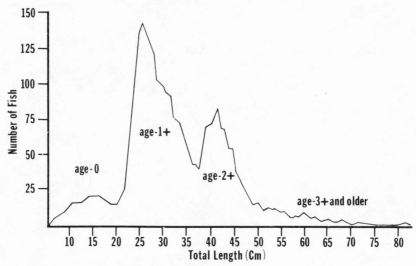

Figure 16.12. The length frequency distribution of a catch of haddock, showing the different size groups of fish caught and corresponding year classes [after Lux (1971) with permission].

represented the modes of year classes (cohorts). Fish hatched in the same year tend to be in the same size range, with most fish being close to an average size. There tends to be a statistically normal distribution of sizes around a modal (most frequent) size. There should be recognizable peaks in the size distribution of a population sample as long as the sample is unbiased and some other assumptions are met that are discussed later in this section.

Table 16.2 Graphical methods for separating polymodal size frequency distributions into cohorts

Methods	Limitations, Advantages	Reference
"Petersen method": simple inspection of modes	- quick-and-dirty estimation - often unreliable (e.g., a small year class may be missed entirely)	
Probability-paper plotting plus trial-and-error estimation of parameters to fit plot	- most often used in fisheries research - assumes normal distribution of size-at-age	Harding 1949; Cassie 1950, 1954
Semi-logarithmic paper plotting plus fitting of parabolas to plot	- assumes normal distribution of size-at-age	Tanaka 1962
Natural logs of size frequencies plotted, straight lines fitted to first significant differences	- assumes normal distribution of size-at-age	Bhattacharya 1967
"Method of successive maxima": modal classes split from left to right side of size distribution plot	- assumes only symmetrical distribution of size-at-age - could introduce bias in mean size-at-age - provides no statistical data for abundance of cohorts	Gheno and LeGuen 1968

Table 16.3 Statistical methods (computer-assisted) for separating polymodal size frequency distributions into cohorts

Methods	Limitations, Advantages	References
"Maximum likelihood" using NORMSEP (FORTRAN program): provides calculated mean size-at-age and standard deviations for each mode	- requires initial estimate of number of size groups, points of overlap (cutoff points) - assumes normal distribution of size-at-age - most often used in fisheries research	Hasselblad 1966; McNew and Summerfelt 1978
Method similar to above using ENORMSEP (to be used with NORMSEP)	- does not require input of initial estimates of number of size groups or cutoff points	Yong and Skilliman 1975
Maximum likelihood in an interactive program (from PDMM; FORTRAN program)	- requires initial estimate of number of size groups, other parameters - can constrain parameters (e.g., to conform to biologically plausible patterns such as Von Bertalanffy growth curve)	MacDonald and Pitcher 1979; Schmute and Fournier 1980
Method tracing age group through sequential data using ELEFAN I (BASIC program)	- does not require input of initial estimates of number of size groups or other parameters (therefore, assumes objectivity) - can be used on minicomputer	Pauly and David 1981

An example from Lux (1971) illustrates this graphical method, called the "Petersen method" (Fig. 16.12). This curve was drawn from a fall catch sampling a haddock population and shows that the larger of the age-0 fish were just beginning to be caught in otter trawl nets. The peaks correspond to modal sizes of about 12 cm for age-0 fish, 22 cm for age-1+ fish, and 36 cm for age-2+ fish. Size overlap begins to obscure the peak at age 2 and, beyond that, prevents the appearance of any obvious peaks (called the "damping" of modes).

Damping of modes because of overlapping sizes of different cohorts restricts the usefulness of the basic Petersen method to the first 2-4 years for most fish populations. More sophisticated graphical procedures have since been developed to separate cohort size distributions from population size distributions (*see* Table 16.2). These procedures may be utilized without computer assistance, which can be an advantage in some fishery facilities. However, these graphical techniques are often not reproducible. Different conclusions may result from analysis of data by different methods or authors (Pauly and David 1981), especially if the size frequency distribution does not have clear modes (MacDonald and Pitcher 1979).

Several computer-assisted procedures have recently been adapted to separate overlapping of normal size frequency distributions in fishery research (*see* Table 16.3). These methods are statistically superior to graphical ones but require large sample sizes to be used effectively. Also, the estimates that result may still have large errors in cases of overlapping modes (McNew and Summerfelt 1978).

All graphical and most statistical methods assume that size distribution within each cohort is normally distributed and that there is some discernible separation between vear-class size distributions. The usefulness of either type of age determination can be limited if growth is largely uniform throughout the year (as in tropical fish), if spawning seasons are prolonged or intermittent, if individuals of a particular species tend to school according to size or maturity, or if size variability among individuals or cohorts is extreme. Older age classes with smaller sample sizes and higher degrees of overlap will be especially difficult to differentiate. Furthermore, none of these methods are free from subjectivity; different techniques for grouping and analyzing data can produce different results. Within their limitations, however, these methods are useful for aging populations if validation of other aging techniques is desired or if catch statistics are the only data available for analysis.

16.8 REFERENCES

Bagenal, T. B., and F. W. Tesch. 1978. Age and growth. Pages 101-136 *in* T. B. Bagenal, editor. *Methods for assessment of fish production in freshwater,* 3rd edition. Blackwell Scientific Publications, Oxford, England.

Barkman, R. 1978. The use of otolith growth rings to age young Atlantic silversides, *Menidia menidia. Transactions of the American Fisheries Society* 107:790-792.

Bhattacharya, C. G. 1967. A simple method of resolution into Gaussian components. *Biometrics* 23:115-135.

Bilton, H. 1974. Effects of starvation and feeding on circulus formation on scales of young sockeye salmon of four racial origins and of one race of young kokanee, coho, and chinook salmon. Pages 40-62 *in* T. Bagenal, editor. *The ageing of fish.* Gresham Press, Old Woking, England.

Blacker, R. W. 1974. Recent advances in otolith studies. Page 67-90 *in* F. R. H. Hardin Jones, editor. *Sea fisheries research.* John Wiley and Sons, Incorporated, New York, New York, USA.

Brothers, E. B. 1979. Age and growth studies on tropical fishes. Pages 119-136 *in* S. B. Saila and P. M. Roedel, editors. *Stock assessment for tropical small-scale fisheries.* International Center for Marine Resource Development, University of Rhode Island, Kingston, Rhode Island.

Brothers, E. B. 1982. Aging reef fishes. Pages 3-22 *in* G. R. Hunstman, W. R. Nicholson, and W. W. Fox, Jr., editors. *The biological bases for reef fishery management.* National Marine Fisheries Service, National Oceanic and Atmospheric Administration Technical Memorandum, Southeastern Fisheries Center-80, Miami, Florida, USA.

Brothers, E. B., C. P. Mathews, and R. Lasker. 1976. Daily growth increments in otoliths from larval and adult fishes. *Fishery Bulletin* 74:1-8.

Carlander, K. D. 1950. Some considerations in the use of fish growth data based upon scale studies. *Transactions of the American Fisheries Society* 79:187-194.

Carlander, K. D. 1981. Caution on the use of the regression method of backcalculating lengths from scale measurements. *Fisheries* 6 (1):2-4.

Carlander, K. D. 1982. Standard intercepts for calculating lengths from scale measurements for some Centrarchid and Percid fishes. *Transactions of the American Fisheries Society* 111:332-336.

Carlander, K. D., and L. L. Smith, Jr. 1944. Some uses of nomographs in fish growth studies. *Copeia* 3:157-161.

Cassie, R. M. 1950. The analysis of polymodal frequency distributions by the probability paper method. *New Zealand Science Review* 8:89-91.

Cassie, R. M. 1954. Some uses of probability paper in analysis of size frequency distributions. *Australian Journal of Marine and Freshwater Research* 5:513-522.

Dahl, K. 1910. *The age and growth of salmon and trout in Norway as shown by their scales.* Translated from Norwegian by Ian Bailee. The Salmon and Trout Organization, London, England.

Fawell, J. K. 1974. The use of image analysis in the aging of fish. Pages 103-107 *in* T. B. Bagenal, editor. *Ageing of fish.* Gresham Press, Old Woking, England.

Gheno, Y., and J. C. Leguen. 1968. Determination de l'age et croissance de Sardinel eba dans la region de Point-Noire. *Orstom Seriel of Oceanography* 6 (2):69-82.

Gilbert, P. W., R. G. Mathewson, and D. P. Rall. 1967. *Sharks, skates, and rays.* The Johns Hopkins Press, Baltimore, Maryland, USA.

Harding, J. P. 1949. The use of probability paper for the graphical analysis of polymodal frequency distributions. *Journal of the Marine Biological Association of the United Kingdom* 28:141-153.

Harrison, E. J., and W. F. Hadley. 1979. A comparison of the use of cleithra to the use of scales for age and growth studies. *Transactions of the American Fisheries Society* 108:431-444.

Hasselblad, V. 1966. Estimation of parameters for a mixture of normal distributions. *Technometrics* 8:431-444.

Hile, R. 1970. Body-scale relation and calculation of growth in fishes. *Transactions of the American Fisheries Society* 99:468-474.

Holden, M. J., and M. R. Vince. 1973. Age validation on the centra of *Raja clavata* using tetracycline. *Journal of the International Council for the Exploration of the Sea* 35:13-17.

Ishiyama, R. 1951. Studies on the rays and skates belonging to the family Rajidae, found in Japan and adjacent regions, on the age determination of Japanese black skate. *Bulletin of the Japanese Society of Scientific Fisheries* 16:112-118.

Jensen, A. C. 1965. A standard terminology and notation for otolith reader. *International Commission of the Northwest Atlantic Fisheries Research Bulletin* 2:5-7.

Koo, S. Y. 1962. Age designation in salmon. Pages 37-48 *in* S. Y. Koo, editor. *Studies of Alaska red salmon.* University of Washington Press, Seattle, Washington, USA.

Lagler, K. F. 1956. *Freshwater fishery biology,* 2nd edition. William C. Brown Company, Dubuque, Iowa, USA.

Lough, R. G., M. Pennington, G. Boltz, and A. Rosenberg. 1982. Age and growth of larval Atlantic herring, *Clupea harengus L.,* in the Gulf of Maine-Georges Bank region based on otolith growth increments. *Fishery Bulletin* 80:187-200.

Lutz, R. A., and D. C. Rhoads. 1980. Growth patterns within the molluscan shell, an overview. Pages 203-254 *in* R. A. Lutz and D. C. Rhoads, editors. *Skeletal growth of aquatic organisms.* Plenum Press, New York, New York, USA.

Lux, F. 1971. *Age determination in fishes.* United States Fish and Wildlife Service, Fishery Leaflet Number 637, Washington, District of Columbia, USA.

MacDonald, P. D. M., and T. J. Pitcher. 1979. Age groups from size-frequency data: a versatile and efficient method of analyzing distribution mixtures. *Journal of the Fisheries Research Board of Canada* 36:987-1001.

Marzolf, R. C. 1955. Use of pectoral spines and vertebrae for determining age and rate of growth of channel catfish. *Journal of Wildlife Management* 19:243-349.

Mayo, R. K., V. M. Gifford, and A. Jearld, Jr. 1981. Age validation of redfish from Gulf of Maine-Georges Bank region. *Journal of the Northwest Fishery Science* 2:13-19.

McNew, R. W., and R. Summerfelt. 1978. Evaluation of a maximum-likelihood estimator for analysis of length-frequency distributions. *Transactions of the American Fisheries Society* 107:730-736.

Menon, M. D. 1950. The use of bones other than otoliths, in determining age and growth-rate of fishes. *Journal du Conseil, International Council for Exploration of the Sea* 16:311-333.

Mosher, K. 1968. Photographic atlas of sockeye salmon scales. *Fishery Bulletin* 67:243-280.

Nichy, F. E. 1977. Thin sectioning fish ear bones. *Sea Technology* 2:27.

Pannella, G. 1971. Fish otoliths: daily growth layers and periodical patterns. *Science* 173:1124-1126.

Pannella, G. 1974. Otolith growth patterns: an aid in age determination in temperate and tropical fishes. Pages 28-39 *in* T. B. Bagenal, editor. *Ageing of fish.* Gresham Press, Old Woking, England.

Pauly, D. and N. David. 1981. ELEFAN 1, a basic program for the objective extraction of growth parameters from length-frequency data. *Meeresforsch* 28: 205-211.

Pycha, R. L. 1955. A quick method of preparing permanent fin ray and spine sections. *Progressive Fish-Culturist* 17:192.

Quinn, S. P., and M. R. Ross. 1982. Annulus formation by the white sucker and the reliability of pectoral fin rays for ageing them. *North American Journal of Fisheries Management* 2:204-207.

Radtke, R., and J. Dean. 1981. Increment formation in the otoliths of embryos, larvae, and juveniles of the mummichog *(Fundulus heteroclitus)*. *Fishery Bulletin* 79:360-366.

Richards, S. W., D. Merriman, and L. H. Calhoun. 1963. Studies on the marine resources of southern New England. *Bulletin of the Bingham Oceanography College, Yale University* 18 (3):5-67.

Ricker, W. E. 1975. *Computation and interpretation of biological statistics of fish populations.* Fisheries Research Board of Canada Bulletin 191, Ottawa, Canada.

Ricker, W. E. 1979. Growth rates and models. *Fish Physiology* 8:677-743.

Ropes, J. W., and L. O'Brien. 1979. A unique method of ageing surf clams. *Bulletin of the American Malacology Union* (1978):58-61.

Schmute, J., and D. Fournier. 1980. A new approach to length-frequency analysis: growth structure. *Canadian Journal of Fisheries and Aquatic Sciences* 37:1337-1351.

Skidmore, W. J., and A. W. Glass. 1953. Use of pectoral fin rays to determine age of the white sucker. *Progressive Fish-Culturist* 7:114-115.

Smith, S. H. 1954. Method of producing plastic impressions of fish scales without using heat. *Progressive Fish-Culturist* 16:75-78.

Sneed, K. E. 1951. A method for calculating the growth of channel catfish, *Ictalurus lacustris punctatus*. *Transactions of the American Fisheries Society* 80:174-183.

Tanaka, S. 1962. A method of analyzing a polymodal frequency distribution and its application to the length distribution of the porgy. *Journal of the Fisheries Research Board of Canada* 19:1143-1159.

Taubert, B. D., and J. A. Tranquilli. 1982. Verification of the formation of annuli in otoliths of largemouth bass. *Transactions of the American Fisheries Society* 11:531-534.

Turner, P. R. 1980. Procedures for age determination and growth rate calculations of flathead catfish. *Proceedings of the Annual Conference Southeastern Association of Fish and Wildlife Agencies* 34:253-262.

Waring, G. T. 1980. *A preliminary stock assessment of the little skate in the northwest Atlantic.* Master's thesis. Bridgewater State College, Bridgewater, Massachusetts, USA.

Whitney, R. R., and K. D. Carlander. 1956. Interpretation of body-scale regression for computing body length of fish. *Journal of Wildlife Management* 20:21-27.

Witt, A., Jr. 1961. An improved instrument to section bones for age and growth determination of fish. *Progressive Fish-Culturist* 23:94-96.

Yong, M. Y. Y., and R. A. Skilliman. 1975. *A computer program for analysis of polymodal frequency distributions.* United States National Marine Fisheries Service, Southwest Fisheries Center Laboratory Document, Honolulu, Hawaii, USA.

Chapter 17
Quantitative Description of the Diet

STEPHEN H. BOWEN

17.1 INTRODUCTION

This chapter presents a summary of considerations and techniques necessary to plan and carry out an efficient study for quantitative description of the diet. It is absolutely essential that the goals of the diet study be clearly established before the details of the study are planned. The filing cabinets of biology are replete with "food habits" data. Although collected at great expense, these data add little to our knowledge of fish ecology because they were collected without a particular goal in mind. By carefully planning the goals of a diet study, it is possible both to produce data to answer specific questions and to plan an efficient study in which only useful data are collected. An efficiently planned study allows the researcher to focus the available resources on the important questions and, thus, to increase greatly the probability of the study's success. Appropriate techniques can be selected only after the goals are established.

17.2 TECHNIQUES FOR COLLECTING FISH

Of the many fish collection techniques, some are more useful than others for collecting specimens to be examined for gut contents. Some capture techniques result in loss of information as a result of regurgitation. Although a systematic study has not been made, various reports suggest that capture techniques that stress fish severely, such as rotenone treatment, electroshocking, gill netting, and trawling at depth, are most likely to cause regurgitation. In some cases, a gentler technique, such as seining, may cause less regurgitation. In other cases, a very sudden shock may be useful. One worker found that a single-pronged spear allowed fish to regurgitate, but a three-pronged spear stunned fish on contact and regurgitation was eliminated. Some fish are more likely than others to regurgitate. Piscivorous predators have large distendable esophaguses which make regurgitation fairly easy, and these fish frequently regurgitate. Fishes that feed on small prey, however, are much less likely to regurgitate. Investigators who plan to study the diet of a piscivore should keep this potential problem in mind. If preliminary collections over a 24-hour cycle yield mostly specimens with empty stomachs, a different collection technique should be tried.

Post-capture digestion may also result in loss of valuable dietary information. Gill nets and traps of various sorts may hold fish for long hours during which much of the diet can be digested beyond the point where identification is possible. Consequently, these capture techniques are most useful where catch rates are high and specimens can be removed from the gear soon after capture. Some post-capture digestion may be tolerable if gill nets or traps are the only effective means by which the target species can be collected.

A third potential difficulty of a diet study stems from fish in traps feeding on foods they would not normally eat. For example, catfish, which are not normally piscivorous, may feed on small fishes during the time they and the small fishes are trapped together. Because of the problems of post-capture digestion and atypical feeding, traps are of limited utility in diet studies.

17.3   SAMPLING STRATEGIES

There are many factors that potentially influence both the amount of food and the type of food found in the digestive tracts of fish. The effects of the diel cycle, seasonal changes, the size and territoriality of fish, and differential digestion rates must be considered if an efficient and accurate sampling program is to be planned.

17.3.1   Diel Effects

The behavior of most fishes is highly structured within the diel cycle. Diel changes in habitat, feeding intensity, and the diet itself must be considered. During nonfeeding hours, fish often seek shelter in low risk environments. Just prior to feeding, they move into habitats where suitable food is available. Some fish feed intensively during a single extended feeding period lasting six hours or longer. Others may feed intensively for only one or two hours, followed by low levels of feeding activity for several hours. Still other fish are known to feed intensively at dusk and dawn with little evidence of feeding at other hours. Although given patterns of diel feeding behavior are commonly associated with particular species, fishes can radically alter their diel cycle in response to changes in food availability or predator distribution (Bowen and Allanson 1982). In order to maximize the amount of information gained per fish collected, fish must be collected when their stomachs are fullest. Therefore, it is often useful to begin a study of diet with a 24-hour series of collections made at 2-hour intervals. This will yield data on both catch per effort and stomach fullness to help plan an efficient sampling program.

Diet may vary with time of day. Salmonids in large streams may feed on invertebrate drift during the night and small fish during the day. I once found a population of golden shiners that fed on periphyton at night, Cladocera at dusk and dawn, and on sessile invertebrates during the day. Comparison of stomach contents from a 24-hour series of collections is necessary to determine if the diet varies according to time of day. If no such variation is apparent, then other factors discussed above will determine the best time of day to sample. If diel variation is detected, it is important to sample at a constant level of effort throughout the feeding period to avoid biasing the diet data.

17.3.2   Seasonal Effects

Fishes are highly responsive to seasonal changes in food availability. Changes within trophic categories, for example, from mayflies to chironomids, are common as invertebrate populations mature and emerge. Changes from one trophic category to another are also know to occur. Bluegills may switch from a diet of invertebrates to a diet of algae as invertebrate numbers decline toward the end of summer (Kitchell and Windell 1970). Thus, to adequately describe the trophic resources utilized by a fish population, it is often necessary to sample at frequent intervals throughout the year.

17.3.3 Effects of Fish Size and Territoriality

Diets selected by fish vary according to size and sex. As fish grow larger, they often select larger prey. The change may be abrupt and associated with a change in habitat as when salmon smolts change from invertebrate diets in freshwater to fish diets at sea. In many cases, the transition is gradual, as with lake herring which change slowly from a diet of nauplii to a diet of adult copepods as they grow. If the change is abrupt, the goals of the study may be served by analysis of gut contents from fish divided into two size groups. When the diet changes more gradually, ten or more size groups may be required to document the change, and it may be necessary to expand the range of habitats sampled in order to obtain statistically significant numbers of individuals for all size groups.

Adult male fishes often establish and defend breeding territories that they rarely leave. They must depend on food resources within the territories, and, consequently, their diets may be considerably different from those of the females and subadult males. If brightly colored males are captured, their gut contents should be compared to those from other specimens. It may be necessary to modify the sampling program so that enough territorial males are collected to permit description of their diet.

17.3.4 Differential Digestion Rates

Stomach contents may not accurately reflect the diet of a fish for two reasons. First, some prey, such as tubificids (Kennedy 1969) or protozoa, may be digested rapidly and leave little recognizable trace in the digestive tract. The importance of these prey, therefore, may be underestimated or overlooked entirely. If such a bias is suspected, it may be useful to hold the fish in an aquarium, watch how they feed, and then collect samples of their food for comparison to gut contents. A second data bias can result from differential rates of digestion for various prey. Slowly digested prey may accumulate relative to rapidly digested prey and, thus, be over-represented in the gut (Mann and Orr 1969; Gannon 1976). This effect can be minimized if fish are collected at the peak of daily feeding intensity when all food items in the anterior gut have been ingested recently.

17.4 REMOVING, FIXING, AND PRESERVING GUT CONTENTS

17.4.1 Removing Gut Contents

Many workers have developed methods for removal of stomach contents from fish that are to be kept alive. Most methods employ a stomach pump to flush stomach contents from the fish with one or more volumes of water (Seaburg 1957; Giles 1980). Tests in which the stomachs were dissected after pumping indicate that this procedure is a highly effective method of removing prey from a fish stomach. Emetics have also been used to obtain stomach contents from live fish (Jernejcic 1969). Both of these methods are useful only with the more robust fishes such as the Percidae, Centrarchidae, and the many catfishes. Delicate fishes such as the Clupeidae and many Cyprinidae are unlikely to survive the experience. Although this approach requires construction of special apparatus, valuable fish are not destroyed, and, in many cases, it is faster than removal of gut contents by dissection.

Often it is preferable to remove gut contents by dissection. This allows the worker to examine the intestine for fullness and to be confident that all prey are removed from the stomach or anterior gut segment. Data on sex and gonad maturity may be collected at the same time (*see* chapter 15). If data on intestinal contents are needed, dissection is essential.

Before dissection, kill the fish be severing the spinal column. For small fishes, scissors are adequate, but a knife may be necessary for larger fishes. Open the coelom to expose the viscera. Using blunt scissors, sever the esophagus, the last few millimeters of the intestine, and the mesentary at its dorsal point of attachment. This allows the visceral mass to be lifted out of the coelom and into a dissection pan for more careful manipulation. Separate the digestive tract from the rest of the viscera and divide it into segments as dictated by the needs of the study. Open each segment by slitting it lengthwise with fine scissors. For piscivorous fishes, prey items can be lifted directly from the stomach. For fish eating smaller prey, it is often useful to hold the slit segment with forceps over a petri dish or small beaker and then to wash the food items from the segment with tap water in a wash bottle. If quantitative recovery of gut contents is not required, the food may be extruded by sliding a blunt probe along the length of the segment. This may, however, also extrude the gut mucosa. Do not mistake this tissue as part of the diet.

17.4.2 Fixing and Preserving Gut Contents

Ten-percent formalin is fully adequate as a fixative (*see* chapter 14 for dilution method). Because it rapidly hardens prey tissues, partly digested prey are more likely to stay intact and, thus, are easier to identify. Before samples are examined in detail, remove excess formaldehyde by soaking the samples in several changes of water. After excess formaldehyde is removed, preserve the samples in a 45-70% aqueous solution of alcohol. Methanol, ethanol, and isopropanol are all satisfactory. Wear plastic gloves and work in a fume hood to minimize exposure to residual formaldehyde. Formaldehyde is suspected of being a carcinogen and can cause permanent damage to fingernail growth areas and sinus tissues.

The availability of field time and facilities usually determines when gut contents are removed and fixed. It is best to remove and fix gut samples immediately after capture. This minimizes post-capture digestion and avoids difficulties associated with dissection of hardened fish tissues (Borgeson 1963). When tens or hundreds of fish are captured at one time, however, it is often necessary to fix the gut contents *in situ* with the least investment of time possible. In warm climates, hold fish on ice if more than a few minutes will elapse before fixation. Fishes of 100-g live weight or less can be fixed whole. In order to halt post-capture digestion, slit the coelom or inject formaldehyde directly into the coelom (Emmett et al. 1982). For larger fish, it is usually more efficient to fix only the digestive tract.

The goals of the study will determine the extent to which individual samples must be kept separate and identified. When gut contents are removed in the field, they may be stored in small numbered vials or plastic bags, and concomitant data (length, weight, sex, etc.) may be recorded on data sheets. When intact digestive tracts are fixed, they may be stored in separate containers. Several digestive tract samples may be stored in a single container if they are wrapped individually in cheese cloth, each with a paper label written in pencil. For some studies it may be adequate to pool gut samples according to size groups of the fish. Although this results in loss of data on variation among individuals, it requires less handling and greatly simplifies subsequent analyses.

17.5 IDENTIFYING THE DIET

17.5.1 Identifying Partly Digested Prey

Identifying stomach contents is often made difficult by digestion. Even recently ingested food items may be ground by jaw or pharyngeal teeth to a point at which recognition is difficult. Consequently, it is usually necessary to identify food items with reference to some characteristic part of the organism that is resistant to digestion. For invertebrates, various parts of the exoskeleton, such as eye capsules, head shields, and tarsal claws, have been commonly used. Otoliths have been used to identify fish prey (Whitfield and Blaber 1978). Macrophytes often can be identified by a characteristic shape or sculpturing along the edge of the leaves. In contrast, algal cells are usually found intact in the anterior digestive tract, and identification of these presents no special problems.

17.5.2 Choosing the Level of Identification

Many fish biologists have invested huge amounts of time identifying at the species level. In a few cases this has yielded interesting data, but often the data are too complex for meaningful interpretation. Competing predators are rarely locked in a struggle for single-species prey resources. In the last ten years, those studies that have provided the greatest insights into the trophic ecology of fish have been those that identified prey at only the higher taxonomic levels, such as order or family, and that included data on prey size. Unless there is a specific need for identification to species, higher taxonomic levels of identification are adequate to produce the most useful data per unit of time invested.

17.6 QUANTITATIVE DESCRIPTION OF THE DIET

There are three simple approaches to quantitative description of the diet: frequency of occurrence, percent composition by number, and percent composition by weight. Each approach provides distinctly different information, and, consequently, it is necessary to discuss each one in some detail.

17.6.1 Frequency of Occurrence

Frequency of occurrence is the fastest approach to quantitative analysis of fish diets. The worker makes up a list of all likely food types and records the presence or absence of each food in each specimen collected. When all specimens have been examined, the proportion of the fish that contained one or more of a given food type is calculated as the frequency of occurrence for that food type. For example, if 18 out of a sample of 22 bluegills contained one or more chironomids, then the frequency of occurrence of chironomids in the diet would be 0.82 or 82%.

These results indicate the extent to which fish in the sample functioned as a single feeding unit. If nearly all fish contained the same set of prey types, this will be clearly documented with uniformly high frequencies of occurrence. Under some circumstances, individual fish concentrate on a single food type, and this will be reflected in low frequencies of occurrence for nearly all foods. A group of omnivorous fish, however, that share a common source of vegetable matter, such as filamentous algae, but specialize individually in selection of invertebrate prey will have high frequency of occurrence values for most algae types and low values for most invertebrate types.

It is important to note what frequency of occurrence data do not show. High frequency of occurrence does not mean that a given food type is of nutritional importance to the consumer. It may be consumed with great regularity but in very small quantities. For example, a fish that feeds on benthic invertebrates may unavoidably ingest a small quantity of sedimented algae. Because these algal cells are small and tend to be evenly distributed throughout a water body, the frequency of occurrence for each algal type may be 100%. The frequency of occurrence for individual invertebrate prey, in contrast, may not excede 75%. This does not suggest that algae are more important than invertebrates in the consumer's diet. Frequency of occurrence data, therefore, describe the uniformity with which groups of fish select their diet but do not indicate the importance of the various types of food selected.

17.6.2 Percent Composition by Number

In this approach, the number of food items of each food type is determined for each fish examined. The percent composition by number is the number of items of a given food type expressed as a percentage of the total number of food items counted. After a preliminary examination of a few samples, it is often necessary to give some thought to the appropriate unit to be counted (Fig. 17.1). If the diet consists of invertebrate prey that are largely disarticulated by the pharyngeal apparatus, some characteristic fragment must be counted. To minimize counting time, it is best to choose some fragment that is found only once per prey. Not all foods, however, are well suited to a counting approach. Detrius and higher plants that are ingested bit-by-bit are not found in discrete units of uniform size, and, thus, counts of these particles have little meaning.

For fish that feed on large prey, it is possible to count all food items in the stomach. However, most fish feed on small prey, and their stomachs often contain such great numbers that counting all items directly is impractical. Under these circumstances, subsampling is necessary. Suspend the gut contents in a known volume of water using a beaker and a magnetic stirrer. When invertebrate prey are to be counted, a large-bore pipette (5-ml graduated pipette with the tip cut off) is used to transfer a known volume of the suspension to a petri dish for counting at the required magnification. When algal cells or other microscopic foods ($<100 \mu$ m maximum dimension) are to be counted, the subsample is a single water drop of known volume which is put on a slide and counted by the drop-transect method (Edmondson 1974). Adjust the volume of the suspension and/or the volume transferred so that 50 to 100 items are counted for each prey type of interest. Subsampling errors become significant if fewer than 50 are counted, but counts above 100 do little to reduce subsampling errors (Allanson and Kerrich 1961). The number of items counted for each food type can be multiplied by the suspension volume and divided by the transfer volume to estimate the total number of items of that food type in the gut sample.

The importance of bacteria in the diets of some fishes is becoming increasingly apparent (Moriarty 1976; Bowen 1979). Counting of bacteria requires an epiillumination microscope. Suspend the diet in particle free water and pass a small subsample through a 0.2-μ m pore-size membrane filter. Stain the bacteria trapped on the filter with a flurochrome; acradine orange is most commonly used. Under the microscope, the light beam from the epiillumination unit is focused on the membrane filter and causes the flurochrome in stained bacteria to fluoresce. Using this approach,

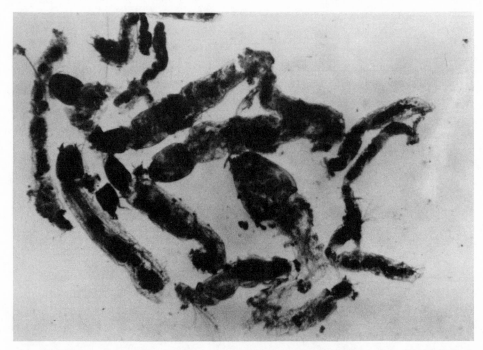

Figure 17.1. Fragments of chironomids teased from detritus in stomach contents of a small white sucker. Head capsules stay intact as softer tissues are digested and can be counted to estimate numbers consumed. Measurements of head capsule width can be used to reconstruct the weight of chironomid prey eaten.

bacteria can be distinguished quantitatively from detritus. For details of bacterial counting methods, *see* Zimmerman et al. (1978).

Unlike the frequency of occurrence approach, which provides a quantitative description only for the entire sample of fish, calculating percent composition by number describes the diet of an individual fish. To summarize results for the entire sample, calculate mean values. Remember that percentage data are not normally distributed and must be transformed for calculation of confidence intervals.

Percent composition by number data are of little value on their own, but they are useful as a step toward other calculations. The raw data from which percent composition by number are derived can be combined with estimates of feeding rate to assess the impact of predators on prey population dynamics. More commonly, percent compostion by number figures are combined with estimates of prey weight.

17.6.3 Percent Composition by Weight

In this approach to quantitative analysis of the diet, the weight of items in each food type is expressed as a percentage of the total weight of ingested food for an individual fish. If prey are large enough to be handled individually and are only slightly digested, they may be weighed directly. Both wet and dry weights have been used. Wet weight is obtained by gently blotting surface fluid from the item and weighing it (*see* chapter 15). Dry weight is obtained by drying the item to constant weight. Preweighed aluminum foil pans are useful for drying and can be purchased commercially or made easily in the laboratory (Fig. 17.2). Because these pans do not

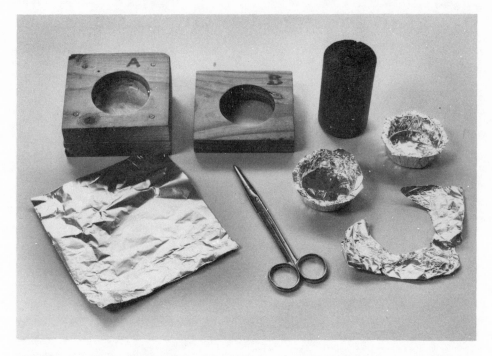

Figure 17.2. Foil pans can be made inexpensively using this simple press. A piece of foil is placed over the press base *(A)*, and then the top *(B)* is placed over the foil. The top is held lightly in place as the cylinder is used to press the foil into a dish shape. Then the top is held down firmly as the cylinder is pressed against the sides to form the top edge of the pan. Finally, the pan is trimmed with scissors. These pans hold 50 ml, but presses for pans of 0.1 - 100 ml volume work just as well. This press was made by Dudley Forsythe, Department of Zoology, Rhodes University.

absorb moisture and are light in weight, they add little weighing error. A wide range of drying temperatures has been recommended, but at 105 C drying is rapid and bacterial action is stopped. Dry weight is more precise than wet weight and should be used when a drying oven is available.

If prey are disarticulated or partly digested, it is necessary to estimate their weight at ingestion by reconstruction. In this procedure, some hard structure such as an operculum, vertebrae, or head capsule is measured and compared to a graph of hard structure size versus whole prey weight (Pearre 1980). The graph should be prepared using prey collected from the fish's feeding habitat. This type of estimate may be quite accurate for fish, but not for invertebrates. Some insect prey, especially chironomids, have vastly different weights with very similar head capsule sizes. In such cases and if there is no evidence that fish select prey according to size, it may be as accurate to sample the feeding habitat and use the average weight of a chironomid as the average weight of chironomid in the diet.

Because many invertebrates and most algae are too small to weigh directly, weight must be estimated from measurements made through a microscope. Although this approach is potentially labor-intensive, a microcomputer combined with a digitizer table (graphics tablet) can be used to reduce the time required and to increase the accuracy of measurement. I have used a camera lucida to trace the outlines of food particles from several microscope fields onto a single sheet of paper. The paper is then put on the digitizer table and the length (long axis) and width (perpendicular to the long axis) are recorded for each particle. Some microscopes are equipped so that the depth of a particle (Z-axis) can be read from the focus knob by focusing first on the top, then on the bottom of the particle. This measurement can be recorded in the computer along with length and width. If depth is not measured, it may be reasonable to assume depth is some constant proportion of width. The volume of the food item is estimated by calculation of the volume of a geometric solid of similar size and shape. The shape approximation need not be complex since this source of error is small relative to subsampling error. Estimated wet volume is converted to dry weight as 1 mm^3 = 0.1 mg. The exact conversion varies among taxa, but this figure is precise enough for most studies (Cummins and Wuycheck 1971).

It is important to cross-check calculations for small particle diets to see if the results from reconstructed weight and indirect weight estimates are realistic. Weigh the whole sample (dry weight, corrected for subsampling), and compare that weight to the sum of estimated weights of all prey. Unless extensive digestion is expected, the sum of the individual prey weights should not exceed the total sample weight. In many cases when fish feed on benthic invertebrates, algae, or macrophytes, the sum of estimated prey weights may be much less than total sample weight. Inorganic sediment and organic detritus are usually responsible for this.

Biologists have tended to ignore material in diets that they cannot recognize as part of an organism. This is clearly a mistake since some fish can derive most or all of their nutritional needs from detritus (Bowen 1979, 1980). To distinguish between detritus and inorganic sediment, combust the sample in a muffle furnace to constant weight at 550 C. Times required to reach constant weight range from 30 minutes to 12 hours depending on the size and composition of the sample. Accuracy can be improved by cooling the samples in a desiccator prior to weighing. The weight loss on combustion provides a very close approximation to total organic weight (Dean and Gorham 1976). For the purpose of these calculations it may be reasonable to assume that organisms are 100% organic matter (with the exception of mollusks for which special calculations must be made). Organic weight in the sample that cannot be accounted for as prey weight must be present in the form of detritus. The weight of detritus may also be estimated directly from measurements made through the microscope as described above.

Percent composition by weight is the only one of the three widely used methods that begins to identify the food or foods important for the fish's nutrition. Although percent composition by number does describe the diet quantitatively, the weight differs for each food type, and, thus, no comparison of types is possible. Which is more important, 10,000 bacteria or 1 amphipod? Numbers alone do not provide an answer. Percent composition by weight quantifies food types in directly comparable weight units. With the obvious exception of mollusks for which special calculations of organic weight must be made, food value is roughly proportional to weight. Thus, percent composition by weight data do suggest the relative importance of individual food types in the nutrition of the fish.

17.6.4   Net Food Value

Biologists using the percent composition by weight approach have developed new techniques which greatly increase our ability to assess the value of different foods. Although the details are beyond the scope of this chapter, persons contemplating a study of fish diets must be aware of these developments to realistically evaluate percent composition by weight data. These new approaches fall into two categories: those applied in the study of animal diets, and those applied in the study of vegetable and detritus diets. Although animal prey are very similar in their food value for consumers, the cost of acquiring individual prey may vary greatly. The net nutritional significance of a prey item depends not only on its weight, but also on how much time and energy it takes to search for, pursue, capture, and eat that prey. A cost-benefit concept has been applied by Werner (1974, 1977), Werner and Hall (1974, 1976), and Stein (1977) among others to account for the role of acquisition cost in diet selection by fishes. Consult these and subsequent work when you want to consider possible differences in the net food value per unit weight of different animal prey.

Food items that make up vegetable and detritus diets differ relatively little in acquistion costs but greatly in the extent to which they are assimilated. Quantities of protein and energy assimilated from each type of food in the diet must be determined before the quality of a particular food type can be assessed. For details of this approach *see* Moriarty and Moriarty (1973), Caulton (1978), Bowen (1979, 1980, 1981) and Buddington (1979).

17.7   DIET INDICES

Biologists have sometimes worried about biases in information provided by frequency of occurrence, percentage by number, and percentage by weight data. In the scientific literature there are statements that percentage by number overemphasizes the importance of smaller prey since they weigh so much less than larger prey, but that percentage by weight overemphasizes the importance of large prey because it takes more time to capture and eat a large number of small prey (Pinkas et al. 1971; George and Hadley 1979; Hyslop 1980). Because of this dissatisfaction with a single data approach, biologists have created several formulas that combine frequency of occurrence, percentage by number, and/or weight data into a hydrid index intended to compensate for the perceived biases of the individual indices. Although these indices may at first offer some intuitive appeal, there is no identifiable biological basis for their interpretation. The fault lies not with frequency of occurrence, percentage by number, or percentage by weight data, but with the investigator's failure to define clearly the objective of the analysis. For example, if the investigator wishes to determine the impact of the predator on a prey's population dynamics, then percentage by number will provide useful data without regard to the size of different prey types. If the investigator wishes to measure the contribution of prey to the predator's nutrition, then percentage by weight is fully adequate. A sample of 25 bass stomachs which contains a total of 5,000 Cladocera weighing 1 g and one fish weighing 99 g indicates that 99% of the bass population's gross energy comes from fish, in spite of the relative numbers of prey consumed. It may be that Cladocera are digested more rapidly than fish or that the single fish prey is not representative of the true diet because the sample size was too small. These problems are not biases attributable to percentage by weight data but to the sampling design. Use of indices that combine frequency of occurrence, percentage by number, and/or percentage by weight data does not solve these problems.

There is a second group of indices which has basis in scientific theory and may be of use in analysis of diet data if the limitations are kept well in mind. Electivity indices can be used to compare the diet selected by fish to the available resources in the feeding environment. The index value indicates the extent to which the observed diet differs from a diet selected at random. This concept was developed by Ivlev (1961), who compared diets and resources for fish feeding in precisely controlled aquarium experiments. Electivity indices have been applied in field studies by a number of workers (*See* Strauss 1979). Application of the concept in field situations makes it necessary to determine accurately the relative abundances of all potential prey. Because of the patchy distribution of most prey, this is a formidable sampling problem (Strauss 1979, 1982). Once electivity has been established, however, it can provide invaluable clues to environmental factors that influence the nutrition of the fish.

Indices which quantify diet overlap are based on niche theory. This area of ecology is evolving rapidly, and at least ten different measures of diet overlap have been proposed. At present, opinions are divided as to which measure has the most desirable mathematical properties, the most statistical reliability, and the most biological relevance (Hurlbert 1978; Jumars 1980; Wallace 1981). Extensive diet overlap is not conclusive evidence for competition because the resources may not be in short supply. Measures of diet overlap may be useful if diets of two similar species are to be compared across a wide range of different water bodies. An inverse correlation between diet overlap and growth rate may suggest that competition for food resources limits growth in some waters. Alternative explanations should be considered as well.

The advantage of these indices is that they may reveal relationships in the data that are not otherwise apparent. The disadvantages are that they possess poorly understood statistical properties and that they shift attention away from the original data. Indices that remove the complexities of original data sometimes let us think that we know more than we do. For these reasons, indices of electivity and diet overlap must be used with full recognition of their limitations.

17.8 REFERENCES

Allanson, B. R., and J. E. Kerrich. 1961. A statistical method for estimating the number of animals found in field samples drawn from polluted rivers. *Verhandlunger Internationale Verein Theoretisch Angewandte Limnologie* 14:491-494.

Borgeson, D. P. 1963. A rapid method for food-habits studies. *Transactions of the American Fisheries Society* 92:434-435.

Bowen, S. H. 1979. A nutritional constraint in detritivory by fishes: the stunted population of *Sarotherodon mossambicus* in Lake Sibaya, South Africa. *Ecological Monographs* 49:17-31.

Bowen, S. H. 1980. Detrital nonprotein amino acids are the key to rapid growth of *Tilapia* in Lake Valencia, Venezuela. *Science* 207:1216-1218.

Bowen, S. H. 1981. Digestion and assimilation of periphytic detrital aggregate by *Tilapia mossambica*. *Transactions of the American Fisheries Society* 110:239-245.

Bowen, S. H., and B. R. Allanson. 1982. Behavioral and trophic plasticity of juvenile *Tilapia mossambica* in utilization of the unstable littoral habitat. *Environmental Biology of Fishes* 7:357-362.

Buddington, R. K. 1979. Digestion of an aquatic macrophyte by *Tilapia zillii*. *Journal of Fish Biology* 15:449-456.

Caulton, M. S. 1978. The importance of habitat temperatures for growth in the tropical cichlid *Tilapia rendalli* Boulenger. *Journal of Fish Biology* 13:99-112.

Cummins, K. W., and J. C. Wuycheck. 1971. Caloric equivalents for investigations in ecological energetics. *International Association for Theoretical and Applied Limnology, Special Communication* 18:1-158.

Dean, W. E., and E. Gorham. 1976. Major chemical and mineral components of profundal surface sediments in Minnesota lakes. *Limnology and Oceanography* 21:259-284.

Edmondson, W. T. 1974. A simplified method for counting phytoplankton. Pages 14-15 *in* R. A. Vollenweider, editor. *A manual of methods for measuring primary production in aquatic environments.* Blackwell Scientific Publications, Oxford, England.

Emmett, R. T., W. D. Muir, and R. D. Pettit. 1982. Device for injecting preservative into the stomach of fish. *Progressive Fish-Culturist* 44:107-108.

Gannon, J. E. 1976. The effects of differential digestion rates of zooplankton by alewife, *Alosa pseudoharengus,* on determinations of selective feeding. *Transactions of the American Fisheries Society* 105:89-95.

George, E. L., and W. F. Hadley. 1979. Food and habitat partitioning between rock bass *(Ambloplites rupestris)* and smallmouth bass *(Micropterus dolomieui)* young of the year. *Transactions of the American Fisheries Society* 108:253-261.

Giles, N. 1980. A stomach sampler for use on live fish. *Journal of Fish Biology* 16:441-444.

Hurlbert, S. H. 1978. The measurement of niche overlap and some relatives. *Ecology* 59:67-77.

Hyslop, E. J. 1980. Stomach contents analysis - a review of methods and their application. *Journal of Fish Biology* 17:411-429.

Ivlev, V. S. 1961. *Experimental ecology of the feeding of fishes.* Yale University Press, New Haven, Connecticut, USA.

Jernejcic, F. 1969. Use of emetics to collect stomach contents of walleye and largemouth bass. *Transactions of the American Fisheries Society* 98:698-702.

Jumars, Peter A. 1980. Rank correlation and concordance tests in community analyses: An inappropriate null hypothesis. *Ecology* 61:1553-1554.

Kennedy, C. R. 1969. Tubificid oligochaetes as food of dace. *Journal of Fish Biology* 1, 11-15.

Kitchell, J. F., and J. T. Windell. 1970. Nutritional value of algae to bluegill sunfish, *Lepomis macrochirus. Copeia* 1:18-190.

Mann, R. K., and D. R. Orr. 1969. A preliminary study of the feeding relationships of fish in a hard-water and a soft-water stream in southern England. *Journal of Fish Biology* 1:31-44.

Moriarty, D. J. W. 1976. Quantitative studies on bacteria and algae in the food of the mullet *Mugil cophalis* L. and the prawn *Metapenaeus bennettae* (Racek and Dall). *Journal of Experimental Marine Biology and Ecology* 22:131-143.

Moriarty, D. J. W., and C. M. Moriarty. 1973. The assimilation of carbon from phytoplankton by two herbivorous fishes: *Tilapia nilotica* and *Haplochromis nigripinnis. Journal of Zoology* 171:41-55.

Pearre, S., Jr. 1980. The copepod width-weight relation and its utility in food chain research. *Canadian Journal of Zoology* 58:1884-1891.

Pinkas, L., M. S. Oliphant, and I. L. K. Iverson. 1971. Food habits of albacore, bluefin tuna and bonito in Californian waters. *California Fish and Game* 152:1-105.

Seaburg, K. G. 1957. A stomach sampler for live fish. *Progressive Fish-Culturist* 19:137-139.

Stein, R. A. 1977. Selective predation, optimal foraging and the predator-prey interaction between fish and crayfish. *Ecology* 58:1237-1253.

Strauss, R. E. 1979. Reliability estimates for Ivlev's electivity index, the forage ratio, and a proposed linear index of food selection. *Transactions of the American Fisheries Society* 108:344-352.

Strauss, R. E. 1982. Influence of replicated subsamples and subsample heterogeneity on the linear index of food selection. *Transactions of the American Fisheries Society* 111:517-522.

Wallace, R. K., Jr. 1981. An assessment of diet-overlap indexes. *Transactions of the American Fisheries Society* 110:72-76.

Werner, E. E. 1974. The fish size, prey size, handling time relation in several sunfishes and some implications. *Journal of the Fisheries Research Board of Canada* 31:1531-1536.

Werner, E. E. 1977. Species packing and niche complementarity in three sunfishes. *The American Naturalist* 111:553-578.

Werner, E. E., and D. J. Hall. 1974. Optimal foraging and size selection of prey by the bluegill sunfish *(Lepomis macrochirus). Ecology* 55:1042-1052.

Werner, E. E., and D. J. Hall. 1976. Niche shifts in sunfishes: experimental evidence and significance. *Science* 191:404-406.

Whitfield, A. K., and S. J. M. Blaber. 1978. Food and feeding ecology of piscivorous fishes at Lake St. Lucia, Zululand. *Journal of Fish Biology* 13:675-691.

Zimmerman, R., R. Iturriaga, and J. Becker-Birck. 1978. Simultaneous determination of the numbers of aquatic bacteria and the number there of involved in respiration. *Applied and Environmental Microbiology* 36:926-935.

Chapter 18
Field Examination of Fish

RICHARD J. STRANGE

18.1 INTRODUCTION

18.1.1 Definition of Health and Illness in Wild Fish

Healthy fish are relatively free from pathogens and environmental conditions that increase mortality or reduce growth and fecundity. Healthy fish may, however, harbor pathogens and probably have encountered unfavorable environmental conditions. Disease organisms and environmental stressors are normal in the lives of fish, and fish usually have the stability to withstand occasional exposures to these challenges without suffering ill effects. There are times when the pathogens become so numerous or the environment sufficiently stressful as to threaten the lives of the fish or, short of that, slow their growth and inhibit their reproduction. When this happens, we can term the fish diseased. Fisheries managers, researchers, and aquaculturists must be capable of detecting overt disease in fish because serious health problems in fish populations can affect the success of their work.

Pathogens are not the only causes of fish disease. In fact, adverse environmental conditions are often more important causes of ill health in fish than pathogens (parasites, bacteria, and viruses). Environmental conditions that can harm fish include degraded water quality (low dissolved oxygen, high ammonia, excessive siltation) and toxic pollutants (chlorine, pesticides, heavy metals). Other factors such as overcrowding and excessive competition can reduce the ability of fish to obtain sufficient food and reproduce normally. Even when pathogens are directly responsible for disease, the infections are usually secondary to degraded water quality or other environmental stressors.

Widespread pathogenic disease is rarely detected in wild fish. Disease organisms are, of course, present in the wild and wild fish certainly die of infections, but the majority of mass fish kills are caused by adverse environmental conditions (*see* chapter 13). In fish culture, however, pathogens often cause substantial mortalities because fish in culture ponds and raceways are generally crowded. Resulting degradation of water quality stresses the fish and encourages the transmission of pathogens.

18.1.2 Limitations of Field Diagnosis

Fish health cannot be completely assessed through the interpretation of clinical signs. Obvious indications of disease can be detected during field examination, but even a specialist in fish health cannot diagnose disease in a fish without thorough laboratory analysis. The causative agent of disease, be it a pathogen, toxicant, or adverse environmental situation, must be identified before a positive diagnosis can be made. The goal, then, is to rule out or confirm specific, suspected disease agents. A biologist without fish health training, even with a microscope and other diagnostic equipment, can only be expected to recognize obvious indications of ill health and be able to correctly take, package, and send samples for further analysis. In the following sections, references to clinical signs and specific disease organisms are intended only as examples, not as a key to definitive diagnosis.

18.2 EXAMINATION OF THE FISH

18.2.1 Behavioral Signs

Even before a fish is in hand, we can observe behavioral indications of environmental stress or ill health. In early morning during the summer, it is not unusual to see fish gulping at the surface, or "piping," in highly eutrophic lakes and ponds. This is an adaptive response to low dissolved oxygen; by pulling the surface film over their gills, the fish are respiring the water with the highest oxygen content. Fish can tolerate occasional, short-term oxygen deficiencies without ill effect, but piping late in the day or piping by fish tolerant of low dissolved oxygen, such as catfish or carp, is indicative of a dangerous situation.

A behavioral indication of external parasites, though more often observed in hatchery raceways than in the wild, is the fish rubbing itself against the bottom. This is termed "flashing" because when one looks down into the water, one sees the silvery side of the fish turn up. Feeding behavior of fish in streams eating aquatic insect drift can be mistaken for flashing. Fish near death often abandon normal behavior and may be seen swimming aimlessly near the surface or finning quietly in shallow water without their normal wariness. Such individuals can usually be dip-netted without difficulty. On rarer occasions, fish may be seen in convulsions (sometimes as a result of pesticide poisoning) or whirling (a sign of certain infectious diseases).

18.2.2 External Signs

Body conformation and color. If at all possible, one should select living fish for examination. Fish undergo rapid post-mortem changes that complicate evaluation of both external and internal signs. It is helpful to first observe fish alive and swimming. To do this, place the living fish in a clear plastic bag of water. Signs of ill health include excessive mucus production on body and gills (often the result of toxicant irritation or ectoparasites), clamped fins and shimmying (a general sign of illness), and faded or blotchy coloration (often a sign of stress).

For further examination, kill the fish with a blow to the head, unless brain tissue samples or cranial examination are anticipated. Does the fish's body conform to what is considered normal for the species? The body should be fairly robust with a head proportional to the body and eyes proportional to the head. An emaciated fish with a large head and eyes is malnourished. Such fish are often found in overcrowded conditions where food is scarce; for example, stunted bluegills occur in an out-of-balance pond (Fig. 18.1). Protruding eyeballs, "pop-eye" or exophthalmia, may be the result of parasites or gas bubble disease caused by water supersaturated with nitrogen.

Fins. The fin membranes should be intact to the end of the rays, free of slime or cottony fungus (e.g., *Saprolegnia* sp.), and without hemorrhagic areas. In the spring, it is normal to occasionally see the lower lobe of the caudal fin frayed in centrarchids because of nesting activity. Otherwise, frayed fins (Fig. 18.1) may indicate the attack of bacteria such as *Flexibacter* sp., especially when slime or hemorrhaging is evident. In

Figure 18.1. An apparently healthy bluegill (upper) compared to one showing a number of signs of overt disease (lower).

advanced cases of fin "rot," fins may be entirely eroded. Also inspect fin membranes for encysted parasites which appear as small white (e.g., *Ichthyoptherius multifilis*) or black (e.g., *Neascus* sp.) spots; these may appear elsewhere on the body, as well.

Skin, scales, and mucus. Scales should lie flat and be firmly attached. A thin, clear, evenly distributed mucus film should cover the fish and the surface of the fish should be free of reddened areas, bloody sores (hemorrhagic lesions), nodular growths, and fungus (Fig. 18.1). Reddened areas and lesions are evidence of systemic (widespread, internal) infections of bacteria (e.g., *Aeromonas* sp.) or superficial bacterial infections (e.g., *Flexibacter columnaris*). Skin lesions may be complicated by parasite infestations (e.g., *Epistylus* sp.) or fungus. Skin lesions in wild fish are seen most often during the early spring when rising water temperatures encourage bacterial growth at a time when fish are least resistant to it. An increased prevalence of skin lesions also has been associated with fish from water with a high organic load, such as below a sewage outfall, and a correspondingly high bacterial community. Nodular growths are a typical host reaction to some parasites. Large ectoparasites of the skin (e.g., *Lernea* sp.) can be seen themselves.

Gills. Pull back or cut off the operculum and carefully inspect the gills. In a freshly killed fish, the gills should be bright red and without a thick mucus covering. The gills are frequently a site of ectoparasites; they are also a sensitive area constantly exposed to the water and, as such, are often the first tissue to show an adverse reaction to unfavorable water quality. Nearly all cultured fish and many wild fish display some gill damage. Though a microscope is necessary to verify a healthy gill, the unaided eye can easily detect pale color, excess mucus, and erosion of the posterior edge of the respiratory tissue (holobranch).

18.2.3 Internal Signs

Opening the fish. After carefully examining the outside of the fish, begin the internal examination, or necropsy, by laying the fish on its side and making an incision from just above the vent, along the top of the rib cage, and forward through the pectoral girdle and fifth non-respiratory gill arch that marks the rear of the opercle cavity. Scissors work best for fish up to 0.5 kg. Take a shallow bite with the scissors so that the internal organs are not damaged. Pull the flap downward to open the body cavity, freeing the forward edge and any clinging mesenteries. The body wall can now be cut along the bottom and removed to fully expose the internal organs (Fig. 18.2). This method of opening a fish is useful in general examination and inspection for parasites. An aseptic technique for opening a fish to obtain material for bacterial isolation is described in section 18.3.3.

Digestive tract. Carnivorous fish, which include most game species, have a gut that is a short, simple S-shaped tube. Beginning anteriorly, the esophagus extends from the oral cavity to a muscular, elastic stomach. The stomach empties into either a straight or simply convoluted intestine that terminates at the vent or anus (Fig. 18.2). The liver, which among other vital functions produces bile, lies around the anterior portion of the stomach. It is connected to the gut by the gall bladder and bile duct. Pyloric caeca, fingerlike out-pocketings of the digestive tract in the region of the pylorus, are present in many fish (e.g., trout, centrarchid bass), but not in all (e.g., catfish). Pyloric caeca often appear to be draped over the front of the stomach because the digestive tract's S-shaped curve places the pylorus anterior to the stomach. Although not part of the digestive tract, the spleen is easily recognized as a small, darkly colored organ located near the posterior portion of the stomach.

Although normal livers usually have a rich, reddish-brown color, they may appear somewhat pale because of species and dietary differences. Such color variation does not necessarily reflect ill health. The liver, however, should be firm, uniformly colored, and free of spots. Bacterial diseases sometimes produce abcesses in the liver that appear as light colored areas; small tumors also may have a similar appearance. Larvae of trematode parasites (small white "grubs") can be seen in the liver without magnification. Inflammation in or around the gut or anywhere in the peritoneal cavity is a frequent sign of a systemic bacterial infection. Large amounts of mucus in an otherwise empty intestine is associated with some viral diseases. Large parasites (e.g., nematodes, cestodes) can be seen easily when present inside the gut. A stomach and intestine containing food items means the fish has fed recently, usually a sign of good health. Absence of food in the gut, however, may only mean that the fish has not eaten recently.

Kidney. The kidney lies along the backbone above the gas bladder. In systemic

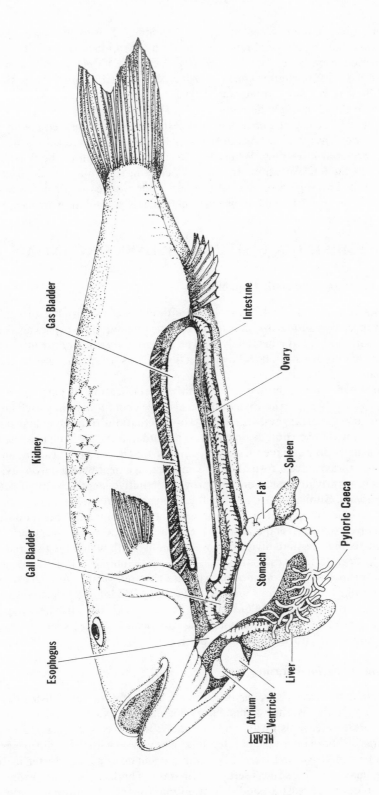

Figure 18.2. A trout dissected to show normal internal organs.

infections, bacteria tend to accumulate in the kidney. This explains why the kidney often shows signs of bacterial infections and why the kidney is the best organ from which to isolate pathogens (section 18.3.3). A healthy kidney is uniformly dark in color and has a flat or slightly concave surface. In contrast, a kidney with white spots and puffiness is indicative of advanced systemic infection or toxicosis.

Muscle. Slices through the white muscle of fish may reveal hemorrhagic areas and free or encysted larvae of trematodes and cestodes that can be seen readily without magnification. Cavitylike lesions in the muscle occasionally are caused by chronic, systemic bacterial infections. While most fish pathogens cannot be transmitted to humans and the few exceptions are killed by cooking, obviously parasitized fish are usually rejected by anglers for aesthetic reasons. There are cases, then, where parasite occurrences not severe enough to be harmful to the fish population are harmful to the fisheries resource.

18.3   SAMPLING A FISH FOR DISEASE ORGANISMS

18.3.1   Locating Diagnostic Capabilities

As stated in section 18.1.2, a biologist without extensive fish health training can only be expected to spot gross indications of ill health and to take, package, and send samples correctly for further analysis. The previous sections outlined methods of detecting obvious disease. The following sections discuss the proper way to obtain and handle samples.

Where can samples be sent for definitive identification of pathogens or toxic residue analysis? Because there is a different arrangement for handling such samples in each state and the arrangements change with fluctuations in funding and staffing, a universal list would be cumbersome and soon outdated. While state fish and wildlife agencies usually do not have their own complete diagnostic facilities for fish health, they almost always have a formal or informal arrangement with a university, federal installation, or private practitioner to provide definitive identification of pathogens when necessary. Similarly, fish and wildlife agencies usually depend on other groups for toxic residue analyses after the more obvious environmental causes of fish kills (e.g., dissolved oxygen, temperature) have been ruled out. Often, pollutant analyses are conducted by the state division of water quality with which the fish and wildlife agencies cooperate in investigating fish kills (*see* chapter 13). Individuals with fish health problems can find diagnostic expertise through their state fish and wildlife agencies or allied organizations such as the Fish and Wildlife Service, the Agriculture Extension Service, or the Soil Conservation Service. Also, many fish health specialists are active in the Fish Health Section of the American Fisheries Society, which is an excellent source of information and expertise.

18.3.2   Sampling for Parasites

Selection and care of specimens. Although field examination of fish is best begun with living fish, this is almost mandatory for parasite examination because many external parasites leave fish within a few minutes after the host's death. Many of the larger parasites, both internal and external, can be viewed with the unaided eye or with the help of a hand lens. An external examination and necropsy conducted in the field, therefore, may provide some useful information. The basic parasite examination, however, requires at least a good dissection microscope. A thorough examination,

---

Box 18.1.   Some Common Pathogen-induced Fish Diseases

Disease	Causative Agent	Typical Host
Parasites		
ich or white spot	*Ichthyoptherius multifilis*	freshwater fish
Epistylus infection	*Epistylus* sp.	freshwater fish
Trichodina infection	*Trichodina* sp.	freshwater fish
whirling disease	*Myxosoma cerebralis*	trout
gyro	*Gyrodactylus* sp.	freshwater fish
yellow "grub" of muscle	*Clinostomum marginatum*	freshwater fish
tapeworm	*Ligula intestinalis*	minnows, suckers
fish "louse"	*Argulus* sp.	freshwater fish
anchor parasite	*Lernea* sp.	freshwater fish
fungus	*Saprolegnia* sp.	freshwater fish
Bacteria		
furunculosis	*Aeromonas salmonicida*	trout
motile *Aeromonas* septicemia	*Aeromonas hydrophila*	freshwater fish
columnaris disease	*Flexibacter* sp.	freshwater fish
vibriosis	*Vibrio anguillarum*	saltwater fish
enteric redmouth	*Yersinia ruckeri*	trout
Edwardsiella septicemia	*Edwardsiella tarda*	catfish
bacterial kidney disease	*Renibacterium salmoninarum*	trout
Viruses		
infectious pancreatic necrosis	ether stable virus	young trout
infectious hemopoietic necrosis	rhabdovirus	salmon
viral erythrocytic necrosis	blood cell virus	saltwater fish
channel catfish virus	herpesvirus	catfish

---

which leads to a definitive inventory of parasites, requires a compound microscope and a person with fish health training. It is necessary, therefore, to transport living fish suspected of having a parasite problem to a laboratory where they can be more fully examined.

Basic parasite examination. While fish health training is necessary to make a complete parasitological report on a fish, a studious biologist who routinely examines fish for parasites can soon learn to recognize the important ones and make judgments regarding the severity of the infection. A basic parasite examination (Hoffman 1967) is begun by killing the fish. If the fish is small, submerge the fish in a petri dish of water, and examine it under a dissection microscope. If the fish is large, keep the body surface moist, remove portions of the fins, and examine them under the dissection microscope. Next, take mucus scrapings from the body of the fish and examine them using a slide, a drop of clean water, and a cover slip, under a compound microscope.

Cut away a gill arch and examine it under a dissection microscope. Then, place a bit of gill tissue in a drop of water between a slide and cover slip and examine it under the compound microscope. Open the fish as described in section 18.2.3, remove the viscera, and examine the organs submerged in water (or preferably in physiological saline) in a petri dish under the dissection microscope. While examining the viscera, tease apart the internal organs with forceps and scrape the inside of the gut to reveal parasites. Tease apart and examine some of the body musculature and check the eyes and brain under a dissection microscope as well.

18.3.3 Sampling for Bacteria

Selection and care of specimens. It should be recognized that a number of fish must be screened for bacterial pathogens in order to make a valid conclusion about the health of a population. While methods of determining the actual sample size necessary are outside the scope of this chapter, it should be obvious that a larger sample of fish is necessary to rule out an active pathogen in a larger population of fish or when incidence of disease is low (McDaniel 1979). Usually, when fish are transported to fish health diagnostic laboratories for bacterial isolation, the first step is bacterial identification. There is no reason, however, that bacterial cultures cannot be begun in the field, as long as an aseptic technique is followed carefully. All fish must be alive when collected and, ideally, freshly killed when bacterial isolation is attempted. If fish cannot be kept alive, they may be enclosed individually in sealed plastic bags and stored on wet ice for preferably less than eight hours and absolutely not more than 48 hours.

Bacterial isolation. While identification of bacterial pathogens requires specialized training as well as access to a variety of stains, media, and laboratory equipment, taking an initial culture is fairly simple and requires only a few basic tools. After conducting the external examination (section 18.2.2), kill the fish with a blow to the head. Sear the area immediately behind the dorsal fin with a glowing red spatula to kill any bacteria on the surface. Dip a pair of scissors in high proof alcohol and heat them in a flame. Cut down through the seared area, severing the spinal cord but not going deeper. Bend the head and tail together to expose the kidney, a dark red mass below the spinal cord (Fig. 18.3). Insert a flamed innoculating loop down into the kidney, extracting a loop of tissue. Streak the tissue on about a quarter of a sterile, nutrient agar plate (blood agar and trypticase soy agar are good for isolation). Flame the loop and streak across the plate at right angles from the edge of the initial streaks. Flame the loop again. Streak the tissue a third time from the edge of the second streaks. This procedure will spread out the bacteria, permitting the growth of

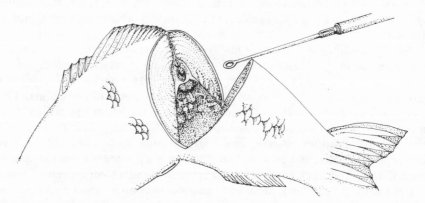

Figure 18.3. Aseptic technique for obtaining kidney tissue for bacterial culture.

individual colonies. Isolations from fish may be incubated at room temperature (20-25 C), although faster growth of colonies will occur in incubation at 37 C.

18.3.4   Sampling for Viruses

Diagnostic work with viruses requires specialized procedures and equipment. Select fish suspected of viral disease and transport them to a fish health diagnostic center as described in the section for bacterial sampling (18.3.3). A biologist who routinely transports samples for viral analysis may want to learn to make the initial tissue homogenates, which, if buffered, may be stored for up to a week under refrigeration (McDaniel 1979).

18.4   SAMPLING BLOOD AND TISSUE

18.4.1   When and Why to Collect Blood and Tissue Samples

As mentioned in section 18.1.1, adverse environmental conditions are more often the cause of widespread ill health in wild fish than are pathogens. Often the specific stressor can be readily identified in the environment, as in the case of an acutely depressed dissolved oxygen level. At other times, however, overt signs of disease and mortality may occur in fish populations when no obvious adverse environmental conditions appear and pathogens are not suspected. In these instances, it is often useful to obtain blood and tissue samples from fish for pathology studies and residue analysis. Such analyses also may be useful in evaluating sublethal stress (Wedemeyer and Yasutake 1977) and often are employed in toxicological and physiological research on fish.

18.4.2   Sampling Blood

Techniques of obtaining blood. Although there are a number of ways to obtain blood samples from fish, perhaps the easiest and most consistent method involves tapping the blood vessels that run in the hemal arches of vertebrae between the anal fin and caudal peduncle. Fish less than 15 cm long must be sacrificed to obtain even a minimal blood sample for microanalysis (0.1 ml). Immobilize the fish with anesthetic (e.g., tricaine methanesulfonate) or by a blow to the head, though excessive force will cause cranial hemorrhaging and a reduction in blood flow. Wrap the fish in a paper towel for ease of handling, cut the tail off just above the caudal peduncle with a razor blade, and immediately place a heparinized (or otherwise anticoagulant-treated) capillary tube with a volume of about 0.25 ml against the severed caudal vessel. The tube will fill with blood and additional tubes may be collected if possible and necessary, though they must be filled without delay since fish blood coagulates quickly.

Blood samples from fish longer than 15 cm may be obtained without killing the fish. Anesthetize the fish (50 to 100 mg/1 of tricaine methanesulfonate acts rapidly and short exposures are not harmful; *see* chapter 5), and prepare a heparinized syringe or Vacutainer blood sampler. Syringes may be heparinized by rinsing with a heparin solution (commercial, injectable preparations of 10,000 units/ml work well). Vacutainers are available with a variety of anticoagulants. While one person holds the fish ventral side up, another inserts the needle (1.5 inches, 21 gauge for small fish, 18 gauge for large fish) at the midline between the anal fin and the caudal peduncle until the needle stops against the vertebral column (Fig. 18.4). Apply slight back pressure

Figure 18.4.   Technique for obtaining a blood sample from a large fish without killing the fish.

with the syringe or break the seal on the Vacutainer; if blood flow does not begin, slight twisting and movement of the needle may initiate it. When an adequate sample is obtained, withdraw the syringe, remove the needle, and expel the blood into a microcentrifuge tube (1.5 ml).

Preservation of samples. Depending on the analysis to be performed, whole blood may be stored up to several hours on ice. Hematocrit (packed cell volume), however, should be run within minutes after blood collection because of the swelling of red blood cells that occurs in blood exposed to the air. Most blood characteristics are determined on plasma, the fluid portion of the blood without the cells. It is best, therefore, to centrifuge immediately after collection, separating the plasma from the cells, and to freeze the sample. This can be accomplished in the field by using a portable generator for the centrifuge and a cooler with dry ice.

18.4.3   Sampling for Histology

Post-mortem changes in histological features (microscopic tissue structures) occur rapidly, making it mandatory that histological samples be taken from fish collected alive. Immediately after killing the fish, remove the tissues of interest. Make small, thin slices, and drop them into labeled vials of fixative. The necessity of taking thin (2 mm) slices deserves emphasis because this promotes rapid penetration by the fixative and minimizes later concern about artifacts caused by poor fixation. Bouin's is an excellent fixative for fish tissue although buffered, 10% formalin is nearly as good and has the advantage of being compatible with routine procedures for slide preparation in large, automated, histopathology laboratories. The volume of fixative in the vial should be at least ten times the volume of tissue to be fixed.

While most fish tissues have some diagnostic value in a histopathology study (Roberts 1978), gill, liver, and kidney tissues are almost always included. Kidney tissue is difficult to dissect from fish because it adheres to the body wall and is soft. In small fish, thin slices can be taken through the vertebral column to include the kidney. In large fish, localized freezing of the kidney with small cans of freon hardens the kidney tissue so it can be separated easily from the body cavity and sliced.

18.4.4   Sampling for Residue Analysis

When sampling for residue analysis, the Environmental Protection Agency suggests that a minimum of five gamefish, five roughfish, and five catfish be collected at each station. Record species, length, weight, and sex of each fish collected.

Tissue for most residue analysis may be taken from dead fish as long as the tissue is not putrid. Wrap tissue for heavy metal analysis in plastic, and wrap tissue to be analyzed for organic compounds in aluminum foil. Keep wrapped tissues on ice until they can be frozen. It is important to sample white muscle because it consititutes the edible portion; take white muscle tissue from the back of the fish in front of the dorsal fin. Analyze organ tissues as well because they tend to accumulate toxins to a greater degree than muscle. Blood plasma and brain tissue are important because they reflect physiologically available residues and do not include an inactive, bound component that may occur in other tissues.

18.5 REFERENCES

Amlacher, E. 1970. *Textbook of fish diseases.* T. F. H. Publications, Neptune City, New Jersey, USA.

Hoffman, G. L. 1967. *Parasites of North American freshwater fishes.* University of California Press, Berkeley, California, USA.

Leitritz, E., and R. C. Lewis. 1976. *Trout and salmon culture.* California Department of Fish and Game. Fish Bulletin 164, Sacramento, California, USA.

McDaniel, D., editor. 1979. *Procedure for the detection and identification of certain fish pathogens.* Fish Health Section of the American Fisheries Society, Bethesda, Maryland, USA.

Plumb, J. A., editor. 1979. *Principal diseases of farm-raised catfish.* Auburn University, Southern Cooperative Series No. 225, Auburn, Alabama, USA.

Roberts, R. J., editor. 1978. *Fish pathology.* Baliere Lindall, London, England.

Sindermann, C. 1970. *Principal diseases of marine fish and shellfish.* Academic Press, New York, New York, USA.

Snieszko, S. F., editor. 1970. *A symposium on diseases of fishes and shellfishes.* American Fisheries Society, Special Publication Number 5, Bethesda, Maryland, USA.

United States Department of the Interior. 1960's to present. Fish Disease leaflets (continually updated series). United States Fish and Wildlife Service, Fish Health Research Laboratory, Kearneysville, West Virginia, USA.

Wedemeyer, G. A., F. P. Meyer, and L. Smith. 1976. *Environmental stress and fish diseases.* T. F. H. Publications, Neptune City, New Jersey, USA.

Wedemeyer, G. A., and W. J. Yasutake. 1977. *Clinical methods for the assessment of the effects of environmental stress on fish health.* United States Fish and Wildlife Service, Technical Paper Number 89, Washington, District of Columbia, USA.

Chapter 19
Underwater Methods

GENE S. HELFMAN

19.1 INTRODUCTION

Much of the information that a fisheries scientist requires is traditionally collected indirectly and this manual focuses on such methods. The emphasis in this chapter is on direct observational techniques, particularly techniques that involve the use of divers.

Direct underwater observation is a valuable and frequently neglected tool in the repertory of a fisheries program. Divers can count and estimate the sizes of organisms in areas where nets or cameras are inappropriate, such as in dense macrophyte beds, under ledges, in holes, or around and under coral heads or rocks. Divers can make observations of spawning, feeding, movement, and other kinds of behavior in a non-destructive manner, a considerable advantage over nets, angling, or poisons. Divers can measure environmental variables (light, temperature, current, water chemistry, etc.) in the exact place of interest rather than in a general area. Divers can also place underwater monitoring gear exactly where surface workers want the equipment and can locate and recover such equipment without the use of costly and elaborate hoisting gear. Underwater methods can serve as a primary means of collecting data or as a complementary tool for supplementing or verifying data collected indirectly or directly through the use of acoustic or biotelemetric methods (chapters 12 and 20).

19.2 DIVING

19.2.1 Basic Diving Methods

Skin Diving. Snorkeling (skin diving) is the least expensive, least disruptive, and logistically simplest way of observing fishes in the field. Skin divers use a mask, snorkel, and fins; a wetsuit is necessary to increase observation time in all but the warmest water. Many fishes that are frightened by the bubbles and sound associated with scuba equipment will allow close approach by a snorkeler. Snorkeling requires little training or energy expense: the snorkeler usually floats at the surface, recording information on a writing slate or cuff (*see* below). Snorkeling is effective for obtaining data on the abundance, distribution, habitat preferences, and behavior of fishes in many habitat types given sufficient water clarity (Keenleyside 1962; Hobson 1968; Ebersole 1977; Hall and Werner 1977).

Snorkeling has obvious disadvantages and limitations. Because observations are made from the surface, water clarity should usually exceed water depth. In water with vegetated or otherwise dark bottoms, many fishes are difficult to observe. Bottom-dwelling and small fishes in water more than a few meters deep are difficult to see, thereby biasing population estimates in favor of large, mobile forms. Behavioral observations are also limited to what can be observed effectively from the surface simply because dives without a scuba tank are limited to less than one minute. Other problems include rough surface conditions which make systematic counts or observations difficult. For basic water safety considerations, two persons are necessary for any diving project.

Scuba. The solution to many of the disadvantages of snorkeling lies in the use of scuba (self-contained underwater breathing apparatus). Information on required and supplementary scuba equipment is available in several publications (Woods and Lythgoe 1971; Empleton et al. 1974; Anonymous 1975; Anonymous 1978; Drew et al. 1976). All of these should be read in combination with completing a diving certification course taught by a licensed instructor before diving begins. The potential hazards of scuba diving are sufficiently great that certified instruction is mandatory. Scuba tanks are filled to high pressure (140 to 210 kg/cm²) with compressed air. Air is delivered on demand to the diver, who breathes through a mouthpiece that regulates air flow so that air is delivered at a comfortable rate regardless of depth. In addition to tank and regulator, divers wear a wetsuit, inflatable vest (bouyancy control device), depth gauge, pressure gauge, watch, weightbelt, and weights. Both scuba divers and skin divers should also carry a dive flag and dive knife. A dive flag warns boaters of the presence of the divers in the water (assuming the boaters know what a dive flag means), and a knife is needed in part to free a diver from monofilament fishing line, which is essentially invisible underwater. Government agencies and universities commonly have diving councils or regulations concerning diving. A prospective research diver should be aware of such regulations.

Scuba permits a diver to submerge to most depths at which fish are abundant and to remain there for extended periods (e.g., more than one hour at less than 20 meters). This allows inspection of cryptic habitats, as well as observation, enumeration, or collection of benthic and small fishes or large, mobile forms that may be frightened by the commotion of a snorkeler's descent. Rough surface conditions can be a problem in getting to and from a study locale, donning equipment, and tending the boat but do not usually affect divers in the water. One person should remain in the boat while divers work in the comparative calm that usually prevails a meter or more below the surface. Divers can operate almost any type of research equipment that terrestrial researchers use if it has been housed or modified for submersion—often at considerable expense.

Scuba diving has definite liabilities. Turbidity makes fish difficult to observe and also disorients the diver. Most fish are at least initially frightened by the noises and bubbles associated with scuba gear. Scuba equipment is heavy, bulky, and expensive, and it is dangerous when used by untrained persons or if it is not kept in good repair. A diver must either have an air compressor to refill scuba tanks or must own or rent several tanks which have to be filled commercially. Full scuba tanks have a considerable explosive potential if punctured, and care must be taken when handling, transporting, and storing them. The time required to prepare for, drive to and from, and clean up after a dive often exceeds the actual time spent diving. Many medical conditions that are only discomforts on the surface, such as respiratory tract or ear problems, can become quite serious when diving or breathing air under pressure. A reputable diving instructor will require a complete medical examination before training a person in scuba diving.

Box 19.1   Useful Equipment for the Fisheries Scientist Working Underwater
The following items may be used in addition to the usual equipment required for diving:

1. Underwater compass. Obtain one with a rotating bezel, marked in at least 5° intervals.

2. Dive watch/stop watch. Digital underwater watches with liquid crystal displays (LCD) have recently become available. A popular model is the Casio Alarm Chronograph. This watch includes a stopwatch, countdown alarm (interval timer), regular alarm, and light. It is water-resistant to 100 m.

3. Underwater lights. For snorkeling, most inexpensive, floating plastic flashlights can withstand submersion to 2 to 3 m. For scuba, the smaller, nonfloating, narrow beam, rechargeable lights (such as the Super Q-Lite, Underwater Kinetics, San Diego, California) are preferable to bulky and/or floating lights.

4. Cameras. Underwater photography can be useful in data collection, record keeping, or for creating visual aids for presentations. Still and movie cameras are available as self-contained, submersible units, such as the Nikonos IV 35 mm camera or the Eumig Nautica Super-8 movie camera. The alternative is to place a terrestrial camera inside a watertight housing, obtainable from a dive shop. Several books are available on underwater photography (e.g., Schulke 1978), and the subject is generally discussed in most scuba certification courses.

5. Writing slate, writing cuff, or writing scroll.

6. Microspear for collecting small fishes.

7. Snap-clip for weight belt. Get the largest size available; serves as a third hand. Keep lightly lubricated with silicone spray.

8. Net bag. Use a large bag with a metal, latching top. Can be attached to clip on weight belt.

9. Thermometer. Use a rugged, protected aquarium thermometer. If using a glass thermometer, set it inside rigid, clear plastic tubing with silicon cement. Calibrate it regularly against a reliable laboratory thermometer.

10. Hand tally counter. Useful for recording numbers of fishes or behavioral events. Usually made from red plastic, with three or four channels. Spray *lightly* with silicone spray to prolong life of metal springs inside.

11. Colored flagging. Use either international orange or yellow. Cut it into 30-cm strips, number it with an indelible marker, and push ten-penny or larger nails through a double section of flagging. A convenient carrier can be made from a short length of bicycle inner tubing with a loop of rope at one end that can be slipped over a wrist or clipped on a weight belt clip. Punch holes in the inner tubing, number the holes with acrylic paint, and push the flagging nails into their respective holes. The same type of inner tubing holder is useful for carrying plastic bags for specimens.

12. Fiberglass measuring tape. Reels that hold more than 50 m are cumbersome.

13. A half-meter stick with thickened black lines at 10-cm intervals for estimating fish sizes. Another technique is to draw outlines of different sizes of particular species on a separate sheet of waterproof paper attached to the back of the dive slate or clipboard.

14. Cyalume lightsticks. These 13-cm by 1-cm, fluid-filled plastic cylinders

Continued

> produce a cold, green light. Useful as an emergency eight-hour backup light or for illuminating a small area with a weak, diffuse light during nocturnal observations.
> 15. Underwater Tape Recorder. Allows recording of abundant data without loss of eye contact with subject and eliminates the need for a clipboard. At present, the only commercially-made underwater tape recorder is the Wet Tape (Sound Wave Systems, Incorporated, Costa Mesa, California). The microphone is incorporated into the scuba regulator mouthpiece and produces a recording that takes practice to understand.
> 16. Gardening gloves, and work pants or overalls. Essential for protection of the diver from stinging or scratching structures and the pockets are useful for storage. Lubricate zippers with silicone spray.

Hookah. Sources of air other than scuba tanks are also available for diving (*see* general references, Scuba). An example of these air sources is the umbilical or hookah rig, which consists of an air pump above water that delivers a continual flow of air to the diver via a hose. Characteristics of this hookah rig, such as constant noise and bubbles and tangling of hoses, have made it more applicable to salvage and repair work than to fisheries-related activities.

19.2.2 Exceptional Diving Conditions

Recreational divers can generally pick their places and times for diving; when conditions are less than ideal, they can cancel a dive. The research diver, however, operates under a different set of constraints and objectives and, as a result, may have to dive where and when conditions merit extra caution and preparation. Such situations can only be mentioned briefly here, but they have been treated extensively in general references and in the specific publications cited below. In all the following situations, divers should carefully plan and even rehearse their actions and must guard against confusion and panic, which become more likely in unfamiliar or stressful circumstances.

Rivers and strong currents. Even weak currents (much less than one knot) are difficult to swim against. Flowing water is often turbid and creates problems associated with reduced visibility (*see* below). In addition, hazards include the potential for (1) exhaustion if a diver attempts to fight a current; (2) being swept too far from the support boat or exit point to make return possible; and (3) becoming entangled or wedged in bottom obstructions, particularly when visibility is poor. Diving under these conditions usually requires (1) additional surface support; (2) a long, highly visible, floating line that trails far behind the surface vessel and terminates in a large, visible object, preferably a raft or boat; (3) extra diving weights to slow downstream movement along the bottom; (4) gloves; and (5) a diving tool with a sharp edge that can be used for prying and cutting. (Strap the knife to the inside of a thigh to prevent it from being entangled and to ensure accessibility by either hand.) If conditions permit, divers should work upstream of the support vessel or exit point and move downstream at the completion of the dive.

Night diving. Simple daytime tasks become surprisingly difficult in the dark. Hazards of nighttime diving include potential for (1) becoming entangled, lost, or disoriented in both horizontal and vertical space; (2) misplacing important equipment;

and (3) flashlight failure. Divers should minimize the equipment carried and simplify the objectives of night dives as compared to day dives. A diver should have a knife, spare flashlight or Cyalume Stick, compass, and gloves that can be removed easily. A strobe light, flashlight, or Cyalume Stick should be suspended below the support boat, and the deck of the boat should be well illuminated to aid the diver's return. Decompression dives at night are extremely dangerous because any problems that require a rapid return to shore are complicated by the probability of getting lost (*see* Boehler 1979).

Turbidity. Turbid conditions, caused by silt, algae, or pollution, prevail in most of the marine and freshwater environments of North America. Turbidity can range from an inconvenient reduction in underwater visibility to completely obscured vision and total darkness. Many of the problems inherent in night diving also apply to turbid situations. However, in very turbid water, available light is extinguished and flashlights provide no additional illumination. All work must be done by feel. Maintaining contact between divers may require a "buddy line," a 1-m length rope connecting both buddies. In very turbid water, exhaled bubbles are no longer visible, and it is easy for a diver to become disoriented to the point of nausea. Although little useful research can be accomplished in low visibility conditions, research divers often find themselves involved in gear recovery in turbid water. Water clarity can also deteriorate rapidly as a result of shifts in tides or currents or when moving through a thermocline. Divers should obtain local knowledge about the probability of such changes and be prepared to handle them.

Ice diving. Surface ice predominates through much of the year in most lakes in northern North America, and winter conditions strongly affect the biology of many fishes. In addition to such problems as hypothermia associated with cold weather (chapter 3), ice diving includes such hazards as: (1) regulator freeze-up in air; (2) disorientation because of cold water in ears; (3) losing contact with the surface hole because the safety lines become unattached at the surface or at the diver or the hole becomes blocked when diving under broken, drifting ice; and (4) inability to manipulate equipment with thick gloves. Surface support, back-up divers, extra scuba tanks and equipment, wind protection, and access to a warm area at the completion of a dive are essential. Suspend a strobe or other light from the surface hole to aid divers in finding their way back. Clear snow away from the area immediately around the hole to improve light penetration through the ice, creating a bright area that helps a diver find the hole. Shovel snow along straight lines radiating out from the hole, thus creating light paths that can be followed by a lost or disoriented diver. If snowmobiles or other traffic are likely, mark the dive hole with a highly visible dive flag. After a dive, replace the ice wedge in the hole, and mark the site to prevent anyone from falling in before the hole refreezes. Before attempting a dive, a diver should receive instruction in ice diving techniques as discussed in Jenkins (1973) and Somers (1974).

Caves and caverns. Cave diving offers little to attract a fisheries biologist, particularly because of the number of fatalities that have occurred in caves. Caverns, where the diver is continually in view of surface light, contain many more organisms and fewer associated hazards than do caves. Major problems of cavern diving involve (1) reduced visibility because of silt suspended by the diver's activities, (2) the need for reliable lights, and (3) the potential for the bends because many caverns have very clear water into which sunlight penetrates to considerable depth. Precautionary practices include carrying extra lights and Cyalume Sticks and paying careful attention to depths and the limits of no-decompression diving. Information on cavern diving can be found in Mount (1973).

Decompression dives. Decompression stops are necessary when divers go too deep for too long and, consequently, accumulate excess nitrogen in their bloodstreams; they cannot come directly to the surface at the end of a dive. Most experienced research divers plan their dives to avoid the need for decompression stops. Careful planning becomes increasingly important when more than one dive is made in less than a 12-hour period. Such repetitive dives lead to build-up of residual nitrogen. The ability to avoid decompression after each deep dive is a major attraction of "saturation" diving from undersea habitats. Unique problems associated with decompression occur when diving in mountain lakes or other high altitude situations where atmospheric pressure is reduced and no-decompression limits are, therefore, shorter and shallower. For similar reasons, many laboratories and government agencies forbid diving during the day prior to flying. *See* Cross (1970) and Smith (1976).

Hypolimnetic diving. Many lakes in North America are strongly stratified, with a layer of warm water overlying cooler water for much of the year. The thermocline or transition region between these two areas can be quite abrupt. In addition to encountering a water mass of much colder water with reduced visibility, the diver may encounter water with a different chemical nature from the water above. A hypolimnion may contain large quantities of hydrogen sulfide (H_2S), ammonium (NH_4), and methand (CH_4) that can be extremely toxic to divers and also cause equipment failure. Lakes that remain stratified throughout the year often contain the strongest concentrations of potentially toxic hypolimnetic chemicals. A diver should always obtain information about conditions beneath the thermocline before diving in unfamiliar waters.

19.2.3 Recording Data While Diving

The most useful way to record information underwater is to use a tape recorder, but such equipment is costly, requires transcription and translation after a dive, and tends to jam. Research divers, therefore, commonly use underwater writing slates. Four types are commonly used.

Erasable slate. An erasable slate is generally made from a thin (2 mm or less) piece of white plastic. A small loop of surgical tubing is attached to one corner for carrying the slate, and a No. 2 pencil sharpened at both ends is attached by a length of surgical tubing or string to another corner. (Holes are drilled through the slate just large enough to slip tubing through.) The pencil can be stored in a loose end of the surgical tubing to keep it from becoming tangled in other gear and/or breaking. The surface of the slate should be roughened with light sandpaper for easier writing. A ruler can be glued to one edge, or centimeters can be marked off with a Sharpie or engraved with a metal engraving tool and filled in with indelible ink. After each dive, notes are transcribed, and the slate is erased using scouring powder. The slate can be any size desired, although 21 cm x 28 cm is convenient. The disadvantages of an erasable slate are that: (1) no permanent record of original data exists, unless notes are photocopied; (2) plastic can break if bent sharply; (3) extra slates must be carried if more data are needed than can be recorded on both sides of one slate; (4) the white surface occasionally alarms fish when it is brought up to writing position; (5) some fish will attack pencils or surgical tubing; (6) both hands are needed for writing, which is particularly inconvenient when diving gloves are being used.

Clipboard with underwater paper. This is probably the most popular underwater writing material. A standard clipboard, either 32 cm or 40 cm long, is fitted with a surgical tubing loop and pencil as described above. Two large rubber bands are slipped around the bottom of the board. The actual writing material is plasticized, waterproof paper. A common brand is Polypaper (Nalge Corporation, Rochester, New York), which comes in standard 21 x 28 cm sheets. The paper should be air dried after the dive, and the notes transcribed, providing the diver with a permanent record of the original data. Information that is regularly needed during dives can be permanently recorded on a piece of waterproof paper with a Sharpie and glued to the back of the clipboard with fiberglass resin. The disadvantages of the clipboard are that: (1) grommets on the clipboard can rust and break free (if this happens underwater while the clipboard is attached to the diver's wrist, the clipboard and data can be lost); (2) the paper has a tendency to lift up in currents, which is why rubberbands are necessary to hold it in place; (3) turning over a piece of paper with gloves on can be surprisingly difficult; (4) although waterproof paper costs only about 15¢ per sheet, this cost can add up quickly in an active diving program; and (5) if the paper is rubbed vigorously, some writing may be erased leaving the diver with the task of transcribing notes from the pencil impressions in the paper.

Writing cuff. When a relatively small amount of data is to be collected, when data are extensively coded and can be fit onto a small area, when extensive written communication between divers is likely, and when divers will need both hands free for other tasks, a useful primary or secondary recording device is a writing cuff (Fig. 19.1). The cuff is a 20-cm long section of 10-cm diameter white polyvinylchloride (PVC) pipe that has been lightly roughened with sandpaper. Three pairs of 7-mm holes are drilled in one end of the pipe, each pair equidistant from the other two pairs. A 40-mm length of surgical tubing is threaded through the holes forming an equilateral triangle inside the pipe through which the diver's arm is passed. A double-pointed pencil is attached by a string just long enough to permit the diver to write on any part of the cuff. The diver records, transcribes, and erases information on the cuff as with the erasable slate. The disadvantages of the writing cuff are the same as those of the erasable slate, except that the surface of the cuff is smaller, i.e., usable surface area is roughly equivalent to one side of a piece of underwater paper. In addition, soreness or cramping in the arm carrying the cuff may result from surgical tubing being too tight, keeping the arm bent in a writing position, or cutting into the elbow region if the cuff is too long (particularly a problem when wearing a wetsuit).

Scroll. A more versatile underwater writing apparatus than the cuff is a continuous scroll (Fig. 19.2), as described by Ogden (1977). The writing surface is a long strip of translucent, matte surface polyester drafting film which is fed from one roller to another across a flat piece of plastic (usually LEXAN). The polyester film is attached to rollers made from solid 13-mm diameter PVC rods that are threaded at either end and to which hexagonal PVC nuts are attached to facilitate hand manipulation of the rollers. By advancing the film, a new writing surface is presented. Tabular data forms, maps, photographs, or lines can be attached to the flat surface, and data can be recorded on the overlying film. The scroll eliminates the need for multiple sheets of paper, provides a large quantity of writing surface, can be photocopied for a permanent record, or can be stored when full and replaced by another roll of polyester film.

The disadvantages of the scroll are that: (1) construction time and costs are greater than for the three methods above, (2) the scroll is relatively bulky, and (3) rollers may bind and be difficult to operate when wearing heavy diving gloves or mittens.

Figure 19.1. *(A)* Writing cuff, made from 20-cm diameter PVC pipe. *(B)* Note pattern of surgical tubing threaded through holes in the end of the cuff. Cuff based on design of R. C. Carpenter, University of Georgia.

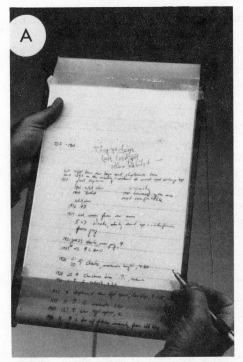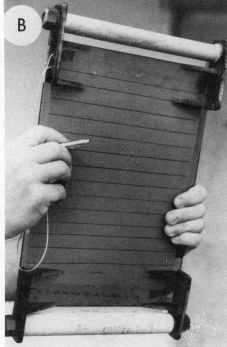

Figure 19.2. Continuous writing scroll, *(A)* front view and *(B)* back view. Design shown is based on modifications by J. C. Ogden from the original scroll described in Ogden (1977). Scroll shown was constructed by E. B. Bermingham.

For all four devices, always carry additional, short pencils sharpened on both ends and a small twist pencil sharpener. Place these in a closeable plastic bag which is weighted sufficiently to keep it and the bouyant pencils from escaping everytime the dive bag is opened.

19.3   OTHER METHODS OF DIRECT OBSERVATION

19.3.1   Video Cameras and Television

Scuba divers are limited to relatively short working times and shallow depths because of problems of nitrogen narcosis, decompression, exhaustion, and heat loss. These problems can be overcome by using underwater videotape or television equipment, which non-destructively samples habitats or populations that may be unavailable to or wary of divers. A video camera can be placed at a chosen locale and monitored with a shoreline or shipboard television, while the image is recorded and stored for future reference on video tape. Such video equipment can be carried by a diver, mounted on a stationary base, or towed by a boat.

Diver-carried video systems are useful for counting fish in a prescribed area, such as around natural or artifical reefs. Diver-held video cameras are particularly useful for recording fish behavior to be analyzed later. Video tapes are less expensive than black and white films, are re-usable, and the investigator knows immediately whether the activities or information have been recorded successfully.

Stationary systems have been used to count fish passing a given point or have been set to swivel at a fixed rate and used to estimate the numbers and species in an area (Smith and Tyler 1973). The lack of a time and depth limit makes this approach particularly useful for determining activity patterns of common fishes in an area or for observing the investigatory, entrapment, and escape responses of organisms around baits, traps, and gill nets (High 1967; Powles and Barans 1980).

Towed video systems have found application in surveys of large areas that are inaccessible to divers because of the depth and size of the area in question. Cameras can be mounted on a sled that is towed along the bottom by a research vessel. The video tapes produced by this method can help locate harvestable aggregations of valuable species, determine the distribution of particular species and habitat types, and provide background data for comparisons in the event a fishery is opened, closed, or otherwise regulated (Uzmann et al. 1977; Henry et al. 1981). Information on analysis of videotape data can be found in Lehner (1979).

The primary drawback to a video system is the initial cost. Depending on the level of sophistication desired (black and white versus color, high resolution of small objects and motion, low light capabilities), underwater video systems at present cost at least $10,000. Any instrument used underwater can leak or break, and repairs are costly and time-consuming. For diver-deployed systems, such as diverheld or stationary cameras, conditions must be appropriate for putting a diver overboard. Diver-carried units have limited mobility because the diver carries a camera with its attached wiring, which can get caught on bottom structure. The mass of the camera and wiring creates problems if currents are present. Water clarity limits the visibility of objects by a camera as it does by the human eye. Stationary systems can only record what moves into their field of view.

19.3.2 Observation Platforms

In situations where a small area (less than 100 m²) is under observation and where the water is relatively clear, shallow, and protected from wind and afternoon sun, a useful means of making extensive, detailed observations of fish behavior and movement is to construct an observation platform above the surface. Such platforms have been used to study the activities of large fishes such as sharks, small fishes and invertebrates in pools, and brown trout in streams (Major 1978; Bachman 1982). Observers can spend as much time as necessary recording the distribution, seasonal and diel movements, details of activities, and effects of experimental manipulation on known individual fish. Location and movement data can be recorded accurately if the study site has been mapped and/or marked off into quadrats visible from the platform. Observation platforms are appropriate to a limited set of situations, but where they are applicable, they can be exceedingly powerful tools for obtaining accurate information on fish behavior and ecology.

19.3.3 Glass-bottom Boats and Lookboxes

Other means of viewing fish that deserve brief mention include glassbottom boats and lookboxes. Glass (or heavy plexiglass) bottom boats are useful for surveying areas for particular habitats or species concentrations or for viewing activities of fishes attracted to baits or devices immediately below the boat (Kalmijn 1978). Limitations are primarily associated with water clarity, depth, and potential for breakage or scratching. Glass must be cleaned often if the boat is left in the water.

A lookbox is a long box with a glass bottom; the glass is kept just below the water surface, with the viewer looking in from the top. Lookboxes are inexpensive to construct but have limited application in fisheries work. They can be used in very shallow water, such as for finding benthic fishes or invertebrates in a stream. Lookboxes have their widest application in surveying for relatively sessile, benthic organisms (e.g., molluscs) in cold water or other situations where a mask and snorkel might be undesirable or inconvenient.

19.4 APPLICATIONS OF DIRECT OBSERVATION

19.4.1 Population Enumeration

A common application of underwater methods is in estimating population abundance (Russell et al. 1978). Counts made by divers can be used as estimates in themselves or as an aid in planning or verifying counts made in an indirect manner (Barans 1982). An initial qualitative survey by divers can help determine if a target species occurs in schools or individually, which is important with respect to the assumption of random distribution of marked fish in a mark-recapture program. Differences in species' locales by day and by night can be rapidly assessed after a few short dives. This can prevent over or underestimation of population size if only one area, depth range, or time period is sampled by an indirect method (Loesch et al. 1982).

Many of the techniques used in making quantitative underwater counts of fishes are modifications of wildlife techniques that are commonly employed for counting small mammals and birds. They are, therefore, subject to the biases and limitations of such terrestrial methods, as well as to problems unique to underwater work (*see* Burnham et al. 1980; Quinn and Galluci 1980). The actual technique used underwater depends on the depth, size, and type of bottom in the area being studied and on the acceptable statistical variance in the data (*see* chapter 1).

Subsampling a large area. A frequently used means of estimating population abundance in a large area is to lay a transect line, count the fishes along the line, and then multiply the density along the transect by the area or volume of the habitat type in question (Hobson 1974; Jones and Chase 1975; Ebeling and Bray 1976; Keast and Harker 1977).

After initial surveys of the area to determine the species present and the approximate distribution of major habitat types, lay the transect. The transect line should be highly visible and made of a floating material (5- to 10-mm diameter, yellow polypropylene rope is often preferred). Weight the rope to keep it near the bottom. Small (5 cm x 10 cm) numbered styrofoam floats can be attached to the weights to orient the diver as to location. Numbered markers also break the transect into sections which permit more quantitative sampling. Subdivision is also necessary if the habitat type is not uniform and information on habitat preferences is desired. Numbered floats should be no farther apart than twice the average horizontal visibility. Thus divers can verify their location at any point along the transect.

In its simplest form, a transect can be laid and taken up after each count. This method is often employed in environmental impact assessment, where divers count fishes and other organisms within a predetermined, estimated distance from the line at several representative sites in the area in question. If a more permanent transect is to be laid, such as one that is to be used for counts of seasonal differences in distribution and abundance, more accurate data can be obtained by counting the fishes occurring between two parallel lines.

The optimal length of a transect is determined in part by bottom and visibility characteristics. Coral reef fishes are usually counted along a 50 to 200-m transect line (e.g., Sale 1980). These lengths are also popular in other habitat types; counts usually require 10 to 30 minutes to complete, depending on fish abundance and recording method. Transect width is generally determined by fish abundance and water clarity. Transects should be no wider than twice the long-term average visibility in the water body. A narrower transect may result in lost information on clear days, but it minimizes inaccuracies when visibility is poorer than average. Most studies have used transects no wider than 10 m (5 m on either side of a diver), probably because small fishes become harder to see and identify beyond that distance and accuracy decreases if fish are abundant in the area.

Standardization of the counting method is crucial for accurate results during any transect count. The diver should swim at a slow, constant speed that is roughly the same each time a count is made. The diver should note the time at the start and end of a count. To minimize the need for slowing or stopping to write, many investigators prepare a data sheet in advance that contains the names of all expected species. The diver then records the number or presence of a species as each section is entered. Counting active fish in a school is difficult. As an aid, divers often carry a small, plastic hand tally counter (available in most grocery stores) to count fish. Underwater tape recorders greatly increase the diver's ability to make accurate counts or other observations of fishes. Tape recorders, however, are expensive and analysis of taped data is very laborious.

A complete census of fishes in an assemblage requires both day and night counts, particularly if nocturnally active species are to be assessed accurately. Many nocturnal fishes, including some of the most abundant species in both marine and freshwater environments such as grunts and crappie, aggregate near structure by day and then disperse at dusk (see Hobson 1968; Helfman 1978, 1981). Daytime counts of these fishes tend to be higher than counts made at night. Other nocturnal fishes seek shelter by day and move about freely at night (e.g., catfishes, squirrelfishes, cardinalfishes). Counts of these species tend to be higher by night than by day. Some active diurnal fishes, such as schooling cyprinids, are more easily counted when dispersed and resting on the bottom at night.

Nighttime presents special problems for fish and invertebrate censuses. Species identification is often more difficult because diurnal color patterns fade and many species assume convergent nocturnal shadings (Emery 1973). Nocturnal counts require flashlights unless the diver has a light amplification device, which is extremely expensive. Flashlights tend to bias counts against fishes that avoid lights and favor fishes that are transfixed by them (Starck and Davis 1966; Luckhurst and Luckhurst 1978). Divers, therefore, should not shine their lights more than a few meters ahead to avoid frightening uncounted fish. Many nocturnal fishes are relatively insensitive to deep red lights. Red wave lengths, however, penetrate water poorly, and red filtered lights are only useful for illuminating nearby objects, i.e., those less than one meter away. Because diver's eyes are also dark-adapted, shining a dive light on a white writing surface is painful. This can be prevented by placing a small amount (approximately 1 cc) of activated Cyalume fluid in a 10-cc translucent plastic vial and attaching the vial to the writing slate. The fluid provides sufficient illumination for writing.

Differences in habitat within the sampling area affect the accuracy of the data. To increase accuracy of abundance estimates, transects should be laid both parallel to and across depth contours (Hixon 1980). Relative abundances of different habitat types should be estimated, using longer transects, survey dives, or towed divers (Box 19.2). Counts of fish along transects can then be stratified for different habitat types. If fish move several meters up in the water column, density calculations should account for such volumetric factors.

Unless water clarity is constant from day to day, horizontal visibility should be measured prior to each transect count. A Secchi disc or some other highly visible object is attached to a rope marked at 0.5-meter intervals. The disc is propped on the bottom, and the diver approaches from the end of the rope and records the distance at which the disc is just visible. Fishes that are countershaded and otherwise camouflaged are generally visible at much shorter distances than the disc, but the diver at least has an idea of the maximum viewing distance on that particular day. This information often proves useful when data from several days are compared.

Counts can be validated by having two divers swim side by side or by making repeated swims along a transect. Two divers are a greater disturbance than one and may cause some fish to flee that might otherwise remain in the area. Repeated counts usually yield lower numbers because many fish will leave an area during the first count. The time interval necessary to correct for this flight factor must be determined empirically for each habitat type. Counts at the same site may also vary as a function of the daily movements of fishes. Counts, therefore, should be made at the same time each day (Colton and Alevizon 1981), the optimal time being determined by test counts at the beginning of the study.

Box 19.2   Towing a Diver

A very effective means of counting aquatic organisms or of locating target species or habitats is to tow divers behind a small boat. A thick (greater than 1 cm diameter) floating rope is tied to the rear of the boat and trailed far enough behind the boat that the diver's vision is not obscured by bubbles from the propeller. A loop is tied in the end of the rope that the diver uses as a stirrup. The diver usually does not use flippers, although neoprene boots and gloves are recommended. Other equipment includes mask, snorkel, and wet suit. The diver puts his foot in the stirrup and grabs the rope with an extended arm. At slow speeds, this technique is surprisingly non-tiring, and the limits to how long a diver can be towed are generally set by comfort factors other than fatigue. While being towed, the diver can plane to either side of the wake by merely pointing his body in the desired direction. This technique can be extended to scuba divers using inexpensive or elaborate underwater sleds, but its use requires more training, careful coordination between boat and diver, and knowledge of possible underwater obstructions with which the diver might collide (Anonymous 1975).

The utility of this technique depends on several factors, such as boat size, water clarity and depth, hazardous biting or stinging aquatic life, and diver confidence. Semiquantitative or comparative data can be collected while towing by traveling along straight-line compass courses for fixed times (e.g., 5 to 10 min). A driver and observer are necessary during tows as support personnel, as in any activity where a person is towed behind a boat. Once the diver overcomes the initial sensation of being trolled, towing can be an extremely efficient underwater method.

The validity of transect counts also can be tested by incorporating a mark-recapture program into the sampling regime. Fish are captured along the transect, tagged, and then released at the capture site. Tagged fish sighted during transect counts are noted, and a second estimate of population size is calculated using standard mark-recapture methods. Although two estimates based on data collected the same way (i.e., by a diver along a transect) are not completely independent, similarity in the results from both population estimates can increase an investigator's confidence in their accuracy.

Alternative subsampling methods. Two other methods used for subsampling fish populations in a large area are cinetransects and species-time counts. The cinetransect is conducted in the same manner as the transects above, but the diver swims along the transect using a motion picture or video camera (Alevizon and Brooks 1975; Powles and Barans 1980). Counts and distributions of fishes are determined upon analysis of the film. This method has the advantage of giving a permanent record of both fishes and habitat. However, it costs more than the standard visual transect, depends on camera resolution for identification of small fishes, does not sample cryptic, secretive, or nocturnal fishes, and is influenced strongly by water clarity and light availability.

The species-time technique is a method for determining the relative abundance of fishes when comparing regions with very high species diversities, such as coral reefs (Jones and Thompson 1978). A diver swims for a fixed time, e.g., 50 minutes, and records all fish species. The diver is free to investigate any habitat type in the study area instead of being limited to movement in a transect or quadrat. Relative abundance within and among areas is determined by how early in the counting period a particular species is seen, based on the assumption that the more abundant species are likely to be seen earlier. Counts are repeated until 90% or more of the common species in that assemblage have been encountered. The species-time technique can also be useful as a means of increasing the accuracy of a transect count. Jones and Chase (1975) found that 30% more species were frequently observed if a visual transect was followed by a species-time count.

Visual censuses of small areas. Diving has one of its most useful applications in determining the numbers and activities of fishes in a small area, such as an artifical reef or other fish attractor or any habitat type sufficiently small to be sampled by a diver swimming around it (Hammond et al. 1977; Prince et al. 1979). The major goal in a visual census is to count all individuals present in the study area. A diver swims around the structure, investigating all potential hiding places. Care is taken when approaching the structure because epibenthic species may swim off with the approach of a diver. Counts should be replicated by the same diver on different days, with the number of replications determined by local conditions and by the desired precision of the estimates. Sale and Douglas (1981) found that the completeness of a census increased continually as the number of counts increased. In the first three enumerations, they saw 82% of the species and 75% of the individuals that were seen after six more counts were performed. Sale and Douglas emphasized that not all species and size classes are counted with equal precision or accuracy.

An alternative, highly accurate method for counting aggregations of relatively stationary fishes is to photograph them. Numbered meter sticks with small weights at one end and floats at the other are placed at approximately one meter intervals around the structure or around the perimeter of a stationary school of fish. The diver swims around the structure and photographs the aggregation, attempting to overlap the photographs by using the meter sticks as orientation marks. The photographs are analyzed with the meter sticks serving both for orientation and for determining fish sizes. Using this method, we have been able to count 95% of 420 fishes in a school that was later captured by net and counted.

19.4.2 Estimating Home Ranges, Habitat Preferences, and Activity Patterns

Knowledge of the area normally used by a species during its activities, of the habitat types used most often by a species, and of the times when a species is most active are all valuable data for the fisheries manager. Accurate estimates of these three biological attributes are almost impossible without some direct measure of fish movement, such as that obtained by underwater observers or through biotelemetry (chapter 20).

Information on movements and distribution can be obtained by divers working with fish that are individually recognizable in an area that contains uniformly distributed underwater markers. Special tagging methods are needed (Box 19.3) because traditional methods (chapter 11) do not allow identification of individual fish underwater, at least not at the distances that most fishes will permit approach by a diver.

Most fishes of interest to fisheries personnel are large and mobile, and their movements encompass more than a few square meters. For these fishes, a grid system should be laid out on the bottom of the study area. To do so, a diver swims along a compass course, unrolling a sinking rope that has been marked at intervals determined by water clarity and the desired scale of areas for the study. Intervals of 5 to 10 m are often used. Numbered floats (markers) are tethered near the bottom at each mark, the rope is rolled back up, and the diver moves over one mark interval and repeats the process. When the grid is completed, markers will be distributed uniformly throughout the study area, and a diver will be able to see a numbered marker at any location under normal visibility conditions.

Home range and activity pattern data can be collected in one of four ways by a diver following an individually recognizable fish. The diver can (1) note the time and number of every marker that is passed, (2) drop strips of consecutively numbered, weighted orange flagging at fixed time intervals, or (3) note the time at which each numbered flag is dropped. In the fourth method, a diver swims through an area on repeated occasions and notes the locations of tagged fish. This last method is better for wary species. When the data from any method are plotted on a map of the grid system and the dots are joined, the pattern gives an initial estimate of home range. The actual formula used for calculating home range depends on the number and pattern of sightings for each animal (Winter 1977; Lehner 1979). Activity patterns can be expressed as rates of movement as determined from the dropped markers and from records of the times at which particular events or activities occur while the fish is being tracked. Care should be taken to avoid following the fish that are wary of divers. If there is sufficient water clarity, this can be tested by initially observing an individual fish from several meters away to establish its undisturbed movement patterns. Observers should also stop periodically to determine whether the fish moves and stops in synchrony with the divers.

Information on habitat preferences can be obtained if the distribution of major habitat types in the gridded study area has been mapped. Such mapping can be done on the basis of percent cover, relative abundance of some habitat characteristics, or some other determination, depending on the desired use of the information (*see* Whittaker 1975). To determine habitat preferences, the diver records fish occurrences

Box 19.3   Tagging Fish for Individual Identification Underwater

For identifying individual fish underwater, the choice of a tag depends on the size of the fish and how close the fish will allow a diver to approach. For large, approachable species, numbered pieces of plastic can be attached to standard dart tags (chapter 11) and inserted behind the dorsal fin of the fish (Heard and Vogele 1969). Such "flag tags" are visible from up to four meters away, depending on lettering size and underwater visibility. Small fish require coded rather than numbered tags. The three common techniques use acrylic paints, colored beads, or band-coded tags.

When a few fish are to be tagged and will be viewed from short distances (less than one meter), acrylic paints injected subcutaneously with insulin syringes are effective (Lotrich and Meredith 1974; Thresher and Gronell 1978). Paints can be injected in fish as small as 20 mm long, and the marks may last up to 16 months. An advantage to this technique is that paint can be injected by one diver underwater. The major disadvantage to the paint injection method is the limited number of colors that can be consistently and accurately discriminated underwater, particularly for small fish or if the water is colored by algae or other substances.

Fish can be identified individually with small plastic beads (available from a hobby or crafts store) attached with very fine monofilament line sewn through the dorsal musculature. Beads provide distinct visual targets, as compared to the blurred edges of injected paints, and they can be seen from farther away. Most persons have difficulty resolving and recording more than four or five such beads, and water color can affect the perceived color of the bead. Another disadvantage of bead tags is that they are attractive to other fishes, including both predators and "cleaner fishes." Cleaner fishes will pull at the beads; this causes tagged fish to lose beads or slows the healing of the wound.

When a large number of fish (more than 50) are to be differentiated, an alternative method is the use of band-coded tags. This method is a modification of a tag described by White and Beamish (1972) which involves attaching banded pieces of plastic tubing to a fish with monofilament fishing line. The most useful tubing types are clear Tygon 5D tubing (inside diameter 0.8 mm, outside diameter 2.4 mm, wall thickness 0.8 mm) and Floy #20 spaghetti tubing (outside diameter 2 mm) in white, yellow, or international orange. Cut the tubing into either 2- or 3-cm lengths, the smaller size for fishes shorter than 15 cm. Mount a large sewing needle in a hand-drill, and slide the piece of tubing over the needle so that the tubing rotates when the drill is turned. Mark thin (2- to 3-mm) lines on the rotating tubing with an indelible marker. Up to four black lines can be marked on each half of a 2-cm tubing piece, allowing for 24 possible combinations of zero to four marks (e.g., 0,1; 2,4; 4,1). Up to three black lines can be marked on each third of the 3-cm tubing (four marks are more difficult to resolve on the three fields of the longer tags), allowing 64 combinations (e.g., 1, 2, 3; 3, 0, 2; etc.). Using all four colors (clear, white, yellow, orange) and both tag sizes, 350 different tags can be made.

Clear Tygon tags are preferable for population counts based on mark-recapture-methods; these tags are less conspicuous than colored ones and, therefore, minimize bias associated with marking.

Attach the tags with 4-kg breaking strength monofilament line. Slide the tubing over a 30-cm length of line that is knotted at one end to keep the tag from

Continued

slipping off. Tie a loose, single overhand knot in the line, just ahead of the tag. Sew the free end of the line through the dorsal musculature of the fish once, usually behind the dorsal fin, and then pass it again through the overhand knot. Tighten the knot. Pull the free end of the line, bringing the now triple overhand knot to within 2 to 3 cm of the fish, and clip off the free end of the line just above the knot. Push the tag over the triple knot so that the knot sits at least 1 cm inside the tag; this prevents tag loss. Cut off the loose end.

Using this technique of band-coded tags, we have tagged more than 1000 marine and freshwater fishes ranging in size from 11 to 70 cm (*see* Helfman 1979, 1981). Tag retention is generally six months to two years. Band-coded tags can be read from up to three meters away. They suffer from the same problem as beaded tags in that they attract the attention of predators and cleaner fish and the colored tubing can be difficult to discriminate in colored waters. The fabrication and application of the tags is also time-consuming compared to paint injections or bead tags, and tags appear cumbersome on most fishes smaller than 12 cm long.

as the diver swims along transects that cross representative habitat types, or the diver remains in areas of uniform habitat for specified times and records the frequencies with which different species visit each area. Habitat preference data also can be obtained incidental to transect counts or by noting the amount of time fish spend in different habitat types while following fish during home range or activity pattern studies.

The primary advantage of direct observation in the above determinations is that it is non-destructive and produces meaningful activity data more quickly than does capturing series of fishes with nets or angling. Fishes that are inappropriate for study by direct observation include species that are either attracted to or repelled by a diver. Direct observation is unlikely to yield meaningful information on activity patterns in turbid water, for benthic or otherwise lurking predators that may be motionless but still seeking prey, or for fishes that may be mobile but not feeding, such as at twilight (Helfman 1981).

19.4.3 Spearfishing for Diet Data

Determinations of food habits form the basis of many fisheries investigations (chapter 17). Major drawbacks of most conventional capture methods include obtaining a desired number of specimens, particularly of very small or large but rare fishes, and preventing regurgitation of food between the time of capture and fixation of the specimen. Spearing is "the most effective way to collect fishes for study of food habits" (Hobson 1974, p. 916). Small fishes can be captured with a microspear (Fig. 19.3), whereas standard slings or spearguns are used to capture large fishes, including those that will not take lures or bait. Spearing is particularly effective when specimens of many species with divergent feeding habits are desired and when differences in feeding between day and night are of interest.

When engaged in spearing, carry an assortment of opaque, numbered plastic bags. Kill and immobilize the captured fish immediately by pushing a spear prong through the brain. Place the specimen in an opaque plastic bag. These actions reduce visual, olfactory, and acoustic cues for predators. Note the time of capture and bag number on a writing cuff. While still underwater or immediately after emerging, inject

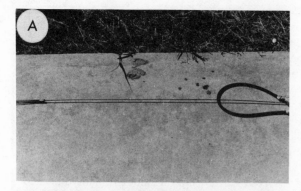

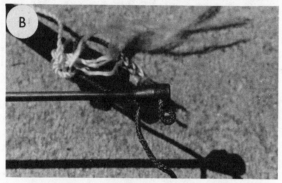

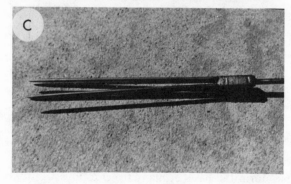

Figure 19.3. A 5-prong microspear for collecting small fishes. The spear is used slingshot style, instead of in the manner that pole spears or Hawaiian slings are used. *(A)* Complete spear, constructed by W. N. McFarland, Cornell University. Spear is 80 cm long. Shaft is constructed from 3-mm diameter, moderately hard stainless steel. *(B)* Top of spear. Nylon cord from surgical tubing is passed through a hole drilled in the brass top. Shaft is inset in brass top, which is then soldered with eutectic stainless-type solder. *(C)* Tines are made from 1.6-mm very hard stainless steel, such as that used in throttle control cables for outboard motor boats. Tines are attached to the shaft with copper wire and soldered in place with eutectic solder. Slippage of tines during soldering can be reduced by first mounting tines in a 5-hole holder or jig.

the gut cavities of the captured fishes and the contents of each plastic bag with formalin. Discard specimens that have been wounded in the abdomen.

The problems of spearfishing must be fully considered before this technique is used. Divers intending to capture fishes with spears should be aware of strong public sentiment and legal restrictions concerning their use. Because diving is a two-person activity, extra precautions must be taken when spears are used. For this reason, friction-type spears (microspears, pole spears, Hawaiian slings) are preferable to spearguns. The presence of large predators in an area should also be considered as a hazard to divers because a substantial number of fishes that are speared are not captured and are, therefore, likely to attract such predators.

19.5 CONCLUSION

The methods outlined in the preceding sections represent only a sampling of the available underwater techniques that are applicable to fisheries questions and to the situations in which they can be applied. No mention has been made of techniques and goals in studies of fish behavior, such as breeding activities, foraging behavior, predator-prey interactions, or territoriality (*see* Keenleyside 1979). All of these activities can be best studied through direct observation, and all can be useful in determining relevant attributes of the biology of commercial and sport fishes. Direct observation of behavior has an additional, important advantage because it often provides valuable insight into other aspects of the biology of fish species.

Every method has advantages and drawbacks, and direct observation is certainly no exception. A well-planned investigation recognizes the limitations of each methodology and attempts to supplement one technique with others. The most satisfactory results usually come from combining the information obtained by different methods (Uzmann et al. 1977; Barans 1982). The training of fisheries personnel in the future will increasingly include underwater techniques and their application to fisheries methods. Technological advances in the design and manufacture of diving gear and visual monitoring devices have reduced the costs and increased the safety and accessability of this equipment. A modern fisheries organization can now afford underwater methods and should incorporate them into its program.

19.6 REFERENCES

Alevizon, W. S., and M. G. Brooks. 1975. The comparative structure of two western Atlantic reef-fish assemblages. *Bulletin of Marine Science* 25:482-490.

Anonymous. 1975. *The NOAA diving manual: diving for science and technology.* United States National Oceanic and Atmospheric Administration, Manned Undersea Science and Technology Office, Stock No. 003-017-00283, Washington, District of Columbia, USA.

Anonymous. 1978. *United States Navy Diving Manual.* United States Government Printing Office, NAVSHIPS 0994-001-9010, Washington, District of Columbia, USA.

Bachman, R. A. 1982. *Cost-minimizing behavior of free-ranging wild brown trout in a stream.* Doctoral dissertation. Pennsylvania State University, University Park, Pennsylvania, USA.

Barans, C. A. 1982. Methods for assessing reef fish stocks. Pages 105-123 *in* G. R. Huntsman, W. R. Nicholson, and W. W. Fox, Jr., editors. *The biological bases for reef fishery management.* United States National Oceanic and Atmospheric Administration, NOAA Technical Memorandum NMFS-SEFC-80, Charleston, South Carolina, USA. (Available from C. A. Barans, South Carolina Marine Resources Institute, P.O. Box 12559, Charleston, South Carolina 29412, USA.)

Boehler, T. 1979. *Night diving.* Deep Star Publishing, Crestline, California, USA.

Burnham, K. P., D. R. Anderson, and J. L. Laake. 1980. Estimation of density from line transect sampling of biological populations. *Wildlife Monographs* 17:1-202.

Colton, D. E., and W. S. Alevizon. 1981. Diurnal variability in a fish assemblage of a Bahamian coral reef. *Environmental Biology of Fishes* 6:341-345.

Cross, E. 1970. Technifacts, high-altitude decompression. *Skin Diver* 19 (11):17.

Drew, E. A., J. N. Lythgoe, and J. D. Woods. 1976. *Underwater research.* Academic Press, London, England.

Ebeling, A. W., and R. N. Bray. 1976. Day versus night activity of reef fishes in a kelp forest off Santa Barbara, California. *Fishery Bulletin* 74:703-717.

Ebersole, J. P. 1977. The adaptive significance of interspecific territoriality in the reef fish *Eupomacentrus leucostictus. Ecology* 58:914-920.

Emery, A. R. 1973. Preliminary comparisons of day and night habits of freshwater fish in Ontario lakes. *Journal of the Fisheries Research Board of Canada* 30:761-774.

Empleton, B. E., E. H. Lanphier, J. E. Young, and L. G. Goff, editors. 1974. *The new science of skin and scuba diving,* 4th revised edition. Association Press, New York, New York, USA.

Fager, E., A. Flechsig, R. Ford, R. Clutter, and R. Ghelardi. 1966. Equipment for use in ecological studies using SCUBA. *Limnology and Oceanography* 11:503-509.

Hall, D. J., and E. E. Werner. 1977. Seasonal distribution and abundance of fishes in the littoral zone of a Michigan lake. *Transactions of the American Fisheries Society* 106:545-555.

Hammond, D. L., D. O. Myatt, and D. M. Cupka. 1977. *Evaluation of midwater structures as a potential tool in the management of the fisheries resources of South Carolina's artificial fishing reefs.* South Carolina Marine Resources Center Technical Report Number 15, Charleston, South Carolina, USA. (Available from South Carolina Wildlife and Marine Resources Department, Recreational Fisheries Section, P.O. Box 12559, Charleston, South Carolina 29412, USA.)

Heard, W. R., and L. E. Vogele. 1968. A flag tag for underwater recognition of individual fish by divers. *Transactions of the American Fisheries Society* 97:55-57.

Helfman, G. S. 1978. Patterns of community structure in fishes: summary and overview. *Environmental Biology of Fishes* 3:129-148.

Helfman, G. S. 1979. Twilight activities of yellow perch, *Perca flavescens. Journal of the Fisheries Research Board of Canada* 36:173-179.

Helfman, G. S. 1981. Twilight activities and temporal structure in a freshwater fish community. *Canadian Journal of Fisheries and Aquatic Sciences* 38:1405-1420.

Henry, V. J., C. J. McCreery, F. D. Foley, and D. R. Kendall. 1981. Ocean bottom survey of the Georgia Bight. Chapter 6 *in* P. J. Popenoe, editor. *Environmental studies in the southeastern U.S. outer continental shelf, fiscal year 1977-1978.* United States Geological Survey Open File Report 81-582-A, Savannah, Georgia, USA. (Available from V. J. Henry, Skidaway Institute of Oceanography, P.O. Box 13687, Savannah, Georgia 31406, USA.)

Herrnkind, W. F. 1974. Behavior: *in situ* approach to marine behavioral research. Chapter 3 *in Experimental marine biology.* Academic Press, San Francisco, California, USA.

High, W. 1967. *Scuba diving, a valuable tool for investigating the behavior of fish within the influence of fishing gear.* Food and Agriculture Organization of the United Nations conference on fish behavior in relation to fishing techniques and tactics, Bergen, Norway, 19-27 October, 1967. Food and Agriculture Organization of the United Nations, Rome, Italy.

Hobson, E. S. 1968. *Predatory behavior of some shore fishes in the Gulf of California.* United States Bureau of Sport Fisheries and Wildlife, Research Report 73:1-92, Washington, District of Columbia, USA.

Hobson, E. S. 1974. Feeding relationships of teleostean fishes on coral reefs in Kona, Hawaii. *Fishery Bulletin* 72:915-1031.

Jenkins, W. 1973. *A summary of diving techniques used in polar regions.* Naval Coastal Systems Laboratory, Panama City, Florida.

Jones, R. S., and J. A. Chase. 1975. Community structure and distribution of fishes in an enclosed high island lagoon in Guam. *Micronesica* 11:127-148.

Jones, R. S., and M. J. Thompson. 1978. Comparison of Florida reef fish assemblages using a rapid visual technique. *Bulletin of Marine Science* 28:159-172.

Kalmijn, Ad. J. 1978. Electric and magnetic sensory world of sharks, skates, and rays. Page 507-528 *in* E. S. Hodgson and R. F. Mathewson, editors. *Sensory biology of sharks, skates, and rays.* Office of Naval Research, Arlington, Virginia, USA.

Keast, A., and J. Harker. 1977. Strip counts as a means of determining densities and habitat utilization patterns in lake fishes. *Environmental Biology of Fishes* 1:181-188.

Keenleyside, M. H. A. 1962. Skin-diving observations of Atlantic salmon and brook trout in the Miramichi River, New Brunswick. *Journal of the Fisheries Research Board of Canada* 19:625-634.

Keenleyside, M. H. A. 1979. *Diversity and adaptation in fish behaviour.* Springer-Verlag, Berlin, West Germany.

Lehner, P. N. 1979. *Handbook of ethological methods.* Garland STPM Press, New York, New York, USA.

Loesch, J. G., W. H. Kriete, Jr., and E. J. Foell. 1982. Effects of light intensity on the catchability of juvenile anadromous *Alosa* species. *Transactions of the American Fisheries Society* 111:41-44.

Lotrich, V. A., and W. H. Meredith. 1974. A technique and the effectiveness of various acrylic colors for subcutaneous marking of fish. *Transactions of the American Fisheries Society* 103:140-142.

Luckhurst, B. E., and K. Luckhurst. 1978. Nocturnal observations of coral reef fishes along depth gradients. *Canadian Journal of Zoology* 56:155-158.

Major, P. F. 1978. Predator-prey interactions in two schooling fishes, *Caranx ignobilis* and *Stolephorus purpureus*. *Animal Behaviour* 26:760-777.

Mount, T. 1973. *Safe cave diving*. National Association for Cave Diving, High Springs, Florida 32643, USA.

Ogden, J. C. 1977. A scroll apparatus for the recording of notes and observations underwater. *Marine Technology Society Journal* 11:13-14.

Powles, H., and C. A. Barnes. 1980. Groundfish monitoring in sponge-coral areas off the southeastern United States. *Marine Fisheries Review* 42 (5):21-35.

Prince, E. D., O. E. Maughan, D. H. Bennett, G. M. Simmons, Jr., J. Stauffer, Jr., and R. J. Strange. 1979. Trophic dynamics of a freshwater artificial tire reef. Pages 459-472 *in* H. Clepper, editor. *Predator-prey systems in fisheries management*. Sport Fishing Institute, Washington, District of Columbia, USA.

Russell, B. C., F. H. Talbot, G. R. V. Anderson, and B. Goldman. 1978. Collection and sampling of reef fishes. Pages 329-345 *in* D. R. Stoddart and R. E. Johannes, editors. *Coral reefs: research methods*. United Nations Educational, Scientific, and Cultural Organization, Paris, France.

Quinn, T. J., II, and V. F. Gallucci. 1980. Parametric models for linear transect estimations of abundance. *Ecology* 61:230-302.

Sale, P. F. 1980. The ecology of fishes on coral reefs. *Oceanography and Marine Biology Annual Review* 18:367-421.

Sale, P. F., and W. A. Douglas. 1981. Precision and accuracy of visual census technique for fish assemblages on coral patch reefs. *Environmental Biology of Fishes* 6:333-339.

Schulke, F. 1978. *Underwater photography for everyone*. Prentice-Hall, Englewood Cliffs, New Jersey, USA.

Shenton, E. 1972. *Diving for science*. W. W. Norton and Company, Incorporated, New York, New York, USA.

Smith, C. L. 1976. *Altitude procedures for the ocean diver*. National Association of Underwater Instructors, Montclair, California, USA. (Available from NAUI Headquarters, 4650 Arrow Highway, Suite F-1, Post Office Box 14650, Montclair, California 91763, USA.)

Smith, C. L, and J. C. Tyler. 1973. Population ecology of a Bahamian suprabenthic shore fish assemblage. *American Museum Novitates* 2528:1-38.

Somers, L. 1974. *Cold weather and under ice scuba diving*. National Association of Underwater Instructors, Montclair, California, USA. (Available from NAUI Headquarters, 4650 Arrow Highway, Suite F-1, Post Office Box 14650, Montclair, California, USA.)

Starck, W. A., and W. P. Davis. 1966. Night habits of fishes of Alligator Reef, Florida. *Ichthyologia* 33:313-356.

Thresher, R. E., and A. M. Gronell. 1978. Subcutaneous tagging of small reef fishes. *Copeia* 1978:352-353.

Uzmann, J. R., R. A. Cooper, R. B. Theroux, and R. L. Wigley. 1977. Synoptic comparison of three sampling techniques for estimating abundance and distribution of selected megafauna: submersible vs. camera sled vs. otter trawl. *Marine Fisheries Review* 39 (12):11-19.

White, W. J., and R. J. Beamish. 1972. A simple fish tag suitable for long-term marking experiments. *Journal of the Fisheries Research Board of Canada* 29:339-341.

Whittaker, R. H. 1975. *Communities and ecosystems*, 2nd edition. MacMillan, New York, New York, USA.

Winter, J. D. 1977. Summer home range movements and habitat use by four largemouth bass in Mary Lake, Minnesota. *Transactions of the American Fisheries Society* 106:323-330.

Woods, J. D., and J. N. Lythgoe, editors. 1971. *Underwater science*. Oxford University Press, Oxford, England.

Chapter 20
Underwater Biotelemetry

JIMMY D. WINTER

20.1 INTRODUCTION

Advances in the electronics industry have enabled scientists to develop sophisticated telemetry methods to monitor the locations, behavior, and physiology of free-ranging aquatic animals. Underwater biotelemetry involves attaching to an aquatic organism a device that relays biological information. The information is relayed via ultrasonic or radio signals to a remote receiving system (Fig. 20.1). Telemetry provides a means to monitor the biology of animals not readily visible, to collect data with a minimal influence on the animal's behavior and health, to collect more data than are normally gathered by techniques such as mark-and-recapture, and to compare physiological and behavioral data collected in the laboratory and in natural systems. In addition, telemetry often provides the only means to solve some biological problems. This chapter describes telemetry systems, methods of attaching transmitters, methods of tracking free-ranging aquatic animals, and data collecting and processing.

If the device attached to the animal emits a signal, it is called a transmitter. If the transmitter returns a signal in response to one sent to it, it is called a transponder. Electronic oscillator circuits produce the signals by inducing high frequency vibrations in the water or air. Signals usually are ultrasonic in the 20-300 kilohertz (kHz) frequency range or radio signals in the 27-300 megahertz (MHz) range. A hertz (Hz) is a measure of the frequency of the signal and is equal to one cycle per second. For comparison, the standard AM radio uses frequencies from 550-1600 kHz and standard FM radio uses 88-108 MHz. Ultrasonic signals are received by a microphone or hydrophone submerged in the water. Radio signals, on the other hand, are received by a variety of antenna types located above the water. Both receiving systems pass the signals to a receiver that converts them to a form that is audible via headphones or that can be electronically processed and automatically recorded. The transmitted signal allows the investigator to locate the animal with a wide variety of methods: by boat, on foot, by truck, from a fixed receiving station, by airplane, etc. If the transmitter has an electronic sensor, it can also relay information on the animal's environment (e.g., water temperature) or on the animal's physiology (e.g., heart rate).

There are many factors to consider before beginning a telemetry project. The first step is to compile simple yes or no questions that can be readily tested (chapter 1). Too many telemetry projects have proceeded without forming testable hypotheses, producing a collection of maps of movements that cannot be meaningfully analyzed. Another important consideration is whether the problem can be solved by means other than telemetry. There is a tendency to adopt techniques because they are novel and flashy, not because they are the best means to answer a question. For example, netting probably would be more efficient than telemetry to determine whether a

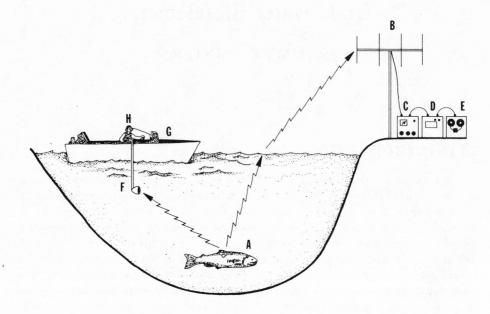

Figure 20.1. A generalized picture of two of the many types of underwater biotelemetry systems. On the left is an ultrasonic tracking system used to locate a fish equipped with a transmitter. On the right is a fixed radio receiving station used to record environmental or physiological data (e.g., temperature) from the transmitter. *(A)* transmitter emitting ultrasonic and radio signals (usually only one type of signal is emitted). *(B)* radio receiving antenna (Yagi). *(C)* radio receiver. *(D)* a signal decoder. *(E)* data recorder. *(F)* ultrasonic hydrophone. *(G)* ultrasonic receiver. *(H)* Headphones, manual data collecting, and recording.

species is found at a particular depth, especially if no other information is desired like diel vertical movement patterns or temperature selection. Since telemetry equipment is expensive and tracking is labor-intensive, the cost per unit of information is high unless large amounts of data are collected. In addition, some problems might be more appropriate to study in the laboratory because it is too difficult to obtain sufficient observations or to account for the many variables in the field.

20.2 TELEMETRY SYSTEMS

Underwater biotelemetry had its origins in the late 1950s. Initially, ultrasonic transmitters were developed because their low-frequency, long-wave signals travel well through water and are minimally affected by the water's conductivity. While ultrasonic transmitters were being developed for aquatic animals, radio transmitters were being developed for terrestrial use (Slater 1963, 1965; Brander and Cochran 1971). In the late 1960s, radio transmitters were modified to work underwater.

The development of underwater biotelemetry has been detailed in the *Underwater Telemetry Newsletter*. The newsletter has been published since 1971 and is available by microfiche and free subscription from Dr. C. C. Coutant (Environmental Sciences Division, Oak Ridge National Laboratory, Oak Ridge,

Tennessee 37830). Excellent reviews of underwater biotelemetry systems have been written by Stasko and Pincock (1977), Ireland and Kanwisher (1978), Nelson (1978), and Winter et al. (1978). This section describes the common features of ultrasonic and radio systems, the advantages of each system, the types of ultrasonic and radio equipment and sensing transmitters, and the selection of an equipment supplier.

20.2.1   Features Common to Ultrasonic and Radio Systems

Transmitter components. Ultrasonic and radio transmitters ("tags") have many common components and features. Both may emit continuous-wave or pulsing signals. Continuous ("whistling") signals are distinguished more easily from background noise and are detected at greater distances than pulsing signals ("beeps"). In addition, animal activity is more easily detected and recorded with continuous signals than pulsing signals (Winter et al. 1978). An active animal causes the intensity of the signal to waver by changing the orientation of the transmitter. Pulsing signals use less energy and thus increase transmitter life. The pulse length (on-time), pulse rate (number of pulses per minute), and pulse interval (time between pulses) can be used to identify individual transmitters and to transmit (code) environmental measurements in sensing tags. If the pulse interval is much longer than one second, it is difficult to determine signal direction. Furthermore, pulse lengths less than 20 milliseconds are difficult to detect by ear (Cochran 1980). The ratios of transmitter on-time to total time (duty cycles) are usually 2% to 4%. Since pulsing signals conserve battery life and can be easily used to transmit biological parameters, pulsing signals rather than continuous signals are most commonly used.

Transmitter encapsulation. Simple expendable tags are encapsulated with epoxy, wax, urethane, silicone, or dental acrylic by pouring (potting) the material over the components in a mold or by dipping the components into the material. In ultrasonic tags, the density of the potting material should be close to that of water for best signal transmission. Although potting transmitters is simple and produces a minimum package size, it does not produce minimum weight or allow easy replacement of components. Complex, expensive, recoverable tags may be incorporated into tubes filled with oil (e.g., dehydrated castor oil) or air. Compared to potted transmitters, tubes are larger, weigh less in water, give easier access to components, and give better transmission of ultrasonic signals between the signal transducer and water. Although most commercial firms will supply the tags encapsulated, it is possible to construct a special attachment method if one knows how to encapsulate tags.

Transmitters are usually turned on by soldering exposed wires or by activating magnetic-reed switches. After soldering the wires, they must be sealed well by epoxy, etc., to prevent water leakage into the transmitter. Magnetic-reed switches are convenient because they are activated by removing a magnet taped to the transmitter over the switch. If transmitters are stored close together, the magnets may cancel each other and drain the batteries. Magnetic-reed switches also occasionally malfunction. One should use magnetic-reed switches when they do not increase the size of the transmitter greatly and when soldering and potting cannot be done easily.

Batteries. Since the battery generally represents more than 50% of the volume and up to 80% of the weight of the transmitter (Campbell and Stoneburner 1980), choice of the battery is critical. Choice of batteries for transmitters is largely determined by their energy per unit weight or volume (volt/gm or ml). Other considerations are battery cost (minor), shelf-life, initial voltage per cell, voltage drop during discharge, performance under environmental conditions, and available sizes and shapes (Nelson

1978). Trade-offs are necessary among transmitter life, size, and range. Larger batteries have greater storage capacities and permit longer transmitter life but increase transmitter size. Increasing the current drain may produce a slight increase in signal range but the transmitter life is greatly decreased. Transmitter life in hours can be estimated by dividing the manufacturer's milliampere-hour (mah) rating for a battery by the average current drain (ma) of the transmitter. However, transmitter life can be considerably less than estimated depending on battery freshness, variability in battery production, and environmental factors. Transmitter reliability increases with certified cells, which are manufactured to rigid specifications and inspected at each stage of construction. Certified cells are more expensive and only available in certain sizes and large quantities.

Five types of batteries are frequently used in transmitters and receivers: lithium, mercury oxide, silver oxide, alkaline, and rechargeable nickel-cadmium. Battery types and sizes available for biotelemetry are listed by Kuechle (1967), Campbell and Stoneburner (1980), and Cochran (1980). Lithium batteries are becoming widely used in transmitters because they produce the highest voltage ($\simeq$ 3.04v) per unit weight and volume. Furthermore, lithium batteries have excellent low temperature performance, good high temperature tolerance, long shelf-life, good efficiency over a wide range of voltage, and low cost (Campbell and Stoneburner 1980). Although the sizes available have been limited, new sizes are developed periodically.

Receivers. A biotelemetry receiver filters input signals, amplifies them, and converts them to a form that is audible to an investigator or is processed by an electronic signal detector. Whether a signal is detected depends largely on the receiver sensitivity, receiver bandwidth, electronic detector, and human hearing. To detect a faint transmitter signal, a receiver must have good sensitivity, i.e., the minimum level of signal that can be detected above the receiver's internal noise. A receiver must also have a narrow frequency bandwidth to exclude as much ambient noise as practical. Bandwidth is the range of frequencies that will be passed through the narrowest filter to the listener or electronic detector. The transmitter signal must be within the bandwidth to be detected. If receiver bandwidths are too narrow, there is an increased chance of not detecting a signal when the tuning dial is off center frequency (Nelson 1978). Furthermore, the frequencies of some transmitters change slightly with different water temperatures and may not be detected if the bandwidth is very narrow. Since most telemetry systems rely on human hearing, the real system bandwidth is the human ear or about 50 Hz. Although electronic detectors have narrower bandwidths, the human ear is a near optimal receiver with high sensitivity (Urick 1975). For example, with moderate pulse lengths, aural detection levels are 10-15 dB lower than for electronic detection. Therefore, human hearing is usually used to search for faint signals while electronic detectors are often used in recording data from transmitters with strong signals.

Other characteristics are also important in choosing receivers. First, the frequency selector should be accurate, and the receiver should be capable of separating many transmitter frequencies in the same locale. Second, a receiver should permit quick searching of frequencies, especially in radio-tracking from an airplane. Specific frequencies can be programmed into some receivers, and the frequencies can be manually or automatically scanned for quick searches or for data recording systems. Although an initial project may not require portable receivers, consider the possible need for portable receivers in future projects before making purchases. If the receiver is to be hand-carried, it should be small, light, and sturdy. In addition, a portable receiver should use rechargeable batteries and get long use per charge. Field receivers also should be capable of working over a wide range of temperatures (-25 C to 50 C). Since fish tracking is often conducted in rain and fog, a receiver should have waterproof switches and be moisture resistant.

20.2.2   Advantages of Ultrasonic and Radio Systems

The advantages and disadvantages of ultrasonic and radio telemetry systems need to be weighed with respect to characteristics of the study area and the animal. Ultrasonic telemetry is well suited for studies in salt water, fresh water with high conductivity, and deep water because these habitats cause little reduction in signal strength. Animals can be located very accurately (< 3-4 m) with ultrasonic systems from a boat.

Since the hydrophone or receiving unit in ultrasonic systems must be submerged in the water, searching for highly mobile animals over great distances or through holes in the ice is difficult and time-consuming. Ultrasonic signals are adversely affected by macrophytes, algae, thermoclines, water turbulence, raindrops, and boat motors. Dense macrophytes, high concentrations of particulate matter, suspended algae, or algae attached to hydrophones can reduce signal range from several hundred meters to a few meters. Thermoclines or temperature gradients reduce range because ultrasonic signals are refracted downward from the warmer water (Fig. 20.2) (Brumbaugh 1980). Ultrasonic telemetry is not suitable for use in turbulent water near dams or rapids because trapped air bubbles attenuate the signals. Since ultrasonic transmitters are usually built without crystals that minimize frequency changes, fewer individuals can be distinguished by different frequencies than with radio telemetry. Individual ultrasonic transmitters are coded by differences in pulse-rates, which are measured by

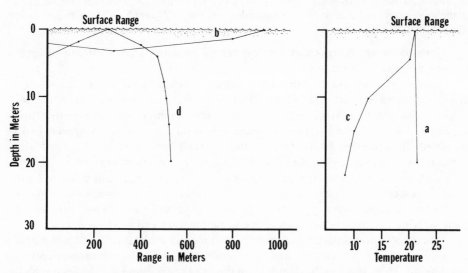

Figure 20.2. Effect of thermal stratification on the signal range of ultrasonic transmitters. Sound travels faster in warm water than in cold water which results in bending of sound energy away from the warmer water. When the water is isothermal (curve *a*), sound is bent towards the surface (curve *b*) because of pressure effects on sound velocity; the range is excellent. If the body of water is stratified (curve *c*), sound is bent downward by the warmer water more than the pressure effects bend the signal upward (curve *d*); the range is considerably reduced. (Adapted from Brumbaugh 1980, with permission.)

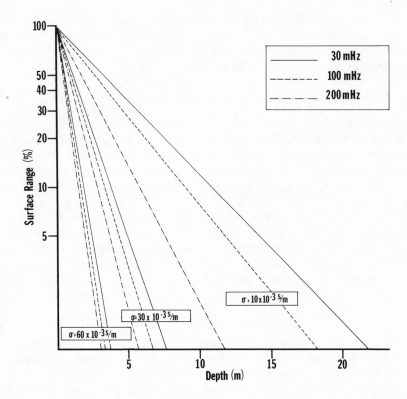

Figure 20.3. The attenuation of different frequencies of radio signals with increasing depth and water conductivities. Curves are calculated from theoretical data on properties of water using a plane wave equation. Seimens *(s)*/m is equivalent to mho/m. (From Kuechle, with permission, unpublished data.)

a stopwatch or an electronic pulse counter or by using a code on one of several channels in a sensing transmitter.

Radio telemetry, on the other hand, is suited well for shallow, low conductivity fresh water and for turbulent water. Since radio receiving antennas do not require contact with water, radio telemetry is excellent for searching large areas to find highly mobile species (e.g., salmon) or for shore-based data recording. Antennas can be mounted on airplanes, boats, trucks, and portable towers or hand-carried along streams (Winter et al. 1978). Radio signals are little affected by vegetation, algae, or thermoclines. Signals can be received through the ice, but they can be considerably reduced under slushy conditions. Since radio transmitters use crystals to minimize frequency changes, each tag can transmit on a different frequency and provide easy identification of many individuals at the same location and time.

The disadvantages of radio telemetry are that it cannot be used in salt water unless the animal surfaces periodically and that signals are reduced by increasing depth and conductivity (Fig. 20.3). If the depth is usually less than 5 m and the species not highly migratory (e.g., largemouth bass), high conductivity may not be a problem. However, with depths greater than 5 m and conductivities greater than 400 μmho/cm, radio telemetry may not work well for highly mobile species (e.g., walleye). Radio signals are also deflected by metal objects, terrestrial vegetation, and terrain; however, experienced trackers can usually surmount these difficulties.

Some features are similar for ultrasonic and radio systems. When used in environments for which they are suited, both types of transmitters have approximately the same battery life and signal ranges. Ultrasonic and radio transmitters can be built in similar sizes and to sense the same parameters. In addition, both receiving systems may receive interference from powerlines, unshielded ignition systems, lightning, and citizen's band radio. Finally, the cost of the equipment is approximately the same. The cost of a receiver and an antenna or hydrophone in 1983 was about $1200 to $2200. Transmitter prices begin at about $100 each and become more costly for environmental sensing or other capabilities.

20.2.3 Ultrasonic Telemetry Systems

Transmitters. Ultrasonic transmitters usually operate from 20-300 kHz with 50-100 kHz being the most common. An oscillator circuit is used to induce vibrations in a transducer which sends ultrasonic vibrations through the water. Transducers are devices that transform energy from one form to another; e.g., at signal levels, electrical energy is changed into mechanical energy or vice versa. The frequency of the vibrations is determined by the transmitter's mechanical resonances. Transducers are usually radially vibrating, electric-to-acoustic ceramic cylinders because this type is inexpensive and approximately omnidirectional at resonance (Stasko and Pincock 1977). Transducers are often the second largest transmitter component.

Hydrophones. The most important component of an ultrasonic system is the underwater microphone or hydrophone (Fig. 20.4). Ultrasonic signals produce vibrations in transducers in the hydrophone and are converted to electrical impulses which are sent to the receiver. When purchasing equipment, obtain hydrophones and

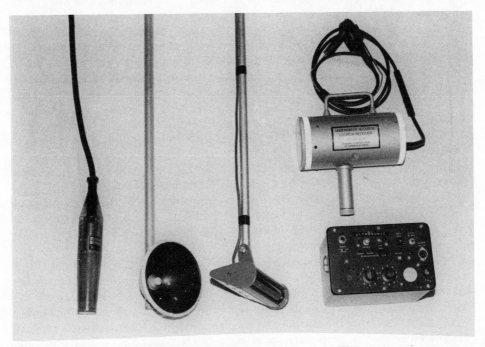

Figure 20.4. Ultrasonic telemetry equipment. Left to right: omnidirectional, cone, and linear array hydrophones. Bottom right: standard ultrasonic receiver. Top right: an ultrasonic receiver and headsets for use underwater.

receivers with compatible electronic specifications. If they are purchased from the same manufacturer, they probably will be matched. Better quality hydrophones often can be obtained from other sources, but you must match hydrophone and receiver specifications or test the units together.

Hydrophones have different angles (beamwidth) of signal reception and receiving patterns. Omnidirectional (360°) hydrophones are useful for searching and for fixed-receiving stations. In quiet environments, they have almost the same signal-to-noise ratios as directional hydrophones. However, in noisy environments you can obtain a better signal-to-noise ratio by using directional hydrophones which block out noise from most directions. Smaller beamwidths allow more accurate bearings, but one must be more careful in orienting the hydrophone to detect faint signals. Directional hydrophones have linear arrays of transducers or a single transducer surrounded by a cone that directs signals to the transducer. Linear array hydrophones have good sensitivities and receive vertically flattened, fan-shaped patterns. Cone hydrophones, on the other hand, have a very low acoustic-to-electric efficiency, receive a cone-shaped signal pattern, and are inexpensive. To reduce ambient electrical interference, good hydrophones have a shielded, twisted pair lead cable instead of a coaxial cable and a low-noise preamplifier near the hydrophone to boost the signal through the cable (Stasko and Pincock 1977).

Directional hydrophones are suspended in the water usually on rotatable, vertical pipes. If the hydrophone is mounted on a boat, the housing should be streamlined to reduce bubbles and permit monitoring at slow speeds. Hydrophone mounting systems which can withstand moderate boat speeds have been described by Stasko and Polar (1973) and Stasko (1976).

Signal transmission. The choice of frequency depends on the size of the transducer, which, in turn, affects the package size that an animal can carry and the acoustic output of the transducer. Small transducers produce high frequencies and, therefore, require more power to produce a given signal range; large transducers produce low frequencies and require less power output (Stasko and Pincock 1977). Since ultrasonic transmission characteristics are well known in salt water (Urick 1975), transmitter range can be mathematically estimated for deep water environments (Nelson 1978). Transmission characteristics are not well known for fresh water, but transmitter ranges are greater in fresh water than in salt water.

Ultrasonic signals are affected by spreading, absorption, noise, and other factors. As sound radiates outward from a source, its energy is spread over an increasing area, and the signal strength decreases. In deep water, spreading is approximately spherical, and the sound level decreases about 6 dB each time the distance doubles (Nelson 1978). Sound spreading is independent of frequency, temperature, and salinity. Some of the sound energy is also absorbed by water and converted to heat. For typical ultrasonic transmitter signals, absorption increases with increasing frequency, increasing salinity, and decreasing temperatures.

Noise can interfere with signal recognition even though the signal level is strong. Noise can be caused by wind, wave action, boat engines, boat movement, rain drops, ice cracking, and biological sources such as snapping shrimp. Other factors that reflect, scatter, or refract signals cause signal losses that are difficult to calculate. Signal reflections from the bottom and vegetation result in less transmitter range in shallow water than in deep water. Scattering of the signal can be caused by small suspended particles, plankton, fish, and air bubbles from waves. Warm surface water causes downward refraction of the signal that can result in the signal not being detected at the surface; however, sometimes the signal can be detected by submerging the hydrophone to a deeper level. In addition, submerged objects block signals, causing a sound shadow on one side of the object. Since sources of noise and reflections can render ultrasonic systems ineffective, consider or monitor these factors before beginning ultrasonic telemetry projects.

20.2.4 Radio Telemetry Systems

Transmitters. Radio telemetry systems usually operate from 27-300 MHz. In underwater radio transmitters, an oscillator circuit produces electromagnetic vibrations at a frequency determined by a crystal. Signals are transmitted through either circular (loop) or straight wire (whip) antennas. Transmitters with loop antennas are simpler to make because they work in air or water with no additional tuning. Whip antennas, on the other hand, require that the transmitter be tuned in the water during construction by selecting the proper capacitor or by trimming the antenna length until the maximum signal is produced. Since the signal transmission pattern of a loop antenna is considerably more directional than a whip antenna, signal levels from loop antennas on moving animals fluctuate greatly, making location of animals more difficult. For a given size battery, whip antennas result in a longer, narrower package than loop antennas and are better suited for implanting in narrow body cavities. A disadvantage of whip antennas is that the wire bends when implanted in the body cavity, and the whip's efficiency decreases.

Receiving antennas. Receiving antennas are usually loop, Yagi, or omnidirectional antennas. Each antenna has special characteristics, but all depend on increasing the height above the water for greater signal detection range. Loop antennas are usually small, circular or diamond-shaped metal tubes with a handle for hand-carrying or mounting on an airplane (Fig. 20.5). A loop antenna is tuned in the

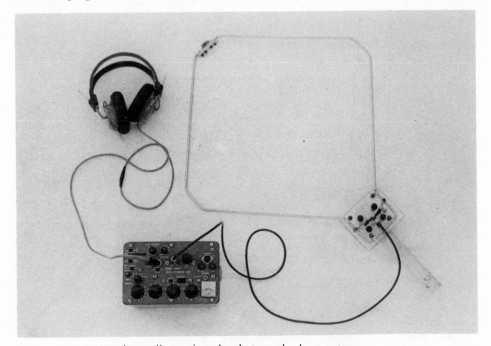

Figure 20.5. A scanning radio receiver, headsets, and a loop antenna.

field by turing a screw on a variable capacitor until the antenna resonates at the correct frequency and produces the strongest signal. Loop antennas are bidirectional with maximum signals occurring when either edge is pointed at the transmitter. When the plane of the antenna is perpendicular to the transmitter direction, a minimum or no (null) signal occurs. Since the null is easier to distinguish than the maximum signal, listeners generally use the null to take bearings. The signal detection range of a loop antenna is usually less than one kilometer, which makes it useful for maneuvering close to an animal or recovering lost transmitters.

Yagi antennas consist of a series of small diameter metal tubes (elements) attached perpendicularly to a long metal tube (boom) (Fig. 20.1 and 20.6). The lengths of the elements and boom increase with decreasing frequency and increasing wave length. Yagi antennas are widely used because they are inexpensive and have excellent directionalities, sensitivities, and ranges. They can be mounted on towers, trucks, and boats (Winter et al. 1978). Because they are small, high frequency (> 100 MHz) Yagis can be hand-carried or mounted on airplanes. Since underwater radio waves are usually vertically polarized (oriented), mount Yagi antennas with the elements vertical to receive maximum radio waves. Yagis are directional with two null points on either side of a peak signal. Since the nulls are sharper than the peak, bearings are determined by taking the mid-point between the two nulls. Null-peak systems are two Yagis connected to a receiver with the antennas spaced 1 or 2 wavelengths apart and pointed in the same direction (Cochran 1980). When the signals from both antennas are received in phase, there is an increase in gain over a single Yagi. By using a phasing switch, the signals can be made out of phase. This produces a sharp null because of signal cancellation. Precision on bearings to a transmitter are within ± 3° for a single Yagi and within ± 1° for null-peak systems; however, the distance error increases with distance from the transmitter.

Omnidirectional antennas are usually ground-plane antennas or straight-wire

Figure 20.6. A 53 MHz Yagi radio antenna mounted on a boat.

whips. A ground-plane antenna consists of a short, vertical, 1/4 wavelength element with four slightly longer elements perpendicular to the base of the vertical element and spaced equally along it (Amlaner 1980). A whip antenna can be a straight, stiff wire or coaxial cable. Omnidirectional antennas are used for quickly finding a transmitter signal or for recording telemetry data where null points must be avoided. Short coaxial whip antennas can be used to pinpoint transmitters, locate fish along tailraces of dams, and record their passages through fish ladders (Johnson 1981).

Signal transmission. Transmission of radio signals is dependent on water depth, water conductivity, signal frequency, reflection, and noises. As the depth of the transmitter increases, transmitter range decreases almost exponentially. Only those signals that travel almost vertically through the water will emerge and travel through air. At the water-air interface, signals from a typical horizontally-polarized underwater antenna are changed to a vertically-polarized radio waves. Since transmitter range also decreases with increasing conductivity, radio telemetry is suitable for conductivities above 400-600 µmho/cm only if the animal remains close to the surface. The effects of depth and conductivity (Fig. 20.3) are greater for frequencies higher than 100 MHz, which are commonly used in terrestrial telemetry, than for frequencies of 30-60 Mhz, which are commonly used in underwater telemetry. Losses in signal strength and errors in determining signal direction also occur when signals are reflected off vegetation, terrain, and buildings. Furthermore, noise from powerlines, ignition systems, and citizen band radios can interfere with signal detection. Therefore, one should monitor the study area for noises before beginning radio telemetry projects and should avoid using transmitters with the same frequencies as the noises. In addition, antennas should be installed in an open area if possible and oriented with the directional end away from sources of noise.

20.2.5 Sensing Tags and Data Recording

Transmitters that sense environmental and physiological parameters provide valuable tools for understanding animal adaptations and requirements. Since these transmitters generate huge amounts of data, there has been simultaneous development of automatic data recording and processing systems. Transmitters which measure temperature and pressure (depth) are readily available from commercial sources. Many other types have been built, but their designs and production methods need improvement. Sensing transmitters (except those which measure temperature) and recording equipment are expensive; however, the cost per unit of information obtained can be lower than that of location telemetry.

Monitoring water temperature is valuable because it can be used to relate animal movements to temperature and to reveal animal depth distribution in stratified waters during the summer. A temperature-sensing transmitter can be produced by replacing a fixed resistor in a location tag with a thermistor whose resistance changes with temperature (Nelson 1978). Temperature is usually coded by changes in pulse rate or interval between pulses. Since the circuitry is simple, temperature tags can be produced reliably and at a cost comparable to location tags.

Before attachment, each transmitter must be calibrated at known temperatures in a water bath. Athough the temperature-pulse relationship is non-linear, curves or mathematical equations can be developed to convert each tag. The usual tag accuracy is about ±0.2 C; however, the critical factor is how much the tag will drift from the calibration curve with time. Water temperatures can be measured by an external transmitter (Ross and Siniff 1980) or by an implanted transmitter with a thermistor extending through the body wall (Rochelle 1974). In addition, the temperatures of specific body locations can be determined by inserting probes (Standora 1977; Prepejchal et al. 1980).

Transmitters that measure water pressure give a direct indication of an animal's depth (Luke et al. 1973; Ross et al. 1979). A strain gauge transducer changes circuit resistance with changing water pressure; thus, pressure can be converted to depth because water pressure increases about 0.10 kg/cm^2 per meter depth. Each tag must be calibrated in a pressure chamber. Since the pressure transducer is expensive and the circuits are more complex, pressure tags are quite expensive. Pressure tags are usually large and have short lives (< 30 days) because the pressure transducer requires great power.

Many other parameters have been or can be monitored from free-ranging aquatic animals. Some of these include: salinity (Mani and Priede 1980), light (Nelson 1978), swimming speed (Nelson 1978), tail-beat frequency (Dalen 1977), swimming activity from sensors in muscles (Luke et al. 1979), swimming direction (Nelson 1978), heart beat (Dalen 1977), and electrocardiogram (Frank 1968).

Sensing-transmitters can be built to transmit several parameters, i.e., multichannel systems (Nelson 1978). The most common method is to transmit the information from each sensor (channel) sequentially with one transmitter frequency. One channel is used for a reference code. Switching of channels (multiplexing) can be rapid or slow. In rapid multiplexing, each sensor transmits for one pulse interval. In slow multiplexing, each sensor transmits for a fixed time interval, e.g., 5 seconds. Multichannel transmitters have been described by Ferrel et al. (1974) and Prepejchal et al. (1980).

The simplest but most time-consuming method for recording data from sensing transmitters is to time a certain number of pulses or pulse intervals with a stopwatch. The signal also can be recorded on magnetic recording tape or stripcharts at preselected intervals controlled by a timer. Although these systems reduce labor in data collecting, they require considerable manual transcribing of data for analysis.

Digital recording systems are used to reduce the amount of manual data transcription (Pincock 1980). The systems consist of a receiver, decoder, and digital recorder. A decoder accepts signals from one or more receivers, extracts sensor data (e.g., time between pulses) and passes it, usually along with the date, time, and other data not obtained from the tag (e.g., water temperature) to the digital recorder. If the data are recorded on computer tapes or discs, the data can be transferred to a large computer for subsequent data analysis. Decoders with a calculator-type keyboard allow the investigator to pass new instructions to the decoder; however, these decoders have little ability to analyze data while data are being collected. By using a small general purpose computer, data can be decoded, displayed, stored, and periodically analyzed while data collection is still in progress.

20.2.6 Choosing a Supplier

Choosing a supplier of telemetry equipment is a crucial decision because equipment, performance, and service by telemetry manufacturers varies greatly. The first step is to review the literature or the *Underwater Telemetry Newsletter* for researchers who are highly involved in telemetry. They can recommend reliable equipment and manufacturers.

Researchers often consider whether to make their equipment or to buy it from an established telemetry company. Although commercial equipment seems expensive, an investigator's time is better spent collecting data than building equipment. Avoid contracting equipment development to companies, friends, colleagues, or students not engaged in the business of animal tracking. This has too often led to reinventing the wheel and usually does not pay off in reliability or service. If the equipment does not exist, most commercial telemetry companies are interested in contracting to develop it. One must also be careful of companies that adapt equipment developed for some other purpose (e.g., marking underwater salvage objects or tracking terrestrial wildlife) with little modification to aquatic animals. Reliable firms usually build a large number of transmitters specifically for aquatic animals and consider aquatic animal tracking an important part of their business.

There are several things to evaluate about a telemetry company based on their reputation and what they promise to do. A firm should fill orders quickly and repair the equipment quickly or loan replacement equipment. In addition, technical problems arise during planning or conducting a project, and a good company will provide advice and travel to the site to solve a problem. Good firms also should be willing to adapt or build equipment for the project rather than change the project to fit their equipment. Finally, the supplier should provide instructions and circuit diagrams with the equipment. Most problems which arise in the field are simple and can be solved with basic information. Some electronic problems can be easily solved by a local radio shop if they have circuit diagrams to follow.

20.3 METHODS OF ATTACHING TRANSMITTERS

The transmitter attachment method depends on the morphology and behavior of the species, the nature of the aquatic ecosystem, and the objectives of the project. Basic methods are external, stomach, and surgical implantations. Before attaching transmitters to experimental animals, practice attachment methods on dead and captive animals. This should include dissection and autopsy to understand how tag placement relates to the anatomy. Since most underwater biotelemetry projects deal with fish, this discussion concentrates on fish attachment methods. Other methods of attachment and types of organisms tagged are listed in Table 20.1. Fish generally should not be equipped with transmitters that weigh more than 1.25% in water or 2% in air of the fish's weight out of water.

20.3.1 External Transmitters

External attachment is quicker and easier than surgical implantations and can be used for spawning and feeding fish. Since it allows the organism to recover quickly after tagging, it is useful if animals must be released immediately or data must be collected over a short period. External transmitters are also necessary for sensing some environmental parameters like water temperature. Because of the higher center of gravity caused by dorsal external transmitters, animals must make greater compensation for balance than when other methods are used. In addition, external transmitters cause increased drag on swimming organisms. Transmitter drag may not affect movement patterns but could affect swimming speed or energy expenditure. The balance and drag problems can be alleviated by reducing the size of the transmitters. Finally external transmitters can cause abrasions on the animal, and they can snag on vegetation, monofilament fishing line, and other objects.

The external method with the widest application attaches the transmitter

Table 20.1. Methods of attaching transmitters to aquatic animals.

Method	Taxon	References
External		
dorsal fin	walleye, bass, perch	Winter et al. 1978
(pins & wires)	perch, northern pike	Ross and Siniff 1980
	porpoise, dolphin, whale	Leatherwood and Evans 1979
straps and pins	salmon	Hallock et al. 1970
hog ring	salmon	Johnson 1960
saddle	rainbow trout	Winter et al. 1978
	brown trout	Prepejchal et al. 1980
Hook and leader	white bass	Henderson et al. 1966
	rainbow trout	Shepherd 1973
dart	shark	Standora and Nelson 1977
	whale	Leatherwood and Evans 1979
alligator clips	cutthroat trout	McLeave and Horrall 1970
harness	Alaska king crab	Monan and Thorne 1973
	American lobster	Lund and Lockwood 1970
	whale	Leatherwood and Evans 1979
screws, epoxy	turtle	Ireland and Kanwisher 1978
sutures	Weddell seal	Siniff et al. 1971
collar		
flipper	Weddell seal	Siniff 1970
neck	alligator	Standora 1977
head	crocodile	Yerbury 1980
Stomach		
ultrasonic	white bass	Henderson et al. 1966
	salmon, trout	Coutant 1969
radio		
whip antenna out		
gill cavity	Atlantic salmon	McCleave et al. 1978
whip antenna		
anchored in mouth	salmon	Monan et al. 1975
whip antenna out		
behind maxillary	salmon	Haynes 1978
Implantation		
no sutures	white bass	Henderson et al. 1966
sutures		
ultrasonic	flathead catfish	Hart and Summerfelt 1975
external thermistor	bass	Rochelle and Coutant 1974
radio whip antenna		
under skin	northern pike	Winter et al. 1978
radio whip antenna		
in body cavity	walleye	Einhouse 1981
radio loop antenna		
in body	bullfrogs	Crossman 1980
radio under the		
skin	Weddell seal	Siniff 1970

alongside the dorsal fin by means of two stainless steel or teflon-coated electrical wires (Fig. 20.7). A surgical needle can be used to pass the wires through the adipose tissue beneath the dorsal fin and between the pterygiophore bones. An alternative method is to insert large hypodermic needles through the fish from one side and thread the wires through the needles from the other side. After removing the hypodermic or surgical needles, the transmitter is pulled snug against the fish, and the wires are

Figure 20.7. Procedures for attaching an external transmitter adjacent to the dorsal fin. Top, threading an attachment wire through a hypodermic needle; middle, dorsal view of a transmitter and plastic plate; and bottom, lateral view of an external radio transmitter with a whip antenna on a yellow perch.

threaded through holes in a soft plastic plate (*see* chapter 11). When the plate is positioned against the fish, knots are tied on the wires to prevent them from pulling through the fish. A neoprene pad placed between the plate and the fish will prevent abrasion. Furthermore, one should attach a radio transmitter so that the whip antenna does not extend beyond the caudal fin.

20.3.2 Stomach Inserted Transmitters

Internal transmitters do not cause drag, cannot become snagged, and are less likely to cause abrasion. Since an internal transmitter is below a fish's center of gravity, one can use a heavier package internally than externally without creating balance problems. Stomach insertion can be done quickly and probably requires the shortest habituation time by fish. The disadvantages of stomach insertion are that transmitters may be difficult to get into the fish's mouth or past the pharyngeal food-crushing apparatus (Stasko and Pincock 1977), may be regurgitated by many species (Coutant 1969; Hart and Summerfelt 1975; Stasko and Pincock 1977), and may rupture the esophagus or stomach (McCleave and Horrall 1970; Gray and Haynes 1979). In addition, external parameters are difficult to monitor with stomach tags.

Stomach tags are inserted through the mouth into the stomach with tubes or with ingested bait. Use a plunger system consisting of a tube within a tube or a veterinarian's bolling-gun. These tubes can be coated with glycerin to slide through the gullet smoothly, but one must be careful not to force the tube. Radio whip antennas are extended out the gullet and can be anchored with a barb in the roof of the mouth (Monan et al. 1975), passed through the tissue behind the maxillary (Haynes 1978), or passed out the gill cavity (McCleave et al. 1978).

20.3.3 Surgically Implanted Transmitters

Surgical implantation into the body cavity is excellent for physiological telemetry, e.g., heart rate, and is the best method for long term attachment. However, implantation takes longer to perform and requires a long recovery period. External factors are also difficult to monitor unless a sensor is passed through the body wall (Rochell and Coutant 1974). For the best results, hold animals a day before surgery to ensure that the capture method has not caused much stress and also a day after surgery to ensure that they are recovering properly before release. Surgical implants are more likely to cause infection than other methods, especially in warm water, and incisions can be slow to heal in cold water. Although eggs can be removed easily from an anesthetized fish to create space for a transmitter, implants may not be suitable for some gravid females. Mortality of fish from implantation is higher after spawning than during other times. Therefore, the best times to implant in most fish are in the spring and early fall when the water is cool and after spawning when the fish have recovered their conditions.

Surgical implantation (Fig. 20.8) is not difficult and does not require elaborate equipment; however, it does require some planning (Hart and Summerfelt 1975). Experiment before beginning the field work to find the dosage of anesthetic (chapter 5) that will keep an animal anesthetized for about 15 minutes and allow recovery in 15 minutes or less. Second, learn how to suture (e.g., from a veterinarian) and then practice on a piece of cloth stretched over the open end of a can. Poor suturing is time-consuming and often the cause of lost transmitters.

Fish can be placed in a V-shaped tagging trough and water passed over the gills via a siphon system or squeeze-bottle. An incision is made slightly to one side of the mid-ventral line and just large enough to admit the transmitter. The incision will vary among species but should be in the area where the body is the deepest and not near vital organs like the heart or liver. Keep the instruments sterile by soaking them in Zephiran chloride, alcohol, etc., and keep water out of the incision. One can apply antibiotics into the body cavity, but this is not necessary if reasonably sterile

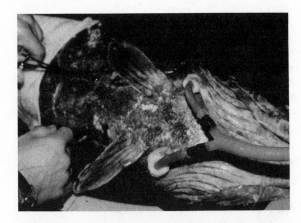

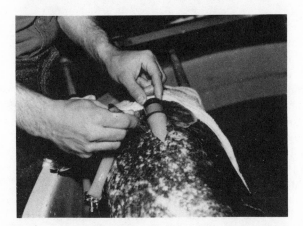

Figure 20.8. Surgically implanting a transmitter into the body cavity of an Antarctic cod. Top, siphon for bathing the gill cavity with an anesthetic or water; middle, inserting the ultrasonic transmitter into the body cavity; and bottom, suturing the incision with monofilament suture.

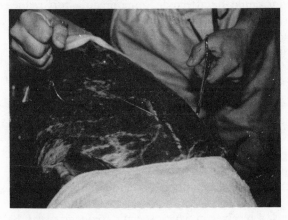

techniques are used. After making an incision, ultrasonic transmitters and radio transmitters with loop antennas are placed into the body cavity. If the transmitter has a whip antenna, the whip should be extended inside the body cavity using a hemostat forceps before the transmitter package is implanted. Whip antennas have been placed under the skin via a tunnel (Winter et al. 1978), but the skin may die and the antenna may pull out. In addition, whip antennas or sensing probes (Rochelle and Coutant

1974) can be passed through the body wall and trailed. Since a radio signal is only slightly attenuated by the fish's body and an external whip may bend, it is questionable if extending a whip antenna through the body wall increases the range much. After implanting the transmitter, the incision is sutured. Although monofilament suture material is available from medical supply houses, 6-8 lb monofilament fishing line works as well. Incisions can be closed by one continuous suture for speed, or each suture can be tied off separately with squareknots for greater reliability. In most fish, the muscle and skin are usually sutured together, but in large fish the muscle should be sutured first and the skin separately with finer material. Care must be taken to pass the needle completely through the muscle so that the skin will completely cover the incision.

20.4 METHODS OF TRACKING

20.4.1 Boat

Aquatic animals can be located by boat, airplane, triangulation, appearance at fixed stations, and signal time delays. Tracking by boat often involves positioning the boat above the animal. This provides a very accurate location (0-10 m depending on transmitter depth) and allows limnological measurements at the site. As a tracker in a boat approaches an animal, the receiver volume and sensitivity (gain) must be continually adjusted downward to obtain nulls. The Yagi or hydrophone will go from direction ahead while approaching the animal, to an omnidirectional signal above the animal, to direction astern after the boat passes over the animal. Hydrophones are more accurate at close distances than a Yagi; however, very accurate (<15 m) radio locations can be made by switching to a low gain coaxial whip antenna. In shallow water, there is a possible data bias because the boat may disturb the animal. If the locations are made infrequently, the disturbance will not alter home range patterns. However, if the movements are being followed continuously, then the animal's direction of movement or behavior could be affected. Some investigators reduce disturbance by attempting to stay a certain distance from the fish as estimated by signal strength, but this can result in considerable location error. The animal's location is usually determined by visual sightings, compass readings, sextant angles, radar readings, and Loran C radio readings on terrestrial locations. Sextants, radar, and Loran C are the most accurate methods.

20.4.2 Airplane Tracking

Tracking from an airplane is useful for locating highly mobile radio-tagged animals. Airplanes provide the greatest detection range for radio signals, usually double or triple that of a Yagi on the ground, but the cost is high. Since airplanes travel rapidly, a tracker can miss an animal while checking many frequencies. In addition, there may not be enough time to detect the signal from a deep transmitter with low range. To minimize these problems, a receiver in which specific frequencies can be programmed and quickly scanned should be used. Gilmer et al. (1981) recommended side-orientation for high frequency Yagis and either side-orientation or a combination of side- and front-orientation for low frequency loop antennas. To detect weak signals, use a plane with the ignition noise suppressed. In addition, use padded monophonic headsets that fit tightly around the ears to reduce interference from noises in the plane.

Transmitter locations are determined by listening for the maximum signal as the plane passes over the transmitter or by flying circles of decreasing circumference until the receiver volume and gain are turned as low as possible. Locations can be determined to within 100 m.

20.4.3 Triangulation

Animals can be located from a distance by taking bearings from two or more locations (triangulation). Bearings can be made by moving the receiving unit between locations or by using several stationary hydrophones or directional radio antennas. Bearings cannot be plotted accurately on a map unless the sites from which the bearings were taken are mapped accurately. Since the most accurate locations occur when bearings intersect at 90°, move mobile antennas to achieve bearings as close to this angle as possible. Fixed stations should be located so that it is not possible for the animal to be near the imaginary line between the stations but very likely that the animal will be near the imaginary perpendicular bisector (most accurate). Since moving animals can cause great error in triangulation, bearings that are taken simultaneously provide more accurate triangulation.

20.4.4 Appearance at Fixed Stations

Receiving systems and recording units can be set up at certain locations such as along rivers (Hallock et al. 1970) or dams (Johnson 1981). The basic system records the presence or absence of a transmitter frequency near the hydrophone or antenna. In streams, narrow beam hydrophones or directional Yagis can be oriented across the stream bed. Where the animal can be in any direction around the station, omnidirectional antennas or hydrophones are used. Omnidirectional hydrophones and antennas can be arranged in a grid or transect; the animals can then be located according to which antenna receives the strongest signal (Zinnel 1980). Ultrasonic signals can also be detected by buoys and relayed by radio signals to another receiving station.

20.4.5 Time Delay Systems

Ultrasonic transmitters can be located by measuring the differences in time that it takes the transmitter signal to reach each hydrophone in an array of three or more fixed hydrophones (Hawkins et al. 1973). This system is very accurate and can be adapted for automatic recording. However, it depends on having a strong signal to trigger the time counter and having animals whose home ranges are similar to the signal detection range of the hydrophones. A similar method can determine the distance to the animal by measuring the time that it takes to send a signal to a tag and receive a signal back from that tag (transponder). If the direction to the signal is also known and the transponder returns information on the animal's depth, the animal can be located accurately in three dimensions. Transponders theoretically save battery life because they transmit information only when needed; however, the receiving portion of the transponder, except in high power output transponders, uses sufficient current to cancel the savings. Methods are currently being developed to locate aquatic animals with satellites by measuring the doppler frequency shift due to the satellite moving closer to or farther from the transmitter (Jennings and Gandy 1980). Since radio signals travel through air much faster than ultrasonic signals travel through water, the development of radio time delay systems for animal tracking has lagged behind that of ultrasonic systems.

20.5 DATA COLLECTING AND PROCESSING

20.5.1 Field Considerations

There are many precautions to take when using telemetry equipment. A good rule for telemetry studies is to have a spare of *every* piece of equipment to avoid data loss because of accidental breakages or failures. Transmitters should be stored in padded boxes because crystals and transducers are very sensitive to jolts. If transmitters are not to be used for months, they should be kept in a refrigerator to keep the batteries fresh. It is a good idea to activate long-lived transmitters a few days before tagging. If any transmitters are defective, the problem is likely to appear during the first few days of activation. Test each transmitter for signal range before attaching to an animal. This will reduce the chance of attaching a poor transmitter. Since the receiver frequency selector seldom exactly matches the theoretical transmitter frequency, determine the correct receiver reading for each transmitter before releasing the animal. This can be determined by reducing the receiver volume and gain controls until the transmitter is detected at one frequency. It is useful to carry a deactivated transmitter to test the receiving system if problems are suspected. However, do not carry an activated transmitter. Its signal may be so strong that it may interfere with reception, or, because it may be received at other frequencies, the signal may cause a mistake in recognizing an animal.

There are several precautions to take with receivers. Moisture will reduce the sensitivity of a receiver, make the gain control noisier, and short out the receiver. Cover receivers with plastic bags to keep moisture out. After use in wet conditions, receivers should be opened and placed in a warm area to dry. Lighter fluid or contact cleaner can be placed on noisy gain control switches. If the receiver is likely to be bounced around, a padded carrying case may prevent damage. In the winter, you can place handwarmers in the carrying bag with the receiver to maintain battery and sensitivity performance. The Nicad batteries in a receiver should be discharged to a low level before charging because they develop a "memory" of how much they usually deliver.

Antennas and, to a lesser extent, hydrophones seldom malfunction. When they do, the most likely problem is the connecting cable. If the signal is discontinuous, check the connections. Since antenna cables get bent or twisted, the wire inside the cable may be broken. Replace cables periodically. Receiving systems can be checked in the field in the absence of a test transmitter. If the antenna or hydrophone is working, an increase in noise level should be heard when the cable is connected to the receiver. Running your finger over the hydrophone also should produce noise.

20.5.2 Number of Animals

The number of tagged animals necessary in telemetry studies depends on cost, labor, availability of animals, type of data desired, and methods of statistical analysis. Studies of home range and migration patterns probably benefit more from long term tracking of individuals than from close daily monitoring. If animals are so widely distributed that they must be located by boat, truck, or airplane, 20-40 individuals are usually the maximum number that can be monitored concurrently. The number can be larger if individuals are located less frequently or if they can be located from fixed receiving stations. Physiological or behavioral studies that necessitate frequent monitoring of individuals, especially from mobile tracking units, usually monitor fewer individuals. Since sample sizes in underwater biotelemetry are often small, investigators must be careful not to scatter their tagging efforts over many sites, different species, and different size classses. It is better to answer one question well than many questions poorly.

20.5.3 Searching

Sound methods of searching for animals equipped with transmitters are important in reducing cost, labor, and bias in data. Most trackers begin searches in the place where they last located the animal. This usually saves much time in searching; however, if animals occasionally disappear, there will be considerable bias in determining habitat utilization because all habitats were not monitored equally. For animals that are very mobile or those that may occupy deep water and have unpredictable movement patterns, e.g., walleyes, the best method is to search along transects.

20.5.4 Sampling Time

The time of the day when the monitoring is done is important. Ideally, all hours should be sampled equally in most studies. Sometimes an investigator is interested only in a certain time period or behavior and, thus, allocates all the monitoring time to that period. Another approach is to choose the days and the hours sampled on a given day from a random numbers table. In practice, this is extremely difficult to do because of investigators' schedules. However, the most practical method is to combine periodic close monitoring of a few individuals over 24-hour periods with the sampling system normally used. This will verify if the usual sampling periods are representative of the animal's typical activity.

20.5.5 Number of Data Points

The number of locations or data points depends on the type of project. Studies on home range and habitat utilization usually monitor individuals throughout a season, and each individual is usually located every 6-48 hours. Procedures are available to determine whether a sufficient number of locations have been made to estimate the area of a home range (Winter 1977; Winter and Ross 1982). Studies on orientation of animals to physical, chemical, or celestial factors usually monitor individuals for a few days or less with locations made minutes to an hour apart. Experiments using sensing transmitters and studies on activity patterns usually record data at evenly spaced intervals of an hour or less. The length of time that an individual is monitored depends largely on transmitter life, which is less for most sensing transmitters than for location tags.

20.5.6 Plotting Data

To process movement data quantitatively, bearings are often drawn on a plotboard that uses a Cartesian coordinate system constructed on a map that has an accurate representation of the shoreline. From a plotboard, each animal location is given X- and Y-coordinate values. These coordinates can be entered into a computer along with information on the fish's identification number, the date, time, weather conditions, water characteristics, habitat type, and any environmental or physiological parameters monitored by the transmitter on the animal. Computer programs such as SPSS or Minitab are available on most computer systems to plot scattergrams (X-Y data) that can be overlaid on a lake map. The scale of the axes must be the same as the map scale. Sometimes, however, it is difficult to get the scales to match. In this case, use the computer to draw a map of the lake to the correct scale from coordinates of the shoreline.

20.5.7 Data Analysis

Since telemetry can generate large amounts of data, data analysis is often done with a computer. One of the most basic parameters for movement studies is the distance between locations. A computer program to determine distance between points on a Cartesian coordinate system can be written easily from a formula found in most algebra books. The program can calculate distances from the previous location to determine swimming speed and activity or from fixed locations such as a stream mouth, underwater structure, or industrial discharge site to determine behavioral responses. Behavioral interactions can be analyzed by calculating the distances between individuals at specific times. Studies on animal orientation usually calculate the angle of movement with respect to the previous locations or some stimulus. Batschelet (1965) has written a report on statistically analyzing data on orientation and biological rhythms. Home ranges are often determined by drawing polygons using the peripheral locations or from the squares on the gridded map that have locations (Winter 1977; Winter and Ross 1982). Habitat utilization can be determined from the number of locations in gridsquares containing each habitat type. Data from sensing transmitters that are coded by pulse interval times can be converted by simple pocket calculator (Coutant 1977) or computer programs (Prepejchal et al. 1980). Sensing transmitter data can be analyzed using standard statistical programs and tests.

20.6 REFERENCES

Amlaner, C. J., Jr. 1980. The design of antennas for use in radio telemetry. Pages 251-261 *in* C. J. Amlaner, Jr. and D. W. Macdonald, editors. *A handbook on biotelemetry and radio tracking.* Pergamon Press, Elmsford, New York, USA.

Amlaner, C. J., Jr., and D. W. Macdonald, editors. 1980. *A handbook on biotelemetry and radio tracking.* Pergamon Press, Elmsford, New York, USA.

Batschelet, E. 1965. *Statistical methods for the analysis of problems in animal orientation and certain biological rhythms.* American Institute of Biological Sciences, Washington, District of Columbia, USA.

Brander, R. B., and W. W. Cochran. 1971. Radio-location telemetry. Pages 95-103 *in* R. H. Giles, Jr., editor. *Wildlife management techniques,* third edition. The Wildlife Society, Washington, District of Columbia, USA.

Brumbaugh, D. 1980. Effects of thermal stratification on range of ultrasonic tags. *Underwater Telemetry Newsletter* 10 (2):1-4.

Campbell, W. B., and D. L. Stoneburner. 1980. Lithium battery technology for biotelemetry. *Underwater Telemetry Newsletter* 10 (2):5-11.

Cochran, W. W. 1980. Wildlife telemetry. Pages 507-520 *in* S. D. Schemnitz, editor. *Wildlife management techniques,* fourth edition. The Wildlife Society, Washington, District of Columbia, USA.

Coutant, C. C. 1969. Temperature, reproduction and behavior. *Chesapeake Science* 10:261-274.

Coutant, C. C. 1977. HP-65 program for calculating basic statistics (X, S.D., S.E.) of temperature selection from pulse interval data. *Underwater Telemetry Newsletter* 7 (2):6-7.

Crossman, E. J. 1980. Radio tracking bullfrogs, *Rana catesbeiana,* in Ontario. *Underwater Telemetry Newsletter* 10 (1):8-9.

Dalen, J. 1977. Underwater telemetry work in Norway. *Underwater Telemetry Newsletter* 7 (1):5-9.

Einhouse, D. W. 1981. *Summer-fall movements, habitat utilization, diel activity and feeding behavior of walleyes in Chautauqua Lake, New York.* Master's thesis. State University College. Fredonia, New York, USA.

Ferrel, D. W., D. R. Nelson, T. C. Sciarrotta, E. A. Standora, and H. C. Carter. 1974. A multi-channel ultrasonic biotelemetry system for monitoring marine animal behavior at sea. *Instrument Society of America Transactions* 13:120-131.

Frank, T. H. 1968. Telemetering the electrocardiogram of free swimming *Salmo irideus. IEEE Transactions on Bio-medical Engineering* 15:111-114.

Gilmer, D. S., L. M. Cowardin, R. L. Duval, L. M. Mechlin, C. W. Shaiffer, and V. B. Kuechle. 1981. *Procedures for the use of aircraft in wildlife biotelemetry studies.* United States Fish and Wildlife Service Resource Publication 140, Washington, District of Columbia, USA.

Gray, R. H., and J. M. Haynes. 1979. Spawning migration of adult chinook salmon *(Oncorhynchus tshawytscha)* carrying external and internal radio transmitters. *Journal of the Fisheries Research Board of Canada* 36:1060-1064.

Hallock, R. J., R. F. Elwell, and D. H. Fry, Jr. 1970. *Migrations of adult king salmon, Oncorhynchus tshawytscha, in the San Joaquin delta.* California Fish and Game Bulletin 151. (Available from Marine Technical Information Center, California State Fisheries, Terminal Island, California 90731, USA.)

Hart, L. G., and R. C. Summerfelt. 1975. Surgical procedures for implanting ultrasonic transmitters into flathead catfish *(Pylodictis olivaris). Transactions of the American Fisheries Society* 104:56-59.

Hawkins, A. D., D. N. MacLennan, and G. G. Urquhart. 1973. Tracking cod in a Scottish sea loch. *Underwater Telemetry Newsletter* 3 (1):1,5-6.

Haynes, J. M. 1978. *Movements and habitat studies of chinook salmon and white sturgeon.* Doctoral dissertation. University of Minnesota, Minneapolis, Minnesota, USA.

Henderson, H. F., A. D. Hasler, and G. G. Chipman. 1966. An ultrasonic transmitter for use in studies of movements of fishes. *Transactions of the American Fisheries Society* 95:350-356.

Ireland, L. C., and J. S. Kanwisher. 1978. Underwater acoustic biotelemetry: procedures for obtaining information on the behavior and physiology of free-swimming aquatic animals in their natural environments. Pages 341-379 *in* D. I. Mostofsky, editor. *The behavior of fish and other aquatic animals.* Academic Press, New York, New York, USA.

Jennings, J. G., and W. F. Gandy. 1980. Tracking pelagic dolphins by satellite. Pages 753-755 *in* C. J. Amlaner, Jr. and D. W. Macdonald, editors. *A handbook on biotelemetry and radio tracking.* Pergamon Press, Elmsford, New York, USA.

Johnson, G. A. 1981. The use of radio tracking to evaluate the powerhouse adult fish passage systems on the Columbia and Snake Rivers. *Underwater Telemetry Newsletter* 11 (1):4-8.

Johnson, J. H. 1960. Sonic tracking of adult salmon at Bonneville Dam, 1957. *Fishery Bulletin* 176:471-485.

Johnsen, P. B., and R. M. Horrall. 1974. A temperature and pressure sensing biotelemetry system. *Underwater Telemetry Newsletter* 4 (1):13-15.

Kuechle, V. B. 1967. *Batteries for biotelemetry and other applications.* American Institute of Biological Sciences/BIAS Information Module M10. Washington, District of Columbia, USA.

Leatherwood, S., and W. E. Evans. 1979. Some recent uses and potentials of radiotelemetry in field studies of cetaceans. Pages 1-31 *in* H. E. Winn and B. L. Olla, editors. *Behavior of marine animals, cetaceans,* volume 3. Plenum Press, Elmsford, New York, USA.

Long, F. M., editor. 1977. *Proceedings of the first international conference on wildlife biotelemetery.* University of Wyoming, Laramie, Wyoming, USA.

Long, F. M., editor. 1979. *Proceedings of the second international conference on wildlife biotelemetry.* University of Wyoming, Laramie, Wyoming, USA. (Available from ICWB, P.O. Box 3295, University Station, Laramie, Wyoming 82701, USA.)

Luke, D. McG., D. G. Pincock, and A. B. Stasko. 1973. Pressure-sensing ultrasonic transmitter for tracking aquatic animals. *Journal of the Fisheries Research Board of Canada* 30:1402-

1404.

Luke, D. McG., D. G. Pincock, P. D. Sayre, and A. L. Weatherly. 1979. A system for the telemetry of activity-related information from free swimming fish. Pages 77-85 in F. M. Long,
editor. *Proceedings of the second international conference on wildlife biotelemetry*. University of Wyoming, Laramie, Wyoming, USA.

Lund, W. A., Jr., and R. C. Lockwood, Jr. 1970. Sonic tag for large decapod crustaceans. *Journal of the Fisheries Research Board of Canada* 27:1147-1151.

Macdonald, D. W., and C. J. Amlaner, Jr. 1980. A practical guide to radio tracking. Pages 143-
159 in C. J. Amlaner, Jr. and D. W. Macdonald, editors. *A handbook on biotelemetry and
radio tracking*. Pergamon Press, Elmsford, New York, USA.

MacKay, R. S. 1970. *Bio-medical telemetry; sensing and transmitting biological information
from animals and man*, 2nd edition. John Wiley and Sons, New York, New York, USA.

Mani, R., and I. G. Priede. 1980. Salinity telemetry from estuarine fish. Pages 617-621 in C. J.
Amlaner, Jr. and D. W. Macdonald, editors. *A handbook on biotelemetry and radio
tracking*. Pergamon Press, Elmsford, New York, USA.

McCleave, J. D., and R. M. Horrall. 1970. Ultrasonic tracking of homing cutthroat trout
(Salmo clarki) in Yellowstone Lake. *Journal of the Fisheries Research Board of Canada*
27:715-730.

McCleave, J. D., J. H. Power, and S. A. Rommel, Jr. 1978. Use of radio telemetry for studying
upriver migration of adult Atlantic salmon *(Salmo salar)*. *Journal of Fish Biology* 12:549-
558.

Mitson, R. B. 1978. A review of biotelemetry techniques using acoustic tags. Pages 269-283 in
J. E. Thorpe, editor. *Rhythmic activities of fishes*. Academic Press, New York, New York,
USA.

Monan, G. E., and D. L. Thorne. 1973. Sonic tags attached to Alaska king crab. *Marine Fisheries Review* 35 (7):18-21.

Monan, G. E., J. H. Johnson, and G. F. Esterberg. 1975. Electronic tags and related tracking
techniques aid in study of migrating salmon and steelhead trout in the Columbia River
Basin. *Marine Fisheries Review* 37 (2):9-15.

Nelson, D. R. 1978. Telemetering techniques for the study of free-ranging sharks. Pages 419-482
in E. S. Hodgson and R. F. Mathewson, editors. *Sensory biology of sharks, skates, and
rays*. Office of Naval Research, Department of the Navy, Arlington, Virginia, USA.

Pincock, D. 1980. Automation of data collection in ultrasonic biotelemetry. Pages 471-476 in
C. J. Amlaner, Jr. and D. W. Macdonald, editors. *A handbook on biotelemetry and radio
tracking*. Pergamon Press, Elmsford, New York, USA.

Prepejchal, W., M. M. Thommes, S. A. Spigarelli, J. R. Haumann, and P. E. Hess. 1980. *An
automatic underwater radiotelemetry system to monitor temperature responses of fish in a
fresh-water environment*. Argonne National Laboratory/Ecological Sciences-108.
(Available from National Technical Information Service, United States Department of
Commerce, 5285 Port Royal Road, Springfield, Virginia 22161, USA.)

Priede, I. G. 1980. An analysis of objectives in telemetry studies of fish in the natural environment. Pages 105-118 in C. J. Amlaner, Jr. and D. W. Macdonald, editors. *A handbook on
biotelemetry and radio tracking*. Pergamon Press, Elmsford, New York, USA.

Rochelle, J. M. 1974. Design of gateable transmitter for acoustic telemetering tags. *IEEE Transactions of Biomedical Engineering* 21:63-66.

Rochelle, J. M., and C. C. Coutant. 1974. Ultrasonic tag for extended temperature monitoring
from small fish. *Underwater Telemetry Newsletter* 4 (1):1, 4-7.

Ross, M. J., V. B. Kuechle, R. A. Reichle, and D. B. Siniff. 1979. Automatic radio telemetry
recording of fish temperature and depth. Pages 238-247 in F. M. Long, editor. *Proceedings of the second international conference on wildlife biotelemetry*. University of
Wyoming, Laramie, Wyoming, USA.

Ross, M. J., and D. B. Siniff. 1980. *Spatial distribution and temperature selection of fish near
the thermal outfall of a power plant during fall, winter and spring*. United States Environmental Protection Agency, Ecological Research Series, EPA-600/3-80-009. (Available
from National Technical Information Service, United States Department of Commerce,
5285 Port Royal Road, Springfield, Virginia 22161, USA.)

Shepherd, B. 1973. Transmitter attachment and fish behavior. *Underwater Telemetry Newsletter* 3 (1):8-11.

Siniff, D. B. 1970. Studies at McMurdo station of the population dynamics of antarctic seals.

Antarctic Journal of United States V (4):129-130.

Siniff, D. B., J. R. Tester, and V. B. Kuechle. 1971. Some observations on the activity patterns of Weddell seals as recorded by telemetry. *Antarctic Pinnipedia* 18:173-180.

Slater, L. E. 1963. *Bio-telemetry.* Pergamon Press, Elmsford, New York, USA.

Slater, L. E., editor. 1965. A special report on bio-telemetry. *Bioscience* 15:81-157.

Standora, E. A. 1977. An eight-channel radio telemetry system to monitor alligator body temperatures in a heated reservoir. Pages 70-78 *in* F. M. Long, editor. *Proceedings of the first international conference on wildlife biotelemetry.* University of Wyoming, Laramie, Wyoming, USA.

Standora, E. A., and D. R. Nelson. 1977. A telemetric study of the behavior of freeswimming Pacific angel sharks, *Squatina californica. Bulletin of the Southern California Academy of Sciences* 76:193-201.

Stasko, A. B. 1975. *Underwater biotelemetry, an annotated bibliography.* Fisheries and Marine Service, Research and Development Technical Report Number 534, Biological Station, St. Andrews, New Brunswick, Canada. (Available from Micromedia Limited, 144 Front Street West, Toronto, Ontario M5J 2L7, Canada.)

Stasko, A. B. 1976. Hydrophone mounting system. *Underwater Telemetry Newsletter* 6 (1):8-11.

Stasko, A. B., and S. M. Polar. 1973. Hydrophone and bow-mount for tracking fish by ultrasonic telemetry. *Journal of the Fisheries Research Board of Canada* 30:119-121.

Stasko, A. B., and D. G. Pincock. 1977. Review of underwater biotelemetry with emphasis on ultrasonic techniques. *Journal of the Fisheries Research Board of Canada* 34:1261-1285.

Urick, R. J. 1975. *Principles of underwater sound,* 2nd edition. McGraw-Hill, New York, New York, USA.

Winter, J. D. 1976. *Movements and behavior of largemouth bass (Micropterus salmoides) and steelhead (Salmo gairdneri) determined by radio telemetry.* Doctoral dissertation. University of Minnesota, Minneapolis, Minnesota, USA.

Winter, J. D. 1977. Summer home range movements and habitat use by four largemouth bass in Mary Lake, Minnesota. *Transactions of the American Fisheries Society* 106:323-330.

Winter, J. D., V. B. Kuechle, D. B. Siniff, and J. R. Tester. 1978. *Equipment and methods for radio tracking freshwater fish.* University of Minnesota Agricultural Experiment Station Miscellaneous Report Number 152. (Available from Communication Resources, Coffey Hall, University of Minnesota, St. Paul, Minnesota 55108, USA.)

Winter, J. D., and M. J. Ross. 1982. Methods in analyzing fish habitat utilization from telemetry data. Pages 273-279 *in* N. B. Armantrout, editor. *Proceedings of the symposium on the acquisition and utilization of aquatic habitat inventory information.* Western Division, American Fisheries Society, Bethesda, Maryland, USA.

Yerbury, M. J. 1980. Long range tracking of *Crocodylus porosus* in Arnhem Land, northern Australia. Pages 765-776 *in* C. J. Amlaner, Jr. and D. W. Macdonald, editors. *A handbook on biotelemetry and radio tracking.* Pergamon Press, Elmsford, New York, USA.

Zinnel, K. A. 1980. *Behavior of walleye pike in experimental channels as monitored by a microcomputer utilizing radio telemetry.* Master's thesis. University of Minnesota, Minneapolis, Minnesota, USA.

Chapter 21
Sampling the Recreational Fishery

STEPHEN P. MALVESTUTO

21.1 INTRODUCTION

This chapter discusses the collection of recreational fishery information using direct interview, on-site, survey sampling techniques. (Mail and telephone surveys will be treated in chapter 23.) Emphasis is placed on marine and large impoundment recreational fisheries, although basic aspects of the sampling techniques apply to all situations. Discussion will center around estimating the traditional fishery descriptors of fishing effort (pressure) and fish harvest. In addition, the chapter will describe the commonly used methods of determining the economic benefit associated with recreational fisheries.

The basic unit of fishing effort is the fisherman-hour or angler-hour, which is one hour of active fishing by a single angler. In certain situations it may be difficult to obtain reasonable estimates of the number of hours spent fishing, and other units of effort, such as the boat-trip or party-trip must be used; boat-trips or party-trips can be converted to angler-trips by knowing the mean number of anglers per boat or party. An angler-trip and an angler-day are equivalent units of effort. Effort over a given time period (usually a year) on a given resource can be expressed on a per unit of surface area basis for comparative purposes. Lambou (1966) gives an overview of the most appropriate ways to express creel survey results for reservoirs.

Benefit derived from recreational fisheries has traditionally been measured as harvest of fishes in number or weight. In the past, the terms harvest and catch have been used synonymously; however, given that recreational benefit may be derived from catching fish and returning them to the water, it is more appropriate to treat harvest as a component of catch and define catch as fish harvested plus fish released. Yield incorporates catch as well as all other benefits derived from the recreational experience (Anderson 1975). Relative comparisons of catch, independent of the magnitude of effort or the size of the water body, are possible by expressing catch on a per unit of effort or on a per unit of surface area basis, respectively. Catch per unit of fishing effort (CPUE), usually measured as harvest per unit of effort, is typically used as an index of stock density (Ricker 1975) where the stock comprises harvestable-sized individuals.

Ultimately, it is desirable to measure the economic value of angling so that it can be related to other segments of the economy. Economic (and social attitude) information can be derived through interviews and, when integrated with catch and effort data, can enhance understanding of the clientele and guide management.

The commonly used expression "creel census" will not be used in this chapter. A census implies a total enumeration of the angling population which is virtually impossible for large recreational fisheries. A survey refers to a sample of the population and is the correct terminology. Given that the creel survey, for cost effectiveness, should be directed at more than just creel data, a more appropriate expression describing this kind of data collection effort is "recreational fishery survey."

21.2 THEORETICAL FRAMEWORK: SAMPLING THE ANGLING POPULATION

There are two basic considerations concerning the collection of recreational fishery information: (1) what statistical survey design provides the best quantitative estimates of the fishery characteristics of interest; and (2) what is the most effective way to contact the anglers to obtain the needed information. This section will address commonly used survey sampling designs and the following section will treat angler contact methods.

Before deciding on a particular sampling design, you must clearly formulate the objectives of the survey. The objectives should identify the anglers of concern (target population), define the temporal and spatial dimensions of sampling, and specify the kind of information to be collected. Objectives may be very broad, for example, if the aim is to describe annual fishing for a state, or quite specific, for example, if information is desired on anglers who fish for one species in one lake for one week a year.

The fishery biologist must divide the time and/or space dimensions of fishing into sampling units (SUs) that should be chosen at random. The primary objective is to give all anglers in the target population some probability of being sampled; although anglers cannot be chosen directly at random, the SUs within which they fish can be (*see* chapter 1). Sampling design concepts are covered in detail in survey technique books; Bazigos (1974) presents these concepts specifically for fisheries surveys in inland waters with emphasis on artisanal fisheries in developing countries. The designs outlined here are simple random sampling, stratified random sampling, and stratified two-stage probability sampling.

21.2.1 Simple Random Sampling

In simple random sampling, the temporal/spatial framework is divided into non-overlapping SUs, a given number of which are then chosen for the sample randomly and with equal probability. Box 21.1 outlines a simple random sampling process used to choose sampling days over a 59-day period. The specific objective of such sampling could be to describe a fishery over the period indicated, on a particular body of water that can be canvassed entirely during one fishing day (for example, 6 A.M. to 6 P.M.), when monetary restrictions permit sampling on only 10 of the 59 possible days.

The random sampling process begins by numbering the SUs (fishing days) from one to 59. Next, 10 random numbers are chosen from a random numbers table by looking at random digits two at a time, for example, and choosing the first 10 between 01 and 59. Sampling days should be chosen without replacement so that once a particular day has been chosen, it cannot be sampled again; therefore, its number is ignored if it appears again in the random numbers table.

21.2.2 Stratified Random Sampling

Stratified random sampling is a technique which reduces sampling variance by controlling variability in the parameter being estimated. A heterogeneous population

Box 21.1 Simple Random Sampling

Outline of a simple random sampling procedure where 10 sampling units (fishing days) were chosen from 59 possible units over a two-month survey period. Randomly chosen days are circled on the calendar.

SAMPLING UNIT DEFINITIONS		RANDOM CHOICES	
SU	DATE	SU	DATE
1	1 Feb	51	23 Mar
2	2 Feb	24	24 Feb
3	3 Feb	09	09 Feb
4	4 Feb	57	29 Mar
		30	02 Mar
		39	11 Mar
57	29 Mar	21	21 Feb
58	30 Mar	36	08 Mar
59	31 Mar	37	09 Mar
		26	26 Feb

FEBRUARY

S	M	T	W	T	F	S
	1	2	3	4	5	6
7	8	(9)	10	11	12	13
14	15	16	17	18	19	20
(21)	22	23	(24)	25	(26)	27
28						

MARCH

S	M	T	W	T	F	S
	1	(2)	3	4	5	6
7	(8)	(9)	10	(11)	12	13
14	15	16	17	18	19	20
21	22	(23)	24	25	26	27
28	(29)	30	31			

is divided into homogeneous sub-populations (strata) which are then subjected to simple random sampling. In order to reduce variability associated with estimates of fishing effort, for example, the days within the survey period are often grouped into weekdays and weekend days. The usual situation is that weekend days receive consistently higher levels of fishing effort than do weekdays. A simple random sample of days would probably include both day-types and would give a highly variable estimate of daily fishing effort. Taking a simple random sample of weekdays and another simple random sample of weekend days, however, would provide two relatively precise estimates of fishing effort that could then be combined to estimate effort for the entire study period.

A stratified random sampling plan is presented in Box 21.2 using the same time period identified in the simple random sampling example. Days within the weekday stratum are enclosed by the dark squares on the calendar. The stratified random sampling plan requires that the sampling units for each stratum be chosen separately. The weekday stratum contains 43 SUs, and the weekend stratum contains 16;

weekdays thus are consecutively numbered from 01 to 43 and weekend days from 01 to 16 so that two separate random samples can be chosen. In the example, 10 sampling days are allocated equally between strata, five days each.

The gain in precision because of stratification depends on the degree of heterogeneity in the total population and the extent to which this heterogeneity is alleviated by the stratification scheme. Thus, if most of the variability (heterogeneity) in daily fishing effort is caused by differences between weekdays and weekend days, then stratification by day-type should decrease sampling variance; if, however, variation in daily fishing effort is caused primarily by other factors (weather, for example), stratification by day-type will do little to improve the precision of the estimate. Also, whereas this stratification would be expected to improve the precision of estimates of fishing effort and probably of catch (since catch is highly correlated with effort), it would not result in a more precise CPUE estimate unless CPUE was found to vary consistently according to day-type.

In Box 21.2, the 10 sampling days were allocated equally between strata. Equal allocation of SUs among strata usually will not be the best design. Cochran (1977) gives the general rule that more samples should be taken within a stratum if: (1) the stratum is larger than others being sampled (more fishing days); (2) the characteristic being measured is more variable in the stratum; and (3) the stratum costs less to sample. These three considerations can be interrelated mathematically (Cochran 1977, section 5.5) to provide optimum allocation of sampling units. Malvestuto et al. (1979), using months as strata with constant cost requirements, showed that for West Point Reservoir, Alabama-Georgia, the sampling variation associated with monthly estimates of fishing effort was highly correlated with fluctuations in air temperature and rainfall; the authors offer a method of more optimally allocating monthly sampling effort based on this information.

Apart from potential gains in precision through stratification, this technique must be used if independent estimates are desired for particular subsets of the total population. Common subsets are geographical regions, habitat types, months or seasons of the year, and fishing methods (bank versus boat, for example). Sometimes multiple stratification is desirable (seasons x day-type x habitat), but given that each stratum must be independently sampled, the number of SUs, and thus the expense, quickly can become too large for practical purposes.

21.2.3 Stratified Two-Stage Probability Sampling

Because of time, cost, and logistical constraints, the fishery may be divided into smaller units for sampling purposes. The stratified random sampling scheme presented earlier, for example, requires the creel clerk to cover the entire body of water and to remain on the water for the entire fishing day; frequently this is impossible. The solution is to subdivide each fishing day into secondary or subsampling units. The stratified sampling procedure is conducted in two stages: (1) fishing days or primary SUs (PSUs) are chosen; (2) within each randomly chosen PSU, one or more secondary sampling units (SSUs) are randomly chosen. Box 21.3 shows a subsampling design where each fishing day is divided into six SSUs (two time periods x three lake sections), and one of these SSUs is chosen from each PSU. The box shows a random sample of SSUs (marked with an X) from the five weekdays and five weekend days previously chosen using stratified random sampling. Note that the time periods are labeled A.M. and P.M. and may realistically represent a morning sampling period of six hours (6 A.M. to 12 noon) and an afternoon sampling period of six hours (12 noon

Box 21.2   Stratified Random Sampling

Outline of a stratified random sampling procedure where five sampling units (fishing days) were chosen from each of two strata, one composed of weekdays, the other of weekend days. The strata are separated by the dark outline on the calendar; randomly chosen days are circled.

SAMPLING UNIT DEFINITIONS

WEEKDAY STRATUM		WEEKEND STRATUM	
SU	DATE	SU	DATE
1	1 Feb	1	6 Feb
2	2 Feb	2	7 Feb
3	3 Feb	3	13 Feb
4	4 Feb	4	14 Feb
41	29 Mar	14	21 Mar
42	30 Mar	15	27 Mar
43	31 Mar	16	28 Mar

RANDOM CHOICES

WEEKDAY STRATUM		WEEKEND STRATUM	
SU	DATE	SU	DATE
43	31 Mar	05	20 Feb
25	05 Mar	13	20 Mar
01	01 Feb	03	13 Feb
22	02 Mar	14	21 Mar
40	26 Mar	12	14 Mar

FEBRUARY

S	M	T	W	T	F	S
	①	2	3	4	5	6
7	8	9	10	11	12	⑬
14	15	16	17	18	19	⑳
21	22	23	24	25	26	27
28						

MARCH

S	M	T	W	T	F	S
	1	②	3	4	⑤	6
7	8	9	10	11	12	13
⑭	15	16	17	18	19	⑳
㉑	22	23	24	25	㉖	27
28	29	30	㉛			

Box 21.3   Stratified Two-stage Sampling

Outline of a stratified two-stage probability sampling procedure where each primary sampling unit (fishing day) was divided into six secondary sampling units (time period/lake section categories), one of which was chosen with nonuniform probability sampling from each primary unit. Chosen secondary units are marked with an "X".

TIME PERIOD PROBABILITIES		LAKE SECTION PROBABILITES	
AM	0.40	1	0.50
PM	0.60	2	0.25
	———	3	0.25
	1.00		———
			1.00

SSU	PROBABILITIES	# RANGES
AM-1	0.20	00-19
AM-2	0.10	20-29
AM-3	0.10	30-39
PM-1	0.30	40-69
PM-2	0.15	70-84
PM-3	0.15	85-99

RANDOM CHOICES

WEEKDAY STRATUM				WEEKEND STRATUM		
PSU	# CHOSEN	SSU		PSU	# CHOSEN	SSU
1 Feb	83	PM-2		13 Feb	44	PM-1
2 Mar	39	AM-3		20 Feb	86	PM-3
5 Mar	09	AM-1		14 Mar	07	AM-1
26 Mar	16	AM-1		20 Mar	50	PM-1
31 Mar	62	PM-1		21 Mar	74	PM-2

Weekdays

	1 Feb. AM PM	2 Mar. AM PM	5 Mar. AM PM	26 Mar. AM PM	31 Mar. AM PM
Lake Section 1			X	X		X
Lake Section 2	X				
Lake Section 3		X			

Weekends

	13 Feb. AM PM	20 Feb. AM PM	14 Mar. AM PM	20 Mar. AM PM	21 Mar. AM PM
Lake Section 1	X		X	X	
Lake Section 2					X
Lake Section 3		X			

to 6 P.M.) in a 12-hour fishing day.

The primary difference in the random, stratified, two-stage probability sampling process relative to the other two designs in the PSUs and SSUs can be chosen with unequal or nonuniform probabilities. In Box 21.3, PSUs were chosen with equal probabilities, but SSUs were chosen with nonuniform probabilities, a desirable approach when SSUs have consistently different fishing pressure (or other characteristic of interest). The A.M. and P.M. time periods were given sampling probabilities of 0.4 and 0.6, respectively, because previous observation indicated that 40% of the fishing effort would occur in the morning and 60% in the afternoon. The unequal probabilities associated with lake sections were based on the same rationale. The sampling probabilities associated with the six SSUs were calculated by multiplying the individual time period probabilities by the lake section probabilities. (Note that the sum of the probabilities equals 1.0.) The random choice for SSUs was based on dividing the number range from 00 to 99 into unequal intervals, according to the probabilities established.

This type of design is similar to that discussed in detail by Malvestuto et al. (1978). The advantage associated with this approach is that, on the average, those sampling units receiving the most fishing effort will occur more often in the sample; more information accrues and precision increases by sampling those units. Fishing effort is typically used to establish probabilities because it is more easily measured (party or boat counts, for example) than other characteristics; however, if information is available on other characteristics which are more closely allied to the objectives of the study (harvest or CPUE), this information should be used to establish sampling probabilities. Additionally, a single set of sampling probabilities (as for SSUs in the current example) may not hold for all strata; a separate set may be desirable for weekdays and weekends, and these, in turn, may change seasonally. The more accurate the probabilities, the better the gain in precision from this type of design. This example was based on a two-stage subsampling design; however, the subsampling process can be extended to three or more stages (multi-stage probability sampling). Pfeiffer (1966) describes a nonuniform probability sampling design for a small lake; Best and Boles (1956) compare the accuracy of four design alternatives to total enumeration data.

21.3 ANGLER CONTACT METHODS

Once the sampling units have been defined and randomly chosen, the creel clerk must contact the anglers and collect the necessary information. Alternatively, anglers can be contacted through household surveys which use telephone or door-to-door canvassing to collect data about earlier trips. In these surveys, the sampling population is usually a list of names (from fishing license receipts or boat registrations, for example), and random sampling is applied directly to the list so that the anglers themselves or their households become the sampling units.

Sampling from lists of names has biases; for example, not all anglers have licenses, and boat registrations allow only boat anglers to be sampled. As Deuel (1980b) points out, the primary advantages of household surveys are that the data can be related to the entire population, response rates are high for telephone and door-to-door interviews, and cost per interview is low for mail and telephone surveys. Household survey information can be unsatisfactory, however, because of the nonsampling errors associated with recall over time—for example, "telescoping," when anglers include events outside the recall period or "omission," when anglers omit events within the recall period (Deuel 1980a). Nonresponse error, as when a particular portion of the target population does not respond to the questionnaire, is an inherent problem associated with mail surveys in particular (although follow-up mailings are possible). Because the mail and telephone surveys are emphasized in chapter 23, the remainder of this section emphasizes on-site anglers contact methods, particularly aerial, roving, access point, and complemented surveys.

21.3.1 Aerial Surveys

Aerial surveys can be loosely categorized as a contact method; anglers are counted from an airplane that is flying low enough and slow enough to get an accurate count of individual anglers in boats and on the bank. This will be difficult if the shoreline is irregular or heavily wooded, and biases will occur if portions of the population cannot be counted. It is also important to establish criteria for deciding if people are engaged in fishing or another recreational activity.

Aerial surveys yield only fishing pressure data, measured as the number of anglers or fishing boats operating over a given time period within a given area (*see* section 21.5.1 on measuring fishing effort). The primary advantage is that large areas can be covered in relatively short periods of time so that total enumeration is possible. Where the sampling system has been divided into spatial and/or temporal sampling units, aerial surveys are particularly useful for establishing sampling probabilities (based on fishing effort counts) and adjusting these probabilities as necessary. Major disadvantages are plane rental costs, which can limit the number of overflights possible, and inclement weather, which may deter pilots but not anglers.

21.3.2 Roving Creel Surveys

The roving creel survey is an on-site intercept survey. The basic advantages of intercept surveys are that response rates are high and most anglers are not required to recall catch information (creel clerks identify species and obtain numbers, lengths, and weights). Disadvantages are the high cost per interview, the difficulty of relating survey results to the entire population, and the logistical problems of contacting a representative sample of anglers in large geographical areas (Deuel 1980a).

With the roving survey method, the creel clerk contacts anglers as he moves through the fishing area along a predetermined route. Creel clerks usually travel by boat, and the following discussion will assume boat travel; in certain situations, shore anglers may be contacted more effectively by automobile or on foot. The original statistical formulation of this approach (Robson 1961) dictates that: (1) the route completely covers the survey area, (2) the clerk begins the route at a randomly chosen point of departure, (3) the clerk randomly chooses one of the two alternative directions of travel, and (4) the clerk travels at a constant speed. In practice, item (2) may be logistically difficult, although boat launching sites can be chosen randomly. In certain instances, item (3) may be impossible because of regulations that restrict direction of boat travel. It is critical that the clerk make a complete circuit of the survey area. Where anglers are too numerous to interview, the clerk should systematically skip parties in an objective manner (every second, or third, or tenth group, for example) so that the circuit is completed; if anglers are too few in number to occupy the full time of the clerk, he should slow down enough to complete the full circuit in the allotted time period.

The primary weakness of the roving survey is that catch and effort information is based on incompleted rather than on completed fishing trips, i.e., anglers are contacted while they are fishing. An incompleted fishing trip is measured from the time an angler begins fishing until the time of the interview; a completed fishing trip is measured from the time an angler begins fishing until he thinks he will finish. Obtaining unbiased estimates of fishing success (measured as CPUE) through the roving survey, thus, requires that catch rate not be dependent on the length of a fishing trip. The literature contains conflicting viewpoints on the validity of this assumption; the preponderance of published information where data on incompleted versus completed trips were actually compared (Carlander et al. 1958; Von Geldern 1972; Malvestuto et al. 1978) suggests that the assumption is reasonable, but it should be checked carefully for every fishery.

Another problem with roving surveys occurs because the probability of contacting an angler is proportional to the length of the trip. Therefore, creel clerks will tend to interview anglers who spend more time on the water, and thus overestimate the mean length of a fishing trip. I found that the arithmetic mean of trip length from the roving survey overestimated actual trip length on West Point Reservoir, Alabama-Georgia, but that the harmonic mean compensated for this positive bias. Other disadvantages associated with the roving survey are that night surveys are generally impossible and, because anglers are interrupted while fishing, that public relation problems may occur (see section 21.4.1 on behavioral protocol).

The primary advantages of the roving survey as outlined by Von Geldern (1972) are that: (1) contact of anglers is more time efficient where multiple access points are present, i.e., waiting time between interviews is limited to travel time between anglers, (2) all angler types (rental boat, private boat, shore, public and private pier) can be contacted in proportion to their actual abundance, and (3) interviews can be combined with angler counts over a large area. With respect to item (3), the creel clerk either makes separate count and interview circuits, randomly choosing which comes first, or combines counts and interviews into a single circuit; the latter is more cost-effective, but precision may increase by keeping count circuits as short as possible (see section 21.5.1 for a discussion of count usage). The suggested count procedure is that the clerk count only those anglers fishing between the shore and the center of the fishing area as he passes them.

21.3.3 Access Point Surveys

The access point survey is another type of on-site intercept survey where the creel clerk is stationed at an access point (boat landing, pier, jetty, beach) so he can contact anglers at the end of their fishing trips. The general advantages and disadvantages of the roving survey previously mentioned hold here as well. Integration of this contact approach with statistical survey techniques dictates that access sites be randomly chosen; it is common to sample access points using nonuniform probabilities proportional to amount of use.

The primary advantage of the access point approach relative to the roving method is that information is based on completed trips rather than on incompleted trips. A basic disadvantage occurs where access points are so numerous that few anglers use any one point, thus contact rates will be low and clerk time is inefficiently used. Also, it is usually impossible to sample all angler types proportional to their level of effort. This is a particular problem with bank anglers who may be widely dispersed along the shoreline and not associated with well-defined access sites. Shoreline facilities may be quite varied, e.g., public and private piers and launch sites, jetties, beaches, parks and other recreational sections of shoreline, so that a very complete sampling design is required. The access point survey is ideal where all anglers must leave from only a small number of points or where anglers must report their catches at a central location (a concession stand, for example).

21.3.4 Complemented Surveys

Complemented surveys are those in which more than one survey method is used. Some of the biases of each individual survey approach can be overcome by using a complemented survey, e.g., a night fishery cannot be monitored reasonably using a roving technique, but it could be surveyed via access points; an access point survey may not adequately sample bank anglers, but a roving survey would, so that both together would cover all anglers.

In a strict sense, a complemented survey is one which has two sampling populations. An excellent example of this is the national marine recreational fishery survey currently being conducted by the National Marine Fisheries Service and described in papers by Deuel (1980a, b). Based on extensive pre-survey evaluation of contact methods, the designers determined that a telephone survey, based on a sampling population of all coastal households with telephones, and an access point survey, based on a sampling population of all fishing sites, together would most effectively provide the information needed to estimate total harvest of marine sport fish. The telephone survey provides estimates of the percentage of anglers in the entire population and of the number of fishing trips by type and location of fishing; the intercept survey provides estimates of harvest per trip by species.

The survey methods discussed here are only a few of many possible types. In many cases, catch is recorded in one form or another, and catch records can be sampled; e.g., charter boat captains keep log books, anglers may be required to record catch on fishing permits, and fishing clubs typically keep records. These surveys usually cover only a specific segment of the fishery and must be interpreted accordingly.

21.4 THE INTERVIEW PROCESS

The verbal interview is a behavioral interaction between an interviewer (creel clerk) and respondent (angler). The data collection instrument used by the interviewer is a questionnaire, often called an interview schedule if questions are verbally read to the respondent. The interview schedule consists of predetermined, exactly worded questions that, ideally, are easily and clearly understood and that elicit responses pertinent to the objectives of the survey. Because anglers are clients as well as respondents, the interview should provide a positive social interaction between the management agency and its clientele.

This section provides a brief overview of the interview process by categorizing the subject into two subtopics, behavioral protocol and questionnaire design and presentation. Detailed consideration is given to social research methodology in books by Babbie (1973), Miller (1977), and Bailey (1978).

21.4.1 Behavioral Protocol

On-site intercept surveys require that anglers be contacted during or after their fishing trips. In either case, the interviewer must realize he is interrupting the respondent's privacy and leisure time to request information. At the same time, the respondent, a resource user, is likely to judge the interviewer as a representative of his management agency. The delicacy of this interaction is readily apparent, and there is a behavioral protocol that will help ensure a successful interview.

Establish contact in as courteous a manner as possible. The situation is especially challenging during on-water roving surveys when you must use a boat to reach and contact anglers. There usually will be entries on the interview form that can be answered prior to verbal contact (*see* top of Fig. 21.1), and there is no need to interrupt the respondent before recording this information. Approach anglers slowly and from far enough away to minimize (if not eliminate) boat wake and to avoid tangling the anglers' gear. (A trolling motor is a handy tool for boat-to-boat interviews.) Call to the anglers from a distance that does not interrupt their fishing. Unless the respondent has harvested fish which you must measure, you can conduct the entire interview with minimum inconvenience to the angler.

Try to gain the anglers' trust from the beginning of the interview. Dress in a manner acceptable to the people being interviewed and be officially identifiable (emblem on shirt or cap, boat label, etc.). After greeting the respondent, provide a brief explanation of the purpose of the survey as soon as possible without shouting. Anglers are not required by law to answer questions and their rights should be respected; ask if they are willing to respond to the questionnaire with the understanding that their answers will remain anonymous. If they do not want to participate, do not pressure them to respond. Emphasize to the angler that his or her responses are very important because only a small portion of all anglers using the resource will be interviewed and that the information collected will be used to improve the fishery.

If accurate harvest data are desired, try to check the creel yourself rather than relying on the anglers' recollections. Measuring fish usually will interrupt fishing activity, especially for boat-to-boat interviews. Do not pressure anglers to allow their fish to be measured but emphasize that the information will be important to fishery managers.

Because of the positive social relationship that the creel clerk seeks to establish with the angler, the objectives of the creel clerk and the law enforcement officer are not complementary. If one objective is to remind anglers when they are in violation of fishing regulations, a standard method should be established so that the creel survey is not jeopardized. The value of the interview as an information exchange mechanism can be enhanced by providing anglers with written information about the survey, including current survey results, if available. Respondents, thus, have a better appreciation of the end-result of donating their time and information.

21.4.2 Questionnaire Design and Presentation

The design and presentation of a questionnaire are critical to the collecting of high-quality data. Questionnaire design refers to the intent, sequence, and wording of questions; questionnaire presentation refers to the interviewer's demeanor, knowledge of question intent, phrasing of questions, and use of verbal probes and visual prompts.

Only include questions in the interview schedule that are relevant to the objectives of the survey. Bailey (1978) suggests that if you cannot decide in advance how the answers will be statistically analyzed and published (or otherwise presented), then you should not ask the question. Take care to avoid two-part questions and ambiguous questions, and whenever possible avoid negatively phrased questions and biased terms or phrases. See Babbie (1973) for examples.

The interview schedule should be well organized, not only to enhance the ease

ROVING CREEL INTERVIEW SCHEDULE

Date_____Time_____Lake Section_____Sample #_____ No. in party____

Fishing From: Bank___ Boat___ With:_ # rods   Sex: M___F___

Location: Open Water ___ Tree Shelter ___ Rip-Rap ___ Pier ___ Bridge ___

"Good morning (good afternoon). My name is _____ and I am
conducting an angler survey for the (affiliation). We are collecting information
that will be used to help manage this resource. Do you mind if I ask you a few
questions about your fishing trip today?"

"What county and state did your fishing trip today originate from? "
          County_____ State_____

"What do you estimate that you will spend on the following items for today's
fishing trip?" Gas $_____ Food $____Bait $_____ Lodging $_____

If boat fisherman: "Which landing did you use to launch your boat ?"_____

"What time did you begin fishing today?" AM_____ PM_____

"What time do you think that you will finish fishing today?"   AM___ PM_____

        " Now I would like to ask you some questions about your catch "

"What kind of fish are you fishing for ?"_____

"How many have you caught and released ?"_____

"How would you rate your fishing success today on a scale of poor, fair, good, or
excellent ?" Poor___Fair___Good___Excellent___

"Would you mind if I record the number and sizes of fish that you have harvested ?"

SPECIES CAUGHT		LENGTH CLASS (SPECIFY)											TOTAL
	No___												
	Wt___												
	No___												
	Wt___												
	No___												
	Wt___												
	No___												
	Wt___												
	No___												
	Wt___												

"That completes the interview. Thank you very much for your time. Do you have
any comments that you would like to make about the management of this resource ?"

Figure 21.1. An example of a roving creel survey interview form. The example is meant to show logical questionnaire construction, not to provide an exhaustive list of questions.

with which anglers can respond, but also to help the interviewer. Place questions in a logical order; for example, the items on the interview form shown in Fig. 21.1 are ordered so that data which can be collected prior to the interview are entered at the top of the form. The actual angler interview begins in the next section of the form with exactly worded questions that progress from asking the place of trip origin to requesting an estimate of the time that the respondent will finish fishing that day (a time sequence). The final section on the form concerns catch information with a space for recording data on harvested fish. The questionnaire ends with an invitation to respondents to ask questions or express opinions. Concentrate on accurately recording the views of the respondents (keeping in mind that responses will have to be categorized for analysis) and refrain from discussion that might destroy an otherwise positive encounter. Some general rules for question order are (1) ask easy-to-answer questions and questions needed for subsequent interviewing first, (2) put sensitive and open-ended questions late in the questionnaire, (3) vary questions in type and length to

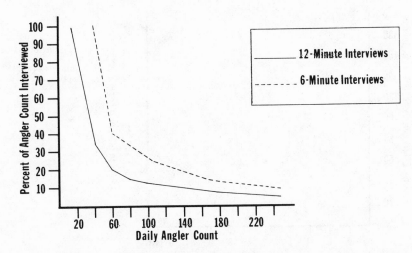

Figure 21.2. Relationship between percent of total anglers interviewed and the total number of anglers counted for two mean interview schedule lengths. The 12-minute interview curve is based on data collected from roving creel surveys conducted on two Alabama-Georgia reservoirs during 1980-81, using four-hour sampling periods; the six-minute interview curve was simulated by doubling the number of people interviewed with the 12-minute schedule.

keep the interest of the respondent, and (4) avoid establishing stereotyped responses. *See* Bailey (1978) for details.

It is tempting to increase questionnaire length because the cost of adding questions is small relative to that of survey equipment and operating costs. However, there is a point of diminishing returns because inconvenience to anglers and data processing time increase to accommodate information of marginal value. Additionally, longer interview schedules reduce the number of anglers that can be interviewed during the sampling period. Figure 21.2 shows the percentage of total anglers interviewed relative to the total number of anglers present during four-hour sampling periods for two interview schedule lengths. In this example, a sample representing 10% of the anglers present during any given sampling period could be collected using either the 6- or 12-minute interview schedule. To sample 20% of the anglers present, however, would require the use of a 6-minute (or shorter) schedule so that the 20% sample could be maintained at pressure counts beyond about 60 anglers per sampling period (Hudgins and Malvestuto 1982). Information trade-offs such as this (more anglers versus more information per angler) must be weighed carefully while designing the survey.

To aid in the consistency of data collection, all questions should be fully expressed on the interview form. Include the introductory remarks, as well as other connecting statements (Fig. 21.1). Interviewers should be trained in questionnaire delivery and should understand the intent of all questions. When a respondent misinterprets a question, the interviewer can provide verbal probes (in a specified manner, as per training) to bring the respondent back on track. Visual prompts, such as holding up a ranking scale (1 = not important, 2 = slightly important, 3 = important, and 4 = very important), can orient the respondent and decrease response error. Data comparability depends on consistent use of the instrument; the interviewer should play a neutral role but facilitate the interview process. Questionnaires should be pretested (Babbie 1973), preferably by conducting a pilot survey.

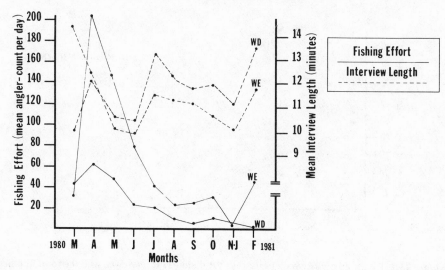

Figure 21.3. Trends in mean length of interview versus mean angler-count for weekdays (WD) and weekend days (WE) from March 1980 - February 1981 on West Point Reservoir. Trends are based on 502 interviews taken on 60 sampling days.

Interviewer inconsistencies can be caused by many factors operating in the system. Hudgins and Malvestuto (1982) show that length of an interview (not including time for collection of harvest data or open-ended comments) was affected by the number of anglers present. Creel clerks tended to shorten interview length as the number of anglers increased in an effort to obtain more interviews and to gain time for pressure counts. The negative relationship was evident both seasonally and between weekdays and weekends (Fig. 21.3). *See* Cannell et al. (1977) for an excellent summary of interviewing methodology.

21.5 OVERVIEW OF QUANTITATIVE PROCEDURES

The primary focus of this section is on collecting information necessary to estimate fishing effort, catch (harvest), and catch per unit of effort (CPUE). Regardless of the sampling design or contact method, catch can be estimated as the product of effort and CPUE so that the primary concern centers on estimating these two components. Effort is estimated using pressure counts, but CPUE is obtained through the interview process. The basic information that must be recorded for each interview is the amount of time spent fishing (unless the chosen unit of fishing effort is something other than the angler-hour) and the number and weight of each species harvested. Taking length measurements will provide additional information on the length structure of the harvestable-sized stocks.

21.5.1 Estimating Fishing Effort

Fishing effort estimates from creel surveys are based on angler count data; the clerk should strive to count all anglers operating within the specified sampling unit. Counts are converted to angler-hours by multiplying the number of anglers by the number of hours in the sampling period (Neuhold and Lu 1957; Lambou 1961). The validity of this calculation rests on the assumption that the number of anglers counted is an unbiased estimate of the number of angler-hours in progress at any given instant, i.e., an "instantaneous count." Twenty anglers fishing during one instant means that after one hour, 20 angler-hours have been expended; after four hours, 80 angler-hours have been expended. This assumption remains the same regardless of how the count is taken, e.g., from a high vantage point, or from a plane or a boat progressively circling the sample section. Field studies show that short counting periods provide better data, and the ideal situation is to take short counts within any given sampling period and average them to obtain an estimate of the true instantaneous count. However, Neuhold and Lu (1957) showed that progressive counts taken over a one-hour sampling period were similar to counts taken from a vantage point, and I found that progressive counts taken over a four-hour sampling period were similar to counts taken over a one-hour sampling period at West Point Reservoir, Alabama-Georgia. Lambou (1961) gives a detailed quantitative discussion of angler count data.

The sampling unit frequently represents only a portion of the angling population present on any given day, as when the day has been divided into sampling periods, a body of water has been divided into sections, or only one of several access points is being sampled. In these cases, angler-hours within the sampling unit are expanded to an estimate of total angler-hours for the entire day by dividing the sampling unit value by the sampling probability associated with the particular unit. Total angler-hours are converted to angler-trips by dividing by the mean time spent fishing per trip. It is usually feasible to count anglers separately by fishing type (bank versus boat), and independent effort estimates can then be calculated. Boat-trips and angler-trips can be converted by knowing the mean number of anglers per boat.

In most instances, it is desirable to partition effort according to the species or group of species sought. This requires that anglers be asked to identify the type of fish they intend to catch. Given a statistically valid sampling design, the sample percentages of effort expended for each class of fish can be multiplied by total effort to estimate "fished for" or "intended" effort. Sample percentages should be obtained using completed trip information.

21.5.2 Estimating CPUE

Creel survey estimates of CPUE are obtained by dividing measured (recorded) harvest by measured effort (usually angler-hours). For the roving survey, measured (incomplete trip) effort is taken as the number of hours from the time the fishing trip began to the time of interview. When fishing days are being subsampled, the calculated CPUE for the subsampling unit is taken to represent the CPUE for the entire day.

There are three major reasons to measure CPUE: (1) to estimate total harvest or harvest by species over a specified time period, (2) to obtain an index of stock abundance for particular species or classes of fish, and (3) to measure fishing quality or fishing success for particular species or classes of fish. To estimate total harvest for a survey period, calculate CPUE by dividing total measured harvest by total measured effort. Then multiply CPUE by the total estimated effort for the survey period. To partition CPUE according to species, divide the recorded harvest of each species by total measured effort; the separate species estimates of CPUE will add to the estimate based on all species combined. Total harvest by species is obtained by multiplying each CPUE by the total estimated effort for the survey period.

It is generally accepted that estimates of CPUE based on total measured effort are not appropriate when interest centers on a particular species or class of fish unless all species are equally vulnerable to all angling techniques or unless the proportional contribution to total effort by anglers fishing for different species is constant over time. The most appropriate measure of CPUE for objectives (2) and (3), therefore, is obtained by dividing the harvest of a given species of fish by the angler-hours directed toward that species (Lambou and Stern 1958; Lambou 1966; Von Geldern 1972; and Von Geldern and Tomlinson 1973). As an index of fishing success, this form of CPUE assumes either that: (1) anglers catch primarily what they seek; or (2) fish caught, but not sought, have little bearing on an angler's perception of his fishing success. *See* Ricker (1975) for a discussion of the use of CPUE as an index of stock abundance.

Catch per unit of effort is a ratio estimate. There are two basic ratio estimators that can be used to calculate it, the mean of ratios estimator and the total ratio estimator. The objectives of the survey will determine which estimator should be used. Given that fishing days are being chosen at random, one way of estimating CPUE for the survey period is to calculate daily CPUE values and take an average of these values over the number of days sampled. This is called the "mean of ratios estimator" and gives equal weighting to each day, regardless of the amount of fishing effort expended. It is the proper estimator if mean daily values of fishing success are desired; the variance for this ratio is calculated as for any set of independent observations.

The other approach to estimating CPUE is to calculate a single ratio by dividing the sum of measured catch over all sampling days by the sum of measured effort over all sampling days. This is called the "total ratio estimator" and is self-weighting, i.e., it is influenced by differences in daily fishing effort and is representative of fishing success for the population as a whole. (*See* Snedecor and Cochran 1980, section 21.12, for the appropriate variance estimator.) The total ratio estimator seems to be the most appropriate for calculating total harvest over the survey period as long as measured effort is proportional to actual effort during the days sampled. Given that there will be a physical limit to the number of people that can be interviewed during a sampling period, measured effort may not be proportional to actual effort over the entire range of possible values. Figure 21.4 is an example of this. With a roving design used on several Alabama reservoirs (four-hour sampling periods), proportionality was

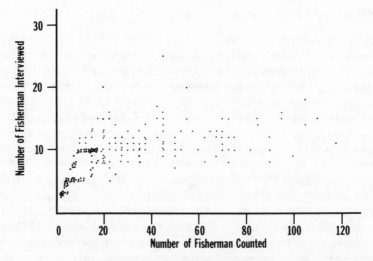

Figure 21.4. Scatter diagram showing relationship between number of anglers counted and number interviewed for roving creel surveys (four-hour sampling periods) on several reservoirs in Alabama.

maintained only when there were fewer than 25 anglers in the sampling section; beyond this density, the number of anglers interviewed was relatively constant regardless of the number counted. In this case, it would be better to use the mean of ratios estimator or to weight daily (CPUEs) by counts of total anglers.

Von Geldern and Tomlinson (1973), who were primarily interested in obtaining an estimate of fishing success applicable to the average angler, suggested a modification of the mean of ratios estimator. They calculated CPUE for each angler (or fishing party) interviewed and a mean of these values for each sample day. They averaged the daily values over the number of days sampled to estimate fishing success on a per angler basis.

The best procedure for estimating effort, CPUE, harvest, and the standard errors of these values depends on the survey sampling design used. Survey design books, such as Cochran (1977), provide the specific statistical formulas. Bazigos (1974) gives formulas and numerical examples. Pinkas et al. (1964) outline the quantitative aspects associated with a stratified random sample of marine sport fishermen using launching and mooring sites and piers and jetties. Malvestuto et al. (1978) provide a step-by-step outline of calculations for a roving survey using stratified nonuniform probability sampling. (*See* section 21.2.3.)

Keep in mind that even if the catch and effort information collected is accurate, it will be of little value for documenting change in the fishery unless it is also precise. Strive to keep relative standard errors as small as possible; values in excess of 20% are not desirable. It may be difficult to increase the number of sampling units because of money and labor restrictions, but appropriate modification of the sampling design can reduce sampling variance.

21.6 MEASURING ECONOMIC BENEFIT

So far, this chapter has focused on estimation of the traditional sport fishery descriptors of effort, catch, and CPUE. These values are the basis of the maximum sustained yield (MSY) approach to fisheries management where the primary objective is usually to improve fishing success (CPUE) for a given species or group of species. Given that yield from a recreational fishery encompasses more than CPUE, the concept of optimum sustained yield (OSY), which incorporates economic and social benefits associated with fishing, is certainly more applicable to the recreational setting. The OSY concept suggests the need for the measurement of economic and social components and the subsequent integration of these values into fishery assessment procedures and management strategies. This section provides an overview of current economic valuation techniques with an emphasis on conceptual problems and data collection. The discussion is based largely on papers by Crutchfield (1962), Knetsch (1963), Gordon et al. (1973), and Dwyer et al. (1977).

21.6.1 The Concept of Economic Benefit

The economic component of yield is referred to as benefit, worth, or value. The gross economic value of a recreational fishery is composed of (1) explicit costs of local, state, and federal services which contribute directly to the fishery; (2) total angler expenditures directed toward the fishery (license and on-site fees, equipment and travel costs, etc.); (3) the value over and above actual expenditures that anglers would be willing to spend, or could be induced to spend, to fish; and (4) the additional value based on nonusers' willingness to pay just for preservation of the fishery. Items (1) and

(4), although contributing to gross economic worth and national or regional welfare, do not contribute directly to the "user-oriented" benefits from the fishery (the value people receive from fishing) and thus are not a part of our focus here. It is the benefits accruing to users of the fishery that are of primary concern. In economic terms, the proper measure of these benefits is simply the total amount of money that people are willing to pay to engage in fishing, i.e., components (2) and (3) above. It should be pointed out that these two components provide a measure of the value of the entire fishing experience to the user; the value of the fish is only a portion of the entire trip value and may be approximated by placing a monetary figure on the harvest as per appropriate market, hatchery, or reimbursement costs (Monetary Values of Fish Committee 1978).

Although actual expenditures and the amount anglers would be willing to spend in excess of actual expenditures both contribute to total willingness to pay, it is the latter component that is theoretically equivalent to the value that we place on items in a market economy. Willingness to pay in excess of actual expenditures is called "net willingness to pay" or "consumers' surplus" and measures the profit, rent, or net value attributable to the fishery, a value theoretically collectable if anglers could be induced to pay this excess amount. Actual expenditures simply measure the costs associated with getting to, using, and returning from the resource and could not be collected as a use fee if, for example, the fishery were privately owned. In other words, if the fishery were to disappear, actual expenditures would be redirected, and the economic loss could not be measured in terms of these expenditures, but rather in terms of the theoretical profit loss associated with abolishment of the fishery. Brown et al. (1964) contrast the two components by pointing out that using total expenditures as a measure of economic worth is similar to stating that the worth of a logging operation is the total monetary outlay by the logger, which is obviously wrong. The primary problem is that although consumers' surplus is the value which must be measured, most recreational fisheries are offered free of charge; therefore, this value cannot be directly estimated. Indirect measurement is possible, however, and two common valuation procedures will be outlined below; Stabler (1980a, b) provides reviews of these techniques and discusses recent developments and future needs. First, however, because actual expenditures do represent a part of angler benefits, a brief overview of expenditure estimation is presented.

21.6.2 Determination of Angler Expenditures

Actual expenditures fall into two general categories: (1) variable or fishing trip costs, which include expenditures for items such as gas, food, bait, and lodging incurred while traveling to, using, and returning home from a fishing trip, and (2) fixed or durable costs for fishing equipment (boats, gear, special clothing), which can be used over a number of years (Gordon et al. 1973). On-site angler surveys can provide estimates of average expenditures per angler-trip and total trip expenditures over a survey period. The quantitative procedures are analogous to those previously discussed for estimating CPUE and total harvest; that is, cost per angler-trip (or angler-hour) is multiplied by the total number of trips (hours) expended over the survey period to give total trip expenditures. Expenditure information is obtained via the interview process and should be itemized (Fig. 21.1). Expenses can be partitioned by type of fishing or type of angler so that the relative contribution of different segments of the fishery to variable cost can be compared.

Fixed expenditures are more difficult to estimate accurately because anglers must recall what they paid for the items (which may have been purchased years previously) and the cost of each item must be depreciated over the period of ownership to determine current value. Another approach is to ask anglers what they have expended recently, over a period of one to two months at most, and then to calculate current fixed expenditures over the survey period. As Gordon et al. (1973) indicate, this method will likely lead to high variance estimates because, for example, one angler may have just spent thousands of dollars on a new boat and others may have purchased nothing. Actual cost values, however, will be less subject to recall bias and depreciation errors. When estimating both fixed and variable costs, care should be taken to determine what portion of these costs are actually spent for fishing. If the primary purpose for visiting an area is to fish, then variable costs can reasonably be attributed to the fishing trip; when other activities are involved, costs should be partitioned, as well as possible, between activities and expressed on a daily basis. Many durable items such as recreational vehicles or clothes can be used for a variety of purposes, and the respondent should be asked to estimate the percentage of time the particular items are used for fishing at the site in question.

Actual expenditures are typically taken to represent the economic worth of a fishery because they can be measured directly and because many people misunderstand what constitutes the true benefit of the fishery. As discussed, expenditures do contribute to the total economic benefit, and consumers value a fishery at least as much as other things that could have been purchased for the same amount (Crutchfield 1962). In addition, expenditure analysis provides insight into how much and for what anglers are spending money, and it can prove useful for allocating personnel and resources toward specific management objectives (Gordon et al. 1973). The true net worth of the fishery, however, is defined as the consumers' surplus, and emphasis should be placed on estimating this value.

21.6.3 Determination of Consumers' Surplus

Fundamental considerations. The following discussion is taken largely from Dwyer et al. (1977) who provide an excellent overview of current recreational resource

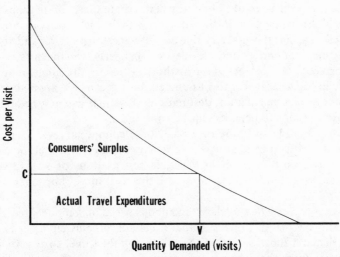

Figure 21.5. Theoretical demand curve showing relation between actual angler expenditures and consumers' surplus.

valuation procedures. The basis of determining consumers' surplus or net willingness to pay is simulation of the demand schedule for the fishery. A demand schedule is simply a curve which indicates the quantity of a good (level of use or visitation) that buyers (anglers) are willing and able to purchase at various prices. Figure 21.5 depicts a hypothetical demand curve. Demand curves generally have a downward slope indicating that decreased amounts of a good are desired at higher prices. Benefits are approximated by the area under the demand curve so that in Figure 21.5, if V number of visits occurred at C cost per visit, then actual angler expenditures are approximated by the area labeled as such under the curve. The area above actual expenditures approximates the amount of money anglers are willing to pay in excess of actual monetary outlay, i.e., consumers' surplus. The entire area under the curve out to point V represents total willingness to pay, or gross consumers' benefits.

The two methods for determination of consumers' surplus, the survey method and the travel cost method, rely on simulation of a demand curve by development of models (equations) that predict (explain) willingness to pay as a function of site and socio-economic characteristics, as well as of location of users relative to the fishery.

Survey method. The term survey method is a misnomer because both this technique and the travel cost method depend on properly designed surveys and interviews (sections 21.2 and 21.4). The basis of the survey method is that anglers are asked what they would be willing to pay over and above actual expenses to fish at a particular site rather than to be deprived of the fishing experience. These dollar values are then quantitatively related to other information obtained via the interview using standard multiple regression techniques. The final outcome is a multiple regression model that allows average willingness to pay (per individual, fishing party, or household) to be predicted using relevant, obtainable information. These predicted values are then used in conjunction with estimates of the number of resource users to simulate the demand function (Fig. 21.5) and calculate consumers' surplus. Dwyer et al. (1977) describe and evaluate several examples of the survey method.

The choice of predictor variables depends on the specific situation at hand but may include standard demographic information (age, sex, income, family size, etc.), number and length of visits to the site, years of experience with the site, distance traveled to the site, a measure of availability of alternate sites, and perhaps a measure of the success of the fishing trip. Basic problems with the method are that respondents may not be able to accurately place a value on willingness to pay over and above actual expenses, they may not understand the questions properly, and the questions may not be phrased correctly. Advantages of the method are that small changes that would not be expected to influence travel costs or number of visits can be evaluated; estimates of the value of a site that is one of many destinations on a single trip may be obtained; and effects of crowding on site benefits can be considered.

The survey method can be used with mail or telephone surveys. Users of specific resource sites are difficult to identify this way. Alternatively, the method can be used on-site. The travel cost method, however, is restricted to on-site surveys. Both methods are ideally suited for integration into the creel survey approaches discussed previously.

Travel cost method. The travel cost method is based on a model predicting use at a particular site by measuring the actual behavior of participants. The method estimates the demand function for a fishery by using the travel costs associated with distances of users from the fishery as a proxy price variable. Because travel cost (as well as travel time) increases as distance from the fishery increases, less participation per unit population will be forthcoming from the more distant locations. This relationship simulates the demand curve that would result if fishing fees were progressively increased—causing fewer people to participate in the recreational activity (Fig. 21.5).

In general, the travel cost method is the appropriate valuation technique when: (1) there is sufficient variation in travel costs (distances) among anglers to estimate the demand function via multiple regression procedures; (2) management strategies being evaluated are significant enough to alter the number of angler-trips that will be made at existing travel costs; and (3) travel expenses have been made primarily for the purpose of fishing at the specific site in question. The method presumes that anglers react to increases in entry fees just as they do to increases in travel cost and that all users place the same value on fishing at the site in question so that they all would be willing to pay the same maximum travel cost per trip. Dwyer et al. (1977) provide a detailed treatment of quantitative advantages over the survey method because the method: (1) explicitly recognizes the price alternatives of the users based on their location relative to the fishery, (2) is based on market behavior of anglers rather than on responses to questions, and (3) does not rely as heavily on the skills of the interviewer.

A simple travel cost model requires information on (1) the total number of visits made from several (not necessarily all) population origins at different distances from the site, where origins are often taken to be towns, counties, or concentric rings around the site; (2) the total population at each origin; and (3) the average trip cost for anglers traveling from each origin, including expenditures for transportation, food, lodging, bait, tackle, and equipment and guide rental. This information allows derivation of a trip demand curve by regressing visits per capita on travel cost. The next step involves derivation of the site demand curve showing how many total visits would be made at various hypothetical fishing fees where differential travel costs simulate these fees. The area under this curve, obtained by integration of the demand function, estimates consumers' surplus.

Angler attendance is partially dependent on factors other than travel cost, and more complex models are usually necessary to accurately predict visitation. Relevant factors include travel distance, travel time, age, sex, family structure, income, and familiarity with the fishery and substitute recreational areas. Recent examples of these are found in Merewitz (1966), Cauvin (1978, 1980), Copes and Knetsch (1981), Palm and Malvestuto (1983), and Weithman and Haas (1982).

21.7 REFERENCES

Anderson, R. O. 1975. Optimum sustainable yield in inland recreational fisheries management. *American Fisheries Society Special Publication* 9:29-38.

Babbie, E. R. 1973. *Survey research methods.* Wadsworth Publishing Company, Belmont, California, USA.

Bailey, K. D. 1978. *Methods of social research.* The Free Press, New York, New York, USA.

Bazigos, G. P. 1974. *The design of fisheries statistical surveys.* Food and Agriculture Organization of the United Nations Fisheries Technical Paper 133, Rome, Italy.

Best, E. A., and H. D. Boles. 1956. An evaluation of creel census methods. *California Fish and Game* 42:109-115.

Brown, W. G., A. Singh, and E. N. Castle. 1964. *An economic evaluation of the Oregon salmon and steelhead sport fishery.* Oregon State University Agricultural Experiment Station Technical Bulletin 78, Oregon State University, Corvallis, Oregon, USA.

Cannell, C. F., K. H. Marquis, and A. Laurent. 1977. *A summary of studies of interviewing methodology.* National Center for Health Statistics, United States Department of Health,

Education and Welfare Publication 77-1343, Rockville, Maryland, USA.

Carlander, K. D., C. J. DiCostanzo, and R. J. Jessen. 1958. Sampling problems in creel census. *Progressive Fish-Culturist* 20:73-81.

Cauvin, D. M. 1978. The allocation of resources in fisheries: an economic perspective. *American Fisheries Society Special Publication* 11:361-370.

Cauvin, D. M. 1980. The valuation of recreational fisheries. *Canadian Journal of Fisheries and Aquatic Sciences* 37:1321-1327.

Cochran, W. G. 1977. *Sampling techniques,* third edition. John Wiley & Sons, New York, New York, USA.

Copes, P., and J. L. Knetsch. 1981. Recreational fisheries analysis: management modes and benefit implications. *Canadian Journal of Fisheries and Aquatic Sciences* 38:559-570.

Crutchfield, J. A. 1962. Valuation of fishery resources. *Land Economics* 38:145-154.

Deuel, D. G. 1980a. Special surveys related to data needs for recreational fisheries. Pages 77-81 *in* J. H. Grover, editor. *Allocation of fishery resources.* European Inland Fisheries Commission, Food and Agriculture Organization of the United Nations, Rome, Italy.

Deuel, D. G. 1980b. Survey methods used in the United States marine recreational fishery statistics program. Pages 82-86 *in* J. H. Grover, editor. *Allocation of fishery resources.* European Inland Fisheries Advisory Commission, Food and Agriculture Organization of the United Nations, Rome, Italy.

Dwyer, J. F., J. R. Kelly, and M. D. Bowes. 1977. *Improved procedures for valuation of the contribution of recreation to national economic development.* University of Illinois, Water Resources Center Report 128, Urbana, Illinois, USA.

Gordon, D., D. W. Chapman, and T. C. Bjornn. 1973. Economic evaluation of sport fisheries - what do they mean? *Transactions of the American Fisheries Society* 102:293-311.

Hudgins, M. D., and S. P. Malvestuto. 1982 (In press). An evaluation of factors affecting creel clerk performance. *Proceedings of the Annual Conference of the Southeastern Association of Fish and Wildlife Agencies.*

Knetsch, J. L. 1963. Outdoor recreation demands and benefits. *Land Economics* 39:387-396.

Lambou, V. W. 1961. Determination of fishing pressure from fishermen or party counts with a discussion of sampling problems. *Proceedings of the Annual Conference of the Southeastern Association of Game and Fish Commissioners* 15:380-401.

Lambou, V. W. 1966. *Recommended method of reporting creel survey data for reservoirs.* Oklahoma Fishery Research Laboratory Bulletin 4, Norman, Oklahoma, USA.

Lambou, V. W., and H. Stern, Jr. 1958. Creel census methods used on Clear Lake, Richland Parish, Louisiana. *Proceedings of the Annual Conference of the Southeastern Association of Game and Fish Commissioners* 12:169-175.

Malvestuto, S. P., W. D. Davies, and W. L. Shelton. 1978. An evaluation of the roving creel survey with nonuniform probability sampling. *Transactions of the American Fisheries Society* 107:255-262.

Malvestuto, S. P., W. D. Davies, and W. L. Shelton. 1979. Predicting the precision of creel survey estimates of fishing effort by use of climatic variables. *Transactions of the American Fisheries Society* 108:43-45.

Merewitz, L. 1966. Recreational benefits of water resource development. *Water Resources Research* 2:625-640.

Miller, D. C. 1977. *Handbook of research design and social measurement,* third edition. David McKay Company, New York, New York, USA.

Monetary Values of Fish Committee. 1978. *Reimbursement values for fish.* North Central Division, American Fisheries Society, Bethesda, Maryland, USA.

Neuhold, J. M., and K. H. Lu. 1957. *Creel census method.* Utah State Department of Fish and Game Publication 8, Salt Lake City, Utah, USA.

Palm, R. C., and S. P. Malvestuto. 1983. Relationships between economic benefit and sport fishing effort on West Point Reservoir, Alabama-Georgia. *Transactions of the American Fisheries Society* 112:71-78.

Pfeiffer, P. W. 1966. The results of a non-uniform probability creel survey on a small state-owned lake. *Proceedings of the Annual Conference of the Southeastern Association of Game and Fish Commissioners* 20:409-412.

Pinkas, L., J. C. Thomas, and J. A. Hanson. 1964. Southern California marine sportfish survey, some methods and preliminary results. *Proceedings of the Annual Conference of Western Game and Fish Commissioners* 44:282-286.

Ricker, W. E. 1975. *Computation and interpretation of biological statistics of fish populations.* Department of the Environment, Fisheries and Marine Service, Fisheries Research Board of Canada Bulletin 191, Ottawa, Canada.

Snedecor, G. W., and W. G. Cochran. 1980. *Statistical methods,* seventh edition. The Iowa State University Press, Ames, Iowa, USA.

Stabler, M. T. 1980a. Estimation of the benefits of fishing on U.K. canals: some problems of method. Pages 346-355 *in* J. H. Grover, editor. *Allocation of fishery resources.* European Inland Fisheries Commission, Food and Agriculture Organization of the United Nations, Rome, Italy.

Stabler, M. T. 1980b. Estimation of the economic benefits of fishing: a review note. Pages 356-365 *in* J. H. Grover, editor. *Allocation of fishery resources.* European Inland Fisheries Commission, Food and Agriculture Organization of the United Nations, Rome, Italy.

Robson, D. S. 1961. On the statistical theory of a roving creel census of fishermen. *Biometrics* 17:415-437.

Von Geldern, C. E., Jr. 1972. Angling quality at Folsom Lake, California, as determined by a roving creel census. *California Fish and Game* 58:75-93.

Von Geldern, C. E., Jr., and P. K. Tomlinson. 1973. On the analysis of angler catch rate data from warmwater reservoirs. *California Fish and Game* 59:281-292.

Weithman, A. S., and M. A. Haas. 1982. Socioeconomic value of the trout fishery in Lake Taneycomo, Missouri. *Transactions of the American Fisheries Society* 111:223-230.

Chapter 22
Sampling the Commercial Catch

ROBERT L. DEMORY AND JAMES T. GOLDEN

22.1 INTRODUCTION

Proper management and conservation of commercial fish and shellfish resources requires collection and summarization of basic harvest data. Researchers and managers need data on catch, effort, and biological characteristics of exploited stocks, as well as on economic data associated with the fishery. These data elements are used to assess stock status and to evaluate impacts of regulatory measures on the fishery and resource.

This chapter describes some of the approaches to obtaining harvest statistics through census and sampling of the commercial catch and the processing and reporting of the data to provide baseline information used by fisheries researchers and managers.

It is best to complement information obtained through census and sampling of the commercial catch with fishery-independent methods such as fishery resource surveys. These are conducted on the fishing grounds and in adjacent areas by research vessels or chartered commercial vessels using standardized fishing gear. Surveys provide information on characteristics of fish and shellfish populations that might otherwise go undetected or that might be distorted by interpreting fisheries data alone. Although these methods are not discussed here, the results of a recent workshop by the Canadian Department of Fisheries and Oceans provides an excellent review (Doubleday and Rivard 1981).

22.2 ESTIMATING THE CATCH

The most elemental data required for analysis of a fishery is the catch. Catch is defined as the total removals of fish or shellfish from the stock(s) by man's efforts. Simple analysis of trends, as well as complex analyses such as cohort analysis (Murphy 1965; Pope 1972), general production modelling (Pella and Tomlinson 1969), and budget simulation, require accurate catch statistics.

The catch can be divided into two categories: (1) the portion of the catch retained for its commercial (or sport) value, and (2) the portion of the catch discarded either at sea or at dockside. Discard rejected at the dock can be easily measured, and some agencies require fish buyers to include this portion in reporting landed weight. This type of discard, called "weighback," is usually deducted from the weight used to calculate payments to the fishermen. If the weighback is not reported for individual species, sampling may be required to determine composition.

Discard can be further subdivided into four components: (1) species which have no commercial value, (2) portions of commercially important species having little or no value (e.g., heads and entrails), (3) under-sized individuals of commercially important species and, (4) commercially important species that have market restrictions or limits. Estimation of the catch of species with no commercial value may be important in establishing baseline data on an underutilized species which may have future commercial value. Estimates of the last three discard components are essential for accurate stock assessments of commercially important species.

Discard of heads and entrails can be accounted for by using conversion factors which translate dressed weight to round weight. The biologist may be required to estimate the discard of whole fish by actual observations aboard commercial fishing vessels on the fishing grounds. Sampling the discard at sea is accomplished by sub-sampling unsorted catches as described by Herrman and Harry (1963). Alternately, fishermen can set aside the discard for later sampling at the dock.

In addition to identifying discarded portions of the catch, biologists may need to sample landings to determine the species composition of the catch. Commercial fishermen and processors frequently lump landed fish into broad market categories with several species in a category. Therefore, independent sampling of the catch may be necessary for scientifically useful data.

Landed weight in United States commercial fisheries usually is reported in pounds. Some stock assessment parameters (such as mortality rates), however, require estimates of numbers of fish caught. Estimating the number of fish in a landing is accomplished directly by counting all fish from a delivery or indirectly by taking a random sample from a bin or cart of fish to obtain average weight. The average weight divided into the total landed weight provides an estimate of the total number of fish landed. If fish are graded by size prior to landing, individual size categories must be subsampled using a stratified sampling scheme. Average weights and, subsequently, total numbers can be determined for each size category (strata) and summed across strata to determine the total numbers landed. Gulland (1965) provides a description of this method along with numerical examples. Remember that estimating total catch in numbers also requires estimating the number discarded.

22.2.1 Determining Where and When Fish are Caught

Determining where fish are caught is important because stock assessments will be carried out on unit stocks whenever possible. The timing of capture may be important in some stock assessments in partitioning fishing effort and mortality inflicted on the unit stock. Knowledge of when and where fish are caught may also be useful in resolving gear conflicts. Seasonal fisheries, crabbing, for example, may be conducted during part of the year in areas used by commercial trawlers. If gear conflicts cannot be resolved by industry, management authorities allocate time or space to each of the fisheries involved. The allocation of area and time is often made on the basis of historical catch summaries.

Fish are usually distributed in a non-random manner determined by environmental conditions such as food supply or terrain preference. Fishing effort, therefore, will also have a non-random distribution. Although it would be ideal if the statistical areas used in reporting catch matched ecological boundaries, the reporting areas or blocks are designed for administrative convenience and conform to state, provincial, or international boundaries. Thus, they tend to be arbitrary with respect to unit stocks and are usually quite large.

Where and when fish are caught can best be determined from logbooks kept by fishermen. In a trawl fishery, for example, an ideal logbook will contain tow-by-tow data that give the specific area fished, the time and depth fished, the gear used, and estimates of weight by species of fish caught. Usually the vessel captain records only the fish retained because this information is used in filling out the market order (sometimes called the "grocery list") and is used to check weights reported by the fish buyer upon delivery.

22.2.2 Determining How Fish are Caught

It is the responsibility of the scientist or manager to describe fishing effort in detail, including types of gear, number of units participating, and fishing power. These data are critical to a complete description of the effects of a fishery on the resource. Different commercial gears fish at different rates and in some cases on different segments of the same stock. Stock assessments usually require partitioning of fishing mortality into its separate components, especially if quantities such as fishing effort are to be reported in equivalent units.

Fishermen using different types of gear are usually grouped together for management. Knowledge of the total number of participants and amount of gear used by each group, as well as of the segment of the stock exploited, may be required to develop optimum management strategies or to settle allocation issues.

A simple form filled out for each vessel and updated periodically simplifies the task of documenting fisheries effort. The form (Fig. 22.1) includes spaces for recording the number of crew, including the captain, for each fishing vessel of a given gear type. Information on the number of individuals making a living in a given segment of a fishery is essential because resource allocation will almost always be based on social,

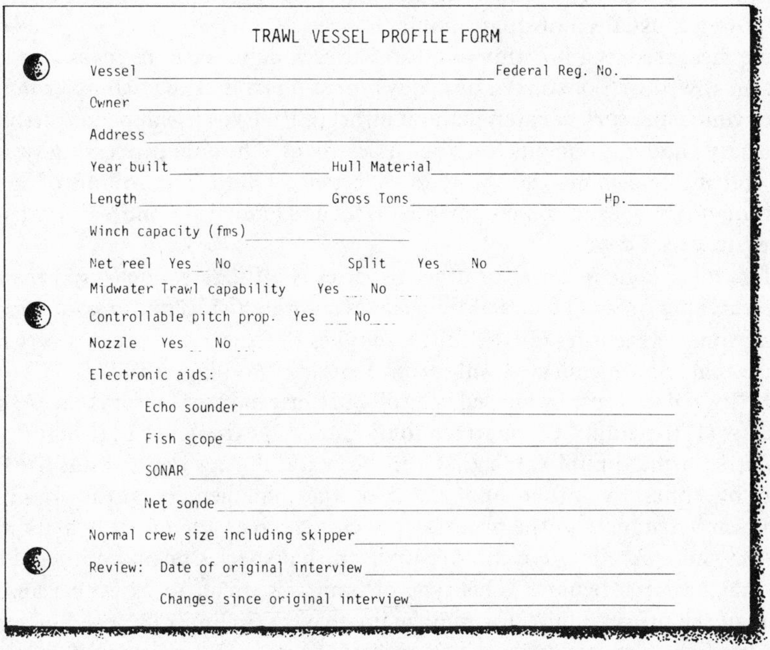

TRAWL VESSEL PROFILE FORM

Vessel_____ Federal Reg. No._____

Owner _____

Address _____

Year built_____Hull Material_____

Length_____Gross Tons_____Hp._____

Winch capacity (fms)_____

Net reel Yes__ No__   Split   Yes__ No__

Midwater Trawl capability   Yes__ No__

Controllable pitch prop. Yes__ No__

Nozzle   Yes__ No__

Electronic aids:

    Echo sounder_____

    Fish scope_____

    SONAR_____

    Net sonde_____

Normal crew size including skipper_____

Review: Date of original interview_____

        Changes since original interview_____

Figure 22.1. Example of a trawl vessel profile form.

economic, and political considerations such as employment rates, rather than on biological reasons.

The amount of effort spent sampling the catch or interviewing participants in different segments of a fishery should be determined by knowledge of the relative importance of the various gear types. Sampling on a basis proportional to the catch by each gear type will probably provide the best coverage for biological data. More detailed information may be required for minor fisheries of historical or ethnic importance to adequately represent these user groups in gear conflict or allocation issues.

In our experience, landings are made in a non-random manner dictated by the processor or markets and further modified by weather, mechanical malfunctions, and price disputes. The fleet catching ability is also influenced by various-sized vessels with different catching ability, skipper expertise, and skipper preference for certain species or gear types. These factors can usually be ignored if sale of fish is recorded on a trip ticket by species or groups of similar species. The use of a trip ticket, which is filled out by the fish buyer, accomplishes a complete census of landings by gear type. Sampling would only be required to account for discard and species composition. Other types of information require more detailed accounting of vessel characteristics.

22.3   ESTIMATING EFFORT AND CATCH PER UNIT EFFORT

There are two aspects of effort—gear and time. Gear is categorized according to the physical elements of the equipment used to catch fish, e.g. motor horsepower, vessel length, and type of electronic aids. The nature of these categories may change slowly over time. Although some of the changes in gear can be quantified (for example, repowering, new Sonar), the impact of new or improved gear on increased fishing power is usually not quantifiable.

Time measured as a function of effort varies widely. It can be measured in hours that gear is towed, set, or soaked or in days spent on trips at the fishing grounds. The choice of which measure of catch per unit effort (CPUE) to use depends on the nature of the fishery and requirements for stock assessment. The data processing system and level of staffing determines the ability to collect effort data. The volume of data is the greatest for effort measured each time gear is set and hauled and the least for trip effort measured in vessel days.

Effort for pelagic fisheries or other fisheries is difficult to measure because they require searching time. The actual time spent catching fish is minor compared to the searching time. Measuring CPUE for these types of fisheries requires a very detailed data base that can be collected only from logbooks (Williams 1977).

All CPUE data must be viewed critically because of three factors that are difficult to quantify: (1) trip limits, (2) new technology, and (3) experience of fishermen. A trip limit, that is, a maximum catch that can be made during one fishing trip, can be imposed by either the processing sector or the management authorities. Unique economic circumstances in the processing sector, such as a temporary market glut of one species, may precipitate a market limit on that species or on another substitute species that has less demand. This type of limit is difficult to assess because not all segments of the processing sector may be involved. Trip limits imposed by managers for conservation purposes are easier to evaluate because they tend to affect all units of fishing effort in a particular area uniformily. CPUE tends to be lower when a trip limit is in force because of under reporting of catch. Logbook data usually contain information on the retained portion of the catch but not on the discard. The end result is an underestimate of fishing mortality and relative abundance.

New technology can increase CPUE due to better efficiency, masking a decline in stock abundance that might be detected by studying CPUE of a standard type of gear. Higher CPUE probably will be short term, however, because the long term result of increased fishing efficiency in an unlimited fishery is declining catch.

The experience of fishermen can impact CPUE because better fishermen will know the right places, times, and ways to fish. Inexperienced trawl fishermen, for example, may stop fishing after several unsuccessful hauls, whereas experienced ones continue because they sense that environmental conditions are about to change. An analysis by the Oregon Department of Fish and Wildlife of CPUE data from trawling showed that 62% of the tow-by-tow variation in CPUE could be explained by horsepower, vessel length, or vessel tonnage (Johnson et al. 1980). The unexplained variance was primarily attributed to a combination of skipper experience, market limits, catchability, and abundance.

The most detailed CPUE data will be obtained from logbooks kept by fishermen. An example of the proposed logbook for west coast trawlers is shown in Figure 22.2. This logbook format is a composite of the logbooks used by California, Oregon, and Washington and was requested by the coastal trawl industry. Note that the area category provides a choice because some fishermen object to supplying specific tow locations but will provide less detailed information on area of catch. Two other unique features are target species and trawl type; most west coast trawlers fish for different species requiring different types of trawls, often on the same trip.

22.4 ECONOMIC DATA

There are many kinds of economic data needed to adequately describe and manage a commercial fishery (*see* chapter 24). The variable costs of fuel, food, and equipment are not easily obtained from logbooks, interviews, fish tickets, and vessel profile forms. Other kinds of data, however, are available from these sources. Number of vessels by gear type is available from tickets and vessel profile forms. Number of people by gear type is available from interviews, vessel profile forms, and logbooks. Total effort and value of the catch by area and gear are available from fish tickets and logbooks. Although these data are collected primarily for stock assessment, they can also provide useful economic and social insights. These data meet only the minimal requirements for economic analysis and must be supplemented by other economic information not addressed in this chapter (*see* Clark 1976, for a comprehensive treatment).

22.5 SAMPLING FOR FISHERIES STATISTICS

Collection of catch, effort, and economic and biological data can be accomplished using various techniques of censusing, sampling, and interviewing. A census or complete examination of a population may be desirable and feasible for some statistics such as the number of vessels participating in a fishery in a given year or the total of each species landed.

Usually a complete census of catch and effort, the amount of fuel consumed to catch all the fish, or length composition of a catch is neither possible nor expedient. In such cases, sampling methods exist to provide data about the population in a more efficient, economical, and timely manner, without having to observe all members of

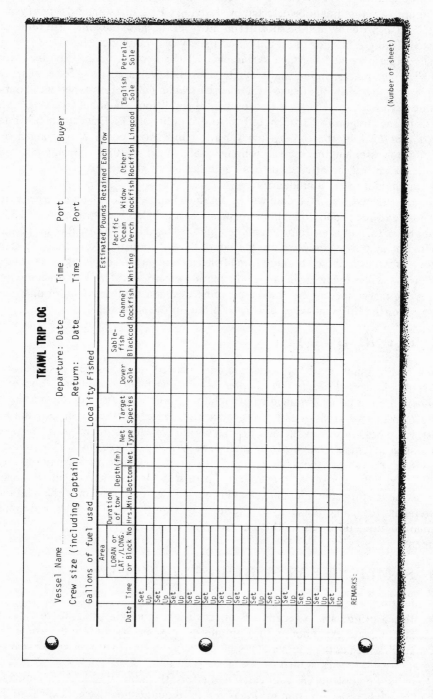

Figure 22.2. Representation of a page from the proposed trawl logbook for the California, Oregon, Washington groundfish fishery. The logbook is designed to use tear-out sheets.

the population (*see* chapter 1). For example, random or stratified random sampling is desirable to estimate size and age composition of species landed (Ketchen 1950). Choosing a survey sampling design requires a detailed analysis of the target population in order to produce accurate estimates. A statistician should be consulted early in the planning stage. Cochran (1977) provides a good review of survey sampling methods.

Stratified sampling is economical when it is impossible to sample all the fish in a landing or all trips into a port. (This is usually the case.) Stratification is the process of partitioning the sampled population (in this case, the fisheries attributes) into units having homogeneous values for the attributes (Cochran 1977). The choice of strata is critical for accurate sampling, and strata should be defined only be someone intimately familiar with the fishery. Some common types of strata for data about fishing trips are seasons, fishing areas, ports where fish are landed, and vessel types.

The choice of a sampling design must not only adhere to sound scientific and statistical principles, but must also be realistic in terms of objectives, personnel, and data processing capabilities. For example, it may be desirable to select strata which can be readily modified to account for unpredictable events such as budget cuts. It is also necessary to set species priorities so that sampling can be designed to optimize data about the most important species. The choice should include a comparison of different sampling designs using the cost per sampling unit as a base (Cochran 1977).

A common use of stratified sampling is for estimating annual catch and effort by statistical area when logbook data are incomplete. The typical stratum consists of the smallest statistical summary unit and one month as the smallest time unit. It is common that catches for such strata from logbook data will be underreported because some logbook records are missing, unuseable, or sometimes unreadable. If an agency has a well-designed trip or landing ticket which includes accurate accounting of all landings, the known logbook information can be applied to all trips including those for which data are unavailable. Trips that have complete logbooks indicating area of catch, hours fished, species pounds retained, etc., make up the knowns. Usually species pounds indicated on the landing record for these trips are prorated to an area and time stratum based on the catch data recorded in the logbook. These intermediate records are summed for all known trips to produce estimated catch and effort by area for each species. All landings from trips without logbook data are then prorated to each area and time stratum in proportion to the known pounds caught in that stratum. Known hours of effort are expanded in a similar manner.

This method is accurate as long as logbook coverage is representative of the activity of the fleet and is consistent from one port to the next. In some cases logbook coverage (or compliance) may be substantially lower in some ports. In such instances stratification by port may be necessary before prorating and expanding the data.

Interviewing methods may be employed for collecting catch and effort data in newly developing fisheries until appropriate data collection forms can be developed. Initially, the fishery scientist may need frequent contact with vessel operators in the new fishery. The fishery manager or scientist may need to ride along on some fishing trips so he can interview the skipper while directly observing the catch and amount and type of effort expended. With detailed observation and interchange with fishermen, an appropriate method of measuring and collecting catch and effort data can be developed. Interview techniques are also important in fisheries where it is difficult to obtain and process large volumes of data through the use of logbook records. In addition, fishermen are more willing to disclose a more condensed account of their fishing activities through personal contact than through a detailed logbook.

Fisheries scientists and managers may employ a variety of methods simultaneously to obtain information required for decision-making. Periodic review of all techniques, especially those used for collecting long term data, is required to insure the adequacy and proper development of the sampling program.

22.6 DATA PROCESSING AND REPORTING

Once fishery data have been collected, a data processing system must exist to tabulate or summarize the data. Data are processed to provide annual summaries of catch and effort by species group. In addition, within-season summaries are required to monitor intensively managed fisheries. Data must also be prepared for use in fishery population dynamics models and stock assessment analysis.

22.6.1 Responsibility for Data

Responsibilities for data processing must be defined clearly at all levels. If several agencies are involved in data collection and processing, one group usually is made responsible for collating the data. The assignment may be permanent or it may rotate among agencies. Data processing personnel should be aware of their function within the hierarchy of authority and responsibility. Priorities and deadlines for collecting and reporting data must be established for within-season and annual data summaries so that multi-agency and international coordination is facilitated.

State fisheries agencies, the National Marine Fisheries Service, and other organizations such as regional fisheries commissions can be assigned ultimate responsibility for collating data. It is useful to have one data committee to coordinate data collations, especially if several state or federal agencies are involved. All agencies involved in the process should further efficiency and timeliness of reporting by reaching common agreement on the data to be collected, the format of reports, and the times of reporting. Standardized output can then be collated easily by the host agency or organization. With the passage of the Magnuson Act of 1976, which fosters coastwide data processing systems, more universal standards are developing.

22.6.2 Processing of Data

The choice of methods used to process fishery statistics is dependent on the type of data, priority of need, time schedule, resources and staff available, and philosophy of a given agency towards data processing. The data processing apparatus of most agencies is a blend of electronic data processing (EDP) and hand tally methods pieced together in an expedient manner.

Data-base management using computers is an appropriate mode of data processing for catch and effort statistics from large fisheries such as the groundfish trawl fisheries. Extensive error checking is required of the data forms and data content. Persons responsible for collecting data should be involved in the error checking cycle to verify intermediate summaries against their knowledge and experience with the fishery. The use of a data-base on a computer allows rapid access to information that otherwise might not be accessible or might require weeks of hand tabulation to produce.

Good EDP systems depend on good software (computer programs and documentation) and hardware (computer equipment) as well as the staffing and material resources to support them. Whenever possible, professional computer scientists and programmers should develop EDP systems, especially complex ones such as data-base management systems. Covvey and McAlister (1982) provide a general treatment of computing choices.

22.6.3 Reporting Data

Standards for data summaries and reports are best handled through committee action whereby participating agencies identify needs and uses of the data and jointly develop a uniform reporting format. The centralized data committee or host agency can then provide annual summaries to participating members.

Statistics used in age-specific models, general production modelling, or trend analysis should be in final form and free of errors. Participating agencies should allocate sufficient time to check and refine annual statistics before submitting them to the collating group. Six months or more may be required to complete housekeeping of statistics from the previous season. The collating group should routinely provide revisions of data sets to participating agencies with instructions to destroy outdated or erroneous data.

Data summaries for within-season management functions, such as monitoring a quota, can be provided by participating agencies at the field level directly to fishery managers. For some fisheries, weekly or daily reporting may be required. Procedures for estimating catches should be standard and one agency or individual should be responsible for collating the data.

22.6.4 Limitations of Fishery Data

Several caveats apply to fishery data and their use. Development of fishery management plans has placed new demands on agencies which collect and use fishery data. In some cases, data are unavailable or not available in an appropriate form for some kinds of analysis. Regulatory analysis directed at assessing impacts of new or proposed management strategies, for example, may be constrained by the lack of data. Such analysis requires linking biological and economic data with trip information. Because local fishing climate, geography, and species distributions contribute to variability among agencies in the way data are collected and processed, trip data required for an analysis may not be in a uniform format. As management plans continue to be developed, new standards for data collection will be required to meet the needs of special within-season assessments.

Depending on budgetary constraints and local data processing development, the timing of data reporting may lag behind a desirable schedule for management use. Priorities may need to be established, and some species may need to be eliminated from detailed consideration.

Finally, interpretation of statistics over a large geographic area requires some caution. Subunit stocks within a larger area are subject to different fishing strategies because of local conditions. Target fishing of a species may occur in some areas and in other areas the catch of that species may be only incidental to the catch of some other species. Contact with local biologists and fishermen can be invaluable in reading between the lines of fishery harvest statistics.

22.7 REFERENCES

Clark, C. W. 1976. *Mathematical bioeconomics. The optimal management of renewable resources.* John Wiley and Sons, New York, New York, USA.

Cochran, W. G. 1977. *Sampling techniques.* John Wiley and Sons, New York, New York, USA.

Covvey, H. D., and N. H. MacAlister. 1982. *Computer Choices: Beware of Conspicuous Computing.* Addison-Wesley, Reading, Massachusetts, USA.

Doubleday, W. G., and D. Rivard, editors. 1981. *Bottom trawl surveys.* Special Publication 58, Department of Fisheries and Oceans, Ottawa, Canada. (Available from Supply and Services Canada, Canadian Government Publishing Centre, Hall, Quebec, Canada K1A059).

Gulland, J. A. 1965. *Manual of methods of fish stock assessment Part 1. Fish population analysis.* FAO (Food and Agriculture Organization of the United Nations) Fisheries Technical Paper 40, revision 1.

Herrman, R. B., and G. Y. Harry, Jr. 1963. *Results of a sampling program to determine catches of Oregon trawl vessels. Part 1. Methods and species composition.* Pacific Marine Fishery Commission 6:40-60.

Hughes, S. E. 1976. System for sampling large trawl catches of research vessels. *Journal of the Fisheries Research Board of Canada* 33:833-839.

Johnson, S. L., W. H. Barss, and R. L. Demory. 1982. *Rockfish assessment studies on Heceta Bank, Oregon, 1980-1981.* Oregon Department of Fish and Wildlife Sub-project Annual Report (NMFS 1-151-R-2).

Ketchen, K. S. 1950. Stratified sub-sampling for determining age distributions. *Transactions of the American Fisheries Society* 79:205-212.

Murphy, G. I. 1965. A solution to the catch equation. *Journal of the Fisheries Research Board of Canada* 22:191-202.

Pella, J., and P. Tomlinson. 1969. A generalized stock production model. *Inter-American Tropical Tuna Commission Bulletin* 13 (3):419-498.

Pope, J. C. 1972. An investigation of the accuracy of virtual population analysis using cohort analysis. *International Commission for Northwest Atlantic Fisheries, Research Bulletin* 9:65-74.

Westerheim, S. J. 1967. Sampling trawl catches at sea. *Journal of the Fisheries Research Board of Canada* 24:1187-1202.

Williams, T. 1977. The raw material of population dynamics. Pages 27-45 *in* J. A. Gulland, editor. *Fish Population Dynamics.* John Wiley and Sons, New York, New York, USA.

Chapter 23
Evaluating Human Factors

COURTLAND L. SMITH

23.1 INTRODUCTION

Evaluating human factors serves a number of needs in fishery science and management. Cultural and social data explain the objectives fishermen seek to achieve. Measuring opinions helps in determining acceptance of policy options. Observing fishermen's behavior explains human factors affecting growth and change in fishing effort. Fishing patterns may complement or confound the efforts of managers trying to protect a resource. Quite often, however, cultural and social data are seen as imprecise and their relevance is questioned (Larkin 1977; Zuboy 1981). Usually this perception has nothing to do with the techniques for gathering data but rather comes from differing basic philosophies on the purpose and objectives of data gathering.

Most analyses are organized according to characteristics describing the fishing experience in which fishing is the focal activity but not necessarily the primary goal. Cultural and social data describe the people who fish, i.e., those who are at the interface of the aquatic-cultural system. The evaluation of human factors usually focuses on those who catch fish, but the same types of research questions are applicable to other people in the fishery such as guides, charter operators, fish buyers, processors, distributors, retailers, and consumers.

Evaluating human factors in many situations has resulted in data serving the practical needs of managers and the public. An example is the development of the Lake Ontario salmon fishery by the New York Department of Environmental Conservation (DEC) in the 1970s (Schuman et al. 1979). In cooperation with the New York Sea Grant Program, initial social and economic impacts of the newly re-established fishery were evaluated. Surveys determined anglers were highly satisfied with the new fishery. There were no complaints of crowding, and fishery-related expenditures more than doubled between 1974 and 1975. The next year, PCB pollution brought about a decline in the use of the fishery. Sea Grant personnel held a conference to inform people about the fish contaminant problem. Based on evaluating human factors, they helped reorient the fishery to minimize impacts from pollution.

Because evaluating human factors is a highly complex task, it, as with all sciences, must be approached from a professional point-of-view. If not done professionally, the data are rarely conclusive. Pacific coast anglers, for example, surveyed commercial gill netters in 1974 to determine the percent of income gill netters obtained from fishing for Columbia River salmon and steelhead trout. Data presented by anglers showed that closing the spring gill net fishery would reduce gill netter incomes by only 8%. These data were not representative of the population of gill netters, were not checked for reliability, and, from the viewpoint of gill netters, were invalid. The object of this chapter is to describe the methods for collecting data about human factors and their biases so that professional fisheries managers can generate meaningful data about their clientele.

23.2 APPROACHES

Approaches for evaluating human factors include surveys to determine basic characteristics of anglers, commercial fishermen, and fishery professionals (Thompson 1979) and highly descriptive accounts of fishing communities. Surveys produce quantitative data while descriptive studies emphasize qualitative information. Opinion and attitude measurement are the most quantitative. Ethnography and network analysis, however, tend to be more qualitative; they focus on the fishery as a cultural or social system. A comparative approach and historical perspective help elaborate the nature of a fishery by contrasting it with other settings or showing change through time.

In most cases fishery management requires data about the characteristics of various aggregates, e.g., anglers, commercial fishermen, steelheaders, trollers, gill netters, or warm water fish enthusiasts. Little emphasis has been placed on how user behavior changes as experience is gained or the conflicts that develop because of differing experience levels (Bryan 1979). Even less concern has been focused on the community or systems relations among groups of fishermen. As managers learn more about the importance of social systems to fishermen, this aspect of human factor analysis will develop.

Research design texts for each social science discipline develop ways to gather cultural and social data (Summers 1970; Miller 1977; Pelto and Pelto 1978; Kerlinger 1979). There is, however, no consistent, standardized approach that can be readily adapted by an outsider.

23.2.1 Opinion Measurement

In opinion measurement the objective usually is to determine if a particular program will be successful. Officials may be interested in angler perceptions or merely concerned about the number of people involved with various fishing activities. For example, in 1977, the Oregon Department of Fish and Wildlife was considering a wild fish policy. The department measured angler opinions on the degree of support for increasing protection of wild stocks of salmon and steelhead trout. The Department contracted a professional survey organization to determine angling patterns, preferences for several management options, and reasons for fishing (Lowry 1978). The survey found that less than 5% of the anglers opposed increased protection for wild fish. Paradoxically, two thirds of the anglers favored increased hatchery production, which biologists linked with the decline of wild stocks.

23.2.2 Attitude Measures

One of the assumptions about human behavior is that how people think about things affects their behavior. If an angler does not like crowding, he is not likely to spend opening day in shoulder-to-shoulder fishing situations. Other anglers may tolerate or even prefer congestion and enjoy the social contact benefits of crowding. This has led to attempts to measure the satisfaction of anglers and commercial fishermen (Moeller and Engelken 1972; Hendee and Bryan 1974; Hampton and

Lackey 1975; Weithman and Anderson 1978; Smith 1980, 1981; Fig. 23.1). In these studies, researchers create a set of attitude statements about the satisfactions obtained from fishing and conduct a survey to learn how people will respond to these attitudes. An analysis of the motivation of anglers fishing on Texas coast charter boats asked respondents to indicate the importance of 36 reasons for fishing (Ditton et al. 1978). Insight was gained on the importance of catching fish, i.e., catching some but not necessarily many fish was acceptable to 63% of those surveyed.

23.2.3 Ethnoscientific Approach

Opinion surveys and attitude measurement assume that people's options and their influence on behavior is known. Ethnoscience is a technique for determining the attitudes of participants from their own point-of-view (Tyler 1969; Smith 1980). It is a technique to determine the questions for which the behaviors observed are answers. This approach assumes a need to find out how the population in question thinks and to predict how it reacts to change.

This technique requires three steps. The first step is an open-ended interview with key informants from the study population to determine their perceptions and to obtain statements about the activity's importance to participants. The second step of this technique requires a broad range of group members rather than merely a representative sample. The objective is to represent as broadly as possible the various points-of-view. The researcher summarizes the observations of the informants and then asks group members to evaluate them. Group members are given a set of cards,

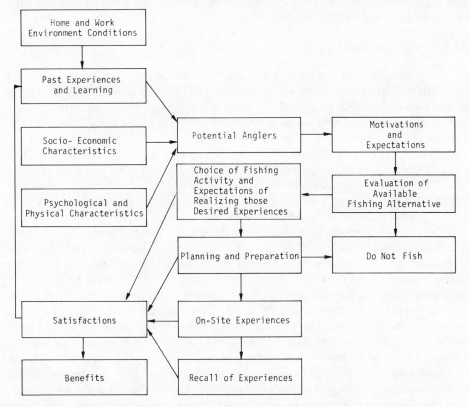

Figure 23.1. Social factors and decisions bearing on satisfaction achieved by anglers [after Dawson and Wilkins (1980:13), with permission].

each of which contains one statement made regarding the activity. Group members then sort the cards with respect to the importance of each attitude. The third step is to select statements that are chosen most often as important and present these statements in a more formalized survey to a representative sample of group members.

23.2.4 Ethnography

Ethnography is the description of a cultural system by an outside observer (Spradley 1980). These qualitative data often provide useful information with respect to how fishermen interact with resources. Ethnographic data are not gathered using predetermined categories; the setting dictates much of what is observed. For example, one of the problems facing tuna seiners is public interest in protecting porpoise from the practices of commercial fishermen. Some species of yellowfin tuna swim under or behind porpoise. Seiners use speed boats to "herd" the porpoise and tuna closely together so they can be surrounded by the seine. Orbach (1977), by living and working with seiners, learned from observing their behavior how these men thought about and acted toward porpoise and also described the rules the fishermen used to allocate the tuna resource at sea.

Observational data are not always qualitative. When Acheson (1975) observed the social organization of lobstermen, he identified two kinds of territories—nucleated and perimeter-defended. By classifying his observations according to type of lobster territory in which they were taken, he found that attitudes toward the resource differed in each territory. In perimeter-defended areas, for example, where access to lobsters was highly restricted, fishermen "cooperate to conserve the lobster stock and raise their own income levels" (Acheson 1975:206).

23.2.5 Networks and Organizational Factors

Ethnography uses a holistic focus and tries to describe the entire fishery or community. Fishermen, however, form groups to improve their effectiveness or to make fishing more enjoyable. Studying the relationships among members of these less formal groups is called network analysis (Barnes 1954; Boissevain and Mitchell 1973; Fischer 1977). Commercial fishermen, for example, organize in social networks, called code groups, for mutual support. The objective of a code group is to restrict the effort of nonmembers and increase member effectiveness. Several boats share information about fish concentrations (Orbach 1977; Anderson 1980; Acheson and Riedman 1982). Members develop secret codes to communicate and prevent outsiders from knowing what the fishing conditions are (Stuster 1978).

On-shore fishermen also form support groups (Stephenson 1980), usually organized around a successful fisherman who is innovative. Through these groups new techniques and information enter the fishery. These groups are used by extension agents in communicating with fishermen and can also be used by managers to contact fishermen regarding management rules and options. Take, for example, the problem of determining the number and activities of New England small-scale commercial fishermen who fish because it provides independence, fulfills an ambition, and gives peace and quiet. Peterson and Smith (1981) contacted shorebased members of fishermen's social networks, e.g., harbor masters, fish buyers, bait store owners, boatyard operators, shellfish wardens, and town registrars, to design a sampling procedure for contacting small-scale fishermen, some of whom did not own a boat.

Most fishing businesses in the United States are family-based. Few large corporations control fishing capacity. In this sense, fishing businesses are like the family farm in agriculture. Dankowski (1981) shows how wives of commercial fishermen play an important shoreside management role in the family unit. A social organization which takes on marketing functions of the fisherman is the cooperative. Cooperatives are often viewed as a vital organizing tool in developing areas (Poggie et al. 1980).

23.2.6   Comparative Approach

Too often studies are designed with only one population as the focus. Much more can be learned about fishermen with a comparative approach (Landberg 1973; Acheson 1981). Using the comparative approach permits the behaviors of one group to be contrasted with those of another, more clearly elaborating the behavior of fishermen. What behaviors, for example, distinguish commercial fishermen from other occupations? This question is investigated by comparing fishermen with millworkers, factory workers, cane workers, and small farmers. Fishermen, for example, who must plan ahead to locate fish, select weather conditions, and save income from good fishing for getting by during bad times display more deferred gratification than other workers (Pollnac and Poggie 1978). Fishing is a dangerous occupation, and this affects the risk taking behavior of commercial fishermen. When compared with millworkers, Poggie and Gersuny (1974) found commercial fishermen were greater risk takers, but their behavior was highly ritualized to reduce the risk of needless hazards.

23.2.7   Historical Approach

Contrasting different groups to obtain a better understanding of fishery participation is one way to gain perspective about the resource. Another way to get comparative insight is an historical perspective, which provides information on trends and patterns of change through time. This involves reviewing documents pertinent to a fishery that elaborate its history. Letters, reports, business records, diaries, photographs, and artifacts all provide useful and important cultural and social data. Many examples of the use of the historical approach exist. The quality and quantity of data in Goode et al. (1887) for United States fisheries often exceed current cultural and social knowledge. These data, gathered with ethnographic techniques, help provide perspective on current fisheries. Groth (1981) used historical analysis of license sales to show patterns in commercial and noncommercial participation in the Gulf shrimp fishery. Smith (1979) suggested the need to review historically the gear efficiency of California anglers to get accurate catch per unit of effort estimates. Failure to do this had led to a serious underestimate of commercial fishing effort for halibut (Skud 1972). Oregon was able to maintain public beach access threatened by private development when old photographs showing a pattern of recreation access were produced.

23.3   TECHNIQUES

Gathering cultural and social data requires choosing the proper variables, appropriate procedures for describing these variables, techniques for checking data quality, and proper methods for data analysis.

23.3.1 Variables

So that the information gathered contributes to meaningful elaboration of a problem, the first question that must be addressed is what variables need inclusion. Should evaluation of human factors strive for completeness or should it adopt a problem-oriented focus? The objectives of the study determine the answer to this question. Is it important to know a great deal about the fishery or to answer a specific question about it?

Ethnographic, network, and historical approaches should be used by those who are interested in completeness and being sure that the broadest range of variables are investigated. Several lists of variables to be considered are available. One developed specifically for fisheries was published in the *Federal Register* 42 (137):36982. Included in this list are age, education, ethnicity, family structure, community organization, employment, and distribution of income. A more complete list for cultural studies is found in *Outline of Cultural Materials* (Murdock 1964). For sociology the compendium assembled by Berelson and Steiner (1960) listing 1065 propositions serves as a resource for the person who does not wish to leave anything out.

To simplify data collected, Burdge and Field (1972) classified six kinds of variables important to leisure research. Christensen and Yoesting (1973) found that two of these kinds of variables, i.e., attitudes and associates (for example, the influence of family and friends), best predicted use of outdoor recreation facilities (*see also* Neulinger and Breit 1969; Murphy 1975; Christensen and Yoesting 1976; Crandall 1976). The other four variables listed by Burdge and Field are angling activities, demographic characteristics of anglers, importance of spatial groupings, and the social organization for leisure activities. A seventh variable relates to environmental qualities (Moeller and Engelken 1972; Hampton and Lackey 1975).

Little agreement exists on the most appropriate variables for evaluating the human factors of fisheries. Completeness is both time-consuming and costly. Ethnographic and network studies are often criticized for being too descriptive and not addressing the problem that needs solution yet too narrow an approach leads to the criticism that an important element has been left out. Usually the problem addressed helps specify the variable of interest.

23.3.2 Data Gathering

Surveys. The investigator's problem, information needs, and available funds help determine the data gathering technique selected for a particular study. The data gathering tool most frequently used is the survey. It is the basis for opinion and attitude measurement. Surveys have the advantage of providing easily quantified data for a user population. Surveys reduce the cost of contacting large numbers of users (Duttweiler 1976). Figure 23.2 summarizes the system for gathering, analyzing, and communicating survey data.

There are three general types of surveys—the mail, telephone, and personal interview surveys. Mail surveys permit contacting large numbers of people at low cost. Telephone surveys also provide a quick and inexpensive way to gather cultural and social data. Personal interviews are usually the most costly. For fishing problems, however, where specific time or area dimensions are important or the dimensions of the fishing population are not known, personal contact may be most effective. This discussion emphasizes mail and telephone surveys; the techniques for conducting

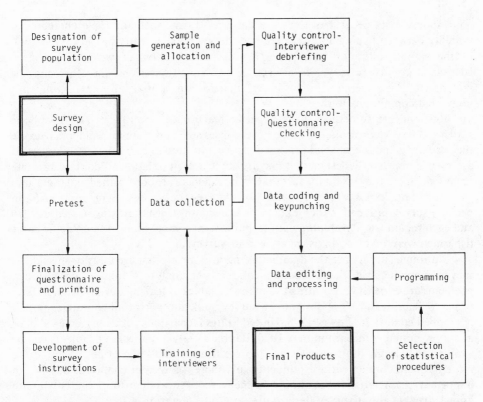

Figure 23.2. The many steps in the design, construction, administration, evaluation, analysis, and reporting of a survey [United States Department of Commerce (1980:4)].

personal interviews are discussed in chapter 21.

Oregon's attempt to determine angler opinions regarding wild fish policy illustrates the design of a typical mail survey. The survey had 17 questions; 16 had a predetermined response format and one was open-ended. The order and wording of questions were perfected after a pretest of the questionnaire with 50 anglers. The technique used was a mail survey with two follow-up mailings. Surveys were sent to a 1%-sample of Oregon's 400,000 anglers; the response rate was 63%.

The failure of some people to return completed questionnaires is a major problem of mail surveys. Heberlein and Baumgartner (1978) reviewed mail survey questionnaires and described procedures for increasing response rates. The critical issue with response rates is not the percentage who return the questionnaire, but rather if a bias in the sample is caused by the nonrespondents. If the respondents do not adequately represent the population being described, the survey results will be biased. If the population is very homogeneous, a low response rate will not necessarily introduce significant bias. Nonrespondents, therefore, should be checked to determine if they differ from the target population (Brown and Wilkins 1978; Brown et al. 1981).

Telephone survey methods are becoming increasingly popular as a data gathering technique as survey costs increase and response rates decline. Telephone surveys are easy to administer, and they enable better control of interviewer effects (Dillman 1978; Groves and Kahn 1979). Telephones have an obvious bias with respect to income. The wealthy have more phones and phone numbers than the poor. Although over 95% of households have phones, telephone surveys of commercial fishermen have very high nonresponse rates because of the work patterns of fishermen and the high degree of mobility between fishing areas. Telephone surveys make identifying the composition of the sample more difficult, and, hence, require complex weighting models. Interviews are fast-paced, and answers to open-ended questions, therefore, may be incomplete. The telephone interviewer, in addition, misses all the respondent's nonverbal communication.

Respondents sometimes have negative reactions to both mail and telephone surveys. Surveys constrain the range of responses solicited from respondents because they usually do not permit elaboration and incorporation of context with responses. If a survey is improperly designed, it may frustrate user populations. People appreciate being asked, but they resent the fact that their responses are constrained and that those gathering data do not care more about how they see the whole system. Users also can become survey fatigued. When extended jurisdiction legislation was passed in 1976 and each region was developing data to build management plans, some Pacific coast fishermen were involved in as many as four surveys.

Although surveys are the data gathering tool most often used by nonspecialists, they are nonetheless difficult to design, conduct, and analyze. Common survey design problems are either incorporating too many variables or leaving out several that later prove to be critical. Before designing a survey, look at reports that contain examples (Campbell et al. 1976; Lowry 1978; United States Department of Commerce 1980) or at survey design guides (Summers 1970; Babbie 1973) so that these problems can be avoided. Although short surveys are more effective than longer ones, they require more frequent contact with respondents, and there is a greater chance of missing an important characteristic that should be measured. Always consult experts who are knowledgeable about survey design, use, and interpretation.

Observation. In many situations, observations of people can be more useful than a survey. For example, an observer located at a critical point in the system can alter observational techniques as needed (Webb et al. 1981). Because human behavior is patterned and the objective is to identify these patterns, observation can provide as much data as asking questions. Much information on effort and factors that affect it can be obtained by observing how people respond to situations. McCay (1978), for example, observed New Jersey fishermen to determine how they arrived at optimum fishing strategies.

Instead of relying on either survey or observation, the use of both can produce the best evaluation. For example, observation of testimony at meetings suggested to Acheson (1980) that fishermen operating in a single-species fishery should favor limited entry because the species would not be exploited by other fishermen. He hypothesized that those fishing multi-species would oppose limited entry. A survey revealed this was not the case. Multi-species fishermen were more in favor of limited entry. Interviews with key informants found that single-species fishermen opposed limited entry because they feared being closed out of their only fishing opportunity.

23.3.3. Checking Data

Because gathering cultural and social data occurs in people-to-people settings, the chances for error are compounded. At every point where people interact, there is a potential for errors in communication. Interviewers may misunderstand or wrongly state the questions of research designers. Respondents may misinterpret the question. The interviewer may not hear the intended response or may record it incorrectly. Coding and analysis errors are possible. Results can be misinterpreted if the wording of the report is not clear. Readers can make erroneous conclusions.

There are several techniques that can be used to spot and prevent errors. First, the use of well-trained or professional interviewers and observers will produce higher quality data. Second, the study designer should review data obtained by interviewers and observers to check these with the study's intent. Third, portions of the data, coding, analysis, and interpretations should be randomly checked. Errors usually increase as the open-endedness of the data gathering technique increases. Fourth, the internal consistency of the data should be checked and where results break from a pattern, they should be rechecked and verified. Three main characteristics of social and cultural data express the value of the information—representativeness, reliability, and validity.

Representativeness. Representativeness is the extent to which the interviews or observations of the sampling program match those of the entire population. The usual way to assure representative data is to collect a random sample, but that may not be appropriate for social and cultural data.

Sample size is extremely important in assuring that the data are representative. Sample size should be set so that the category of data with the fewest expected responses or observations has an error that can be tolerated in response percentages. While problems vary, fewer than 30 observations typically are unacceptable because they will have a wide band of statistical error. More than 300 observations are not useful because they will not appreciably reduce the standard error (Lowry 1978; Neave 1978).

For example, suppose a study was designed to contrast Oregon steelhead anglers who catch many fish and those who catch none. A random sample of angling license holders would be the simplest sampling design, but in this case, it would yield poor data. In most fisheries the distribution of catch is skewed toward large numbers of anglers catching none or very few fish. Over 50% catch no steelhead, and only 1% of those catching fish take over 20 per year. Suppose a random sample of Oregon's 400,000 licensed anglers is constructed. Data on catch are confidential so a stratified sample is not possible. A 1%-sample with 50%-response rate gives 2000 anglers. Of these 2000 anglers, however, over 1000 would have caught no steelhead trout at all. Only 10 would have caught over 20 steelhead trout.

The high catch sample, i.e., 10 anglers, is inadequate to make meaningful comparisons unless it can be shown that this is a homogeneous population whose variability can be represented adequately with 10 cases. If this is not the case, some other sampling technique needs to be used that increases the representation of high catch steelheaders. The solution in this case was to sample the organization of Northwest Steelheaders of Trout Unlimited, who were known to have a higher percentage of successful anglers (Smith 1980). This solution, however, made it very difficult to determine representativeness in terms of such characteristics as age or residence. The more that is done to adequately determine the nature of the sample, the more costly and complex is the data gathering and analysis process.

Sampling license holders can create other problems, as well. Dunning and Hadley (1978) found that over 40% of western New York anglers were under 15 and not licensed. They also found that 25% of the anglers fished illegally! A representative sample of licenses in this situation would severely misrepresent the population of anglers. Some states hold license information to be confidential. This makes it especially difficult for investigators to link catch statistics with license holders. Fishermen are quite mobile. Smith (no date) found an annual Oregon commercial license turnover of 25%. Taking a sample from the previous year's license holders means possibly 25% will be difficult to contact and may not be able to provide data on current fishing activities. Considerable time will be spent chasing elusive respondents.

While representativeness is not always easy to determine, the objective is to save money and effort by sampling a representative portion of the total population (Cohen 1969). Representativeness is checked by comparing known characteristics of the whole population against the group surveyed. The analyst must be prepared to defend his data against allegations that an important segment of the population is unintentionally over or under represented.

Reliability. Reliability describes whether or not the method used to collect data reproduces the same results when used again. Reliability is checked with repeated or alternate observations. For a survey, this would mean administering it more than once to a user population or surveying matched user populations to determine whether the questions provide similar answers with repeated use. Reliability is one good reason for using measures that have been tested on other populations and for which there is a base of comparative data.

For ethnographic studies reliability is checked by observing the same behavior in alternate ways. For example, some trawlmen fill out logbooks or answer surveys in ports indicating where they fished on a particular trip and what they caught. Observing how fishermen behaved at sea, however, suggested to Peterson (1977) that the port sampling data were not very reliable. Fishermen sought to protect hot spots and their favorite grounds by giving incorrect responses. Their responses could be biased against locating the best fishing spots and unreliable regarding the location of good quality fishing.

Validity. Validity tests if the question measures what it is intended to measure. Does the question asked evoke an answer with the correct information? The angler who can cast to the same spot time after time has a high degree of casting reliability. Casting to the same spot or going to the same fishing location does not necessarily mean quality fishing, however, because "good quality fishing" means different things to various anglers. Validity of a question depends upon agreement between those concerned with the topic being evaluated. A measure can be very reliable, but, if it does not adequately gage the behavior or primary interest, it lacks validity.

There are no easy tests of validity. Cultural and social data, however, are most frequently questioned on this issue. What is the measure of good fishing quality? Is it the quantity of fish caught, their size, the spot's seclusion, or the association with friends? If quality fishing is determined by the number of fish caught, then this is a valid measure. Not everyone, however, will agree that this is the appropriate measure, and thus validity is only established to the extent people agree that the measure does indicate a desired quality.

23.3.4 Analysis

The most common error in evaluating human factors is not allocating enough time for data analysis. As with data gathering, analysis will go better if alternative hypotheses are used and if different approaches are taken to data.

Surveys lend themselves to high levels of quantification and automatic data processing. Many studies only report simple descriptive statistics and do not address multivariate questions the data could answer. In analyzing quantitative data, select a statistical package that enables data modification. No matter how carefully coding categories are defined, there is always the need to lump categories together, to recombine responses, and to make new variables out of existing ones in the data. Biomedical programs (BMD), Statistical Package for the Social Sciences (SPSS), and Statistical Analysis System (SAS) have a great deal of analytical flexibility. They are clearly described, relatively easy to learn, and give references to statistical literature (Nie et al. 1975; Engelman et al. 1977; Helwig 1978). SAS is more limited than BMD and SPSS to computer systems having an IBM type architecture, and thus its availability is somewhat restricted.

Various analytic methods are useful. Regression, discriminant, and factor analysis are the most widely used multivariate techniques (Van de Geer 1971; Blalock 1979), although analysts must be careful with the quality and kind of measurement (Siegel 1956).

Regression analysis has the objective of developing an equation that shows how several independent variables predict a single dependent variable (Draper and Smith 1966). It also shows direction and degree of influence of an independent variable. Discriminant analysis is used to determine which variables best distinguish between types of anglers or commerical fishermen. Factor analysis determines the underlying patterns of relationships (Rummel 1970; Harmon 1976; Cattell 1978); it shows the associations between many variables. Factor analysis organizes variables along vectors. It is analogous to a fly caster seeking the best line to avoid overhanging limbs, skirt rocks, and hit the pool where a large trout lies. What is the best line, given these variables? Other types of analysis, such as cluster, path, and canonical correlation are available and, depending on the problem, should be evaluated.

Analysis of data gathered by observation relies more heavily on concept than quantification. To be most useful, the analyst needs to clearly relate observations with scientific and policy issues. Very often this is incompletely done. Code groups, for example, are very interesting to social scientists as a means for fishermen to deal with incomplete knowledge about the environment.

23.4 LEGAL AND ETHICAL RESPONSIBILITIES

When gathering cultural and social data, the researcher has legal and ethical responsibilities to many people. Principal among these are the people who respond to surveys or who allow researchers to observe their activities.

Gathering data may be conceived as objective and in the best interests of science. Yet cultural and social data describe human factors. Data gathering is an exchange; the researcher takes information from people, and they expect something in return. The person providing data usually feels that more has been given than is received. The decisions, resulting from cultural and social data, involve distribution of catch between user groups, limitations on freedom of enterprise, and usually more rules rather than fewer.

Cultural and social data represent a trust that researchers must carefully manage. How and to whom these data are disseminated are key issues. Most reports are prepared for managers and decision makers. In addition, cultural and social data are used to report to colleagues on advances in social science knowledge. Much less effort is given to feedback with respondents and the general public who often feel they have shared information but received little benefit for doing so.

Certain responsibilities to respondents are specified by legislation on the protection of human subjects and administrative rules stemming from the Fishery

Conservation and Management Act. Rules with respect to observation and survey research have relaxed some, but the overriding concern for personal privacy remains. Respondents should be informed of the background reasons for gathering cultural and social data, any hazards that development of this information may cause, and the purpose for which these data are intended.

University committees for the protection of human subjects set requirements for informing respondents of risks and assuring that respondent rights and privacy are protected. State agencies have similar requirements, and surveys conducted with Federal funds need to be reviewed by the Office of Management and Budget. In most cases people know all too well the hazards. Often this affects their responses and may result in biased behavior or answers to questions. Techniques exist for getting information on sensitive issues. One approach is to ask people not about themselves but about what they see others do. Such a technique, however, strikes at the rights to privacy in a free society. Social scientists must continually weigh the issue of whether the information is worth the social cost of getting it.

23.5 REFERENCES

Acheson, J. M. 1975. The lobster fiefs: economic and ecological effects of territoriality in the Maine lobster industry. *Human Ecology* 3(3):183-207.

Acheson, J. M. 1980. Attitudes towards limited entry among finfishermen in northern New England. *Fisheries* 5(6):20-25.

Acheson, J. M. 1981. Anthropology of fishing. *Annual Review of Anthropology* 10:275-316.

Acheson, J. M., and R. Riedman. 1982. Technological innovation in the New England fishing industry: an examination of the Downs and Mohr hypothesis. *American Ethnologist* 9 (3): 538-558.

Anderson, R. 1980. Hunt and conceal: information management in Newfoundland deep-sea trawler fishing. Pages 205-228 *in* S. K. Tefft, editor. *Secrecy.* Human Sciences Press, New York, New York, USA.

Babbie, E. R. 1973. *Survey research methods.* Wadsworth Publishing Company, Incorporated, Belmont, California, USA.

Barnes, J. A. 1954. Class and committees in a Norwegian island parish. *Human Relations* 7 (1): 39-58.

Berelson, B., and G. A. Steiner. 1964. *Human behavior, an inventory of scientific findings.* Harcourt, Brace, and World, Incorporated, New York, New York, USA.

Blalock, H. M. 1979. *Social statistics.* McGraw-Hill Book Company, New York, New York, USA.

Boissevain, J., and J. C. Mitchell. 1973. *Network analysis: studies in human interaction.* Mouton, The Hague, Netherlands.

Brown, T. L., and B. T. Wilkins. 1978. Clues to reasons for nonresponse and its effect on variable estimates. *Journal of Leisure Research* 10(3):226-231.

Brown, T. L., C. P. Davison, D. L. Hastin, and D. J. Decker. 1981. Comments on the importance of late respondent and nonrespondent data from mail survey. *Journal of Leisure Research* 13(1):76-79.

Bryan, H. 1979. *Conflict in the great outdoors; toward understanding and managing for diverse sportsmen preference.* Bureau of Public Administration, Sociological Studies No. 4, The University of Alabama, Tuscaloosa, Alabama, USA.

Burdge, R. J., and D. R. Field. 1972. Methodological perspectives for the study of outdoor recreation. *Journal of Leisure Research* 4(1):63-72.

Campbell, A., P. E. Converse, and W. L. Rogers. 1976. *The quality of American life; perceptions, evaluations, and satisfactions.* Russell Sage Foundation, New York, New York, USA.

Cattell, R. B. 1978. *The scientific use of factor analysis in the behavioral and life sciences.* Plenum Press, New York, New York, USA.

Christensen, J. E., and D. R. Yoesting. 1976. Statistical and substantive implications of the use of stepwise regression to order predictions of leisure behavior. *Journal of Leisure Re-*

search 8(1):59-65.

Cohen, J. 1969. *Statistical power analysis for the behavioral sciences.* Academic Press, New York, New York, USA.

Crandall, R. 1976. On the use of stepwise regression and other statistics to estimate the relative importance of variables. *Journal of Leisure Research* 8(1):53-58.

Dankowski, F. 1981. *Fishermen's wives: coping with an extraordinary occupation.* University of Rhode Island Marine Bulletin Series No. 37., Kingston, Rhode Island, USA.

Dawson, C. P., and B. T. Wilkins. 1980. Social considerations associated with marine recreational fishing under FCMA. *Marine Fisheries Review* 42(12):12-17.

Dawson, C. P., and B. T. Wilkins. 1981. Motivations of New York and Virginia marine boat anglers and their preferences for potential fishing constraints. *North American Journal of Fisheries Management* 1:151-158.

Dillman, D. A. 1978. *Mail and telephone surveys: the total design method.* Wiley and Sons, New York, New York, USA.

Ditton, R. B., T. J. Mertens, and M. P. Schwartz. 1978. Characteristics, participation, and motivations of Texas charter boat fishermen. *Marine Fisheries Review* 40(8):8-13.

Draper, N. R., and H. Smith. 1966. *Applied regression analysis.* Wiley and Sons, New York, New York, USA.

Duncan, D. J. 1978. Leisure types: factor analyses of leisure profiles. *Journal of Leisure Research* 10(2):113-125.

Dunning, D. J., and W. F. Hadley. 1978. Participation of nonlicensed anglers in recreational fisheries, Erie County, New York. *Transactions of the American Fisheries Society* 107: 678-681.

Duttweiler, M. W. 1976. Use of questionnaire surveys in forming fishery management policy. *Transactions of the American Fisheries Society* 105:232-239.

Engelman, J., J. W. Frane, and R. I. Jennrich. 1977. *BMDP-77, biomedical computer programs, p-series.* University of California Press, Berkeley, California, USA.

Fischer, C. S. 1977. *Networks and places.* The Free Press, Glencoe, Illinois, USA.

Groth, P. G. 1981. Effective use of sociocultural data in fisheries management: a case study. *Fisheries* 6(2):11-16.

Groves, R. M., and R. L. Kahn. 1979. *Surveys by telephone: a national comparison with personal interviews.* Academic Press, New York, New York, USA.

Hampton, E. L., and R. T. Lackey. 1975. Analysis of angler preferences and fisheries management objectives with implications for management. *Proceedings of the Annual Conference Southeastern Association of Game and Fish Commissioners* 29:310-316.

Harman, H. H. 1976. *Modern factor analysis.* The University of Chicago Press, Chicago, Illinois, USA.

Heberlein, T. A., and R. Baumgartner. 1978. Factors affecting response rates to mailed questionnaires: a quantitative analysis of the published literature. *American Sociological Review* 43(4):447-462.

Hendee, J. C., and H. Bryan. 1974. Social benefits of fish and wildlife conservation. Pages 234-254 *in Proceedings of the Western Association of Fish and Wildlife Agencies,* San Diego, California, USA.

Helwig, J. T. 1978. *SAS introductory guide.* SAS Institute Incorporated, Cary, North Carolina, USA.

Kelly, J. R. 1978. Situational and social factors in leisure decisions. *Pacific Sociological Review* 21(3):313-330.

Kerlinger, F. N. 1979. *Behavioral research a conceptual approach.* Holt, Rinehart, and Winston, New York, New York, USA.

Landberg, L. C. W. 1973. *A bibliography for the anthropological study of fishing industries and maritime communities.* University of Rhode Island, Kingston, Rhode Island, USA.

Larkin, P. A. 1977. Epitaph on the concept of maximum sustainable yield. *Transactions of the American Fisheries Society* 106(1):1-11.

Lowry, H. M. 1978. *Survey of Oregon resident annual angler license holders: recreational fishery use and preferences among management options.* Survey Research Center, Oregon State University, Corvallis, Oregon, USA.

McCay, B. J. 1981. Optimal foragers or political actors? Ecological analyses of a New Jersey fishery. *American Ethnologist* 8(2):356-382.

Miller, D. C. 1977. *Handbook of research design and social measurement,* 3rd edition. David

McKay Company, Incorporated, New York, New York, USA.

Moeller, G. H., and J. H. Engelken. 1972. What fishermen look for in a fishing experience. *Journal of Wildlife Management* 36(4):1253-1257.

Murphy, P. E. 1975. The role of attitude in the choice decisions of recreational boaters. *Journal of Leisure Research* 7(3):216-224.

Murdock, G. P. 1971. *Outline of cultural materials*, 4th edition. Human Relations Area Files, New Haven, Connecticut, USA.

Neulinger, J., and M. Breit. 1969. Attitude dimensions of leisure. *Journal of Leisure Research* 1(3):255-261.

Neave, H. R. 1978. *Statistics tables for mathematicians, engineers, economists and the behavioral and management sciences*. George Allen & Unwin, London, England.

Nie, N. H., C. H. Hull, J. G. Jenkins, K. Steinbrenner, and D. H. Bent. 1975. *SPSS: statistical package for the social sciences*, 2nd edition. McGraw-Hill Book Company, New York, New York, USA.

Orbach, M. K. 1977. *Hunters, seamen, and entrepreneurs: the tuna seinermen of San Diego*. University of California Press, Berkeley, California, USA.

Pelto, P. J., and G. H. Pelto. 1978. *Anthropological research: the structure of inquiry*. Cambridge University Press, New York, New York, USA.

Peterson, S. B. 1977. Social science research in the New England fishing industry. Pages 72-85 *in* M. K. Orbach, editor. *Report of the national workshop on the concept of optimum yield in fisheries management*. United States Department of Commerce, Washington, District of Columbia, USA.

Peterson, S. B., and L. J. Smith. 1981. *Small-scale commercial fishing in southern New England*. Woods Hole Oceanographic Institution, Technical Report 81-72, Woods Hole, Massachusetts, USA.

Poggie, J. J., Jr., and C. Gersuny. 1974. *Fishermen of Galilee*. University of Rhode Island Marine Bulletin Series No. 17, Kingston, Rhode Island, USA.

Poggie, J. J., Jr., J. Stuster, R. B. Pollnac, F. Carmo, B. J. McCay, J. R. McGoodwin, M. K. Orbach, and J. S. Peterson. 1980. Maritime anthropology. *Anthropological Quarterly* 53(1):1-74.

Pollnac, R. B. 1980. *Continuity and change in marine fishing communities*. University of Rhode Island, International Center for Marine Resource Development, Kingston, Rhode Island, USA.

Pollnac, R. B., and J. J. Poggie, Jr. 1978. Economic gratification orientations among small-scale fishermen in Panama and Puerto Rico. *Human Organization* 37(3):355-367.

Rummel, R. J. 1970. *Applied factor analysis*. Northwestern University Press, Evanston, Illinois, USA.

Schuman, S. P., T. L. Brown, and M. W. Duttweiler. 1979. The roles of research and extension education in the developing Lake Ontario salmonid fishery. *Fisheries* 4(3):6-8.

Siegel, S. 1956. *Nonparametric statistics for the behavioral sciences*. McGraw-Hill Book Company, New York, New York, USA.

Skud, B. E. 1972. *A reassessment of effort in the halibut fishery*. International Pacific Halibut Commission Scientific Report No. 54. Seattle, Washington, USA.

Smith, C. L. 1979. *Salmon fishers of the Columbia*. Oregon State University Press, Corvallis, Oregon, USA.

Smith, C. L. 1980. Attitudes about the value of steelhead and salmon angling. *Transactions of the American Fisheries Society* 109:272-281.

Smith, C. L. 1981. Satisfaction bonus among salmon fishermen: implications for economic evaluation. *Land Economics* 57(2):181-196.

Smith, F. J. No date. *Some characteristics of Oregon fishermen*. Extension Advisory Program, Commercial Fishing Publication No. 12, Oregon State University, Corvallis, Oregon, USA.

Smith, S. E. 1979. Changes in saltwater angling methods and gear in California. *Marine Fisheries Review* 41:32-44.

Spradley, J. P. 1980. *Participant observation*. Holt, Rinehart, and Winston, New York, New York, USA.

Stephenson, G. O. 1980. *"Pushing for the highline": the diffusion of innovations in the Oregon otter trawl fishery*. Master's thesis. Oregon State University, Corvallis, Oregon, USA.

Stuster, J. 1978. Where "Mabel" may mean "sea bass." *Natural History* 87(9):64-71.

Summers, G. F. 1970. *Attitude measurement.* Rand McNally, Chicago, Illinois, USA.

Thompson, T. P. 1979. Job satisfaction among fisheries professionals: a report. *Fisheries* 4 (4): 7-13.

Tyler, S. A. 1969. *Cognitive anthropology.* Holt, Rinehart, and Winston, Incorporated, New York, New York, USA.

United States Department of Commerce. 1980. *Marine recreational fishery statistics survey, Atlantic and Gulf coasts, 1979.* United States Department of Commerce, Current Fishery Statistics, Number 8063. Washington, District of Columbia, USA.

Van de Geer, J. P. 1971. *Introduction to multivariate analysis for the social sciences.* W. H. Freeman and Company, San Francisco, California, USA.

Webb, E. J., D. T. Campbell, R. D. Schwartz, L. Sechrest, and J. B. Grove. 1981. *Nonreactive measures in the social sciences.* Houghton Mifflin Company, Boston, Massachusetts, USA.

Zuboy, J. R. 1981. A new tool for fishery managers: the Delphi technique. *North American Journal of Fisheries Management* 1:55-59.

Chapter 24
Economic Considerations for Fishery Management

FRED J. PROCHASKA AND JAMES C. CATO

24.1 INTRODUCTION

Economic considerations are an integral part of the development and management of any fishery. Fishery producers, handlers, and processors are interested in making a profit both in the short-run and on a long term continuous basis. Fishery managers are concerned with resource management to insure proper utilization of what is often a public resource. Certain economic models and concepts provide useful guidelines in both developing and managing fishery stocks.

One role of economics in fishery management is to help formulate goals for management programs and to determine conditions consistent with goals. Maximum economic yield (MEY) is an acceptable economic goal, but it is only one of many possible goals fishery managers may choose or consider. Economic analysis can be used to measure economic impacts of goals other than maximum economic yield on income, prices, consumption, profits, trade, and employment.

This chapter is presented in four sections. First, basic economic production and demand models are reviewed. The second section contains a review of economic data requirements and associated problems. The third section is designed to teach economics by way of example. Use is made of both economic production and marketing concepts. Economic data collection and its incorporation with biological data is discussed. Management inferences using both data types are made. The final section is a summary and conclusion.

This chapter focuses primarily on considerations of fishery economics management practices that relate to open fisheries (stock harvested from the wild) as opposed to closed systems such as catfish farms or aquaculture operations. Shang (1981) gives a layman's treatment of the basic concepts and methods of analysis of aquaculture systems.

24.2 APPLICATION OF ECONOMIC MODELS IN FISHERY MANAGEMENT

Demand and supply jointly determine the economic value of any fishery stock. This includes the quantity produced and consumed of the stock and the allocation of labor, management, and capital to the fishing industry that can be used to produce fish and shellfish products. Demand reflects what consumers are willing to pay for various amounts of fishery products. Changes in demand encourage more or less effort to be devoted to harvesting fishery stocks and, therefore, must be included in fishery management considerations. Supply reflects the amount of fishery products producers are willing to sell at different prices. As price increases, the quantity offered for sale increases, and as price decreases, the available quantity decreases. The positive relation between price and quantity supplied and shifts in supply are based on production economics while demand relationships between prices charged and quantities purchased are related to consumption parameters. Each of these is discussed in the following sections.

24.2.1 Production Concepts

Production economics models of importance in fishery management are of two types: (1) industry production functions or bioeconomic models and (2) firm production models. Bioeconomic models address the question of maximum economic yield and optimum allocation of resources from a total industry or stock perspective. The firm production model addresses questions of efficiency and profitability from the view of the individual fishing firm.

Bioeconomic models. These models originate with a yield function based on both biological and physical production parameters as demonstrated in Williams and Prochaska (1977). Yield is measured in terms of landings or catch over time. Effort represents the physical production variable(s) and is usually measured in terms of inputs such as fishermen, vessels, and/or gear units (*see* chapter 22). The inputs are indexed to some base period so that changes in technological efficiency or change in input mix are considered.

Total fishery revenue (TR) is expressed as a function of effort by multiplication of the yield function by product price. The total cost (TC) function associated with different levels of effort increases with increased levels of effort. Total cost in an economic model includes normal returns to labor, management, and capital. As long as TR exceeds TC, pure profits exist, and additional effort will be induced into the fishery until TR equals TC. Maximum economic yield occurs at an effort level where the difference in TR and TC is a maximum. This occurs at effort levels less than those required for maximum sustainable yield (MSY) since as additional units of effort are applied, productivity decreases and the cost of each marginal unit of effort is greater than the marginal revenue earned by the product resulting from that unit of effort.

MEY results in maximum economic efficiency in terms of all resources used (fish stock, fuel, manpower, etc.). The pure profits or economic rent may be taxed away to provide a return to the public sector from the commonly owned fish resource. In overfished fisheries a reduction in effort often involves social costs which must be weighed against the gains in economic efficiency. In addition, administrative cost of any management program must be weighed against the expected benefits resulting from a mandated change in fishing effort. The concept of maximum economic yield and associated theory may be examined in detail in Scott (1954), Crutchfield (1965), Prochaska and Baarda (1974), Christy (1975), Nielsen (1976), Anderson (1977), Larkin (1977), and Bell (1978). Prochaska and Cato (1980) also give another overview of the economic considerations in the management of an entire fishery.

Firm production models. Individual firm or boat production models provide the basis for an analysis of costs, revenues, and most profitable resource allocation for that firm. These models are estimated from cross-sectional surveys of fishing firms during a given time or production period. A cross-sectional survey is a sampling of firms concerning production during a given period of time. The stock level is fixed for a given production period and thus is not a factor in determining effort levels. The profit-maximizing firm will allocate individual units of effort in such a way that the marginal value of the product resulting from a dollar's worth of one input is equal to the marginal value of production of every other input used.

Firm production models are appropriate for fishery management considerations when management programs are implemented which hold stock at some desired levels. Firm models are used to analyze profits and costs associated with the management program. More importantly, firm models are used to determine changes in "effective" units when effort is the control or managed variable. Economically motivated fishermen change the mix of individual effort inputs in response to changes in input prices and/or changes in technological efficiency of individual inputs. This substitution of one effort unit for another changes the "effective" units of effort devoted to the fishery. Firm production models aid in the analysis of this problem. For example, restricting the number of firms in the fishery will not limit effort if vessel size and gear units per vessel are allowed to change in response to changing economic conditions. Prochaska and Williams (1978) provide an example of using firm production models.

Budget information or cost and return studies are not derived using either of these production models, but rather through simple accounting or budgeting procedures. These budgets allow easy examination of the impact of a management regulation on the profit of an individual fishing firm and also allow determination of the economic feasibility of participating in certain fisheries. Examples of budgetary analysis can be found in Prochaska and Williams (1978), Cato and Prochaska (1977b), and Cato and Lawlor (1981). Smith (1975) provides a complete overview of a fishing firm's business management considerations.

24.2.2 Demand and Consumption Concepts

Marketing and demand data and analyses are essential for effective fisheries management. Consumer demand parameters associated with factors such as per capita income, population, and price of substitute products play a large part in determining product prices which, in turn, affect the amount of effort devoted to the fishery. Prices of luxury products are highly income responsive. As income levels increase, demand and prices for these products increase causing total revenue (TR) to increase and additional effort to enter the fishery. This is the case for many overfished fisheries. Increases in population levels generally increase demand for all products. Knowledge of these demand parameters may be used to predict potential problems in expanding fisheries, and management measures may be instituted before serious problems arise. Cato (1976) and Lampe (1978) give examples of demand analysis in fisheries.

Marketing and demand considerations are also important for fisheries where management measures are in the process of implementation or evaluation. Management of fisheries usually changes the quantities moving into the market. As quantities change, prices offered for the product by consumers change, and this changes the economic targets of the producers. Thus, demand functions must be considered simultaneously with production and cost functions (supply) to arrive at MEY solutions as shown in Prochaska and Baarda (1974).

Marketing and demand information also are the basis for determining the economic impacts of alternative management programs. Price equations at each market level can be used to determine the value of the fishery for different levels of production. Market channel and structure studies identify individuals affected and help determine impacts at the wholesale, processing, broker, import, and export market levels. Prochaska and Andrew (1974), Alvarez et al. (1976a, 1976b), and Prochaska (1978) all give examples of market channel and structure studies.

24.3 DATA REQUIREMENTS, PROBLEMS, AND APPLICATIONS

24.3.1 Collection Methods and Data Use

Economic research is generally based on one of three types of data: (1) primary, (2) secondary, and (3) experimental. Primary data refers to data collected by the economist at some level in the economic system from basic producer to final consumer. A cross-section survey of industry participants is taken through either personal interviews or mail interviews (*see* chapter 23). For purposes of statistical analysis and testing, a random sample is desired. However, at some market levels, especially the fishermen level, participants are highly mobile, making it extremely difficult and costly to obtain a truly random sample. In addition, adequate population lists are seldom available from which to draw the sample. A stratified sample based on known industry characteristics often becomes the data base. Data of this nature usually accurately reflect industry practices. Firm production, market structure, and socioeconomic studies are based on these types of data.

Secondary data among economists generally refer to statistics published on a continuous basis by data management agencies. Data of importance for economic consumption and demand studies are prices, quantities, per capita income, and data describing socioeconomic characteristics of consumers. Because of the size of the markets and the number of consumers and market outlets, it is usually not economically feasible to collect these data through primary surveys. The shortcomings of these data are that they are often out of date and that individual products cannot be traced through all levels of the market system.

Experimental data are generally of limited use in economic studies with the exception of market development studies. Experimental catch-effort studies reflect how a scientist catches fish rather than how the fisherman fishes. The latter is generally the subject of economic research and is used most often in feasibility studies. A complete description of economics data and research needs in the management of a fishery is given in Table 24.1. Data sources are discussed in chapter 22. The United States National Marine Fisheries Service (1981) provides a comprehensive list of fishery data sources which is updated annually.

24.3.2 Production Data

Landings. Production or landings statistics are most useful when reported on a monthly and annual basis. This allows for trends that can be analyzed both seasonally and cyclically. Pounds landed and the value of the catch are important for use in economic analysis.

Effort. Complete measurement of effort requires data for several individual variables. The number of boats and vessels classified by length, width, tonnage, and propulsion type and size enables determination of the capital investment and of the harvesting capacity in the fishery. Documentation of gear units by specific type deployed in the fishery is also essential. These gear units must be classified by craft characteristics and catch of individual species.

Table 24.1 Data requirements for an effective economic research data base and management system [modified from Cato and Prochaska (1980)]

Data Requirements	Frequency needed		
	Monthly	Annually	3-5 years
I. Economic Data Base Needs			
A. Production Statistics			
1. Landings			
a. Pounds	X	X	
b. Value	X	X	
2. Effort			
a. Boats and Vessels			
i. number		X	
ii. size by length		X	
iii. classified by motor size		X	
b. Fishermen			
i. full-time		X	
ii. part-time		X	
iii. fishing rate	X	X	
c. Gear Units by Specific Type		X	
3. Cross Classification of Landings and Effort		X	
B. Market Statistics			
1. Employment			
a. Number of Wholesalers, Processors, Middlemen, Importers and Exporters		X	
b. Employment in Marketing Sector		X	
2. Prices			
a. Fisherman Level			
i. Prices received from wholesalers	X	X	
ii. Prices received from retailers	X	X	
iii. Prices received from consumers	X	X	
b. Wholesale Level			
i. Price received from retailers	X	X	
ii. Price received from consumers	X	X	
c. Retail Price Received from Consumer	X	X	
d. Import and Export Prices	X	X	
e. Processed Product Prices	X	X	
3. Volume			
a. Fisherman Level			
i. Volume of sales to wholesalers	X	X	
ii. Volume of sales to retailers	X	X	
iii. Volume of sales to consumers	X	X	
b. Wholesale Level			
i. Volume of sales to retailers	X	X	
ii. Volume of sales to consumers	X	X	
c. Retail Volume of Sales to Consumers	X	X	
d. Volume of Imports and Exports	X	X	
e. Volume of Processed Product	X	X	
II. Research Needs			
A. Production			
1. Costs and Returns Budget by Gear Type and Size of Operation			X
2. Industry and Firm Production and Cost Functions		X	
B. Consumption and Demand Equations			X

Continued

Table 24.1 Continued

Data Requirements	Frequency needed		
	Monthly	Annually	3-5 years
C. Marketing			
1. Description of Product Flows			X
2. Descriptive Study of Marketing and Processing Activities			X
3. Feasibility of New Methods (as needed)		X	
D. Social and Economic Profile of Fishermen			X

Finally, the total human effort or man-days required to harvest the catch provides another necessary measure of effort. Fishermen should be reported as full-time and part-time fishermen. The monthly and annual fishing rates are also important. These effort data are obtained through primary surveys and from secondary data sources.

Cross classification. Landings data should be cross classified with effort variables on an annual basis for analysis of the relationship between catch and effort. For example, catch by specific gear type should be reported. Care should be exercised that the data are reported in the same manner each year to provide continuity in the data. This continuity allows for time series econometric analysis of catch-effort data after sufficient years have been reported.

24.3.3 Market Statistics

Employment. Annual reporting of the number of wholesalers, processors, middlemen, importers, and exporters provides information on the size, structure, economic importance, and growth of the marketing sector. Reporting of the number of employees provides a similar measure and also allows a measure of efficiency when annual output rates for employers are recorded. These data are available in statistical publications by the National Marine Fisheries Service.

Prices. Price data are necessary in order to analyze seasonal and cyclical price variations at different marketing levels. Price data permit the estimation of market margins and the effect on price of changes in the volume harvested and marketed. Price data should be collected at least monthly at the fishermen, wholesale, and retail levels by product forms, such as fresh and processed. Finally, import and export product prices allow the determination of how domestic market prices compete in international markets.

Volume. Quantities marketed should be reported at the same levels at which price data are obtained. This allows for analysis of the relationship between price and quantity. Volume data also allow the determination of the economic structure, conduct, and performance of the marketing system. Knowledge of product flows facilitates market development where the resource base is adequate for market expansion. In the event of problems with the resource base, product flow data would allow determination of which segments of the market would most likely suffer hardships.

24.4 LAKE OKEECHOBEE: EXAMPLE OF USING ECONOMICS TO MANAGE A FISHERY

24.4.1 Problem and Objectives

Applied economics methods can be used to determine the economic feasibility of a fish management and utilization program prior to the initiation of the program. The Florida Game and Fresh Water Commission recommended in 1975 that Lake Okeechobee in southern Florida be opened to commercial fishing subject to certain limitations. Biological recommendations indicated that the lake was overpopulated with "rough fish" such as gar and shad and that annual harvest of one-half the standing stock in the lake would make the lake a more successful fishery from both a sport and commercial viewpoint. Fish proposed for harvest included all freshwater fish except largemouth bass and chain and redfin pickerel. Fishing devices proposed included the traps and trotlines which were then legal in Lake Okeechobee and the use of commercial haul seines and trawls subject to certain limitations. All sunfish and crappie (scale fish) harvested were to be tagged before marketing because they are considered gamefish and were not commercially legal in other parts of Florida.

The economic study (Cato and Prochaska 1977a) was designed to indicate prices acceptable to consumers of both scale and rough fish and to determine if these prices would allow adequate economic incentives to fishermen, wholesalers, and retailers to become involved in the production and marketing of the fish. Specific commercial outlets for the scale fish, market potential for the rough fish, and the number of fishing permits that could be issued to support an economically and biologically successful fishery were determined. The following sections summarize the analysis of trawl and haul seines used to catch sunfish, crappie, and rough fish.

24.4.2 Production Potential

Production data for haul seines were based on an average catch of two pounds per yard of net per haul from an earlier study of Lake Okeechobee. The estimated annual fishing season was based on a proposed management program of commerical fishing only on weekdays and considerations of down time for repairs and weather. Total production per haul seine was then estimated to be 226.4 metric tons per year, composed of 37.5, 27.3, 17.7, and 143.9 metric tons, respectively, of catfish, sunfish, crappie, and roughfish.

Total revenue per haul seine was projected by multiplying the estimated production by expected product price. Catfish prices of $683.44 per metric ton were based on commercial market data. Sunfish and crappie prices were estimated to be between $551.16 and $661.39 per metric ton based on primary data surveys of potential markets ($661.39 is used in this example). Market surveys were also used to estimate a price of $66.14 per metric ton for rough fish. With these prices, total annual revenue per haul seiner was estimated to be $63,511.

Cost schedules were developed for haul seine fishing from a primary cross-sectional survey of 19 Florida and 7 out-of-state fishermen using haul seines in other areas. Variable costs for gas and oil, ice, repair and maintenance, crew shares, and miscellaneous cost items represented 84% of the total annual cost per haul seiner. Depreciation, licenses, and permits made up the fixed cost component. Total annual cost was estimated to be $50,178.

Comparison of the estimated costs and revenues showed a net annual return of

$13,333 to a haul seiner. Total investment in the haul seine operation was estimated to be $22,600. Net return per day of haul seining was $85.47.

Similar analyses for trawling showed that total annual revenue per trawler was estimated to be $19,102, and total cost was estimated at $7,122, leaving a net return of $11,980. Total investment of $9,900 was less than half of that required for haul seiners, and variable costs were only 50% of total cost, compared to a 84% for seiners. The main difference in variable cost was that four crewmen were required for seining compared to one crewman for trawling operations. Net return per day to a trawler owner was estimated to be $71.72. With these anticipated prices and estimated harvest rates, it was concluded that fishermen would enter the fishery.

The minimum recommended harvest to reach biological control standards for the biomass in the lake was based on the annual catch rates for trawls and haul seines; 39 haul seines or 170 trawls would be necessary to reach this harvest. Various combinations of haul seine units and trawl units to reach the minimum recommended harvest were also determined for the total catch and for the individual species. Because the population of some species was more limited than others, the minimum harvest would be reached with fewer days of fishing effort for these species. Linear equations were provided to determine the number of haul seines and fishing units for various combinations of fishing units and desired days of fishing effort.

24.4.3 Marketing Potential

The market potential for sunfish and crappie was assessed by determining the expected prices and margins at both the retail and wholesale market levels. Since these species were not previously legal in Florida markets (and in most other states), expected prices were estimated rather than recording existing secondary market prices.

Retail. A cross-sectional primary survey of Florida retail markets resulted in an anticipated average annual sales per retail market of 5.01 metric tons of sunfish and 5.04 metric tons of crappie. At these rates it would only require 225 retail markets out of the 5,000 licensed retail seafood firms in Florida to handle the entire desired catch. Retailers felt crappie would be slightly preferred to sunfish. Expected retail price to consumers was $1,807.80 per metric ton for crappie and $1,653.47 per metric ton for sunfish. The weighted average marketing margin necessary to cover marketing costs and profits was estimated to be $881.85 per metric ton for crappie and $793.67 for sunfish. Retailers were then willing to pay $925.94 and $859.81 per metric ton of crappie and sunfish, respectively.

Wholesale. Twelve potential sunfish and crappie wholesalers who marketed catfish from Lake Okeechobee were interviewed to determine expected prices and marketing margins. Wholesalers projected that they could sell annually nearly 453.59 metric tons of sunfish and crappie in-state and 2,993.70 metric tons out-of-state. The expected demand was nearly twice the recommended harvest of sunfish and crappie and would only meet the demand of 45 Florida retail markets. With the expected national demand to be nearly twice the supply it was concluded that prices would be sufficiently high to insure success of the program.

Wholesale marketing margins were determined in two ways. First, wholesalers were asked their expected sales prices and the prices they would be willing to pay fishermen with the difference being the margin. Second, cost of transportation, boxes, ice, labor, overhead, investment, and management was estimated. The estimated marketing margins ranged from $367.73 to $491.19 per metric ton.

24.4.4 Management Decisions

Expected market price and margins for edible fish and production costs were determined by the above procedures. It was also necessary to determine the cost of administering the program before the final decision to implement the management program could be made.

Each scale fish had to be tagged since sunfish and crappie would only be legal in commercial trade if harvested from Lake Okeechobee. Tagging cost was estimated to include a fee to cover the cost of the tag and its application and to provide enough revenue to cover the state's cost of administering the program. The tagging cost determined was $48.72 per metric ton for labor at the wholesale level and $123.68 per metric ton for the tag and administration.

The process of determining the market feasibility of catching and marketing sunfish and crappie involved adding fishing and marketing costs at each level and comparing these to expected or "reasonable" prices. The costs of production analysis indicated that $661.39 per metric ton would yield fishermen an acceptable level of returns and encourage fishing activity. The tagging fee used was $123.68 per metric ton or 1.87 cents per tag. Addition of production costs, tagging costs, and marketing margins at the wholesale and retail levels gave an estimated retail price of $2,039.28 per metric ton in Florida. This price was consistent with those currently charged for tilapia and catfish in Florida. Consideration of other market outlets also presented a favorable economic potential.

Favorable results of the analysis of out-of-state markets was especially important since wholesalers thought these markets could absorb more than one and one-half times the minimum recommended harvest from the lake. This meant that anticipated Florida prices would be driven up to obtain any of the available supply. There appeared to be sufficient demand to absorb the total catch with adequate returns to all involved in the production, marketing, and administration of the sunfish and crappie segment of the lake fishery.

The Lake Okeechobee Fish Management and Utilization Plan was initiated on October 15, 1976, partly as a result of the favorable economic analysis of production and market potential. In the five months ending on March 4, 1977, a total of five haul seine permits and 200 trawl permits were issued. Once the program was underway, wholesalers were paying fishermen $771.62 to $1,102.32 per metric ton for sunfish and crappie. The fish were selling for up to $1,653.47 per metric ton at the wholesale level and up to $2,623.51 at in-state retail markets. The anticipated excess demand materialized, and resulting market prices exceeded the estimates in the report.

24.5 SUMMARY

When fishery management decisions are based to some extent on economic considerations, nearly all economic concepts are involved. Therefore, this entire chapter should be considered at most an extremely brief summary of economic considerations for fishery management. Each topic could easily be a chapter in itself.

The Lake Okeechobee study was an example of how economic analysis can improve fishery management programs, especially when one of the goals of the program is to pay for itself. The study was also an example of how biological and economic data can be used together to develop fishery management plans which make sense from both the viewpoint of the fishermen who is trying to make a living and the fishery manager who is trying to manage the resource.

24.6 REFERENCES

Alvarez, J., C. O. Andrew, and F. J. Prochaska. 1976a. Dual structural equilibrium in the
    Florida shrimp processing industry. *Fishery Bulletin* 74:879-883.
Alvarez, J., C. O. Andrew, and F. J. Prochaska. 1976b. *Economic structure of the Florida
    shrimp processing industry.* State University System of Florida Sea Grant Report
    Number 9, Gainesville, Florida, USA.
Anderson, L. G. 1977. *The economics of fisheries management.* The Johns Hopkins University
    Press, Baltimore, Maryland, USA.
Bell, F. W. 1978. *Food from the sea: the economics and politics of ocean fisheries.* Westview
    Press, Boulder, Colorado, USA.
Cato, J. C. 1976. Dockside price analysis in the Florida mullet fishery. *Marine Fisheries Review*
    38(6):4-13.
Cato, J. C., and F. L. Lawlor. 1981. *Small boat longlining for swordfish on Florida's east coast:
    an economic analysis.* Florida Cooperative Extension Service, Marine Advisory Pro-
    gram, MAP-15, Gainesville, Florida, USA.
Cato, J. C., and F. J. Prochaska. 1977a. Economic potential for producing and marketing
    underutilized fish: Lake Okeechobee scale and rough fish. *Proceedings of the Second
    Annual Tropical and Subtropical Fisheries Conference of the Americas.* Texas A & M
    University Sea Grant Report 78-101, Gainesville, Florida, USA.
Cato, J. C., and F. J. Prochaska. 1977b. A statistical and budgetary economic analysis of
    Florida-based Gulf of Mexico red snapper-grouper vessels by size and location, 1974 and
    1975. *Marine Fisheries Review* 39(11):6-14.
Cato, J. C., and F. J. Prochaska. 1980. Economic management concepts in small-scale lobster
    fisheries. *Proceedings of the 33rd Annual Gulf and Caribbean Fisheries Institute* 33:301-
    321.
Christy, F. T. 1975. *Alternative entry controls for fisheries. Limited entry into the commercial
    fisheries.* Institute for Marine Studies, Seattle, Washington, USA.
Crutchfield, J. A. 1965. *Economic objectives for fishery management. The fisheries: problems in
    resource management.* University of Washington Press, Seattle, Washington, USA.
Lampe, H. C. 1978. Demand analysis and its implications for fisheries development. *Proceed-
    ings of the 30th Annual Gulf and Caribbean Fisheries Institute* 30:174-180.
Larkin, P. A. 1977. An epitaph for the concept of maximum sustained yield. *Transactions
    of the American Fisheries Society* 106:1-11.
Nielsen, L. A. 1976. The evolution of fisheries management philosophy. *Marine Fisheries Re-
    view* 38(12):12-28.
Prochaska, F. J. 1978. Prices, marketing margins and structural change in the king mackerel
    marketing system. *Southern Journal of Agricultural Economics* 10(1):105-109.
Prochaska, F. J., and C. O. Andrew. 1974. Shrimp processing in the southeast: supply problems
    and structural change. *Southern Journal of Agricultural Economics* 6(1):247-252.
Prochaska, F. J., and J. R. Baarda. 1974. *Florida fisheries management programs: their devel-
    opment, administration, and current status.* Florida Agricultural Experiment Station
    Bulletin Number 768, Gainesville, Florida, USA.
Prochaska, F. J., and J. S. Williams. 1978. Economic analysis of spiny lobster firms at optimum
    stock levels. *Southern Journal of Agricultural Economics* 10(2):93-100.
Prochaska, F. J., and J. C. Cato. 1980. Economic considerations in the management of the
    Florida spiny lobster fishery. *Fisheries* 5(4):53-56.
Scott, G. H. 1954. The economic theory of a common property resource: the fishery. *Journal
    Political Economics* 62:
Shang, Y. C. 1981. *Aquaculture economics: basic concepts and methods of analysis.* Westview
    Press, Boulder, Colorado, USA.
Smith, F. J. 1975. *The fishermen's business guide.* The International Marine Publishing Com-
    pany, Camden, Maine, USA.
United States National Marine Fisheries Service. 1981. *Fisheries of the U.S., 1980.* United
    States Department of Commerce, Washington, District of Columbia, USA.
Williams, J. S., and F. J. Prochaska. 1977. Maximum economic yield and resource allocation in
    the spiny lobster industry. *Southern Journal of Agricultural Economics* 9(1):145-150.

Index